Edited by
Mike Martin
Kwabena Boysen

Carbohydrates – Tools for Stereoselective Synthesis

Related Titles

Koskinen, A.

Asymmetric Synthesis of Natural Products

Second Edition

2012
ISBN: 978-1-119-97668-4

Cornils, B., Herrmann, W. A., Wong, C.-H., Zanthoff, H.-W. (eds.)

Catalysis from A to Z

A Concise Encyclopedia
Fourth, Completely Revised and Enlarged Edition

2012
ISBN: 978-3-527-33307-3

Hanessian, S., Giroux, S., Merner, B. L.

Design and Strategy in Organic Synthesis

From the Chiron Approach to Catalysis

2012
ISBN: 978-3-527-33391-2

Wang, B., Boons, G.-J. (eds.)

Carbohydrate Recognition

Biological Problems, Methods, and Applications

2011
ISBN: 978-0-470-59207-6

Brahmachari, G.

Handbook of Pharmaceutical Natural Products

2010
ISBN: 978-3-527-32148-3

Hudlicky, T., Reed, J. W.

The Way of Synthesis

Evolution of Design and Methods for Natural Products

2007
ISBN: 978-3-527-32077-6

Edited by Mike Martin Kwabena Boysen

Carbohydrates – Tools for Stereoselective Synthesis

WILEY-
VCH

WILEY-VCH Verlag GmbH & Co. KGaA

The Editor

Dr. Mike Martin Kwabena Boysen
University of Hannover
Institute of Organic Chemistry
Schneiderberg 1
30167 Hannover
Germany

All books published by **Wiley-VCH** are carefully produced. Nevertheless, authors, editors, and publisher do not warrant the information contained in these books, including this book, to be free of errors. Readers are advised to keep in mind that statements, data, illustrations, procedural details or other items may inadvertently be inaccurate.

Library of Congress Card No.: applied for

British Library Cataloguing-in-Publication Data
A catalogue record for this book is available from the British Library.

Bibliographic information published by the Deutsche Nationalbibliothek
The Deutsche Nationalbibliothek lists this publication in the Deutsche Nationalbibliografie; detailed bibliographic data are available on the Internet at <http://dnb.d-nb.de>.

© 2013 Wiley-VCH Verlag & Co. KGaA, Boschstr. 12, 69469 Weinheim, Germany

All rights reserved (including those of translation into other languages). No part of this book may be reproduced in any form – by photoprinting, microfilm, or any other means -- nor transmitted or translated into a machine language without written permission from the publishers. Registered names, trademarks, etc. used in this book, even when not specifically marked as such, are not to be considered unprotected by law.

Print ISBN: 978-3-527-32379-1
ePDF ISBN: 978-3-527-65457-4
ePub ISBN: 978-3-527-65456-7
mobi ISBN: 978-3-527-65455-0
oBook ISBN: 978-3-527-65454-3

Cover Design Adam-Design, Weinheim

Typesetting Toppan Best-set Premedia Limited, Hong Kong

Printing and Binding Markono Print Media Pte Ltd, Singapore

Printed on acid-free paper

Contents

Foreword *XI*
Preface *XIII*
List of Contributors *XV*

Part I Carbohydrate Auxiliaries *1*

1 Reactions of Nucleophiles with Electrophiles Bound to Carbohydrate Auxiliaries *3*
Zhiwei Miao
1.1 Introduction *3*
1.2 Strecker Reactions *3*
1.3 Ugi Reactions *8*
1.4 Allylations *13*
1.5 Mannich-Type Reactions *15*
1.6 Addition of Phosphites *19*
1.7 Dynamic Kinetic Resolution of α-Chloro Carboxylic Esters *22*
References *24*

2 1,4-Addition of Nucleophiles to α,β-Unsaturated Carbonyl Compounds *27*
Kiichiro Totani and Kin-ichi Tadano
2.1 Introduction *27*
2.2 1,4-Additions to Acrylic Amides and Acrylic Esters *27*
2.3 1,4-Addition to 4- and 2-Pyridones *40*
References *45*

3 Reaction of Enolates *47*
Noureddine Khiar, Inmaculada Fernández, Ana Alcudia, Maria Victoria García and Rocío Recio
3.1 Introduction *47*
3.2 Aldol Alkylation *48*
3.3 Aldol Addition *56*

3.4	Concluding Remarks	60
	References	62

4 Cycloadditions 65
Mike M.K. Boysen

4.1	Diels–Alder Reactions	65
4.1.1	Dienophiles Attached to the Auxiliary	65
4.1.2	Dienes Attached to the Auxiliary	83
4.2	1,3-Diploar Cycloadditions	93
4.2.1	1,3-Dipoles Attached to the Auxiliary	93
4.2.2	Dipolarophiles Attached to the Auxiliary	101
4.3	[2 + 2] Cycloadditions	101
	References	104

5 Cyclopropanation 107
Noureddine Khiar, Inmaculada Fernández, Ana Alcudia, Maria Victoria García and Rocío Recio

5.1	Introduction	107
5.2	Epoxidation	113
5.3	Construction of Chiral Sulfur and Phosphorus Centers	117
5.4	Chiral Phosphorus Compounds	120
5.5	Concluding Remarks	122
	References	123

Part II   Carbohydrate Reagents 125

6 Hydride Reductions and 1,2-Additions of Nucleophiles to Carbonyl Compounds Using Carbohydrate-Based Reagents and Additives 127
Omar Boutureira and Benjamin G. Davis

6.1	Introduction	127
6.2	Hydride Reductions	127
6.2.1	Modified Aluminohydrides	128
6.2.2	Modified Borohydrides	129
6.3	1,2-Additions of Nucleophiles to Carbonyl Compounds	135
	References	140

7 Aldol-Type Reactions 143
Inmaculada Fernández, Noureddine Khiar, Ana Alcudia, Maria Victoria García and Rocío Recio

7.1	Introduction	143
7.2	Titanium Lewis Acids for Enolate Formation	143

7.3	1,2-Additions of Nucleophiles to Carbonyl Compounds Using Stoichiometric Reagents *149*	
7.3.1	Allyl-Titanium Reagents *149*	
7.3.2	Allylsilicon Reagents *151*	
	References *153*	

Part III Carbohydrate Ligands *155*

8   Hydrogenation Reactions *157*
Carmen Claver, Sergio Castillón, Montserrat Diéguez and Oscar Pàmies

8.1	Introduction *157*
8.2	Hydrogenation of C=C and C=N Bonds *158*
8.2.1	Hydrogenation of C=C Bonds *158*
8.2.2	Hydrogenation of C=N Bonds *158*
8.3	P-Donor Ligands *159*
8.3.1	Phosphinite Ligands *160*
8.3.2	Phosphite Ligands *166*
8.3.3	Phosphine Ligands Including Mixed Donor Ligands *173*
8.4	P–N Donors *177*
8.5	P–S Donors *178*
	Acknowledgments *180*
	References *180*

9   Hydroformylations, Hydrovinylations, and Hydrocyanations *183*
Mike M.K. Boysen

9.1	Hydroformylation Reactions *183*
9.1.1	Diphosphite Ligands *184*
9.1.2	Diphosphinite Ligands *198*
9.1.3	Diphosphine Ligands *200*
9.1.4	Mixed Phosphorus Donor and P-S Donor Ligands *201*
9.2	Hydrovinylation Reactions *202*
9.2.1	Monophosphinite and Monophosphite Ligands *203*
9.3	Hydrocyanation Reactions *206*
9.3.1	Diphosphinite Ligands *206*
9.4	Hydrosilylation Reactions *211*
9.4.1	Monophosphite Ligands *211*
9.4.2	Diphosphite and Phosphite–Phosphoroamidite Ligands *212*
9.4.3	P-S Donor Ligands *212*
9.5	Conclusion *214*
	References *215*

10	**Carbohydrate-Derived Ligands in Asymmetric Tsuji–Trost Reactions** *217*
	Montserrat Diéguez and Oscar Pàmies
10.1	Introduction *217*
10.2	Ligands *220*
10.2.1	P-Donor Ligands *220*
10.2.1.1	Phosphine Ligands *220*
10.2.1.2	Phosphinite Ligands *221*
10.2.1.3	Phosphite Ligands *223*
10.2.1.4	Phosphoroamidite Ligands *227*
10.2.1.5	P-P′ Ligands *228*
10.2.2	S-Donor Ligands *231*
10.2.3	Heterodonor Ligands *233*
10.2.3.1	P-S Ligands *233*
10.2.3.2	P-N Ligands *236*
10.2.3.3	P-O Ligands *240*
10.2.3.4	N-S Ligands *240*
10.3	Conclusions *242*
	Acknowledgments *242*
	References *242*

11	**Carbohydrate-Derived Ligands in Asymmetric Heck Reactions** *245*
	Montserrat Diéguez and Oscar Pàmies
11.1	Introduction *245*
11.2	Ligands *247*
11.3	Conclusions *250*
	Acknowledgments *251*
	References *251*

12	**1,4-Addition of Nucleophiles to α,β-Unsaturated Carbonyl Compounds** *253*
	Yolanda Díaz, M. Isabel Matheu, David Benito and Sergio Castillón
12.1	Copper-Catalyzed Reactions – Introduction *253*
12.2	Copper-Catalyzed Reactions – Ligands *257*
12.2.1	Bidentate P,P-Ligands *257*
12.2.1.1	*Xylo* Ligands *257*
12.2.1.2	*Ribo* Ligands *257*
12.2.1.3	D-*Gluco*-, D-*Allo*-, L-*Talo*-, and D-*Galacto*-Ligands *258*
12.2.2	Bidentate P,N-Ligands *261*
12.2.2.1	*Xylo* and *Ribo* Ligands *261*
12.2.2.2	D-*Gluco* Ligands *261*
12.2.3	Bidentate P,S- and O,S-Ligands *261*
12.2.4	Monodentate P or S-Ligands *263*
12.2.5	Miscellaneous Ligands with Peripheral Donor Centers. N,N-, N,O-, and P,N-Ligands *265*

12.3	Enantioselective Copper-Catalyzed 1,4-Addition of Organometallics to α,β-Unsaturated Carbonyl Compounds *266*	
12.3.1	Substrate: 2-Cyclohexenone *266*	
12.3.1.1	Bidentate Diphosphite and Diphosphinite Ligands *266*	
12.3.1.2	Bidentate Ligands with Different Coordinating Functional Groups: P/P, P/N, S/O, and S/P *267*	
12.3.1.3	Monodentate P- or S-Ligands *280*	
12.3.1.4	Miscellaneous Ligands with Peripheral Donor Centers. N,N-, N,O-, and P,N-Ligands *281*	
12.3.2	Substrate: 2-Cyclopentenone, 2-Cycloheptenone *281*	
12.3.3	Linear Substrates: *trans*-Non-3-en-2-one and *trans*-5-Methyl-3-hexen-2-one *281*	
12.4	Rhodium-Catalyzed Reactions *286*	
12.5	Conclusions *287*	
	References *288*	

13 1,2-Addition of Nucleophiles to Carbonyl Compounds *293*

M. Isabel Matheu, Yolanda Díaz, Patricia Marcé and Sergio Castillón

13.1	Introduction *293*
13.2	Addition of Organometallic Reagents to Aldehydes *294*
13.2.1	Addition of Diethylzinc to Aldehydes. N,O Donor Ligands (Aminoalcohols) *294*
13.2.2	Addition of Trialkylaluminium Compounds to Aldehydes. P,P and P,O-Donor Ligands *302*
13.2.3	Zinc-Triflate-Catalyzed Addition of Alkynes to Aldehydes and Imines. N,O Ligands *305*
13.3	Addition of Trimethylsilyl Cyanide to Ketones. P=O,O Ligands *308*
13.4	Conclusion *311*
	References *312*

14 Cyclopropanation *313*

Mike M.K. Boysen
References *318*

Part IV Carbohydrate Organocatalysts *319*

15 Oxidations *321*

O. Andrea Wong, Brian Nettles and Yian Shi

15.1	Oxidations *321*
15.1.1	Epoxidations with Ketone Catalysts and Oxone *321*
15.1.2	Oxidation of Miscellaneous Substrates *345*
15.2	Conclusion *346*
	References *347*

16 Enantioselective Addition Reactions Catalyzed by Carbohydrate-Derived Organocatalysts *351*
Jun-An Ma and Guang-Wu Zhang

16.1 Introduction *351*
16.2 Strecker, Mannich, and Nitro-Mannich Reactions *351*
16.2.1 Strecker Reactions *352*
16.2.2 Mannich Reactions *355*
16.2.3 Nitro-Mannich Reactions *356*
16.3 Michael Additions *358*
16.3.1 Bifunctional Primary Amino–Thiourea Catalysts *358*
16.3.2 Bifunctional Secondary Amino–Thiourea Catalysts *362*
16.3.3 Bifunctional Tertiary Amino–Thiourea Catalysts *363*
16.4 Miscellaneous Reactions *364*
16.4.1 Radical-Chain Reactions *365*
16.4.2 Aldol Reactions *367*
16.5 Outlook *367*
Acknowledgment *368*
References *368*

Index *371*

Foreword

Among the organic compounds provided by nature, carbohydrates are those which contain the highest density of stereochemical information. In addition, carbohydrates are cheap and readily available in large quantities. Logically, carbohydrates, in particular monosaccharides, emerged as valuable enantiomerically pure starting materials for numerous total syntheses of interesting natural products and drugs. In spite of these attractive properties, carbohydrates have almost been ignored as tools for stereodifferentiation in stereoselective syntheses for a long time. This may be traced back to some frustrating experiences of leading researchers in the field as well as to the widespread impression among chemists that carbohydrates, although they can be sweet, sometimes tasting and even convertible to stimulating liquids, are difficult to handle and to purify if they do not crystallize.

It is a great benefit of the book *Carbohydrates – Tools for Stereoselective Synthesis* and a particular merit of the editor Mike Boysen and the authors he invited that they convincingly show how useful and efficient carbohydrates actually are as stereodifferentiating tools in a broad range of stereoselective reactions. In a number of cases, these stereoselective conversions finally paved an elegant way to access interesting enantiomerically pure products of quite different structures.

In the chapters of the book, the authors describe briefly, but comprehensively, carbohydrates in their function as chiral auxiliaries in diastereoselective reactions (Part I), as stereoselective reagents (Part II), as the decisive chiral ligands of enantioselective catalysts (Part III), and as organo-catalysts in enantioselective syntheses (Part IV). The subdivision according to the type of reaction provides a profound survey over the accomplishments achieved so far. Clear schemes displaying the reactions and all required information concerning conditions, yield, stereoselectivity, and original literature enlarge the profit the reader can gain from this book. In many cases, the interpretation of the stereodifferentiating effects induced by carbohydrate is also outlined, thus stimulating interested chemists to let their own ideas climb up the dense chirality of the carbohydrates. The book is a competent and inspiring source for preparative chemists aiming at demanding chiral target compounds on innovative paths.

Mainz, November 2012 *Horst Kunz*

Preface

Carbohydrates are arguably one of the most important classes of natural products. They serve as important energy sources and energy storage compounds in both animals and plants, and are essential as integral parts of structural fibers for plants, fungi, insects, spiders, and crustaceans. Apart from this, glycoconjugates – carbohydrate structures covalently bound to proteins and lipids – have been recognized to play a fundamental role in biological recognition and signaling processes on a cellular level.

From the point of view of synthetic organic chemistry, carbohydrates are primarily of interest to scientists working in the fields of glycoconjugates and glycomimics. As far as total synthesis of complex natural products and medicinal chemistry are concerned, simple carbohydrates – mainly monosaccharides – are occasionally used as inexpensive enantiopure chiral starting materials. In these ex-chiral-pool syntheses, some or all of the stereocenters of the carbohydrates are incorporated in the target structure. Application of carbohydrates as starting materials for the design of chiral auxiliaries, reagents, complex ligands, or organocatylsts, that is, as synthetic tools for de novo setup of stereocenters, on the other hand, has long been avoided by chemists. This may be due in part to some deeply rooted prejudices against carbohydrate chemistry, which I have encountered myself ever since I started working with carbohydrate compounds during my PhD time: carbohydrates are frequently believed to be "difficult" substrates because of their manifold functional groups, and I have even been asked whether it is at all possible to purify these "sticky" and "over-functionalized" compounds. Thus, carbohydrates have often been regarded as unsuitable or impractical for the design of stereodifferentiating agents for asymmetric synthesis. The foundation of these prejudices is, however, quickly dispelled by proper research of the literature, revealing highly successful examples of all kinds of carbohydrate-based tools for the setup of new stereocenters.

In this context, we set out to collect successful and instructive examples for carbohydrate tools in stereoselective synthesis for this book. It is the first publication to give the reader a comprehensive overview of today's scope and limitations of these tools, which in some areas have already become indispensable supplements to the arsenal for modern stereoselective synthesis. This book covers all four types of carbohydrate tools comprising a furanose- or a pyranose-type scaffold; open chain structures and derivatives of tartaric acid are only included in some exceptional cases. Our aim is not only to bring carbohydrate tools to the awareness of the readers, but also to encourage them to apply these to their advantage in their own synthetic efforts as they often complement the scope of more traditionally used stereodifferentiating agents. Further, we would like to motivate especially young researchers to use their creativity and skills to add their own contributions to the toolbox of carbohydrate-derived stereodifferentiating agents: carbohydrate scaffolds offer unique opportunities for both the design and optimisation of novel synthetically useful tools.

I would like to thank Ms. Elke Maase, who helped me to initiate this book project, and Ms. Lesley Belfit, whose help during the editing and production process was invaluable. Finally, I would like thank all authors who contributed to this venture!

Hannover, October 2012 *Mike Boysen*

List of Contributors

Ana Alcudia
CSIC–Universidad de Sevilla
Investigaciones Químicas
C/Américo Vespucio, s/n, Isla de la Cartuja
41092 Seville
Spain

David Benito
Universitat Rovira i Virgili
Facultat de Química
Departament de Química Analítica i Química Orgànica
C/Marcel·lí Domingo s/n
43007 Tarragona
Spain

Omar Boutureira
University of Oxford
Department of Chemistry
Chemistry Research Laboratory
12 Mansfield Road
Oxford OX1 3TA
UK

Mike M.K. Boysen
Leibniz University of Hannover
Institute of Organic Chemistry
Schneiderberg 1B
30167 Hannover
Germany

Sergio Castillón
Universitat Rovira i Virgili
Facultat de Química
Departament de Química Analitica i Química Orgànica
C/Marcel·lí Domingo s/n
43007 Tarragona
Spain

Carmen Claver
Universitat Rovira i Virgili
Facultat de Química
Departament de Química Física i Inorgànica
C/Marcel·lí Domingo s/n
43007 Tarragona
Spain

Benjamin G. Davis
University of Oxford
Department of Chemistry
Chemistry Research Laboratory
12 Mansfield Road
Oxford OX1 3TA
UK

Yolanda Díaz
Universitat Rovira i Virgili
Facultat de Química
Departament de Química Analítica i
Química Orgànica
C/Marcel·lí Domingo s/n
43007 Tarragona
Spain

Montserrat Diéguez
Universitat Rovira i Virgili
Facultat de Química
Departament de Química Física
i Inorgànica
C/Marcel·lí Domingo s/n
43007 Tarragona
Spain

Inmaculada Fernández
Universidad de Sevilla
Facultad de Farmacia
Departamento de Química
y Farmacéutica
c/Prof. Garcia Gonzalez 2
41012 Seville
Spain

Maria Victoria García
CSIC–Universidad de Sevilla
Investigaciones Químicas
C/Américo Vespucio, s/n, Isla de
la Cartuja
41092 Seville
Spain

Noureddine Khiar
CSIC–Universidad de Sevilla
Investigaciones Químicas
C/Américo Vespucio, s/n, Isla de
la Cartuja
41092 Seville
Spain

Jun-An Ma
Tianjin University
Department of Chemistry
Nankai Wu, Weijin Road 92
Tianjin 300072
China

Patricia Marcé
Universitat Rovira i Virgili
Facultat de Química
Departament de Química Analítica i
Química Orgànica
C/Marcel·lí Domingo s/n
43007 Tarragona
Spain

M. Isabel Matheu
Universitat Rovira i Virgili
Facultat de Química
Departament de Química Analítica i
Química Orgànica
C/Marcel·lí Domingo s/n
43007 Tarragona
Spain

Zhiwei Miao
Nankai University
Research Institute of Elemento-
Organic Chemistry
Tianjin 300071
China

Brian Nettles
Colorado State University
Department of Chemistry
220 W Lake Street
Fort Collins, CO 80523
USA

Oscar Pàmies
Universitat Rovira i Virgili
Facultat de Química
Departament de Química Física
i Inorgànica
C/Marcel·lí Domingo s/n
43007 Tarragona
Spain

Rocío Recio
CSIC–Universidad de Sevilla
Investigaciones Químicas
C/Américo Vespucio, s/n, Isla de
la Cartuja
41092 Seville
Spain

Yian Shi
Colorado State University
Department of Chemistry
220 W Lake Street
Fort Collins, CO 80523
USA

Kin-ichi Tadano
Keio University
Department of Applied Chemistry
3-14-1 Hiyoshi
Kohoku-ku
Yokohama 223-8522
Japan

Kiichiro Totani
Seikei University
Department of Materials and
Life Science
3-3-1 Kichijoji-kitamachi
Musashino-shi
Tokyo 180-8633
Japan

O. Andrea Wong
Colorado State University
Department of Chemistry
220 W Lake Street
Fort Collins, CO 80523
USA

Guang-Wu Zhang
Tianjin University
Department of Chemistry
Nankai Wu, Weijin Road 92
Tianjin 300072
China

Part I
Carbohydrate Auxiliaries

General Remarks on the Use of Pseudo-Enantiomers of Carbohydrate Tools

The application of pseudo-enantiomers is of great importance when carbohydrate-derived chiral tools are used: While d-monosaccharides are easily available from the *chiral pool*, the corresponding l-enantiomers are mostly expensive and in some cases even unavailable. For the preparation of a pseudo-enantiomer of a given carbohydrate tool, a carbohydrate scaffold with opposite configuration at relevant stereocenters is chosen. These relevant stereocenters are usually those directly involved in the events determining the direction of the asymmetric induction, that is, the one(s) carrying the substrate (in the case of carbohydrate auxiliaries), coordinating metal centers (in the case of carbohydrate ligands), or shielding one face of a substrate. The remaining stereocenters, which are further from the reacting sites, are neglected and may have any configuration. Thus the synthesis of a pseudo-enantiomeric tool can start from other l-monosaccharides, which are more readily available, that is, l-rhamnose and l-fucose and even d-carbohydrates may be employed. By this approach the preparation of a real enantiomer from an expensive l-enantiomer of a d-carbohydrate can be avoided altogether. Attractive and powerful as this approach may be it is important to note that choosing suitable a pseudo-enantiomeric auxiliary offering high levels of stereoinduction is by no means trivial. Usually, several tentative pseudo-enantiomers can be envisioned for a given carbohydrate tool but which of them – if any – gives high levels of stereoinduction cannot, unfortunately, be predicted. Therefore, finding suitable pseudo-enantiomers remains a process of trial and error. With this in view, unsurprisingly, some highly efficient carbohydrate tools have remained without any suitable pseudo-enantiomer (e.g., the Duthaler–Hafner reagent, Chapter 7).

1
Reactions of Nucleophiles with Electrophiles Bound to Carbohydrate Auxiliaries

Zhiwei Miao

1.1
Introduction

Carbohydrates are widespread chiral natural products found worldwide and they have been transformed into diverse, interesting chiral products in ex-chiral pool syntheses. However, carbohydrates were not used as chiral auxiliaries in stereoselective syntheses for a long time. About 30 years ago Vasella reported the earliest example of carbohydrate auxiliaries tools in organic synthesis [1]. During the following decades, carbohydrates slowly became recognized as versatile starting materials for chiral auxiliaries in stereoselective reactions, and today a multitude of structures has been developed and applied to various reactions [2].

1.2
Strecker Reactions

The three-component Strecker reaction as well as the hydrocyanation of imines (modified Strecker reaction) are fundamental carbon–carbon bond-forming processes [3], which are efficient methods for preparing α-amino acids (Scheme 1.1).

In 1987 Kunz and coworkers first reported pivaloyl protected D-galactosyl amine **3** as a very useful tool for asymmetric aminonitrile syntheses [4]. Galactosyl amine **3** can be obtained from penta-*O*-pivaloyl-β-D-galactopyranose **1** by reaction with trimethylsilyl azide/tin tetrachloride to give the galactosyl azide **2** followed by hydrogenation (Scheme 1.2) [4].

Condensation of **3** with aldehydes **4** yields galactosyl aldimines **5**, which undergo highly diastereoselective Strecker reactions with trimethylsilyl cyanide (TMSCN) in the presence of Lewis acids (Scheme 1.3). The observed diastereoselectivity is a result of the attack of the cyanide anion on the face of the (*E*)-imine opposite to the sterically demanding 2-*O*-pivaloyl group. Separation of the minor diastereoisomer and subsequent hydrolysis with hydrochloric acid affords the corresponding enantiomerically pure α-amino acid **7** (R = *p*-ClC$_6$H$_4$).

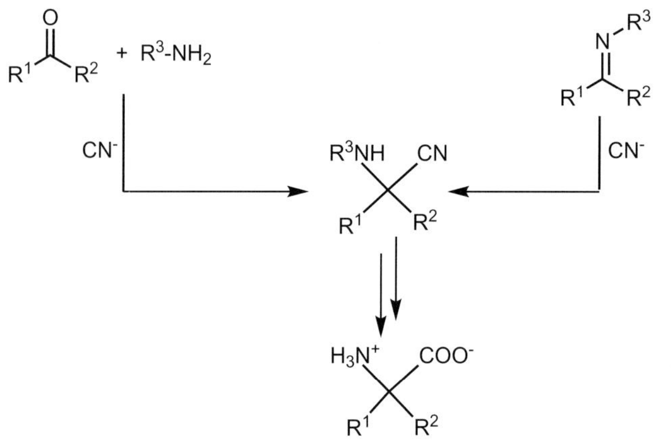

1 Reactions of Nucleophiles with Electrophiles Bound to Carbohydrate Auxiliaries

Scheme 1.1 Strecker-type reactions for the synthesis of α-amino acids.

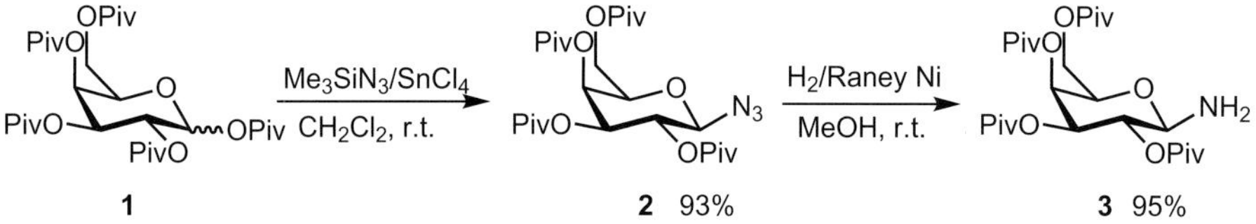

Scheme 1.2 Synthesis of D-galactosyl amine **3**.

The solvent has a strong impact on the direction of the stereoinduction. Stannic chloride in tetrahydrofuran or zinc chloride in isopropanol give α-aminonitriles with the *(R)* configuration with high diastereoselectivity [4a, b], while zinc chloride in chloroform reverses the direction of the asymmetric induction in favor of the *(S)* enantiomer [4c]. Therefore, this method is highly attractive for the preparation of α-amino acid derivatives as by simply changing the reaction conditions the aminonitrile product can be obtained in both configurations from the D-configured galactose auxiliary in a stereodivergent manner (Scheme 1.3).

Kunz ascribed the high selectivity of the Strecker reactions to steric and stereoelectronic effects arising from the carbohydrate auxiliary in combination with the Lewis acid. In the transition state (Figure 1.1) the activating Lewis acid catalyst $ZnCl_2$ is apparently coordinated by the imine nitrogen and the carbonyl oxygen of

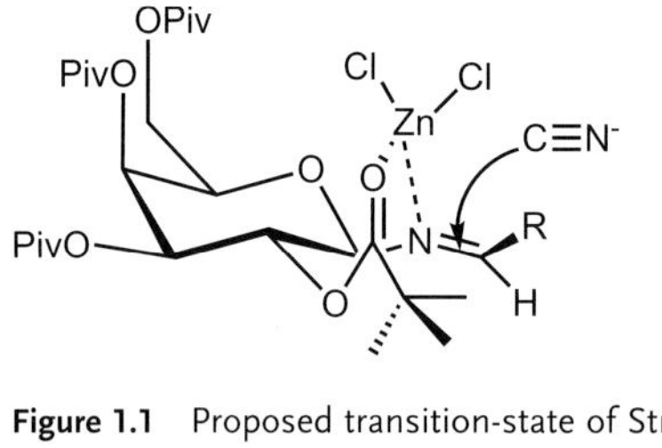

Figure 1.1 Proposed transition-state of Strecker reactions with the galactose auxiliary.

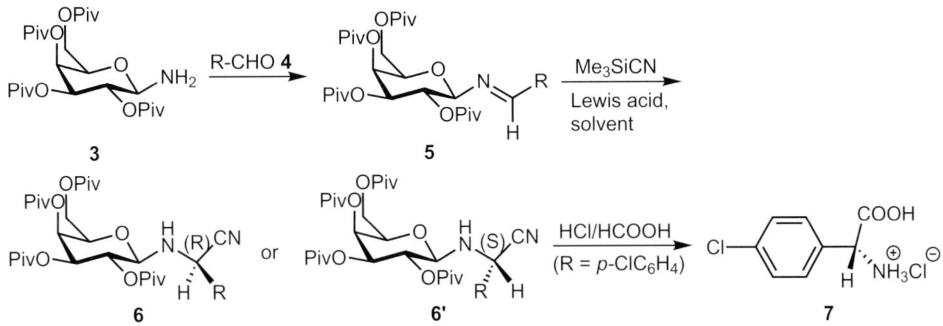

1.2 Strecker Reactions

Scheme 1.3 Kunz's asymmetric Strecker reactions with a galactose-derived chiral auxiliary.

Conditions for **6**: SnCl₄, THF or ZnCl₂, *i*-PrOH

R	dr
p-tolyl	6.5:1 dr (ZnCl₂, *i*-PrOH)
p-tolyl	12:1 dr (SnCl₄, THF)
p-NO₂-C₆H₄	7:1 dr (ZnCl₂, *i*-PrOH)
o-NO₂-C₆H₄	only (R) (SnCl₄, THF)
p-F-C₆H₄	6.5:1 dr (ZnCl₂, *i*-PrOH)
p-F-C₆H₄	10:1 dr (SnCl₄, THF)
p-Cl-C₆H₄	11:1 dr (SnCl₄, THF)
*i*Pr	8:1 dr (SnCl₄, THF)
*i*Bu	13:1 dr (SnCl₄, THF)

Conditions for **6'**: ZnCl₂, CHCl₃

R	dr
Me₃C-	1:9 dr
Me₂CH-	1:5 dr
C₆H₅CH₂CH₂	1:3 dr
p-CH₃C₆H₄	1:4.5 dr
p-F-C₆H₄	1:3 dr
p-Cl-C₆H₄	1:4 dr
m-Cl-C₆H₄	1:6 dr

the 2-*O*-pivaloyl group. This complex is preferably attacked by the cyanide, which is liberated in the polar medium from TMSCN, from the sterically less hindered rear face, that is, the *Si* face of the imine [4a].

After the successful syntheses of D-amino acids via Lewis-acid-catalyzed Strecker reactions with galactosylamine **3** as the stereodifferentiating auxiliary, Kunz has developed the pivaloylated-arabinosylamine **10** as a new chiral auxiliary [5]. Apart from the missing hydroxy methyl group at C5, D-arabinose is a mirror image of D-galactose and therefore arabinosylamine can be regarded as a pseudo-enantiomer of D-galactosylamine **3**. To prepare pivaloylated-arabinosylamine **10**, the peracetylated arabinopyranose is transformed into arabinopyranosyl azide **8**, as has been described by Paulsen and coworkers [6]. After deacetylation and subsequent pivaloylation, **8** gives arabinopyranosyl azide **9**, which is subsequently reduced by hydrogenation with Raney nickel to furnish the auxiliary **10** (Scheme 1.4).

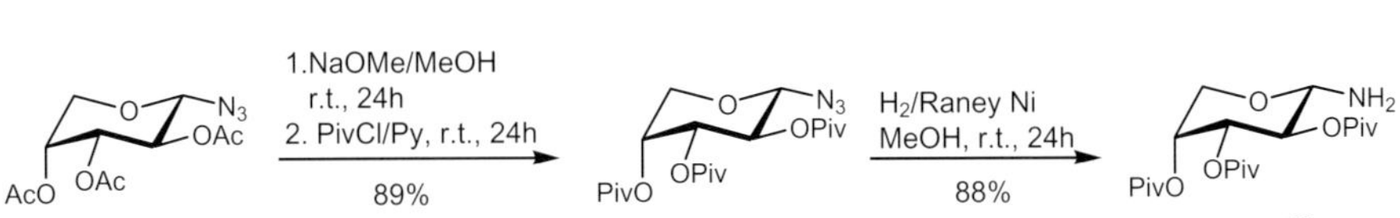

Scheme 1.4 Synthetic pathway to arabinosylamine **10**.

By using the arabinosylamine **10** in the Strecker reaction L-amino nitriles have been successfully obtained. To this end, **10** was condensed with aldehydes to give the N-arabinosylimines **11**, which with TMSCN/tin tetrachloride furnish the α-amino nitriles **12**. The diastereoselectivity was determined as 7–10:1 in favor of the L-diastereomer after hydrolysis and cleavage of the aminonitrile from the auxiliary Hydrolysis of pure **12** with hydrogen chloride/formic acid forms exclusively L-phenylglycine (Scheme 1.5) [5].

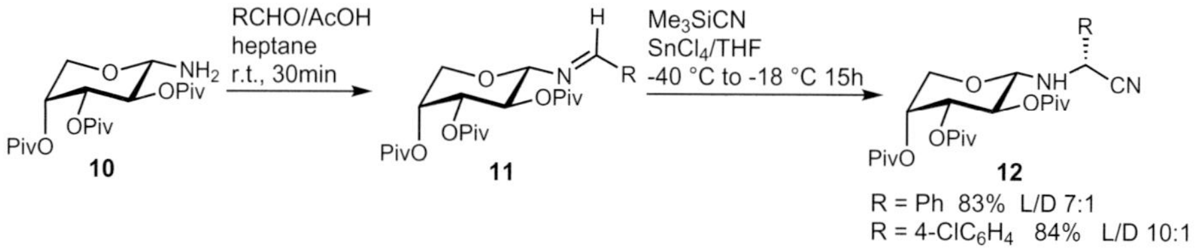

Scheme 1.5 Synthesis of α-L-amino nitriles **12** by Strecker reaction on a D-arabinose template.

The asymmetric Strecker synthesis using carbohydrate auxiliaries has also been studied in some detail by Zhang using a D-glucose-based chiral template [7]. In continuation of Kunz's studies a general protocol for the asymmetric synthesis of α,β-diamino acids involving enantiomerically pure α-amino aldehydes, O-pivaloylated glucopyranosylamine, and TMSCN was developed. The α-aminoaldehydes **14** reacted with glucopyranosylamine **13** in CH_2Cl_2 to give the corresponding imines **15a** and **15b** in high yields. The nucleophilic addition of TMSCN to aldimines **15a** and **15b** employed CuBr·Me2S as promoter to activate the C=N group and afforded α,β-diaminonitriles **16a** and **16b**, respectively. The absolute configuration of the new stereocenter formed in the Strecker reaction is predominantly controlled by the carbohydrate auxiliary, which overrules the stereoinduction by the stereocenter stemming from the amino aldehyde part. The diastereoselectivities were 96% and 82% de, respectively, indicating only a small matched/mismatched effect between carbohydrate auxiliary and the stereocenters from the amino-aldehyde substrates. The bis-hydrochlorides **17a** and **17b** were obtained by hydrolysis of the α,β-diaminonitriles in acidic medium (Scheme 1.6).

Zhang and coworkers also studied the copper(I)-promoted Strecker reaction of sugar-modified α,β-unsaturated imines [8]. Under acidic conditions, the imines **19** were prepared from glucosyl-amine **13** and a series of substituted cinnamic aldehydes **18**. The nucleophilic addition of TMSCN to aldimines **19** afforded the products **20** with the aid of CuBr·Me$_2$S (1 equiv.) as the Lewis acid. In all reactions, only 1,2- rather than 1,4-addition products were observed [9]. This indicates that the carbohydrate auxiliary plays a significant role in controlling the regio- and diastereoselective 1,2-addition of cyanide to the α,β-unsaturated aldimines. The (R)-configured 2-amino-4-phenylbut-3-enoic acids **21** can be obtained by hydrolysis of compounds **20** in acidic medium (Scheme 1.7).

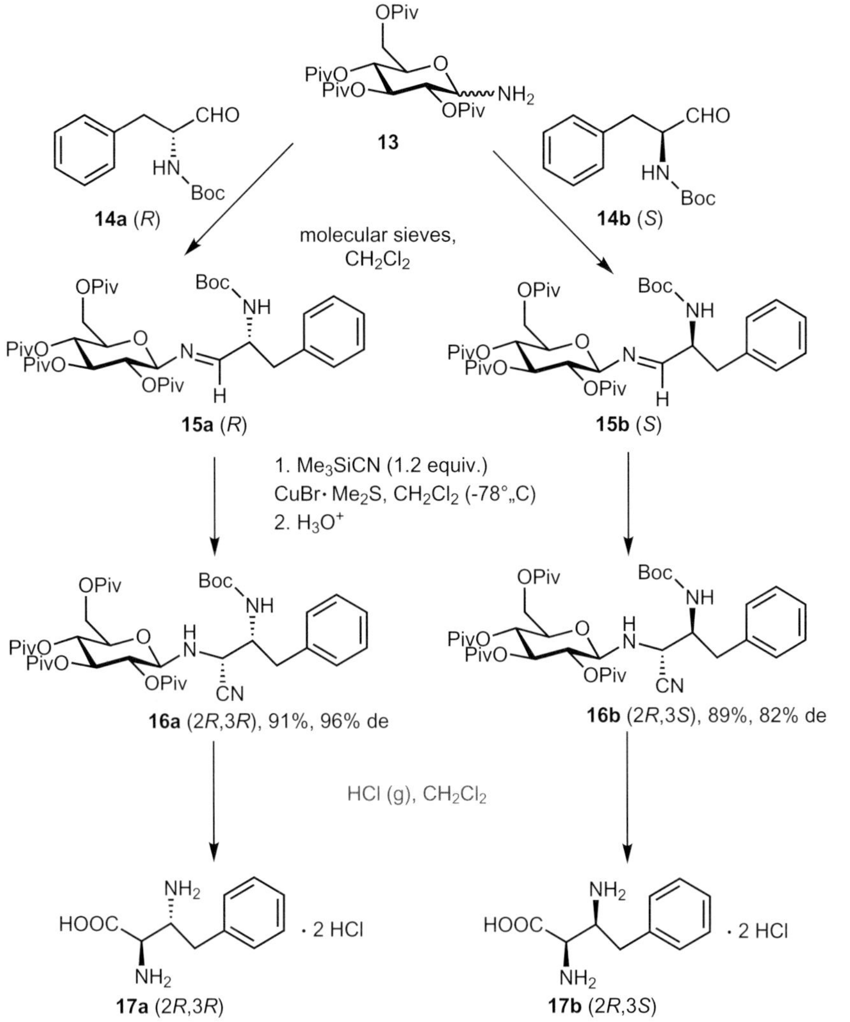

Scheme 1.6 Asymmetric synthesis of α,β-diamino acids via Strecker reaction of aldimines **15**.

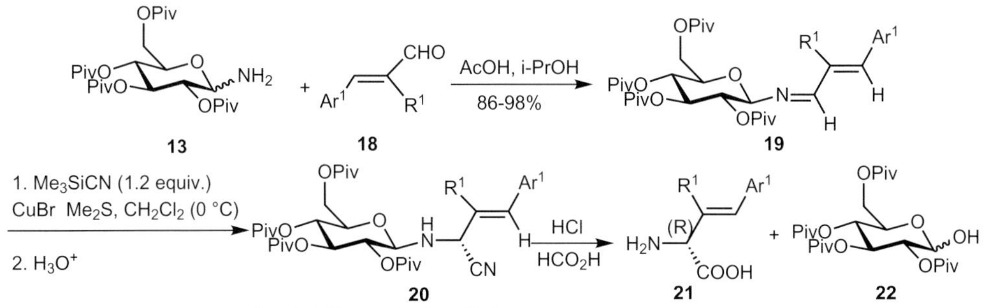

For **20**:
Ar¹ = Ph, R¹ = H 92% 86% de
Ar¹ = 4-NO$_2$-C$_6$H$_4$, R¹ = H, 95% 88% de
Ar¹ = 4-Me$_2$N-C$_6$H$_4$, R¹ = H, 83% 87% de
Ar¹ = 4-MeO-C$_6$H$_4$, R¹ = H, 90% 88% de
Ar¹ = Ph, R¹ = Me, 92% 89% de

Scheme 1.7 Copper(I)-induced regio- and diastereoselective Strecker reaction.

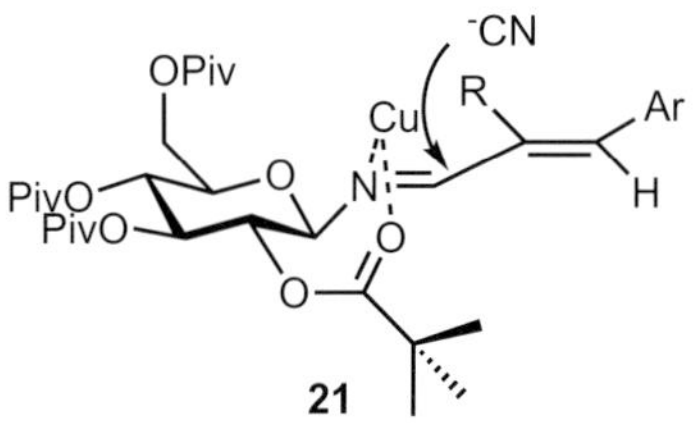

Figure 1.2 Proposed transition-state for Strecker reactions of α,β-unsaturated imines bound to a D-glucose-derived template.

Figure 1.2 shows the proposed transition state **23** leading to products **20**. It is very similar to the one invoked by Kunz for Strecker reactions of galactose-modified imines. The Lewis acid CuBr is coordinated to both the N-atom of the imine and one of the O-atoms of the 2-O-pivaloyl group. This would decrease the electron density at the C-atom of the C=N moiety and direct the attachment of cyanide.

1.3
Ugi Reactions

The terms Ugi four-component reaction (Ugi-4CR) or Ugi four-component condensation (U-4CC) usually refer to the reaction of an amine (usually a primary amine; less frequently ammonia or a secondary amine), a carbonyl compound (an aldehyde), an isocyanide, and a carboxylic acid [10]. In the course of the reaction two peptide bonds and one carbon–carbon bond are formed and a new chiral center is created (Scheme 1.8) [11].

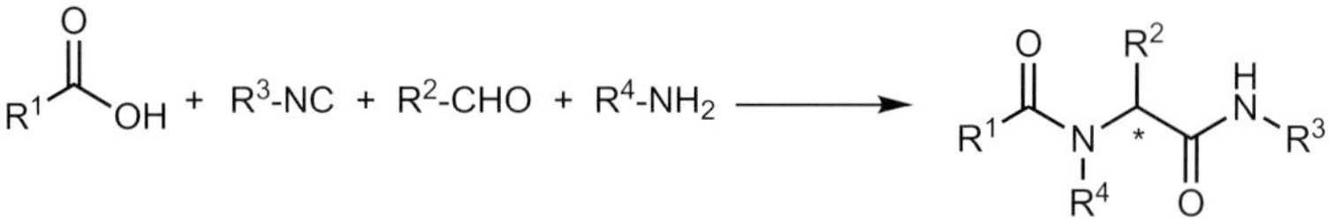

Scheme 1.8 General outline of a four-component Ugi reaction.

A major difficulty in conducting Ugi reactions stereoselectively is that reaction conditions for the transformations vary considerably (e.g., solvent, temperature, and highly diverse starting materials) and consequently the reactions follow different mechanisms. In one successful example Kunz employed his galactosylamine auxiliary as chiral template in the Ugi reaction (Scheme 1.9) [12].

When galactosylamine **3** was allowed to react with an aldehyde, an isocyanide, and a carboxylic acid (preferably formic acid) in the presence of zinc chloride in THF, N-galactosyl amino acid amide derivatives **24** were obtained in almost quantitative yield and high dr. The N-galactosyl amino acid amide derivatives **24** can be transformed into a series of valuable chiral products, for example, 1,2-diamines and β-amino alcohols. At −25 °C (for aliphatic imines −78 °C) D-configured amino acid derivatives **24** were formed with a diastereoselectivity of about 95 : 5

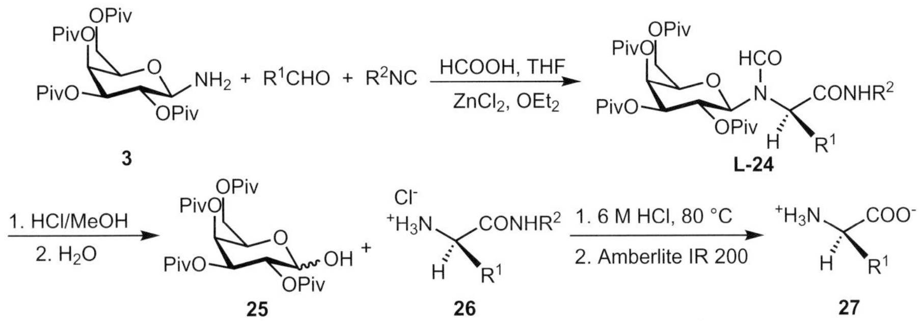

Scheme 1.9 Asymmetric Ugi reaction using a galactose-derived chiral auxiliary.

R' = nPr, R" = tBu, 80% 94:6 dr
R' = iPr, R" = tBu, 86% 95:5 dr
R' = tBu, R" = tBu, 80% 96:4 dr
R' = Bn, R" = tBu, 80% 95:5 dr
R' = 2-furyl, R" = tBu, 90% 95:5 dr
R' = 2-thienyl, R" = tBu, 93% 96:4 dr
R' = Ph, R" = tBu, 81% 91:9 dr
R' = p-Cl-C$_6$H$_4$, R" = tBu, 92% 97:3 dr
R' = p-NO$_2$-C$_6$H$_4$ R" = tBu, 91% 94:6 dr
R' = styryl, R" = tBu, 75% 95:5 dr
R' = N≡C-(CH$_2$)$_3$, R" = Ph, 81% 94:6 dr

(Scheme 1.9). After acidolytic cleavage of the N-glycosidic bond the tetra-O-pivaloyl-galactose **25** is reisolated in quantitative yield. Hydrolysis of the amino acid amides **26** and subsequent deprotonation gives the free α-D-amino acids **27** [12].

The arabinosylamine **10** also was applied in Ugi reaction by Kunz and shows a slightly enhanced reactivity in comparison to the galactosylamine [5, 13]. At −25 °C, **10** reacts with aldehydes, *tert*-butyl isocyanide, and formic acid in the presence of zinc chloride in THF to form the N-formyl-N-arabinosyl amino acid amides **28** in almost quantitative yield. The diastereomeric ratio for the L-amino acid derivatives **28** ranges from 22:1 to 30:1. The free enantiomerically pure L-amino acids **31** can easily be released from the carbohydrate templates by a two-step acidic hydrolysis and the carbohydrate template can be recovered in quantitative yield (Scheme 1.10).

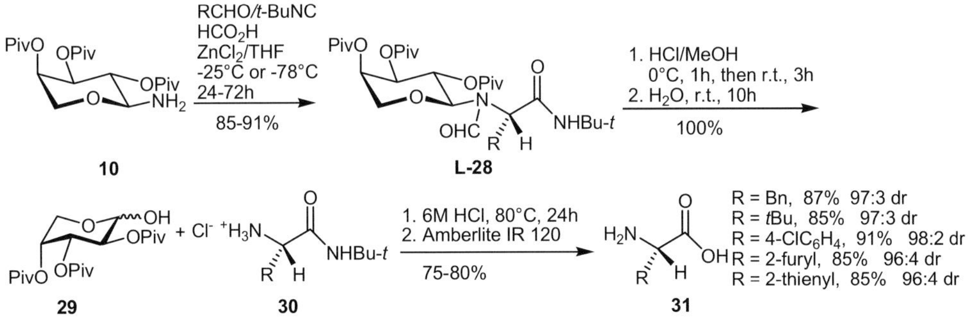

R = Bn, 87% 97:3 dr
R = tBu, 85% 97:3 dr
R = 4-ClC$_6$H$_4$, 91% 98:2 dr
R = 2-furyl, 85% 96:4 dr
R = 2-thienyl, 85% 96:4 dr

Scheme 1.10 Stereoselective synthesis of L-amino acid derivatives **31** using a D-arabinose-derived auxiliary.

Kunz and coworkers introduced their chiral carbohydrate based auxiliaries successfully onto a solid phase [14]. They synthesized 2,3,4-tri-*O*-pivaloylated-β-D-galactopyranosyl azide bearing a hydroxyl-functionalized spacer unit at the C6 position of the galactose and immobilized this on a solid phase by using a polymer-bound chlorosilane. The azide was reduced to the corresponding galactopyranosylamine, which served as a versatile chiral auxiliary in highly diastereoselective Ugi four-component condensation reactions at ambient temperature. Fluoride-induced cleavage from the polymeric support furnished N-glycosylated N-acylated α-amino acid amides **32** (Scheme 1.11).

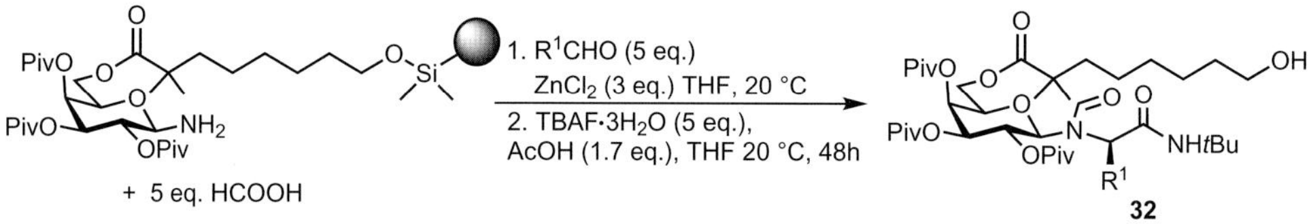

R = *p*-NO_2-C_6H_4, 66%, 76:24 dr
R = *p*-CF_3-C_6H_4, 65%, 74:26 dr
R = Ph, 58%, 96:4 dr
R = *p*-Cl-C_6H_4, 84%, 80:20 dr
R = *p*-MeO-C_6H_4, 67%, 94:6 dr,
R = *i*Pr, 68%, 91:9 dr
R = *n*Pr, 70%, 91:9 dr,
R = *i*Bu, 64%, 90:10 dr

Scheme 1.11 Kunz's auxiliary made available on solid phase and its use in an Ugi reaction.

Pellicciari *et al.* have reported the stereoselective synthesis and preliminary biological evaluation of (+)- and (−)-3-methyl-5-carboxythien-2-ylglycine (3-MATIDA), **36** and **37**. They used chiral sugar based auxiliaries **3** and **10** to prepare the enantiomerically pure unnatural amino acids using a U-4CR [15]. The reaction of thiophenecarbaldehyde **33** with *tert*-butyl isocyanide, formic acid, and D-galactosylamine **3** or D-arabinosylamine **10**, respectively, in the presence of zinc chloride in THF at −25 °C and subsequent cleavage afforded the *N*-formyl-*N*-galactosyl amino acid amide **34** in a 17 : 1 diastereomeric ratio and the *N*-formyl-*N*-arabinosyl amino acid amides **35** in a 32 : 1 diastereomeric ratio, respectively (Scheme 1.12).

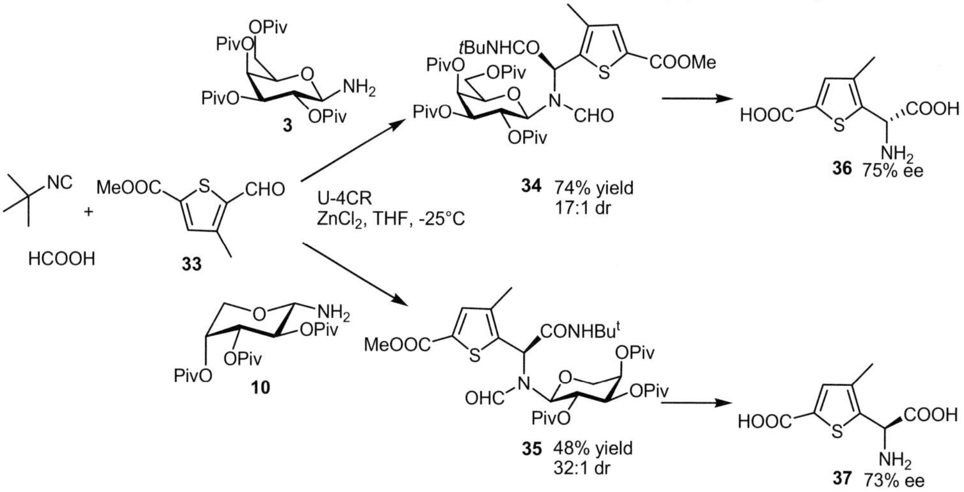

Scheme 1.12 Enantioselective preparation of carboxythiophene α-amino acids by Ugi reactions.

Ugi and coworkers have presented a highly improved sugar derived auxiliary, which was tested as amine compounds for peptide synthesis [16]. Glucopyranosides **39** were prepared from methyl α-D-glucopyranoside (**38**) by methylation. Subsequent acetolysis to give **40** followed by ammonolysis yielded **41**, which was transformed into the auxiliary tetra-*O*-methyl glucopyranosylamine **42** [17] by mesylation and subsequent treatment with gaseous ammonia in a one-pot reaction developed by Vasella (Scheme 1.13) [18].

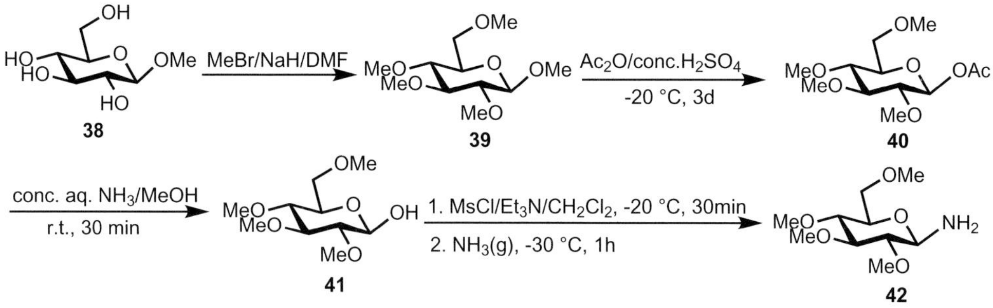
MeBr/NaH/DMF
Ac2O/conc.H2SO4
-20 °C, 3d
conc. aq. NH3/MeOH
r.t., 30 min
1. MsCl/Et3N/CH2Cl2, -20 °C, 30min
2. NH3(g), -30 °C, 1h
38
39
40
41
42

Scheme 1.13 Synthetic pathway to tetra-*O*-methyl-glucopyranosylamine auxiliary **42.**

Glycosylamine **42** has been tested as chiral template in various types of Ugi reaction, and the results show that the major diastereomers of the products **43** have the D-configuration at the newly installed stereocenter [19]. Trifluoroacetic acid (TFA) in combination with a soft base can cleave the Ugi product **43** into peptide **44** and the carbohydrate auxiliary (Scheme 1.14).

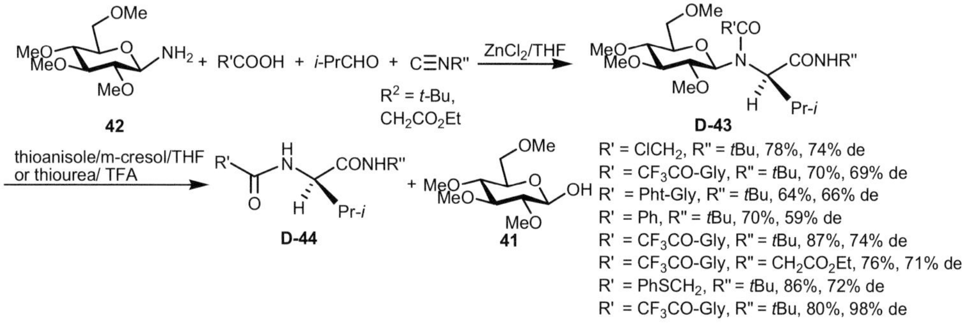
+ R'COOH + i-PrCHO + C≡NR''
ZnCl2/THF
R2 = t-Bu, CH2CO2Et
42
D-43
thioanisole/m-cresol/THF
or thiourea/ TFA
D-44
41
R' = ClCH2, R'' = tBu, 78%, 74% de
R' = CF3CO-Gly, R'' = tBu, 70%, 69% de
R' = Pht-Gly, R'' = tBu, 64%, 66% de
R' = Ph, R'' = tBu, 70%, 59% de
R' = CF3CO-Gly, R'' = tBu, 87%, 74% de
R' = CF3CO-Gly, R'' = CH2CO2Et, 76%, 71% de
R' = PhSCH2, R'' = tBu, 86%, 72% de
R' = CF3CO-Gly, R'' = tBu, 80%, 98% de

Scheme 1.14 Formation of α-acylamino acid derivatives **44** by diastereoselective Ugi reaction on carbohydrate template **42.**

In 1995, Ugi examined the stereoselective syntheses of peptide derivatives with acetylated 1-amino-glucopyranose **45** as the chiral template [20]. The acetylated amino-glucopyranose **45** as auxiliary is prepared from readily available *N*-acetylglucosamine in three steps [21]. Condensation of an aldehyde with the amine **41** yielded glucosyl aldimines **46**, which reacted with isocyanide and acid in the presence of zinc chloride to form the N-glucosyl peptide derivatives **47** in good yields (Scheme 1.15).

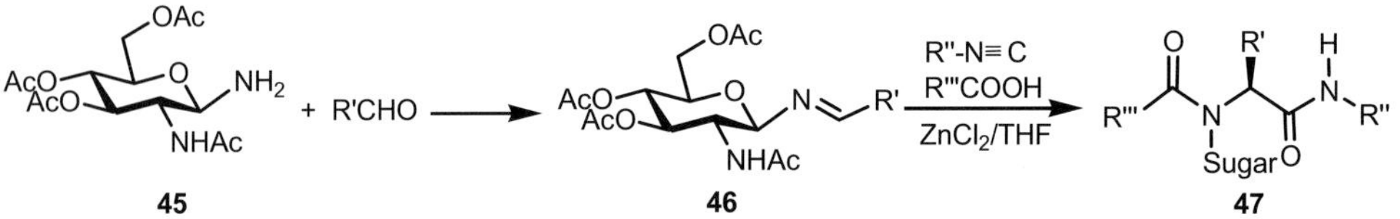

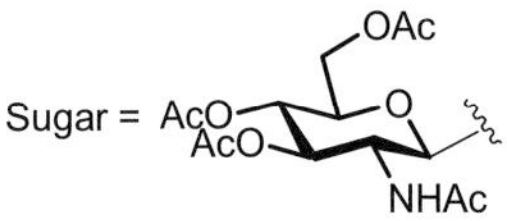

Scheme 1.15 Selection of Ugi reaction products obtained with amino-glucopyranose as auxiliary.

R' = iPr, R" = tBu, R'" = Ph, 85%, >99% de
R' = iPr, R" = CH$_2$COOMe, R'" = Ph, 80%, >99% de
R' = iPr, R" = (iPr)CHCOOMe, R'" = CF$_3$CONHCH$_2$, 85%, >99% de
R' = CH$_3$, R" = tBu, R'" = Ph, 81%, 89% de
R' = iPr, R" = tBu, R'" = tBuOCONHCHCH$_3$, 79%, >99% de
R' = C$_2$H$_5$, R" = Cy, R'" = Ph, 72%, 94% de
R' = iPr, R" = CH$_2$COOMe, R'" = HOCH$_2$CHNH$_2$, 50%, >99% de
R' = Bn, R" = CH$_2$COOMe, R'" = CF$_3$, 50%, 98% de
R' = Bn, R" = CH$_2$COOMe, R'" = Ph, 80%, 99% de
R' = iPr, R" = tBu, R'" = H, 90%, >99% de
R' = iPr, R" = CH$_2$COOMe, R'" = HCONHCH$_2$, 5%, 0% de
R' = CH$_3$, R" = CH$_2$COOMe, R'" = HCONHCHCH$_3$, 17%, 92% de
R' = iPr, R" = TsCH$_2$, R'" = H, 10%, 0% de
R' = p-AcNHPh, R" = CH$_2$COOMe, R'" = tBuOCONHCHCH$_3$, 7%, 0% de
R' = iPr, R" = CH$_2$COOMe, R'" = tBuOCONHCH(p-OHPhCH$_2$), 83%, >99% de

Ugi also reported a thiasugar as a chiral auxiliary for the stereoselective reaction four-component synthesis of amino acids [22]. According to Ingles and Whistler's method [23] 5-desoxy-5-thio-D-xylose **49** can be prepared in six steps from D-xylose **48**. This product can be peracylated to **50** by an excess of isobutanoyl chloride in pyridine. In the presence of tin tetrachloride, **50** can be converted into azide **51** by treatment with trimethylsilyl azide. The anomerically pure β-amine hydrochloride **52** is obtained from the α/β-azide mixture **51** by reduction with 1,3-propanedithiol. During workup, β-amine **53** can be precipitated from an etheric solution as the hydrochloride salt **52** (Scheme 1.16) [24].

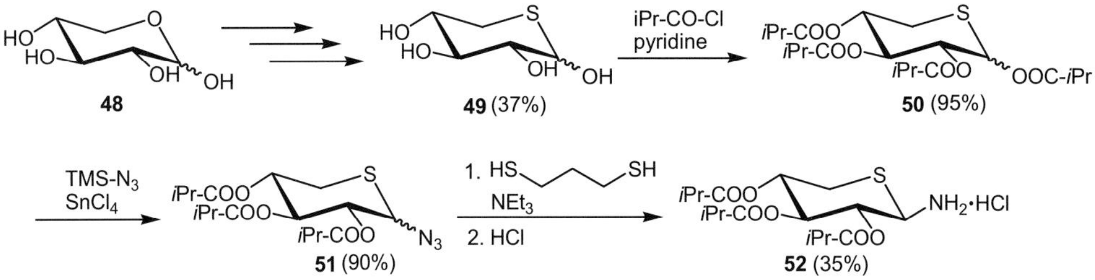

Scheme 1.16 Synthesis of thiasugar auxiliary **52** from D-xylose.

The free amine **53** and isovaleraldehyde are subsequently converted into the imine **54**, which is reacted under Ugi reaction conditions with zinc chloride diethyl etherate, *tert*-butyl isocyanide, and benzoic acid. The product **55** is formed in 92% de (diastereomeric ratio 24:1) and a yield of 92%. The readily crystallizing product **56** is obtained from **55** by removing its *O*-acyl groups by aminolysis with methyl amine. The *O*-deacylated chiral auxiliary **49** can be cleaved off under mild acidic conditions to afford the *N*-benzoyl-D-leucine-*tert*-butylamide **57** (R = *t*Bu) (Scheme 1.17) [25].

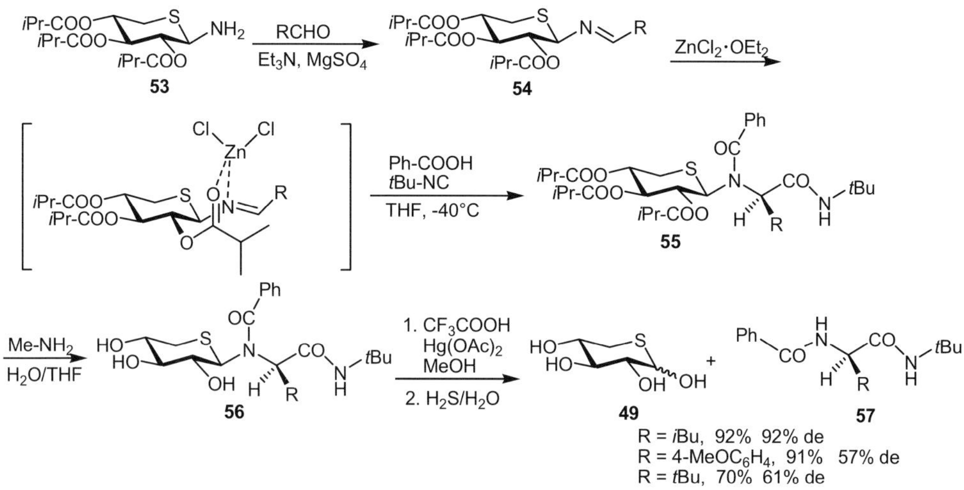

R = *i*Bu, 92% 92% de
R = 4-MeOC$_6$H$_4$, 91% 57% de
R = *t*Bu, 70% 61% de

Scheme 1.17 Stereoselective U-4CR with thiasugar auxiliary **53**.

1.4
Allylations

Homoallyl amines are useful precursors of a various compounds, especially β-amino acids and β-lactam antibiotics, which can be obtained by subsequent functionalization of the double bond. An attractive method for the synthesis of homoallyl amines is the organometallic allylation of chiral imines carrying a chiral template on the nitrogen, which can be successively removed [26].

Kunz reported that *(S)*-configured homoallylamines can be synthesized diastereoselectively by the Lewis acid induced addition of allylsilanes to Schiff bases of tetra-*O*-pivaloyl-galactosylamine **3** (Scheme 1.18) [27a, c] giving moderate to good diastereoselectivity for imines **5** with non-aliphatic residues. The nucleophilicity of the allylic organometallic compound can be improved by changing the metal from silicon to tin [27c]. Thus imine **5** with R = 4-Cl-C$_6$H$_4$ was converted into the corresponding homoallyl amines **58** by using allyltributylstannane instead of allyltrimethylsilane under identical conditions, resulting in an increased yield, but reduced asymmetric induction.

When the reaction is conducted with the *O*-pivaloyl-protected L-fucosylamine **59** instead of *N*-galactosylamine **3**, the *(R)*-configured homoallyl amines **61** can be isolated in high diastereoselectivities [27b, c]. The advantage of this reaction is that most *N*-fucosyl-homoallyl-amines **61** are crystalline and can be obtained as the pure *(R)* diastereomer or as a strongly enriched mixture simply by recrystallization (Scheme 1.19) [27b, c]. It should be noted that allyltributylstannane is used instead

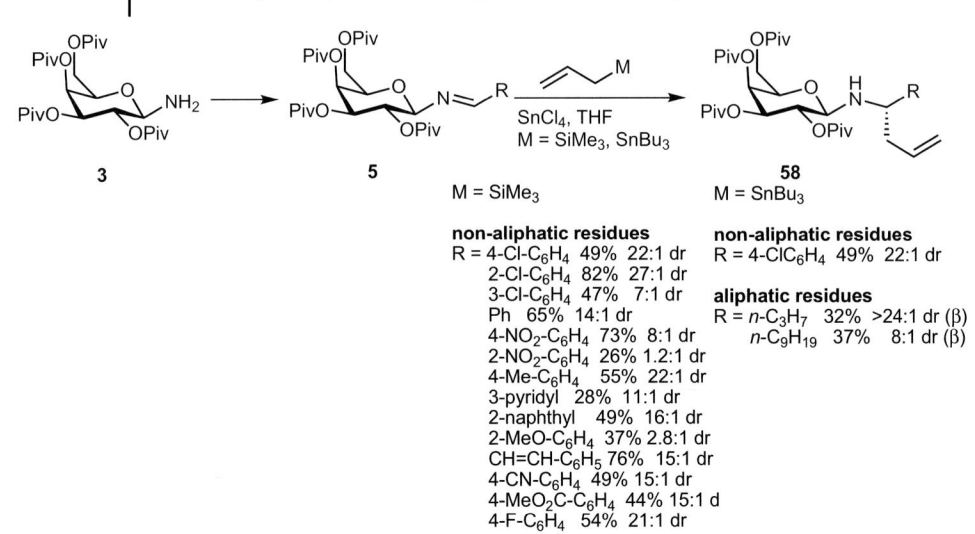

Scheme 1.18 Asymmetric synthesis of (S)-configured homoallyl amines using galactosyl amine **3**.

M = SiMe₃

non-aliphatic residues
R = 4-Cl-C₆H₄ 49% 22:1 dr
2-Cl-C₆H₄ 82% 27:1 dr
3-Cl-C₆H₄ 47% 7:1 dr
Ph 65% 14:1 dr
4-NO₂-C₆H₄ 73% 8:1 dr
2-NO₂-C₆H₄ 26% 1.2:1 dr
4-Me-C₆H₄ 55% 22:1 dr
3-pyridyl 28% 11:1 dr
2-naphthyl 49% 16:1 dr
2-MeO-C₆H₄ 37% 2.8:1 dr
CH=CH-C₆H₅ 76% 15:1 dr
4-CN-C₆H₄ 49% 15:1 dr
4-MeO₂C-C₆H₄ 44% 15:1 d
4-F-C₆H₄ 54% 21:1 dr

M = SnBu₃

non-aliphatic residues
R = 4-ClC₆H₄ 49% 22:1 dr

aliphatic residues
R = n-C₃H₇ 32% >24:1 dr (β)
n-C₉H₁₉ 37% 8:1 dr (β)

Scheme 1.19 Asymmetric synthesis of (R)-configured homoallyl amines using fucosyl amine **58**.

R = Ph 44% 23:1 dr,
4-Cl-C₆H₄ 86% 8:1 dr
2-Cl-C₆H₄ 69% 3.3:1 dr,
4-NO₂-C₆H₄ 68% 20:1 dr
4-MeO₂C-C₆H₄ 53% 17:1 dr,
4-CN-C₆H₄ 56% 13:1 dr
4-Me-C₆H₄ 43% 49:1 dr,
2-naphthyl 40% 12:1 dr
n-Pr 26% 4.5:1 dr

of allyl trimethylsilane in the allylic addition of the corresponding β-L-fucosyl imines **60** (R = nPr).

Schiff bases derived from glucosyl amines and aliphatic aldehydes do not react with allyltrimethylsilane under the same conditions. Even at low temperature (−78 °C), only anomerization and decomposition occurred. However, these imines could be converted into the corresponding homoallylamines using allyltributylstannane instead of the silane at −78 °C, and SnCl₄ (1.2 equiv) was used to activate the imine.

The homoallylamines **62** can be released from the carbohydrate template using aqueous HCl in methanol. Homoallylamine hydrochlorides **62** could easily be N-protected and were subsequently oxidized to yield the N-protected β-amino acid **64**, which was finally deprotected to the corresponding β-amino acid **65** (Scheme 1.20).

A tentative reaction mechanism was proposed by the authors. In the transition state the tin atom of the Lewis acid SnCl₄ has octahedral coordination, with sites occupied by chlorine atoms, the imine nitrogen and the carbonyl oxygen of the

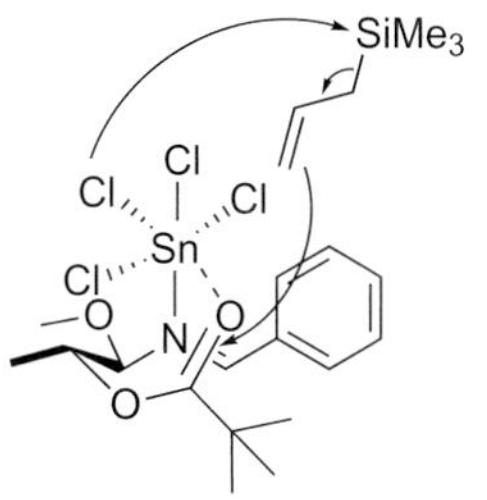

Figure 1.3 Transition state proposed by Kunz for the allylation reaction of glycosyl imines.

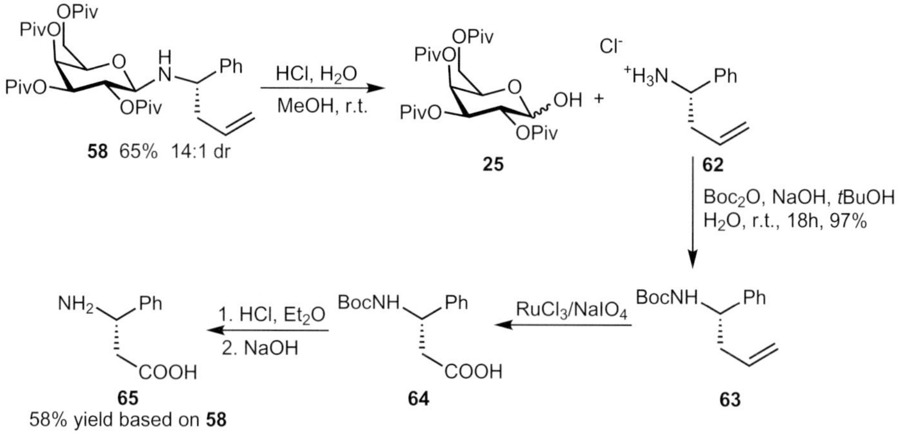

Scheme 1.20 Conversion of N-galactosyl-N-homoallylamines into β-amino acids.

(C2) pivaloyloxy group; one of the four chlorines is removed when allyltrimethylsilane is added. The S_N2'-type attack of the allylic compound occurs preferentially from the rear face of the imine, as the 2-O pivaloyl group effectively shields the front face. The mechanism indicates that the pivaloyl group in the aldimines **5** and **60** plays a significant role in controlling the diastereoselective addition of allyltrimethylsilane (Figure 1.3).

1.5
Mannich-Type Reactions

Mannich-type reactions are among the most important transformations in organic chemistry because they afford synthetically and biologically important β-amino carbonyl compounds [28]. Asymmetric Mannich-type reactions provide useful routes for the synthesis of optically active β-amino ketones or esters that are versatile chiral building blocks in the preparation of many nitrogen-containing biologically important compounds [29].

The first asymmetric Mannich reactions were diastereoselective and involved the addition of preformed enolates and enamines to performed imines using stoichiometric amount of chiral auxiliaries [30]. More recently, direct catalytic asymmetric Mannich-type reactions have been reported [31]. The transformations are catalyzed by both organometallic complexes and metal-free organic catalysis. The different catalysts are highly stereoselective and complementary in their applicability and selectivity.

During investigations on Mannich-type reaction, *N*-galactosyl aldimines **5** were employed as the chiral template [32]. Like α-amino acids generated by the Strecker reaction, β-amino acid derivatives accessible via Mannich reactions are important building blocks for the construction of natural products [33]. The *N*-galactosyl-β-amino acid esters **67** were obtained by the treatment of silyl ketene acetals **62** with the Schiff bases **5** in the presence of zinc chloride at −78 to −30 °C within 24 h. The β-phenyl-β-alanine ester derivatives **68** can be removed from the carbohydrate template almost quantitatively with HCl in methanol (Scheme 1.21).

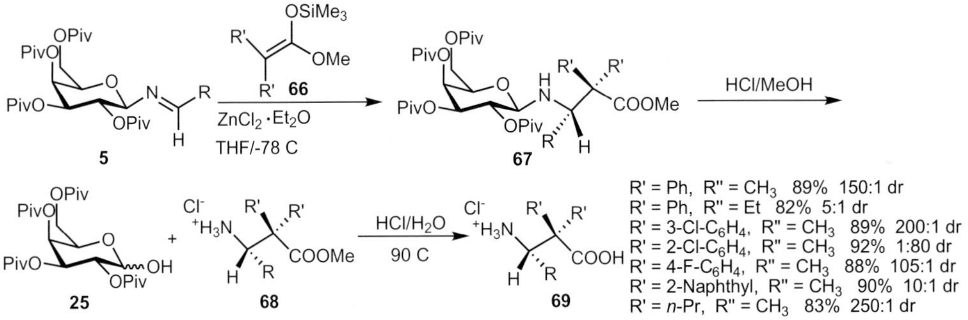
OPiv
PivO
O
N
R
PivO
OPiv
H
5
OSiMe3
R'
OMe
R' 66
ZnCl2·Et2O
THF/-78 C
67
HCl/MeOH
COOMe
25
68
69
HCl/H2O
90 C
COOH
R' = Ph, R" = CH3 89% 150:1 dr
R' = Ph, R" = Et 82% 5:1 dr
R' = 3-Cl-C6H4, R" = CH3 89% 200:1 dr
R' = 2-Cl-C6H4, R" = CH3 92% 1:80 dr
R' = 4-F-C6H4, R" = CH3 88% 105:1 dr
R' = 2-Naphthyl, R" = CH3 90% 10:1 dr
R' = n-Pr, R" = CH3 83% 250:1 dr

Scheme 1.21 Diastereoselective Mannich-type reaction of ketene acetals with imine **5** for the synthesis of β-alanine derivatives.

The diastereoselective Mannich reaction of *O*-pivaloylated *N*-galactosyl aldimines **5** containing two new stereocenters bis-silyl ketene acetals **70**, which was reported by Kunz, proved an efficient stereoselective access to chiral β-amino acid derivatives **71** [34]. The yields and diastereoselectivities of these Mannich reactions are high and only two of the four possible diastereomers are formed. In most cases one of them is obtained in large excess. The *N*-glycosidic bond of compound **71** was readily cleaved under mildly acidic conditions to give enantiomerically pure β-amino acids or their hydrochlorides **72** (Scheme 1.22). To assign the configuration of the β-amino acids **73**, 2,3-diphenyl-β-alanine **72** was released from **71** with 0.01M HCl in methanol. Subsequent reduction with lithium aluminum hydride yields 3-amino-2,3-diphenylpropanol **73**.

To extend the scope of asymmetric reactions using *N*-glycosyl imines to *N*-alkyl or *N*-aryl amino acid derivatives, *O*-pivaloylated galactosyl bromide **74** was employed in Mannich reactions of *N*-alkyl and *N*-aryl aldimines **75** with *O*-trimethylsilyl ketene acetals **76.** The reactions were carried out in a one-pot procedure to give the β-amino acid esters **77** in high yield and with moderate diastereoselectivity (Scheme 1.23) [35].

3,4-dihydroisoquinoline (**78**) reacted with silyl ketene acetal **79** after activation by *N*-galactosylation to give the β-amino acid ester **80** with high diastereoselectivity (Scheme 1.24) [35].

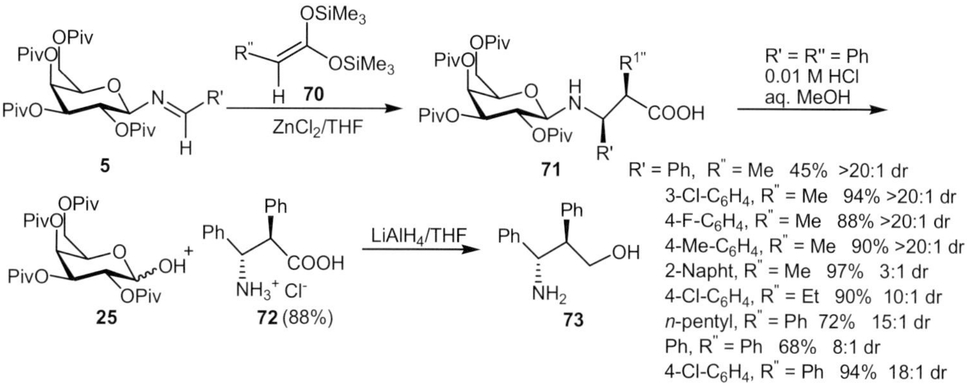

5
70
71
R' = R" = Ph
0.01 M HCl
aq. MeOH
ZnCl2/THF
25
72 (88%)
LiAlH4/THF
73
R' = Ph, R" = Me 45% >20:1 dr
3-Cl-C6H4, R" = Me 94% >20:1 dr
4-F-C6H4, R" = Me 88% >20:1 dr
4-Me-C6H4, R" = Me 90% >20:1 dr
2-Napht, R" = Me 97% 3:1 dr
4-Cl-C6H4, R" = Et 90% 10:1 dr
n-pentyl, R" = Ph 72% 15:1 dr
Ph, R" = Ph 68% 8:1 dr
4-Cl-C6H4, R" = Ph 94% 18:1 dr

Scheme 1.22 Diastereoselective synthesis of β-amino acids **73.**

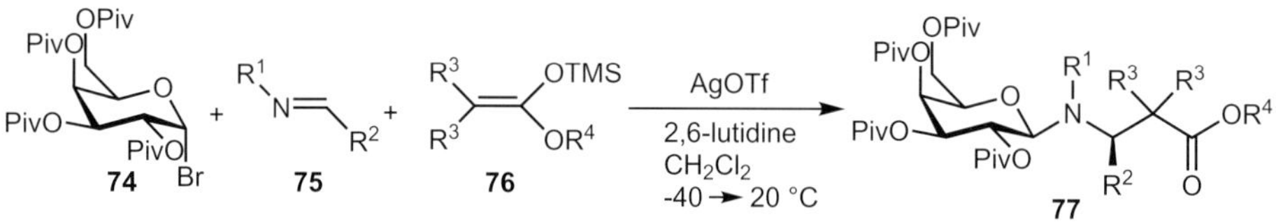

74
75
76
AgOTf
2,6-lutidine
CH2Cl2
-40 → 20 °C
77

R^1 = Et, R^2 = Ph, R^3 = Et, R^4 = Me, 68% (R), 18% (S) 4:1 dr
R^1 = Et, R^2 = Ph, R^3 = Me, R^4 = Me, 48% (R), 17% (S) 3:1 dr
R^1 = Et, R^2 = Ph, R^3 = -$(CH_2)_5$-, R^4 = Et, 63% (R), 23% (S) 3:1 dr
R^1 = Et, R^2 = 4-NO_2-Ph, R^3 = Et, R^4 = Me, 53% (R), 7% (S) 8:1 dr
R^1 = Et, R^2 = 3,4$(MeO)_2$Ph, R^3 = Et, R^4 = Me, 68% (R), 24% (S) 3:1 dr
R^1 = Ph, R^2 = Ph, R^3 = Et, R^4 = Me, 88% 3:1 dr
R^1 = All, R^2 = Ph, R^3 = Et, R^4 = Me, 70% (R), 18% (S) 4:1 dr
R^1 = Bn, R^2 = Ph, R^3 = Et, R^4 = Me, 37% (R), 30% (S) 5:4 dr

Scheme 1.23 Diastereoselective synthesis of β-amino acid esters via an *in situ* glycosylation method.

In 1989, Kunz reported the stereoselective tandem Mannich–Michael reactions for the synthesis of piperidine alkaloids again using galactosylamine **3** as an effective chiral auxiliary [36]. A subsequent publication described how the *N*-galactosyl aldimines **5** react with silyl dienol ether **81** in the presence of zinc chloride in tetrahydrofuran at −20 °C to give the Mannich bases **82/83** with high diastereoselectivities. The Michael addition then occurs to give the dehydropiperidones **84/85** in high yields upon hydrolysis with 1M HCl (Scheme 1.25) [37].

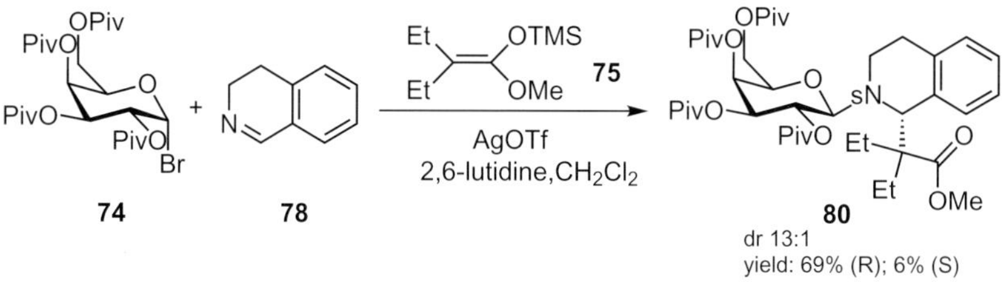

75
AgOTf
2,6-lutidine,CH2Cl2
74
78
80
dr 13:1
yield: 69% (R); 6% (S)

Scheme 1.24 Diastereoselective synthesis of β-amino acid ester **80** from dihydroquinoline **78.**

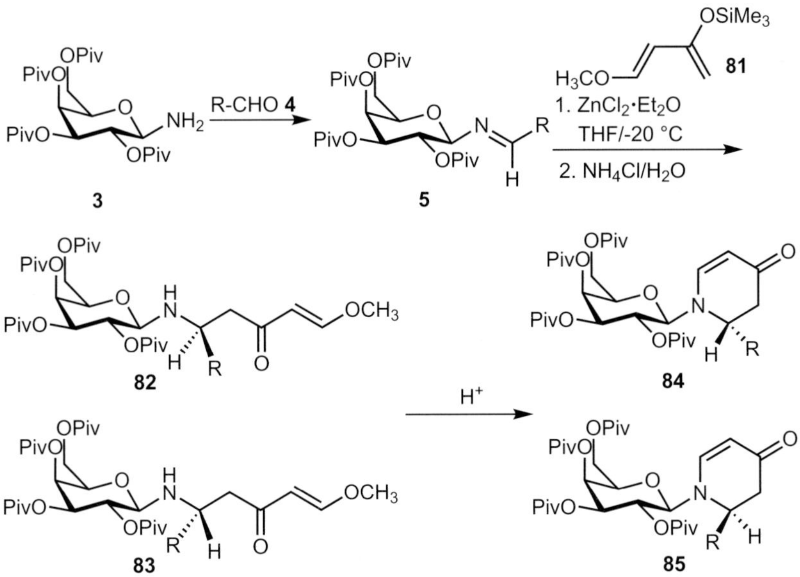

R = *n*-Pr, 96%, 38:1 dr
R = 3-pyridyl, 90%, 34:1 dr
R = *i*-Pr, 58%, >30:1 dr
R = 4-NCC_6H_4, 71%, >20:1 dr
dr = ratio **85**:**84**

Scheme 1.25 Diastereoselective synthesis of *N*-galactosyldehydropiperidones **84** and **85.**

In 2004, Kunz reported the application of arabinosylamine **10** as a suitable *pseudo* enantiomeric auxiliary to the galactosylamine **3** [38a]. *N*-Arabinosylimines **11** react with silyl dienol ether **81** in a domino Mannich–Michael reaction sequence to give 2-substituted 5,6-dehydropiperidinones **86.** The 2-substituted dehydropiperidinones are formed with opposite stereochemistry compared to those from the tandem Mannich–Michael reaction with D-galactosylamine as auxiliary (Scheme 1.26).

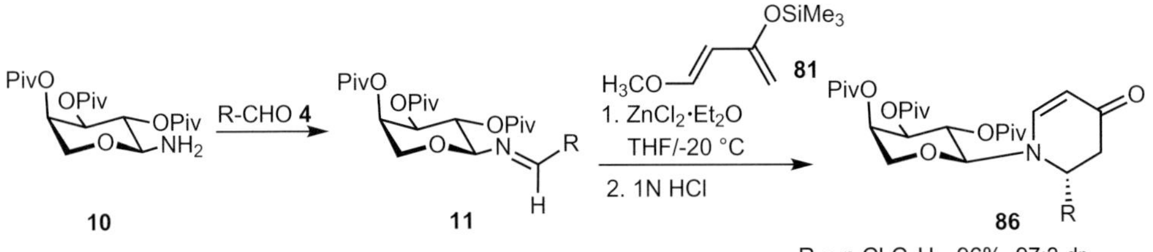

R = *p*-Cl-C_6H_4, 96%, 97:3 dr
R = *p*-NO_2-C_6H_4, 74%, 96:4 dr
R = Ph, 84%, 92:8 dr
R = Me, 49%, 93:7 dr
R = Et, 67%, 96:4 dr
R = *n*Pr, 81%, 87:13 dr
R = iPr, 72%, 99:1 dr
R = $(CH_2)_3$OTBDPS, 65%, 96:4 dr

Scheme 1.26 Diastereoselective synthesis of 2-substituted *N*-arabinosyl dehydropiperidinones **86.**

1.6
Addition of Phosphites

Vasella and coworkers first reported the stereoselective synthesis of α-aminophosphonic acids by means of carbohydrate auxiliaries [39, 40]. In the first experiments N-mannofuranosylnitrones **87** (R = iPr, CH$_2$OBn, Me) were reacted with lithium dialkyl phosphites, affording the corresponding α-aminophosphonic acids with up to 90% de [39]. In a second approach, which was amenable to a wider range of N-mannosylnitrones **87**, tris(trimethylsilyl)phosphite (**88**) was employed under acid catalysis with HClO$_4$, giving (R)-N-hydroxyphenylphosphaglycines **90** in high yield and with an optical purity of 88% after acidic work-up. Hydrogenolysis of **90** gives (R)-phenylphosphaglycines **91**, with optical purities of up to 88% (Scheme 1.27) [40].

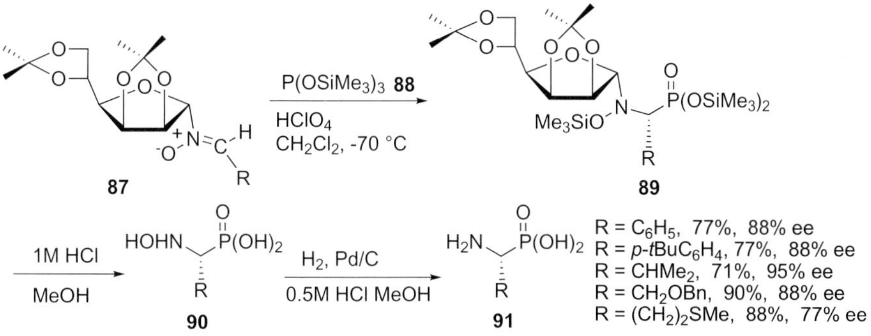

Scheme 1.27 Synthesis of (R)-phenylphosphaglycine **91** using a mannose-derived carbohydrate auxiliary.

In 1992 Kunz and coworkers reported the stereoselective synthesis of α-aminophosphonic acid derivatives from O-pivaloylated galactosylamine as chiral auxiliary [41]. The galactosyl amine **3** was reacted with various aldehydes to give N-galactosyl aldimines **5**, which were reacted with diethyl phosphite to furnish the four diastereomeric N-galactosylphenyl phosphonoglycine esters **92** in high yield by catalysis with tin(IV) chloride in THF (Scheme 1.28). The new stereocenter in esters **92** was preferentially obtained in (S)-configuration, and the anomeric configuration was predominantly β, except for the cases with R = 2-MeOC$_6$H$_4$ and R = Pr, where substantial amounts of the α-anomers were found.

The (R)-configured aminophosphonic acids can be obtained by employing the L-fucose-derived Schiff base **93** as a *pseudo* enantiomeric auxiliary [41]. The diastereomeric mixture of the addition products **94** was treated with 1M hydrogen chloride in methanol at room temperature, giving the carbohydrate template and the α-aminobenzylphosphonate hydrochloride **96** in quantitative yield (Scheme 1.29).

Miao has also reported the diastereospecific formation of α-aminophosphonic acids derivatives in high yield via a Mannich-type reaction [42]. The reaction was

Scheme 1.28 Synthesis arylphosphonoglycine esters using carbohydrate auxiliary **3**.

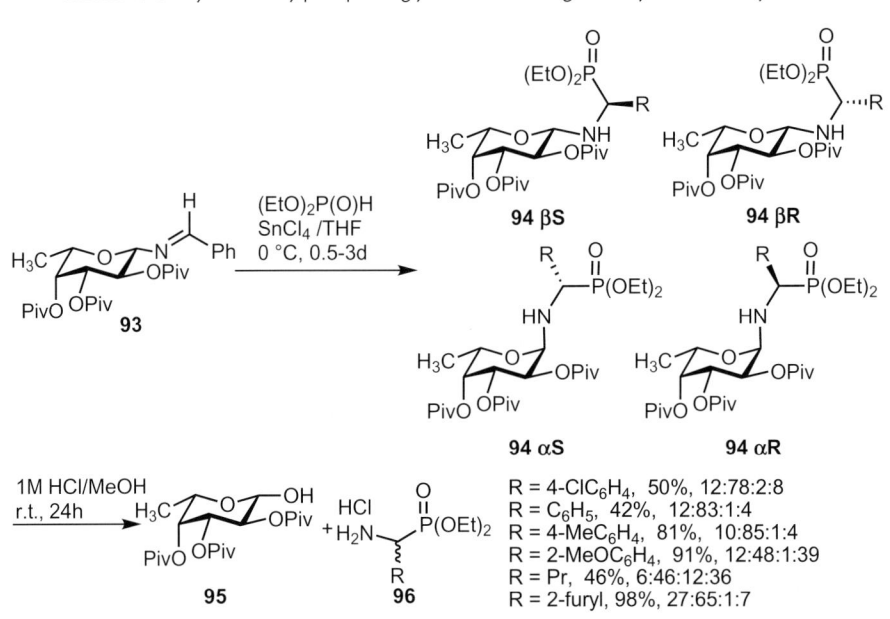

R = 4-ClC$_6$H$_4$, 67%, 79:11:6:4
R = Ph, 50%, 90:7:2:1
R = 4-MeC$_6$H$_4$, 78%, 91:6:2:1
R = 2-MeOC$_6$H$_4$, 80%, 55:1:43:1
R = Pr, 71%, 33:13:37:17
R = 2-furyl, 89%, 69:16:11:4

Scheme 1.29 Diastereoselective synthesis of *(R)*-aminophosphonates using L-fucose auxiliary.

R = 4-ClC$_6$H$_4$, 50%, 12:78:2:8
R = C$_6$H$_5$, 42%, 12:83:1:4
R = 4-MeC$_6$H$_4$, 81%, 10:85:1:4
R = 2-MeOC$_6$H$_4$, 91%, 12:48:1:39
R = Pr, 46%, 6:46:12:36
R = 2-furyl, 98%, 27:65:1:7

performed by using *O*-pivaloylated galactosylamine **3** as a chiral template and boron trifluoride diethyl etherate as a catalyst in THF. Imines **5** [4b] of aromatic aldehydes and diethyl phosphite were converted into *N*-galactosyl α-aminoalkylphosphonates **97** with diastereomeric ratios higher than 19:1 (Scheme 1.30).

1.6 Addition of Phosphites

Scheme 1.30 Synthesis of N-galactosyl α-aminoalkylphosphonates **97**.

R = p-NO$_2$, 88%, 88 de
R = p-Br, 85%, 88 de
R = p-F, 82%, 86 de
R = o-Br, 90%, 62 de
R = p-Cl, 82%, 86 de
R = p-CH$_3$, 85%, 84 de
R = p-OCH$_3$, 80%, 77 de
R = H, 87%, 80 de

The diastereomerically pure compounds **97** were obtained by simple recrystallization from n-hexane and diethyl ether. To determine the absolute configuration of the main isomer of the diethyl phosphite addition to N-galactosyl aldimines **5**, a single-crystal X-ray diffraction study of **97** (R = p-Cl) was performed. The molecular structure of **97** (R = p-Cl), shown in Figure 1.4, proves that the absolute configuration of the main product is *(S)* [42].

Figure 1.5 shows a possible mechanism for the reaction. The preferred formation of the *(S)*-configured diastereomers of **97** can be rationalized by an attack of diethyl phosphite from the *Si* side of N-galactosylaldimines **5**. Initially, the Lewis acid boron trifluoride is coordinated to the imine nitrogen of **5**. The *Re*-face of the imine is shielded by the 2-*O*-pivaloyl group, leaving the *Si*-face exposed. Upon attack of the diethyl phosphite in the transition state, one fluoride may be removed from the Lewis acid and the vacant coordination site may then be filled by the carbonyl oxygen of the 2-*O*-pivaloyl group.

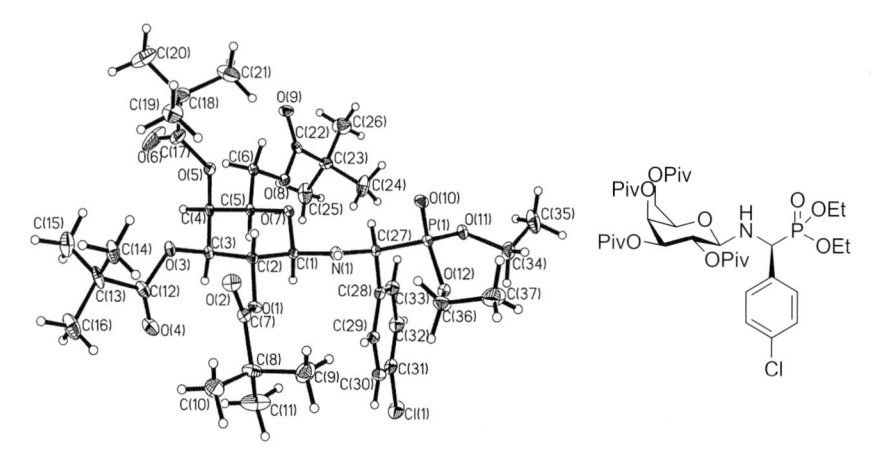

Figure 1.4 ORTEP presentation of the crystal structure of **97** (R = p-Cl).

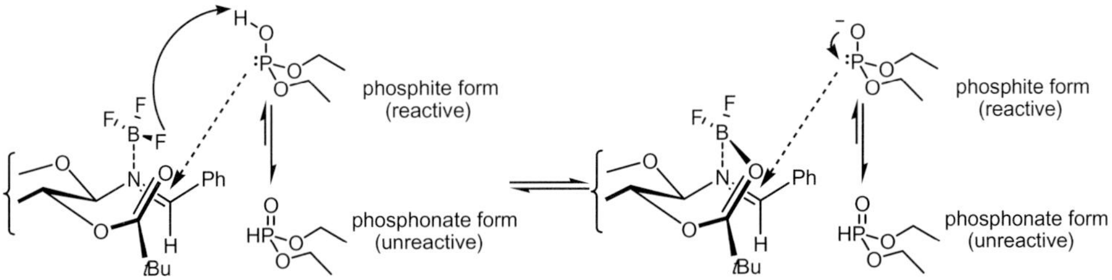

Figure 1.5 Plausible reaction mechanism of the addition of phosphites to galactosyl imines **5**.

In 2009, Miao reported the stereoselective synthesis of α-amino(phenyl)methyl (phenyl)phosphinic acids with D-galactosylamine as chiral auxiliary [43]. Aldimines **5** of aromatic aldehydes and ethyl phenylphosphinate **98** were converted into N-galactosylarylphosphonoglycine esters **99** with diastereomeric ratios higher than 20:1. α-Amino(phenyl)methyl-(phenyl)phosphinic acids **100** can be obtained by treatment with 1M hydrogen chloride in methanol (Scheme 1.31).

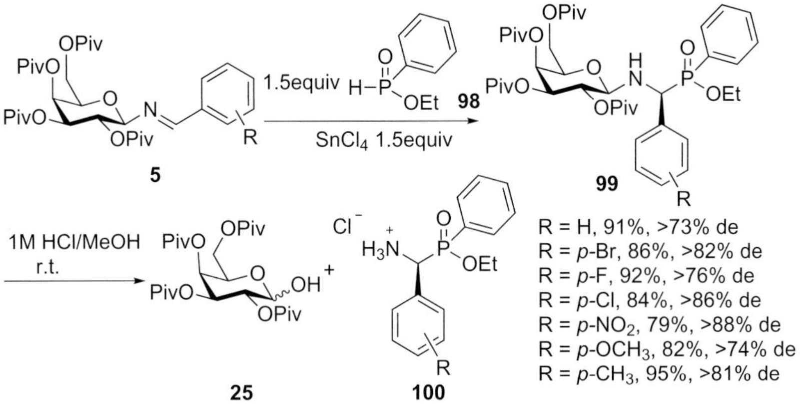

R = H, 91%, >73% de
R = p-Br, 86%, >82% de
R = p-F, 92%, >76% de
R = p-Cl, 84%, >86% de
R = p-NO₂, 79%, >88% de
R = p-OCH₃, 82%, >74% de
R = p-CH₃, 95%, >81% de

Scheme 1.31 Synthesis of N-galactosyl arylphosphonoglycine esters **99**.

1.7
Dynamic Kinetic Resolution of α-Chloro Carboxylic Esters

Another interesting application of carbohydrate-derived auxiliaries is the dynamic kinetic resolution of racemic α-halogenated carboxylic esters [44]. Park reported a D-glucose-derived auxiliary in the dynamic resolution of α-halo esters in an asymmetric nucleophilic substitution [45, 46]. α-Chloro-α-phenyl ester (α*RS*)-**101** was obtained as a diastereomeric mixture by the reaction of diacetone-D-glucose and racemic α-chloro-α-phenylacetyl chloride in the presence of Et₃N. Treatment of (α*RS*)-**101** with various amines, and diisopropylethylamine (DIPEA) in the presence of tetrabutylammonium iodide (TBAI), gave the amino acid derivatives **102** in high yields and high diastereomeric ratios. After treatment of esters **102** in methanol with Et₃N at room temperature, the chiral auxiliary was successfully removed (Scheme 1.32).

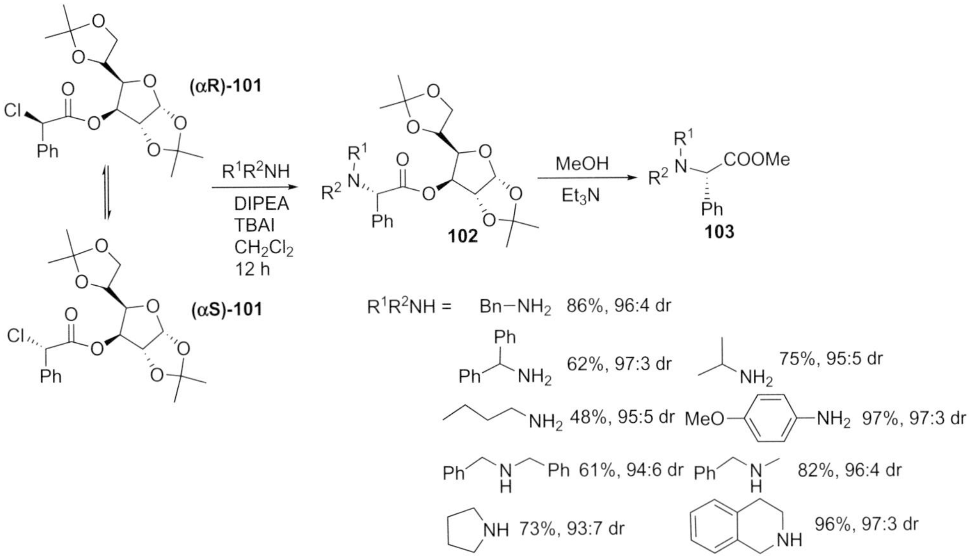

1.7 Dynamic Kinetic Resolution of -Chloro Carboxylic Esters

Scheme 1.32 Dynamic kinetic resolution of α-chloro ester **101** in nucleophilic substitution.

Park also employed D-allofuranose as auxiliary for the dynamic kinetic resolution of α-chloro esters in nucleophilic substitutions [46]. Using the same reaction conditions previously for D-glucose derivative **101** and benzylamine as the nucleophile, dynamic resolution of α-chloro acetate **104** took place with high stereoselectivity, affording **105** in moderate isolated yield with 90:10 dr (αS:αR) (Scheme 1.33).

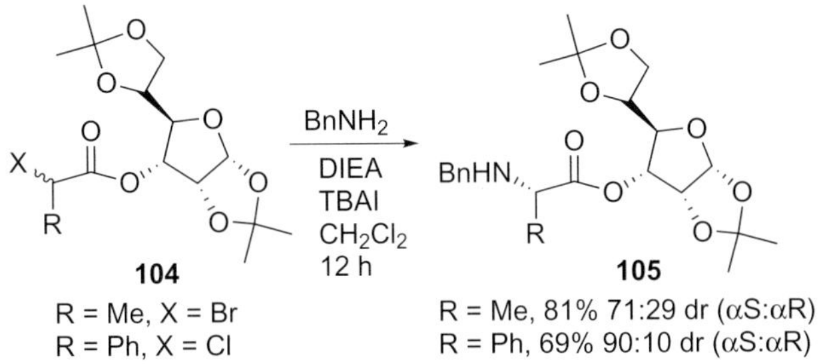

Scheme 1.33 Dynamic kinetic resolution of α-chloro ester **104** in nucleophilic substitution.

Based on the results, a plausible mechanism for the nucleophilic substitutions of D-glucose derivatives and D-allose derivatives has been suggested (Figure 1.6) [46]. The authors proposed two transition states in which the α-R group and the C=O bond in the ester substituent adopt an *s-cis* conformation, while the ester carbonyl group is in an eclipsed position relative to the hydrogen atom at C3 of the furanose. The nucleophilic attack of an amine nucleophile may then be aided by hydrogen bond formation with one oxygen atom from the 5,6-O and 1,2-O

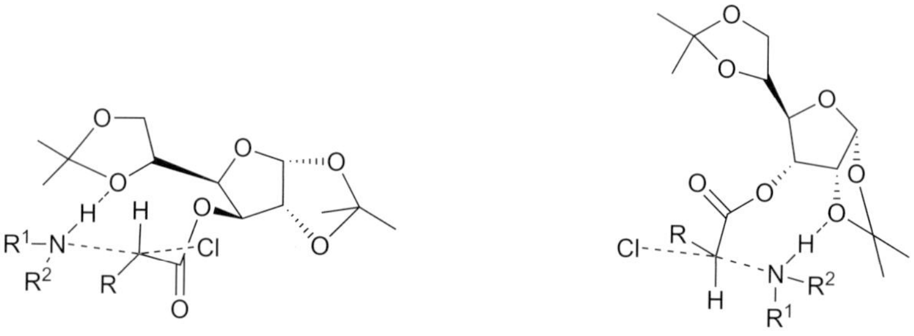

Transition state for D-glucose derivatives   Transition state for D-allose derivatives

Figure 1.6 Proposed transition state structures for dynamic kinetic resolution of α-halo esters of glucose- and allose-derived carbohydrate auxiliaries.

dioxolanes of the chiral auxiliaries in **(αR)-101** and **(αS)-101**, respectively. These tentative transition states explain the *(S)*-configurations of the products observed for both the D-glucose and D-allose derived auxiliary.

References

1 Vasella, A. (1977) *Helv. Chim. Acta*, **60**, 1273.
2 (a) Cintas, P. (1991) *Tetrahedron*, **47**, 6079; (b) Reissig, H.U. (1992) *Angew. Chem. Int. Ed. Engl.*, **31**, 288; (c) Kunz, H. and Rück, K. (1993) *Angew. Chem. Int. Ed. Engl.*, **32**, 336; (d) Hale, K.J. (1993) *Second Supplement to the Second Edition of Rodd's Chemistry of Carbon Compounds*, vol. 1E/F/G (ed. M. Sainsbury), Elsevier, Amsterdam, Chapter 23b, p. 273; (e) Kunz, H. (1995) *Pure Appl. Chem.*, **67**, 1627; (f) Hultin, P.G., Earle, M.A., and Sudharshan, M. (1997) *Tetrahedron*, **53**, 14823; (g) Knauer, S., Kranke, B., Krause, L., and Kunz, H. (2004) *Curr. Org. Chem.*, **8**, 1739; (h) Boysen, M.M.K. (2007) *Chem. Eur. J.*, **13**, 8648; (i) Lehnert, T., Özüduru, G., Grugel, H., Albrecht, F., Telligmann, S.M., and Boysen, M.M.K. (2011) *Synthesis*, 2685.
3 (a) Enders, D. and Shilvock, J.P. (2000) *Chem. Soc. Rev.*, **29**, 359; (b) Merino, P., Marqués-López, E., Tejero, T., and Herrera, R.P. (2009) *Tetrahedron*, **65**, 1219.
4 (a) Kunz, H. and Sager, W. (1987) *Angew. Chem. Int. Ed. Engl.*, **26**, 557; (b) Kunz, H., Sager, W., Schanzenbach, D., and Decker, M. (1991) *Liebigs Ann. Chem.*, 649; (c) Kunz, H., Sager, W., Pfrengle, W., and Schanzenbach, D. (1988) *Tetrahedron Lett.*, **29**, 4397.
5 Kunz, H., Pfrengle, W., Rück, K., and Sager, W. (1991) *Synthesis*, 1039.
6 Paulsen, H., Györgydeak, Z., and Friedmann, M. (1974) *Chem. Ber.*, **107**, 1568.
7 Wang, D., Zhang, P.F., and Yu, B. (2007) *Helv. Chim. Acta*, **90**, 938.
8 Zhou, G.B., Zheng, W.X., Wang, D., Zhang, P.F., and Pan, Y.J. (2006) *Helv. Chim. Acta*, **89**, 520.
9 Porter, J.R., Wirschun, W.G., Kuntz, K.W., Snapper, M.L., and Hoveyda, A.H. (2000) *J. Am. Chem. Soc.*, **122**, 2657.
10 Ugi, I., Offermann, K., Herlinger, H., and Marquarding, D. (1967) *Justus Liebigs Ann. Chem.*, 7091.
11 (a) Dömling, A. (2006) *Chem. Rev.*, **106**, 17; (b) Dömling, A. and Ugi, I. (2000) *Angew. Chem. Int. Ed.*, **39**, 3168.
12 (a) Kunz, H. and Pfrengle, W. (1988) *J. Am. Chem. Soc.*, **110**, 651; (b) Kunz, H. and Pfrengle, W. (1988) *Tetrahedron*, **44**, 5487.

13 Kunz, H., Pfrengle, W., and Sager, W. (1989) *Tetrahedron Lett.*, **30**, 4109.
14 Zech, G. and Kunz, H. (2004) *Chem. Eur. J.*, **10**, 4136.
15 Costantino, G., Marinozzi, M., Camaioni, E., Natalini, B., Sarichelou, I., Micheli, F., Cavanni, P., Faedo, S., Noe, C., Moroni, F., and Pellicciari, R. (2004) *Farmaco*, **59**, 93.
16 Ugi, I., Marquarding, D., and Urban, R. (1982) *Chemistry and Biochemistry of Amino Acids, Peptides and Proteins*, vol. 6 (ed. B. Eeinstein), Marcel Dekker, New York, p. 245.
17 Preobrazhenskaya, M.N. and Suvorov, N.N. (1965) *Zh. Obshch. Khim.*, **85**, 888.
18 Aebischer, B.M., Hanssen, H.W., Vasella, A.T., and Schweizer, W.B. (1982) *J. Chem. Soc. Perkin Trans. 1*, 2139.
19 Goebel, M. and Ugi, I. (1991) *Synthesis*, 1095.
20 Lehnhoff, S., Goebel, M., Karl, R.M., Klösel, R., and Ugi, I. (1995) *Angew. Chem. Int. Ed. Engl.*, **34**, 1104.
21 (a) Micheel, F. and Klemer, A. (1961) *Adv. Carbohydr. Chem.*, **16**, 95; (b) Pfleiderer, W. and Bühler, E. (1966) *Chem. Ber.*, **99**, 3022.
22 Ross, G.F., Herdtweck, E., and Ugi, I. (2002) *Tetrahedron*, **58**, 6127.
23 Ingles, D.L. and Whistler, R.L. (1962) *J. Org. Chem.*, **27**, 3896.
24 (a) Birkofer, L. and Ritter, A. (1965) *Angew. Chem. Int. Ed. Engl.*, **4**, 417; (b) Paulsen, H., Gyorgydeak, Z., and Friedmann, M. (1974) *Chem. Ber.*, **107**, 1568; (c) Strumpel, M.K., Buschmann, J., Szilagyi, L., and Gyorgydeak, Z. (1999) *Carbohydr. Res.*, **318**, 91.
25 Ross, G.F. and Ugi, I. (2001) *Can. J. Chem.*, **79**, 1934.
26 (a) Yamamoto, Y., Nishii, S., Maruyama, K., Komatsu, T., and Ito, W. (1986) *J. Am. Chem. Soc.*, **108**, 7778; (b) Yamamoto, Y., Komatsu, T., and Maruyama, K. (1984) *J. Am. Chem. Soc.*, **106**, 5031; (c) Yamamoto, Y., Komatsu, T., and Maruyama, K. (1985) *J. Chem. Soc., Chem. Commun.*, 814.
27 (a) Laschat, S. and Kunz, H. (1990) *Synlett*, 51; (b) Laschat, S. and Kunz, H. (1990) *Synlett*, 629; (c) Laschat, S. and Kunz, H. (1991) *J. Org. Chem.*, **56**, 5883.
28 (a) Risch, N., Arend, M., and Westermann, B. (1998) *Angew. Chem., Int. Ed. Engl.*, **37**, 1044 and references therein; (b) Tramontini, M. and Angiolini, L. (1994) *Mannich Bases, Chemistry and Uses*, CRC Press, Boca Raton, FL, and references therein; (c) Volkmann, R.A. (1991) *Comprehensive Organic Synthesis*, vol. 1 (eds B.M. Trost and I. Fleming), Pergamon, Oxford, UK, p. 355 and references therein.
29 Vilaivan, T., Bhanthumnavin, W., and Sritana-Anant, Y. (2005) *Curr. Org. Chem.*, **9**, 1315.
30 (a) Seebach, D. and Hoffmann, M. (1998) *Eur. J. Org. Chem.*, 1337; (b) Aoyagi, Y., Jain, R.P., and Williams, R.M. (2001) *J. Am. Chem. Soc.*, **123**, 3472 and references therein; (c) Evans, D.A., Urpi, F., Somers, T.C., Clark, J.S., and Bilodeau, M.T. (1990) *J. Am. Chem. Soc.*, **112**, 8215; (d) Kober, R., Papadopoulos, K., Miltz, W., Enders, D., Steglich, W., Reuter, H., and Puff, H. (1985) *Tetrahedron*, **42**, 1693; (e) Palomo, C., Oiarbide, M., Landa, A., Gonzales-Rego, M.C., Garcia, J.M., Gonzales, A., Odriozola, J.M., Martin-Pastor, M., and Linden, A. (2002) *J. Am. Chem. Soc.*, **124**, 8637 and references therein.
31 (a) Córdova, A. (2004) *Acc. Chem. Res.*, **37**, 102; (b) Loh, T.P. and Chen, S.L. (2002) *Org. Lett.*, **4**, 3647.
32 Kunz, H. and Schanzenbach, D. (1989) *Angew. Chem. Int. Ed. Engl.*, **28**, 1068.
33 (a) Braun, M., Sacha, H., Galle, D., and El-Alali, A. (1995) *Tetrahedron Lett.*, **36**, 4213; (b) Ojima, I., Habus, I., Zhao, M., Georg, G.I., and Jayasinghe, L.R. (1991) *J. Org. Chem.*, **56**, 1681.
34 Kunz, H., Burgard, A., and Schanzenbach, D. (1997) *Angew. Chem. Int. Ed. Engl.*, **36**, 386.
35 Allef, P. and Kunz, H. (2000) *Tetrahedron: Asymmetry*, **11**, 375.
36 Kunz, H. and Pfrengle, W. (1989) *Angew. Chem. Int. Ed. Engl.*, **28**, 1067.
37 Weymann, M., Pfrengle, W., Schanzenbach, D., and Kunz, H. (1997) *Synthesis*, 1151.
38 (a) Kranke, B., Hebranlt, D., Schulz-Kukula, M., and Kunz, H. (2004) *Synlett*, 671; (b) Kranke, B., and Kunz, H. (2006) *Can. J. Chem.*, **84**, 625.

39 Huber, R., Knierzinger, A., Obrecht, J.-P., and Vasella, A. (1985) *Helv. Chim. Acta*, **68**, 1730.
40 Huber, R. and Vasella, A. (1987) *Helv. Chim. Acta*, **70**, 1461.
41 Laschat, S. and Kunz, H. (1992) *Synthesis*, 90.
42 Wang, Y.D., Wang, F., Wang, Y.Y., Miao, Z.W., and Chen, R.Y. (2008) *Adv. Synth. Catal.*, **350**, 2339.
43 Wang, Y.D., Wang, Y.Y., Yu, J., Miao, Z.W., and Chen, R.Y. (2009) *Chem. Eur. J.*, **15**, 9290.
44 For examples using other auxiliaries see: (a) Valenrod, Y., Myung, J., Ben, R.N. (2004) *Tetrahedron Lett.*, **45**, 2545; (b) Nam, J., Lee, S.-K., and Park, Y.S. (2003) *Tetrahedron*, **59**, 2397; (c) Nam, J., Lee, S.-K., Kim, K.Y., and Park, Y.S. (2002) *Tetrahedron Lett.*, **43**, 8253; (d) Lee, S.-K., Nam, J., and Park, Y.S. (2002) *Synlett*, 790; (e) Ben, R.N. and Durst, T. (1999) *J. Org. Chem.*, **64**, 7700.
45 Kim, H.J., Shin, E.K., Chang, J.Y., Kim, Y., and Park, Y.S. (2005) *Tetrahedron Lett.*, **46**, 4115.
46 Kim, H.J., Kim, Y., Choi, E.T., Lee, M.H., No, E.S., and Park, Y.S. (2006) *Tetrahedron*, **62**, 6303.

2
1,4-Addition of Nucleophiles to α,β-Unsaturated Carbonyl Compounds

Kiichiro Totani and Kin-ichi Tadano

2.1
Introduction

In this chapter we describe the 1,4-additions of various nucleophiles to α,β-unsaturated carbonyl groups on carbohydrate templates. Many examples have revealed that carbohydrates serve as effective chiral auxiliaries for realizing useful levels of diastereoselectivity [1–8]. As acyclic 1,4-addition acceptors, acrylic amides or acrylic esters on a carbohydrate template are described. Then, as cyclic acceptors, 2- and 4-pyridones are also exemplified.

2.2
1,4-Additions to Acrylic Amides and Acrylic Esters

As an early example concerning the utility of carbohydrate-based auxiliaries for stereoselective carbon–carbon bond formation, Kunz and Pees reported in 1989 the stereoselective 1,4-addition of the carbon nucleophile generated from diethylaluminum chloride to a D-xylose-derived N-cinnamoyl 2-oxazinone **2** [9, 10]. The substrate **2** was synthesized by N-cinnamoylation of 3,5-O,N-oxazin-2-one derivative **1**, which in turn was synthesized from D-xylose (Scheme 2.1) [9]. The 1,4-addition of diethylaluminum chloride to **2** at −80 °C proceeded with remarkable π-facial selectivity to provide (S)-3-phenylpentanoyl derivative **3** in high yield with 74% diastereomeric excess (de). Hydrolysis of **3** with HCl provided enantioenriched (3S)-phenylpentanoic acid **4**, the absolute stereochemistry of which was confirmed by comparison of its optical rotation to the reported $[\alpha]_D$ (the yield and ee of **4** were not reported).

Later, the Kunz group demonstrated other stereoselective syntheses of chiral β-branched carboxylic acids using the 1,4-addition of organometallic species to carbohydrate-derived N-acyloxazolidinones [10–12]. Thus, oxazolidinone **5** as the auxiliary for the 1,4-addition was synthesized from D-galactose via azidonitration

Scheme 2.1 1,4-Addition of an ethyl nucleophile to D-xylose-derived N-cinnamoyl-2-oxazinone 2.

of 3,4,6-tri-O-pivaloyl-D-galactal, followed by hydrolysis and oxazoline formation via a Staudinger–aza-Wittig reaction cascade (Scheme 2.2).

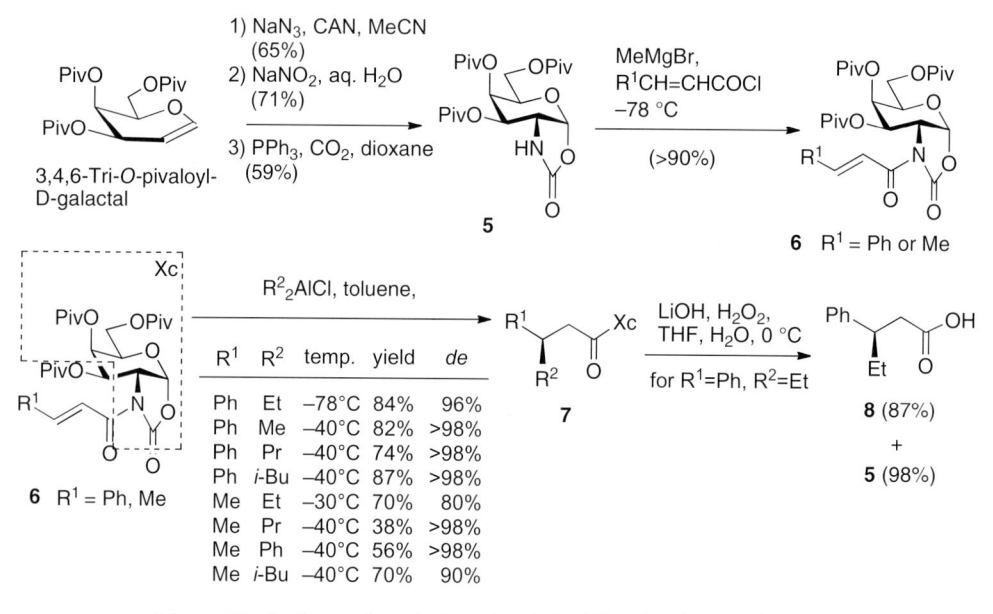

R[1]	R[2]	temp.	yield	de
Ph	Et	−78°C	84%	96%
Ph	Me	−40°C	82%	>98%
Ph	Pr	−40°C	74%	>98%
Ph	i-Bu	−40°C	87%	>98%
Me	Et	−30°C	70%	80%
Me	Pr	−40°C	38%	>98%
Me	Ph	−40°C	56%	>98%
Me	i-Bu	−40°C	70%	90%

Scheme 2.2 Synthesis of D-galactosamine-derived N-acylated oxazolidinones 6 and subsequent 1,4-additions with various organoaluminum species.

Conversion of 5 into the N-acyl derivatives 6 by deprotonation with methylmagnesium bromide was followed by treatment with the two α,β-unsaturated acyl chlorides. The resulting N-acyloxazolidin-2-ones 6, fused with per-O-pivaloylated

D-galactopyranose at C1 and C2, were subjected to 1,4-additions with various organoaluminum chlorides at low temperature. As a result, β-substituted N-acyl derivatives were obtained in high to excellent diastereoselectivity and in moderate to high yield. In all cases, attack of the carbon nucleophiles occurred from the front side, opposite the pivaloyloxy group installed at C3. The configurations of the newly introduced stereocenters at the β-carbon of **7** were determined after removal of the D-galactose moiety by alkaline hydrolysis in the presence of hydrogen peroxide, producing **8**. The oxazolidinone-type template **5** was recovered efficiently. On 1,4-addition of the (diisobutyl)aluminum reagent to **6** with R^1 = Ph, the *(S)*-adduct **7** was obtained in diastereomerically pure form (>98% de). Using the α,β-unsaturated carbonyl derivative **6**, the Kunz group also explored the formation of β-branched α-halo carboxylic acid derivatives via 1,4-addition of dialkylaluminum chlorides followed by treatment with N-chlorosuccinimide [13].

To obtain the opposite enantiomers of **8**, the Kunz group designed another carbohydrate-based template as a chiral auxiliary. For this purpose, they chose D-arabinose-based bicyclic oxazolidin-2-one **10** as the auxiliary, which in turn was synthesized from readily available 3,4-di-O-acetyl-D-arabinal via de-O-acetylation, per-O-pivaloylation, and then azidonitration, providing the 2-azido derivative **9**. By using the analogous route employed for the synthesis of **5**, the 2-azido sugar **9** was efficiently converted into **10** (Scheme 2.3) [12]. Obviously, the two bicyclic oxazolidin-2-one derivatives **5** and **10** have a pseudo-enantiomeric relationship in constituting a chiral environment [14]. A cinnamoyl or crotonyl group was introduced into the oxazolidin-2-one moiety in **10** by the same procedure used for the preparation of **6**. As shown in Scheme 2.3, the 1,4-additions of several carbon nucleophiles to the resulting unsaturated amides **11** proceeded with moderate to high diastereoselectivity to provide the β-branched N-acyl oxazolidin-2-ones **12** with the opposite configuration at the β-carbon, compared to those in the adducts **7**. The auxiliary moiety in the 1,4-addcut **12** (R^1 = Ph, R^2 = Et) was efficiently removed with H_2O_2 in an aqueous solution in the presence of LiOH to provide *(S)*-3-phenylpentanoic acid **13** along with efficient recovery of the oxazolidinone **10**. In some 1,4-additions of a carbon nucleophile to **11**, organoaluminum compounds, prepared by mixing alkyllithium or alkyl Grignard reagents and $MeAlCl_2$, were essential for obtaining higher diastereoselectivity.

On the other hand, the Kunz group has also investigated D-glucosamine-based 1,2-oxazolidin-2-ones as a chiral auxiliaries by exploring the 1,4-additions of diethylaluminum chloride to the corresponding N-cinnamoyl derivative. In this case, the diastereoselectivity of the 1,4-adduct was slightly decreased (91% yield, 80% de) compared to the case of **11** [15].

Hon and coworkers explored the utility of D-xylofuranose derivatives as chiral auxiliaries for stereoselective 1,4-addition to obtain enantioenriched β-branched carboxylic acids [16]. For example, the 5-phenylated derivative **17** was synthesized as a chiral auxiliary from readily prepared 3-O-benzyl-1,2-O-isopropylidene-α-D-glucofuranose (Scheme 2.4). Oxidative cleavage of the 5,6-diol in the staring D-glucofuranose derivative provided *aldehydo*-sugar **14**, which was treated with PhMgBr to provide **15**. Chlorination of **15** followed by reductive dechlorination of

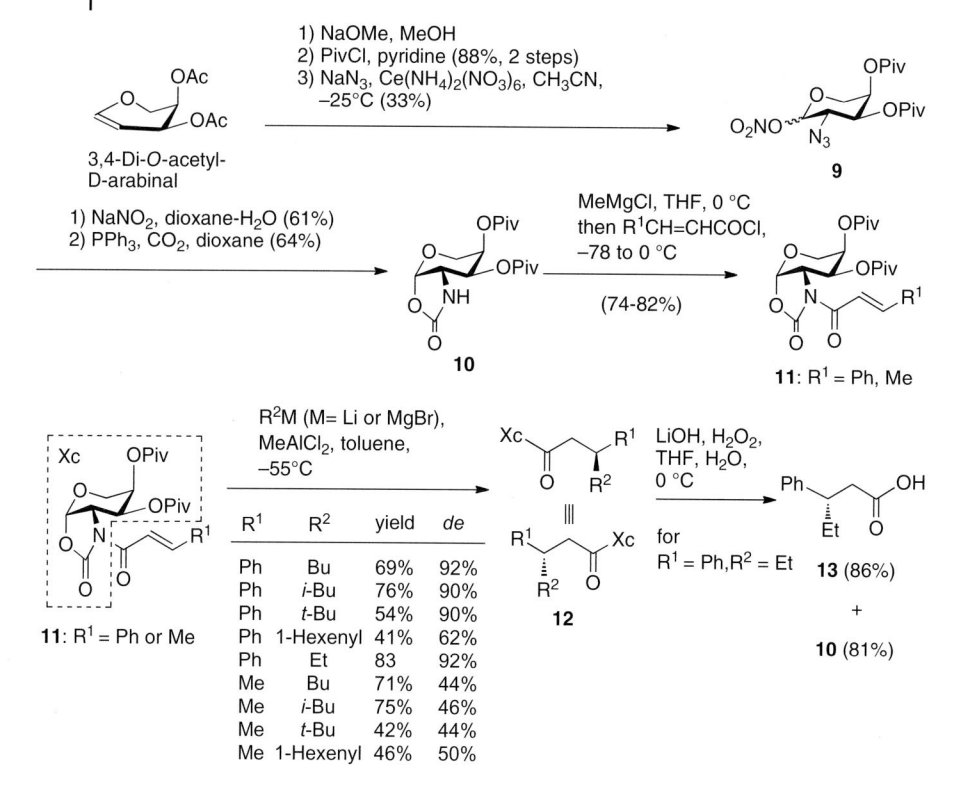

Scheme 2.3 Synthesis of D-arabinal-derived N-acylated oxazolidin-2-ones **11** and subsequent 1,4-additions with various mixed organoaluminum species.

the resulting chloride **16** eventually provided the auxiliary **17**. Crotonylation of **17** provided the 3-O-crotonyl ester **18**. The 1,4-additions of various organocuprates, prepared by mixing alkyl Grignard reagents and CuBr·Me$_2$S, to **18** provided the 1,4-adducts **19** in moderate to high yield with approximately 90% de for R^1 = Ph, vinyl, allyl, heptyl, Et, and i-Pr. The Hon group also explored the 1,4-additions of similar organocuprates to the corresponding 3-O-cinnamoyl derivative. In these cases, the diastereoselectivity was in general less practical. In addition, the 3-O-cinnamoyl and 3-O-crotonyl derivatives of the one-carbon-elongated 1,2-O-isopropylidene-5-deoxy-5-phenylmethyl-α-D-xylofuranose were explored as substrates for 1,4-additions. The 1,4-additions using these compounds provided the corresponding 1,4-adducts in lesser diastereoselectivity. The authors rationalized the high diastereoselectivities obtained using **18** as follows: the addition of the cuprates to unsaturated ester **18** occurs from the side opposite of the bulky phenyl group. As shown in Scheme 2.4, the *s-trans* conformation in the crotonate moiety is the most favorable in this case.

Since the 1990s, Tadano and coworkers have designed, synthesized, and utilized various hexopyranoside-derived templates to verify their potential as efficient

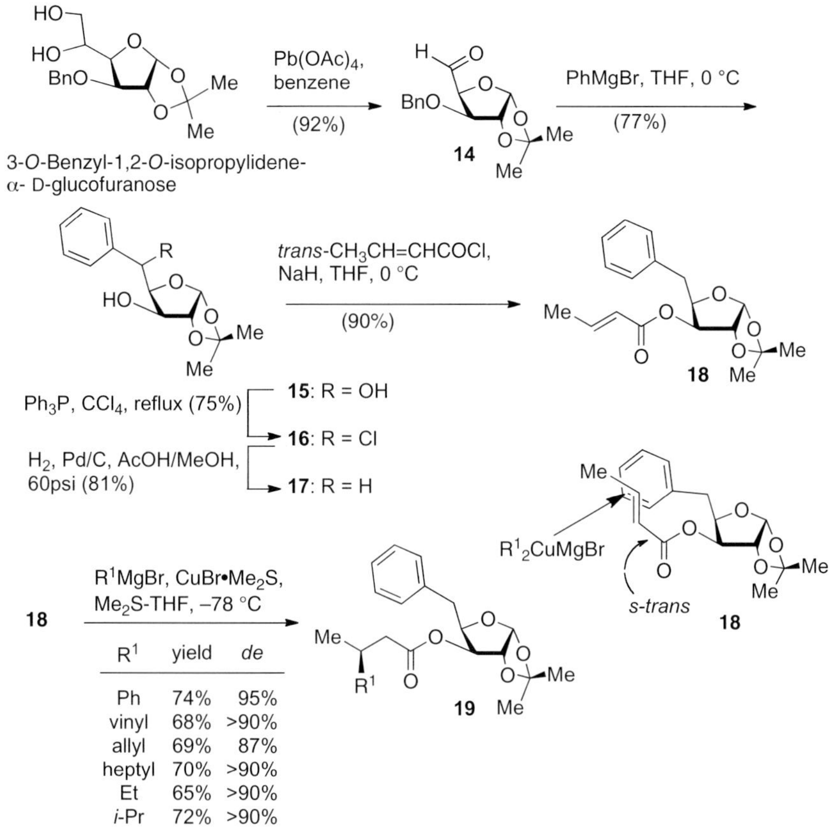

R¹	yield	*de*
Ph	74%	95%
vinyl	68%	>90%
allyl	69%	87%
heptyl	70%	>90%
Et	65%	>90%
i-Pr	72%	>90%

Scheme 2.4 Synthesis of D-xylose-derived 3-*O*-crotonyl ester **18** and subsequent 1,4-additions with various organocuprates.

chiral auxiliaries for various carbon–carbon bond-forming reactions [6]. The Tadano group explored extensively the difference in the spatial environment constituted by each hexopyranoside configuration, which affected the diastereoselectivity of the attempted organic reactions. Consequently, they synthesized various carbohydrate-based templates from D-glucose, D-mannose, or D-galactose, in which an unprotected hydroxyl group at C2, -3, -4, or -6 was used to introduce the reaction site. These substrates were subjected to 1,4-addition reactions, alkylations, cycloadditions, and other reactions. In each reaction, the Tadano group expected that stereodifferentiation would proceed principally by the shielding effect of the protecting group neighboring the reaction site. The sugar template was then removed from each product by basic, acidic, or reductive treatment. As a result, various α- and/or β-chiral carboxylic acids or alcohols could be obtained in enantiomerically pure or highly enriched form. In addition, the used sugar templates could be recovered.

2 1,4-Addition of Nucleophiles to α,β-Unsaturated Carbonyl Compounds

As the typical carbon–carbon bond-forming reaction using carbohydrate templates, the Tadano group explored the 1,4-conjugate additions of organometallic species to α,β-unsaturated esters, such as crotonyl or cinnamoyl esters incorporated at the C4 of methyl 6-iodo-, 6-deoxy-2,3-di-O-protected, or 2,3,6-tri-O-protected or at C6 of 2,3,4-tri-O-proteced α-D-glucopyranosides [17, 18]. As one example, the synthesis of methyl 2,3-di-O-benzyl-6-deoxy-α-D-glucopyranoside **20** and the 1,4-addition of an organocopper reagent to its 4-crotonyl ester **21** are depicted in Scheme 2.5. The carbohydrate auxiliary **20** was synthesized efficiently from known methyl 2,3-di-O-benzyl-α-D-glucopyranoside by a two-step reaction sequence, that is, hydride reduction of the corresponding 6-O-tosylate. The crotonyl ester formation of the hydroxyl group at C4 in **20** provided **21**. The 1,4-addition of a vinylcopper reagent, prepared by mixing vinylmagnesium bromide and cuprous bromide–dimethyl sulfide complex in THF–Me$_2$S (2:1), provided preferentially the 1,4-adduct **22** in 90% yield with 74% de.

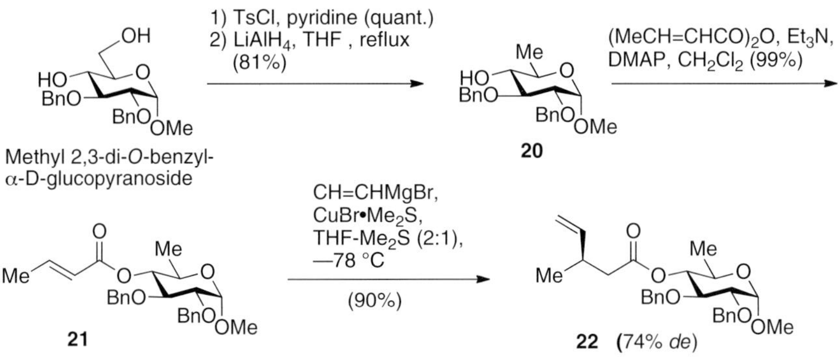

Scheme 2.5 Synthesis of 2,3-di-O-benzyl-6-deoxy-α-D-glucopyranoside **20** and subsequent 1,4-addition of a vinyl cuprate to its 4-O-crotonyl ester **21**.

The Tadano group introduced 6-deoxy-2,3-di-O-(t-butyldimethylsilyl)-α-D-glucopyranoside **26** as the most effective chiral auxiliary to date for the 1,4-conjugate addition of organocuprates to an unsaturated ester incorporated in a hexopyranoside derivative. Compound **26** was prepared efficiently staring from methyl 4,6-O-benzylidene-α-D-glucopyranoside [19] (Scheme 2.6): O-Silylation of the 4,6-O-benzylidene derivative with t-butyldimethylsilyl (TBS) chloride, followed by removal by hydrogenolysis of the benzylidene acetal in the resulting **23**, and successive mono-O-sulfonyl ester formation provided the 6-O-tosyl derivative **24**. After substitution of the O-tosyl group by an iodo group, the resulting iodide **25** was treated with Raney nickel in EtOH to provide the auxiliary **26**. To introduce an unsaturated ester as the reaction site for the planned 1,4-addition, **26** was treated with crotonic acid in the presence of trifluoroacetic anhydride (TFAA) to provide the 4-O-crotonyl derivative **27**. The 1,4-additions of three organocuprates to **27** provided the 1,4-addcuts **28** in high yields with useful levels of diastereoselectivity (90% de) (Scheme 2.6) [20]. All the 1,4-adducts **28** (R^2 = vinyl, ethyl, and phenyl) possess the same configuration at the newly introduced stereocenter at

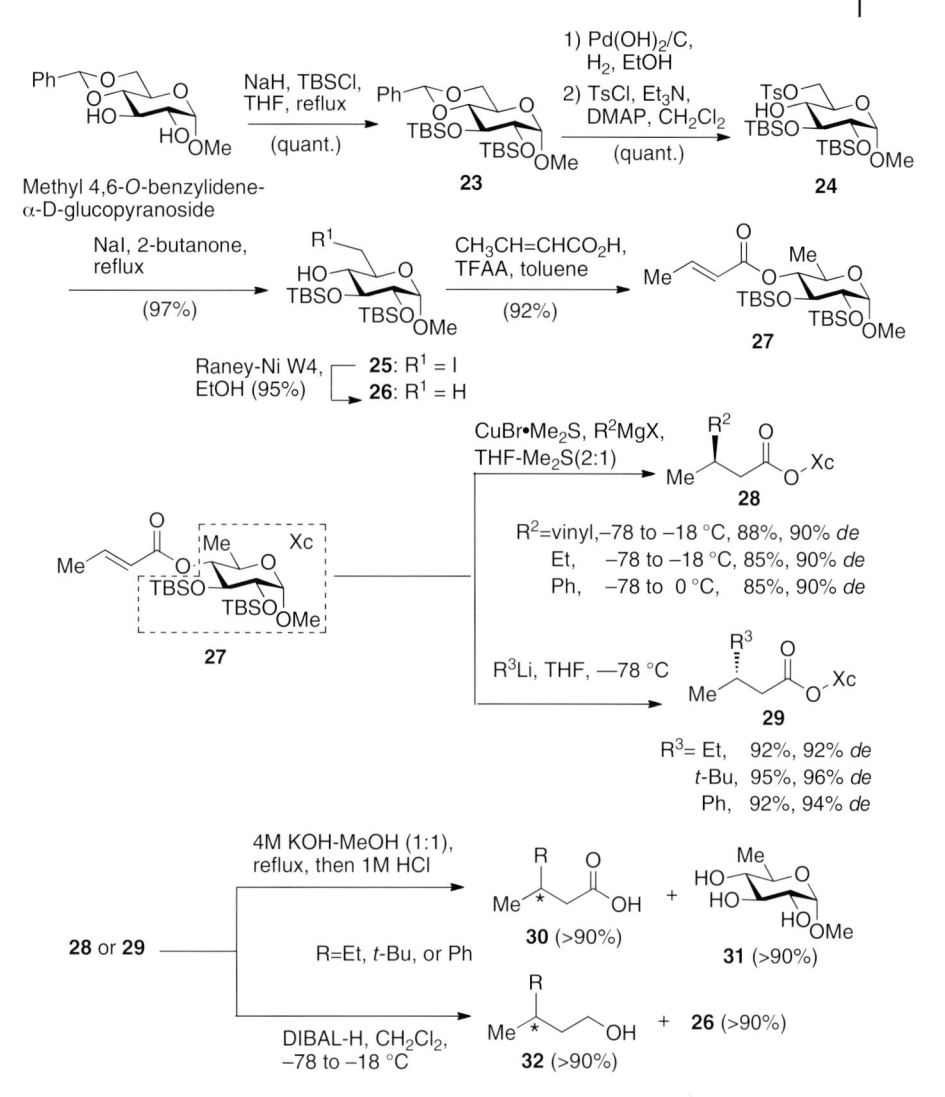

Scheme 2.6 Synthesis of methyl 2,3-di-O-(t-butyldimethylsilyl)-6-deoxy-α-D-glucopyranoside **26** and subsequent 1,4-additions of organocuprates and organolithiums to its 4-O-crotonyl ester **27**.

the β-carbon of the butanoyl ester moieties. In contrast, the 1,4-additions of alkyllithium compounds (ethyl, t-butyl, or phenyllithium) to **27** at −78°C resulted in exclusive formation of the 1,4-adducts **29** (instead of the 1,2-adducts), possessing β-carbon configurations opposite to those in **28**, in both high yields and excellent diastereoselectivities. From the 1,4-adducts **28** or **29**, β-alkylated (or arylated) butanoic acids **30** or 3-alkylated (or arylated) butanols **32**, respectively, were obtained in highly enantioenriched form by alkaline hydrolysis (aqueous KOH) or DIBAL-H

2 1,4-Addition of Nucleophiles to α,β-Unsaturated Carbonyl Compounds

reduction. The carbohydrate moiety was either re-isolated as methyl 6-deoxy-α-D-glucopyranoside **31** after alkaline hydrolysis, or carbohydrate auxiliary **26** was recovered efficiently after hydride reduction [6].

As shown in Figure 2.1, it seems reasonable to assume that the unsaturated ester moiety in **27** exists preferentially in an *s-trans, syn* conformation to avoid the steric congestion incurred by coordination of the organometallic species to the ester carbonyl in the *s-cis, syn* conformation. In the favorable transition state **TS-A**, the neighboring bulky 3-O-TBS group effectively shields the front side of the 4-O-crotonyl ester moiety. Thus, the attack of organocuprates occurs preferentially from the less congested rear side, providing 1,4-adducts **28**. For the addition of alkyllithium reagents to **27**, it seems reasonable that the crotonyl ester exists

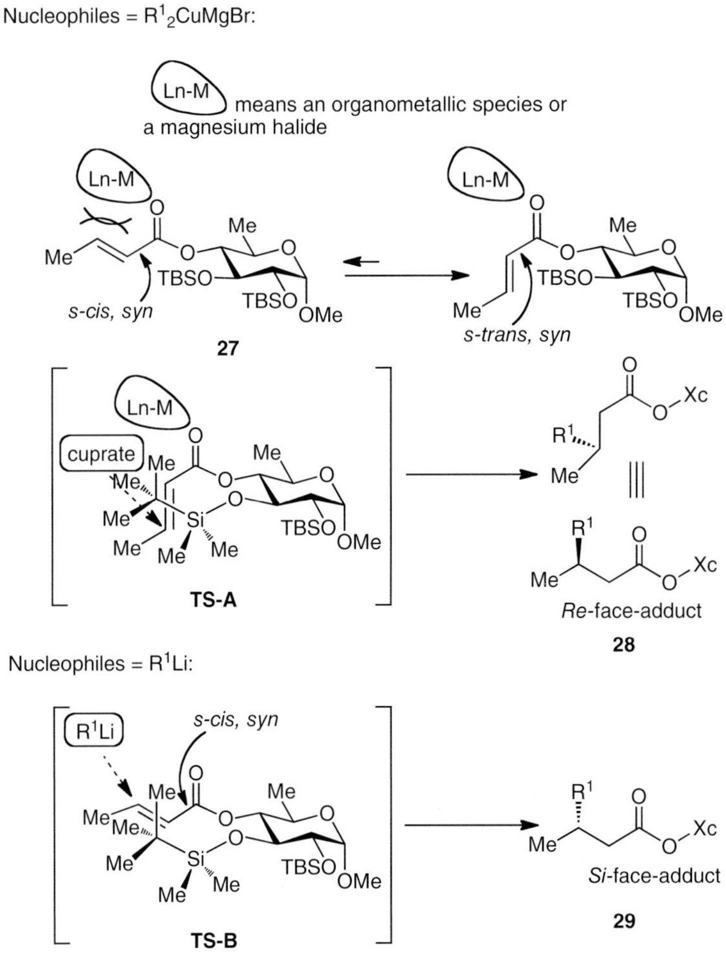

Figure 2.1 Preferred transition states for the 1,4-addition of organocuprates or organolithium to **27**.

predominantly in an *s-cis*, *syn* conformation and that no coordination by the lithium cation occurs. According to these considerations, **TS-B** is the most favorable transition state, leading to an attack of the alkyllithium compound from the less congested rear side, providing **29**.

The 1,4-addition of carbon nucleophiles to the 4-*O*-crotonyl ester **27**, followed by α-alkylation of the enolate intermediate in a one-pot process, was realized with high *anti/syn*-selectivity and good diastereoselectivities (Scheme 2.7) [20]. Namely, the 1,4-addition of a phenyl cuprate, prepared from phenylmagnesium bromide and copper(I) bromide, to **27**, followed by trapping of the intermediate enolate with iodomethane, provided α,β-disubstituted butanoic acid derivative **33** with high *syn*-selectivity in 52% yield along with 39% of the 1,4-adduct **28**. Furthermore, it should be emphasized that the diastereomeric excess of this *syn*-adduct **33** was almost complete. Another interesting finding was observed by using phenyllithium as the carbon nucleophile instead of phenyl cuprate. This one-pot reaction provided the *anti*-product **34** with remarkably high *anti/syn*-selectivity and diastereoselectivity (>98% de). From the two products **33** or **34**, respectively, the almost enantiomerically pure 2-methyl-3-phenylbutanols **35** or **36** were obtained by reductive removal of the carbohydrate auxiliary.

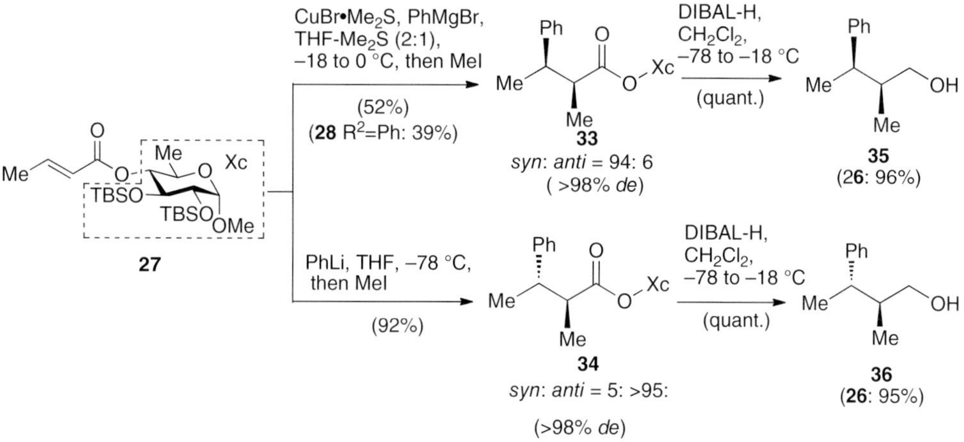

Scheme 2.7 One-pot 1,4-addition/α-alkylation of phenyl-metallic species/MeI to **27** and removal of the carbohydrate auxiliary from the product.

As an application of the carbohydrate-based stereoselective carbon–carbon bond-forming reactions to natural product synthesis, the Tadano group achieved the highly stereoselective total synthesis of (−)-lasiol **41**, an acyclic monoterpene alcohol (Scheme 2.8) [21]. Natural lasiol was isolated as an insect sex-attracting pheromone, which was found in the mandibular gland secretions of the male ant *Lasius meridionalis*. The structure of this pheromone (**41**) is shown in Scheme 2.8. For the synthesis the synthesis of **41** started with auxiliary **26**, which was transformed into 4-*O*-propionyl ester **37**. After deprotonation of **37** at the α-position of the ester moiety in **37**, a highly diastereoselective 1,4-addition to methyl crotonate produced 1,4-adduct **38** almost as a single diastereomer. The product **38** was

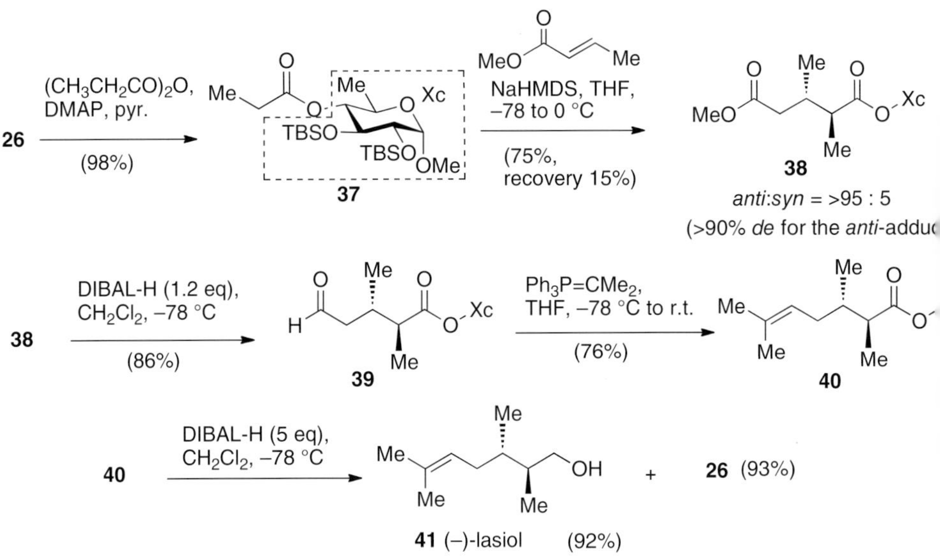

Scheme 2.8 Total synthesis of (−)-lasiol (**41**) via the stereoselective 1,4-addition of the enolate derived from the 4-O-propionyl ester **37** to methyl crotonate.

obtained with *anti*-stereochemistry (*anti*:*syn* = >95:5) for the vicinal methyl substituents at the α- and β-carbons of the 4-O-diester moiety. Furthermore, the diastereoselectivity in favor of the major *anti*-adduct **38** was determined to be more than 90% de. Selective DIBAL-H reduction of the less-congested methyl ester moiety in **38** efficiently provided aldehyde **39**. Wittig olefination of **39** with Ph$_3$P=C(Me)$_2$ provided the 2-methylpropenylated product **40**. Reductive removal of the carbohydrate auxiliary from **40** with excess DIBAL-H provided (−)-lasiol **41**, and the chiral template **26** was efficiently recovered.

The 1,4-additions of radical species to the 4-O-crotonyl derivative of methyl 6-deoxy-2,3-O-protected α-D-glucopyranosides were also explored by the Tadano group (Scheme 2.9) [19]. As the most efficient auxiliary for the 1,4-radical addition, methyl 6-deoxy-2,3-di-O-mesitoyl (2,4,6-trimethylbenzoyl) -α-D-glucopyranoside **45**, prepared from methyl 4,6-O-benzylidene-α-D-glucopyranoside using the analogous route to **26**, was found to be the most efficient auxiliary for this purpose. Thus, 2,3-di-O-acylation of the 4,6-O-benzylidene acetap, followed by debenzylidenation, provided **42**, which was then converted into 6-deoxy derivative **45** via the 6-O-tosyl derivative **43** and 6-iodide **44**. Crotonylation of **45** provided the substrate **46** for the 1,4-radical reaction. The radial reactions – carried out under standard conditions involving a combination of alkyl iodide, tributyltin hydride, and triethylborane/oxygen in the presence of a Lewis acid (AlCl$_3$, BF$_3$·Et$_2$O, or Et$_2$AlCl) – provided various β-alkylated butanoic acid esters **47**. For example, the 4-O-crotonyl ester **46** provided the 1,4-adducts **47** (R^1 = *i*-propyl, ethyl, or *t*-butyl) most effectively when Et$_2$AlCl was used as the Lewis acid. The enantiomerically enriched β-alkylated butanoic acids were isolated as the respective anilides **48** by saponification of the adducts **46**, followed by treatment of the resulting β-alkylated butanoic acids **47** with aniline in the presence of N-(3-dimethylamino)propyl-N′-ethylcarbodiimide (EDCI).

The glycopyranoside-based diastereoselective 1,4-additions were further developed by the Tadano group using other hexopyranosides, such as α-D-manno- and α-D-galactopyranosides. Scheme 2.10 shows representative examples for the D-mannopyranose-based approach [18]. Thus, two 3-O-crotylated-α-D-mannopyranosides, **50** and **52**, were synthesized from known methyl 4,6-O-benzyl-α-D-mannopyranoside.

Selective O-benzylation of the 4,6-di-O-benzyl derivative provided **49**, which was acylated with crotonic anhydride to provide the 3-O-crotonyl ester **50**. On the other hand, treatment of the 4,6-O-benzyl derivative with crotonic anhydride in the presence of $CuCl_2$ provided preferentially the 3-O-crotonyl ester **51**, which was then acylated with pivaloyl (Piv) chloride to provide **52**.

The 1,4-addition of a vinyl cuprate, prepared from vinylmagnesium bromide and cuprous bromide, to **50** preferentially provided the 1,4-adduct **53** possessing the (R)-configuration at the new stereogenic center, in 84% yield. Alkaline hydrolysis of **53** to remove the carbohydrate auxiliary and successive alkylation of the resulting carboxylate benzyl bromide provided β-vinylated butanoic acid benzyl ester **54** in 76% ee. On the other hand, 1,4-addition of vinyl cuprate to **52** provided the 1,4-adduct **55**, possessing a new stereogenic center with (S)-configuration, in 78%

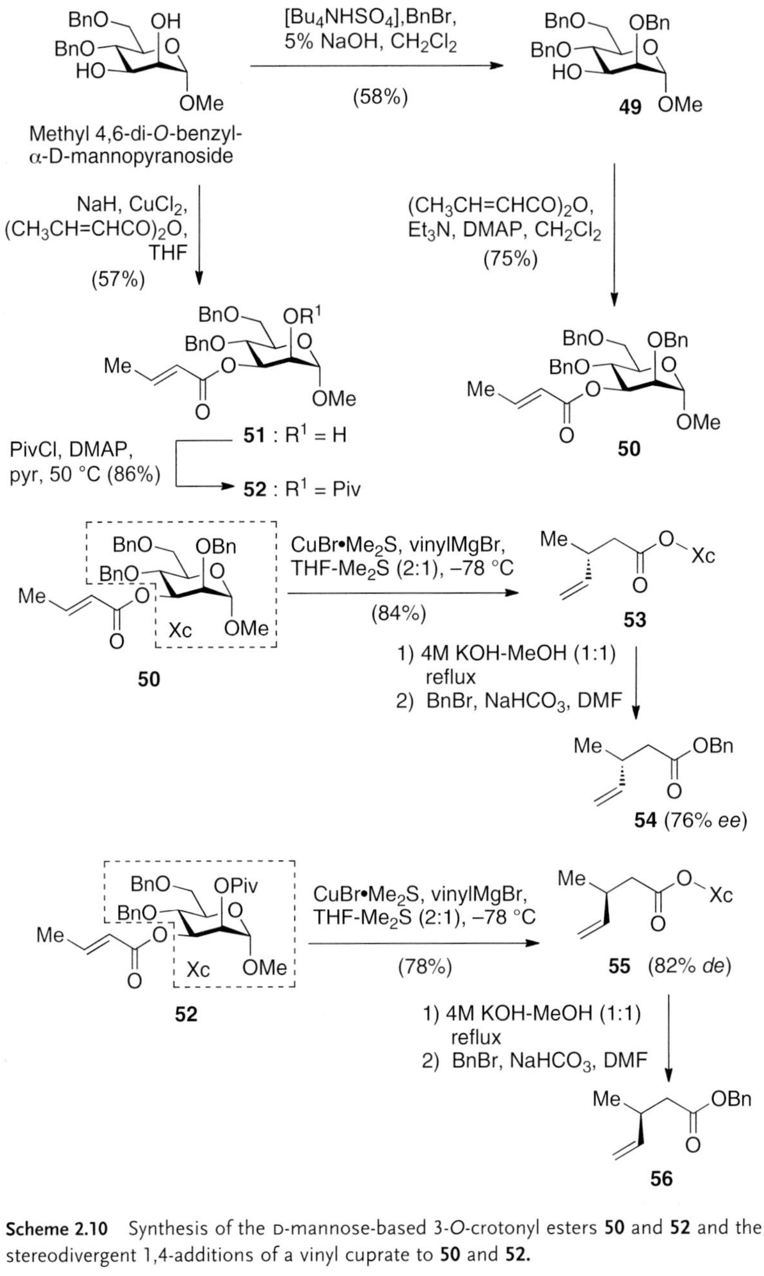

Scheme 2.10 Synthesis of the D-mannose-based 3-O-crotonyl esters **50** and **52** and the stereodivergent 1,4-additions of a vinyl cuprate to **50** and **52**.

yield. Removal of the auxiliary from **55** and alkylation of the resulting carboxylate provided **56**.

Analogously, 3-O-crotonyl-α-D-galactopyranosides **57** and **59** provided two kinds of 1,4-adducts with moderate to high levels of diastereoselectivity [18]. As shown in Scheme 2.11, the 3-O-crotonyl esters **57** and **59** were prepared from known

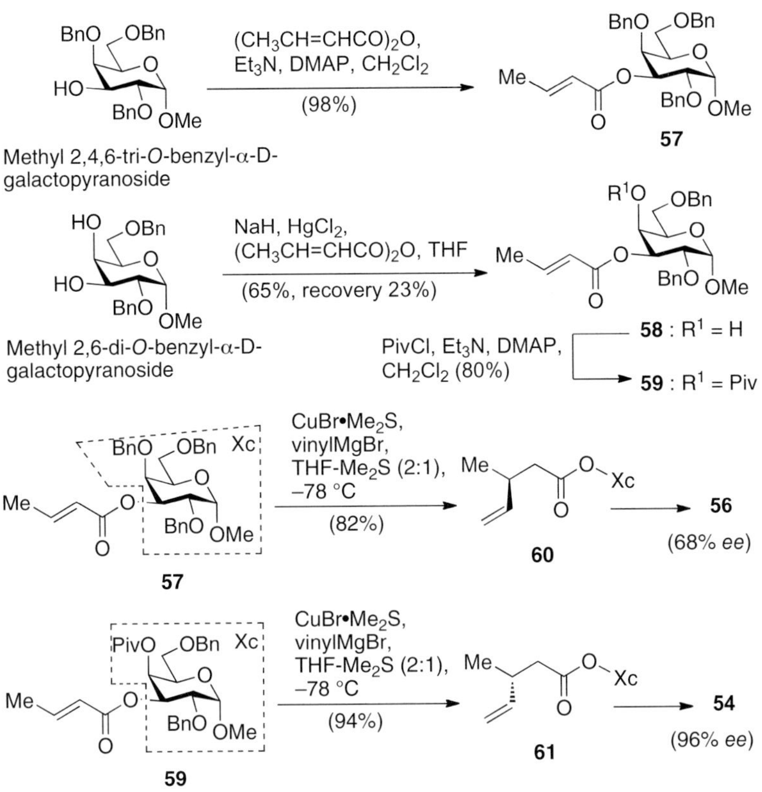

2.2 1,4-Additions to Acrylic Amides and Acrylic Esters | 39

Scheme 2.11 Synthesis of the D-galactose-based 3-O-crotonyl esters **57** and **59** and the stereodivergent 1,4-additions of a vinyl cuprate to **57** and **59**.

methyl 2,4,6-tri-O-benzyl-α-D-galactopyranoside and methyl 2,6-di-O-benzyl-α-D-galactopyranoside, respectively, by procedures analogous to those used for the syntheses of D-mannose derivatives **50** and **52**. The 1,4-addition of a vinyl cuprate to **57** provided the 1,4-adduct **60**, which was converted into **56** with moderate ee of 68%. On the other hand, 1,4-addition of the same vinyl cuprate to **59** provided the 1,4-adduct **61**. The (R)-3-Vinylbutanoic acid benzyl ester **54**, obtained from adduct **61**, was obtained in an excellent ee of 96%.

Enders and coworkers utilized 1,2:5,6-di-O-isopropylidene-α-D-allo-hexofuranose **62** as an effective chiral auxiliary, which was obtained from 1,2:5,6-di-O-isopropylidene-α-D-glucofuranose by a well-known oxidation/reduction procedure (Scheme 2.12). In the context of the asymmetric synthesis of analogues of homotaurine (3-aminopropanesulfonic acids), the Enders group has explored the diastereoselective and enantioselective synthesis of α,β-disubstituted γ-nitro methyl sulfonates using the auxiliary **62** [22, 23]. This asymmetric approach started from 3-O-sulfonyl-D-allofuranoses **63**, which was prepared by sulfonylation of **62** with two (phenylmethyl)sulfonyl chlorides (R^1 = H or R^1 = t-Bu). Deprotonation at the

40 | 2 1,4-Addition of Nucleophiles to α,β-Unsaturated Carbonyl Compounds

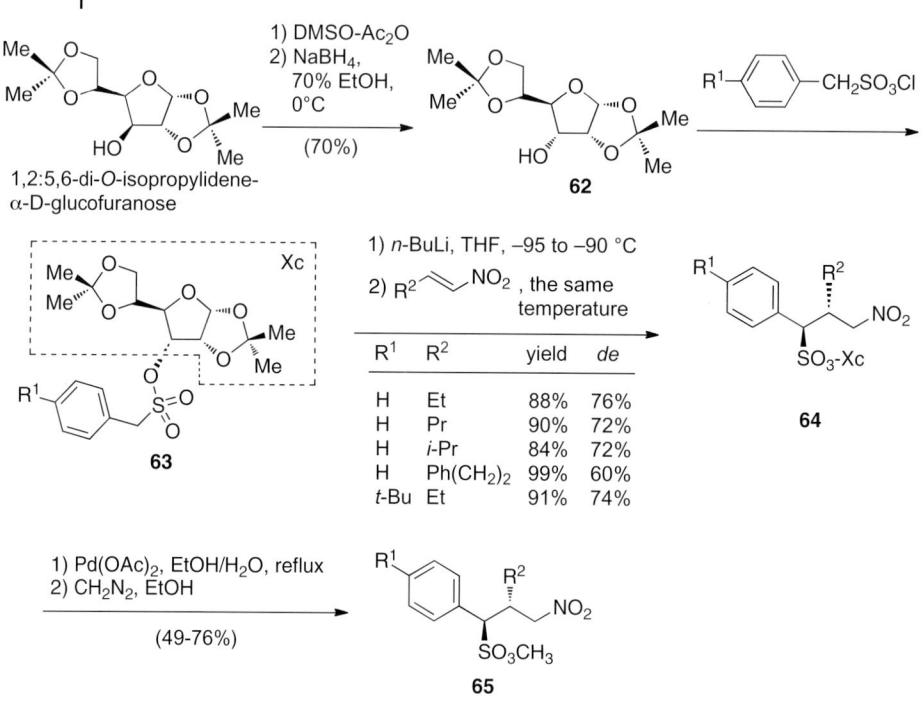

R¹	R²	yield	de
H	Et	88%	76%
H	Pr	90%	72%
H	i-Pr	84%	72%
H	Ph(CH$_2$)$_2$	99%	60%
t-Bu	Et	91%	74%

Scheme 2.12 Synthesis of D-allofuranose-derived phenylmethyl-sulfonate **63** and its 1,4-additions to various β-alkylated nitroalkenes.

phenylmethyl group inside the sulfonyl ester moiety with BuLi at −95 °C produced the α-lithiated sulfonates, which were then reacted with various β-alkylated nitroalkenes. The 1,4-additions proceeded smoothly and produced the 1,4-adducts **64** in good to excellent yields and moderate diastereoselectivities. After improving the diastereomeric excesses by preparative HPLC or recrystallization, various γ-nitrosulfonic esters **64** were obtained in virtually enantiopure form. By treatment of **64** with catalytic Pd(OAc)$_2$ in aqueous EtOH, followed by esterification of the resulting sulfonic acid with diazomethane, the desired α,β-disubstituted γ-nitro methyl sulfonates **65** were obtained.

2.3
1,4-Addition to 4- and 2-Pyridones

The Kunz group has extensively explored the utility of pivaloylated galactosylamine **67** for stereoselective carbon–carbon bond-forming reactions. This compound is readily available from D-galactose as shown in Scheme 2.13 [24]. Pivaloylation followed by treatment of the resulting penta-O-pivaloyl derivative with trimethylsilyl azide in the presence of SnCl$_4$ provided the β-D-galactosyl azide **66** as a single anomer.

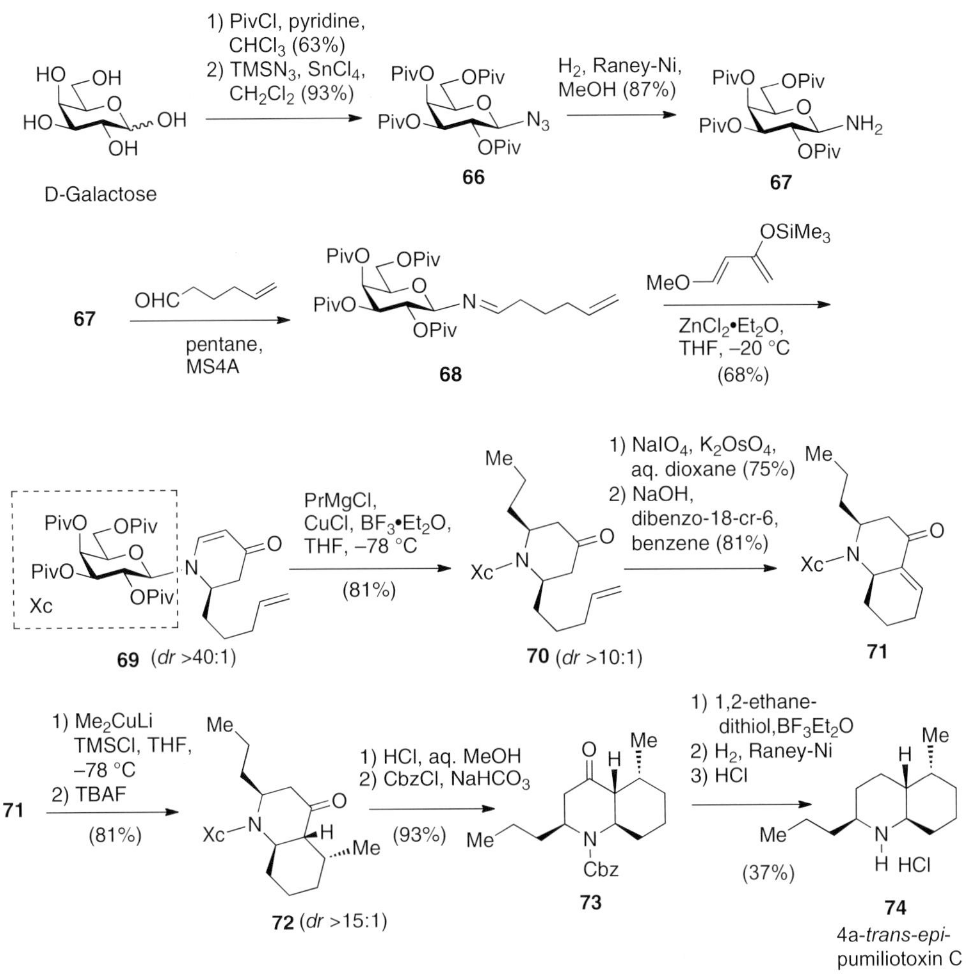

Scheme 2.13 Total synthesis of 4a-*trans* epi-pumiliotoxin (**74**) from per-*O*-pivaloyl-β-D-galactosylamine (**67**).

Hydrogenation of **66** provided the β-glycosylamine **67**. As a representative example of stereoselective carbon–carbon bond-forming reactions using **67**, the Kunz group developed the tandem Mannich/1,4-addition reaction of various glycosylimines prepared from **67**. The products of this reaction sequence are starting materials for the synthesis of structurally simple piperidine alkaloids [25]. Later, the Kunz group has extended this strategy using **67** as a chiral platform for the synthesis of functionalized decahydroquinoline-type alkaloids. As shown in Scheme 2.13, β-D-galactosylimine **68**, prepared from **67** and 5-hexenal, was treated with the Danishefsky–Kitahara diene in the presence of ZnCl$_2$·Et$_2$O to provide in excellent diastereoselectivity (dr >40:1) galactopyranosyl 5,6-dehydropiperidin-4-one **69**, which possesses a 4-pentenyl substituent at C2 with

(R)-configuration in the dehydropiperidine ring [26]. The conversion of **68** into **69** involved an initial Mannich reaction between the glycosylimine **68** and the siloxy diene, followed by 1,4-addition of the intermediate glycosylamine to the resulting conjugate enone and subsequent elimination of MeOH. For construction of 2,6-disubstituted 4-piperidone skeletons, a second 1,4-addition to the enone moiety in **69** was explored. A combination of the organocuprate, prepared from PrMgCl and CuCl, and boron trifluoride worked well for this purpose. Thus cis-2,6-dialkylated piperidin-4-one **70**, which incorporates the carbohydrate template at nitrogen, was obtained with high diastereoselectivity (dr >10:1) and in an acceptable yield of 81%. The Kunz group further demonstrated the synthetic utility of piperidinone **70** for the synthesis of decahydroquinoline alkaloids. Pumiliotoxin C, a representative of the cis-annulated decahydroquinoline alkaloids, was isolated from a glandular secretion of a South American frog of the genus *Dendrobates*. Some trans-annulated decahydroquinoline alkaloids are also known. The Kunz group completed the total synthesis of trans-annulated 4a-epi-pumiliotoxin C (**74**) from **70**: Oxidative cleavage of the vinyl group in the side chain in **70**, followed by base-mediated intramolecular aldol condensation of the resulting aldehyde, produced octahydroquinoline derivative **71**. The 1,4-addition of Me_2CuLi to the bicyclic enone **71** in the presence of TMSCl efficiently provided **72** (dr >15:1) after cleavage of the intermediate silyl enol ether. From trans-annulated decahydroquinoline **72**, trans-4a-epi-pumiliotoxin C (**74**) was synthesized in a five-step reaction sequence, including removal of the D-galactosyl moiety and removal of the carbonyl functionality in the resulting **73** via hydrogenolytic desulfurization of the corresponding dithiolane. More recently, the Kunz group applied these 1,4-additions based on the 4-dihydropyridone scaffold attached at the anomeric carbon of the D-galactopyranose to solid-phase synthesis [27, 28]. Furthermore, the Kunz group has extended this glycosylamine-based synthesis of piperidine alkaloids to a D-arabinosylamine-derived auxiliary, which was prepared from 2,3,4-tri-O-pivaloyl-α-D-arabinopyranosylamine by a reaction sequence analogous to that used for **67**. This compound acts as a pseudo enantiomer to galactosyl amine **67** and thus the reaction with the Danishefsky-Kitahara diene leads to carbohydrate-bound dehydropiperidine derivatives with stereocentres in opposite configuratiion compared to those in galactose-bound dehydropiperidines **69**. The 1,4-addition of various organocopper reagents to these 2-alkylated (or -arylated) N-arabinopyranosyl 5,6-dehydropiperidin-4-ones provided various 2,6-cis-substituted piperidines [29].

Desymmetrization of prochiral compounds is one of the most convenient strategies for supplying enantioenriched chiral compounds. The Kunz group reported desymmetrization of a 4-pyridone tethered to a carbohydrate unit by the nitrogen atom. For this purpose, per-O-pivaloylated α-D-galactopyranosyl fluoride **75**, which was conveniently prepared from D-galactose [24], was reacted with 4-(trimethylsiloxy)pyridine in the presence of a Lewis acid ($TiCl_4$), which provided N-(β-D-galactopyranosyl)-4-pyridone **76** quantitatively (Scheme 2.14) [30, 31]. The 4-pyridone **76**. This compound was subsequently treated with iPr_3SiOTf and 2,6-lutidine to form the corresponding N-glycosyl 4-siloxypyridinium salt, which was reacted with organomagnesium halides (R^1 = Me, Pr, and Ph). The expected 1,4-addition proceeded with good to excellent diastereoselectivities (80–100% de),

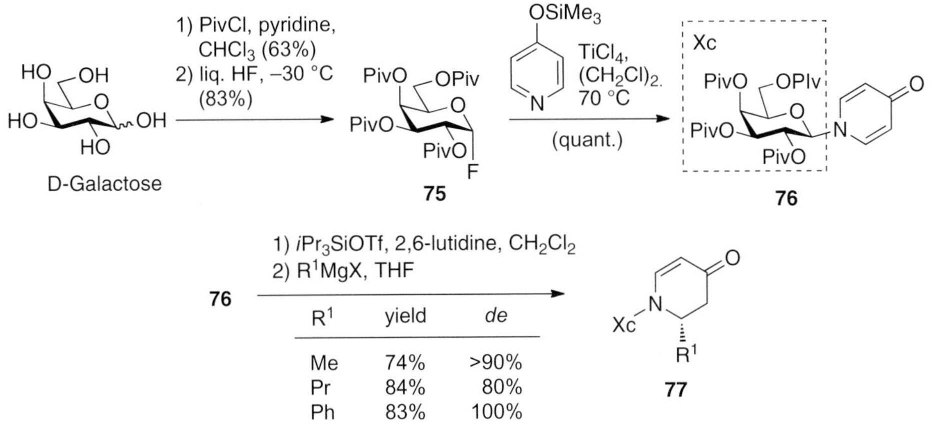

Scheme 2.14 Synthesis of N-(β-D-galactopyranosyl)-4-pyridone **76** and 1,4-additions of Grignard reagents to its 4-pyridone moiety.

providing 2-alkyl-substituted 5,6-dehydropiperidin-4-ones **77** in high yields. The carbohydrate auxiliary mediates a diastereoselective attack of the Grignard reagent at the planar 4-silyloxypyridinium intermediate; this constitutes a selective monofunctionalization and efficient diastereofacial differentiation of the two double bonds of the 4-pyridone in **76**.

The Kunz group also explored the synthetic potential of N-(β-D-galactopyranosyl)-2-pyridone **78**, which was synthesized by the reaction of galactosyl fluoride **75** and 2-(trimethylsiloxy)pyridine in the presence of Lewis acids such as $TiCl_4$ (Scheme 2.15) [15, 31]. The 2-pyridone **78** carrying the D-galactopyranosyl moiety as a chiral auxiliary was treated with TMSOTf and various Grignard reagents (R = Pr, i-Pr, Ph) in the presence of 2,6-lutidine. The 1,4-addition products **79** were obtained in all cases with excellent diastereoselectivity (>98% de). In the case of **78**, two rotamers **78-A** and **78-B**, resulting from the rotation around the glycosidic C–N bond, are stabilized by the *exo*-anomeric effect. However, the rotamer **78-B** is favored owing to the minimized electrostatic repulsion between the 2-O-pivaloyl carbonyl and the pyridone carbonyl group. Consequently, the Grignard reagent attack occurs exclusively at the 4-position of pyridone from the less shielded *Si*-face.

The Kunz group utilized the products **80** formed in this transformation for the stereoselective synthesis of benzomorphan derivatives (Scheme 2.16) [32]. The 4-benzylated 5,6-dehydropiperidin-2-ones **80** were synthesized by the reaction of **78** and benzylmagnesium chloride in the presence of 2,6-lutidine after the formation of O-silylated pyridinium salt with iPr_3SiOTf. As expected, these 1,4-additions of the Grignard reagents occurred stereoselectively in moderate to good yields. Intramolecular amino-alkylation of **80** (R^1 = Bn) in the presence of a mixture of HCl and $SnCl_4$ provided tricyclic benzazocinone **81** via the intermediate N-acyl iminium ion. Further elaboration of the functional groups in **81** eventually provided 7,8-benzomorphane hydrochloride **83**. This transformation included reductive removal of the δ-lactam carbonyl via the thioamide, prepared by treatment of **81** with Lawesson's reagent followed by hydrogenolysis in the presence of Raney

44 | *2 1,4-Addition of Nucleophiles to α,β-Unsaturated Carbonyl Compounds*

Scheme 2.15 Synthesis of *N*-D-galactopyranosyl-2-pyridone **78** and 1,4-additions of Grignard additions to its 2-pyridone moiety.

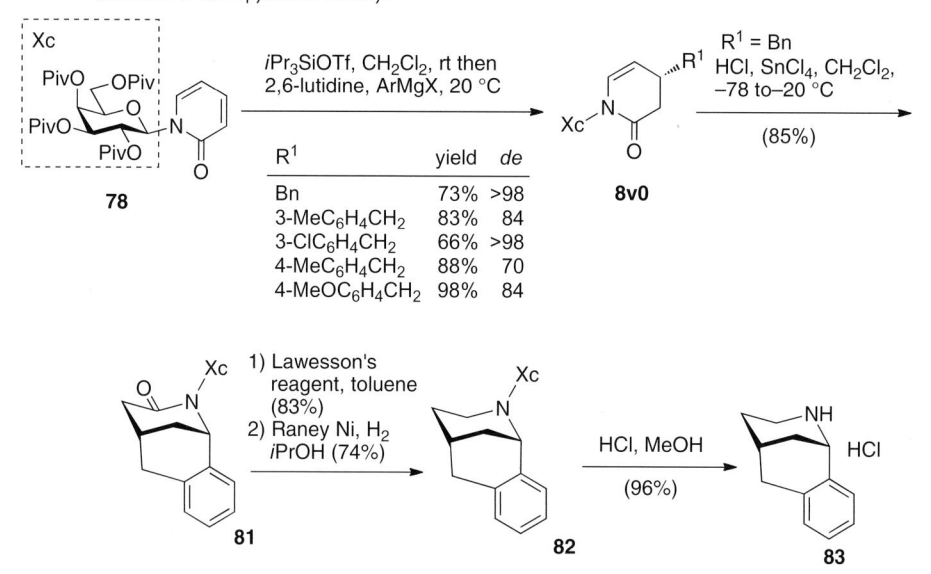

R^1	yield	de
Pr	68%	>98%
i-Pr	88%	>98%
cyclohexyl	79%	>98%
Ph	76%	>98%

R^1	yield	de
Bn	73%	>98
3-MeC$_6$H$_4$CH$_2$	83%	84
3-ClC$_6$H$_4$CH$_2$	66%	>98
4-MeC$_6$H$_4$CH$_2$	88%	70
4-MeOC$_6$H$_4$CH$_2$	98%	84

Scheme 2.16 Synthesis of 7,8-benzomorphane **83** via 1,4-addition of benzyl Grignard reagent to **78**.

nickel, and acid hydrolytic removal of the sugar template in the resulting compound **82**, eventually providing **83**.

References

1. Kunz, H. and Rück, K. (1993) *Angew. Chem. Int. Ed. Engl.*, **32**, 336–358.
2. Kunz, H. (1995) *Pure Appl. Chem.*, **67**, 1627–1635.
3. Seyden-Penne, J. (1995) *Chiral Auxiliaries and Ligands in Asymmetric Synthesis*, John Wiley & Sons, Inc., New York.
4. Hultin, P.G., Earle, M.A.M.A., and Sudharshan, M. (1997) *Tetrahedron*, **53**, 14823–14870.
5. Rück-Braun, K. and Kunz, H. (1999) *Chiral Auxiliaries in Cycloaddition*, Wiley-VCH Verlag GmbH, Weinheim.
6. Totani, K., Takao, K., and Tadano, K. (2004) *Synlett*, 2066–2080.
7. Boysen, M.M.K. (2007) *Chem. Eur. J.*, **13**, 8648–8659.
8. Totani, K. and Tadano, K. (2008) *Glycoscience – Chemistry and Chemical Biology*, 2nd edn, (eds B.O. Fraser-Reid, K. Tatsuta, and J. Thiem), Springer Verlag, Berlin, pp. 1029–1075.
9. Kunz, H. and Pees, K.J. (1989) *J. Chem. Soc. Perkin Trans. 1*, 1168–1169. For preparation of **1**, see: Kunz, H., Mohr, J., Pfrengle, W., and Sager, W. (1988) *Proceedings 2nd Akabori-Conference, Kashikojima 1987* (ed. S. Sakakibara), Protein Research Foundation, Osaka, p. 1.
10. Rück, K. and Kunz, H. (1993) *Synthesis*, 1018–1028.
11. Rück, K. and Kunz, H. (1992) *Synlett*, 343–344.
12. Elzner, S., Maas, S., Engel, S., and Kunz, H. (2004) *Synthesis*, 2153–2164.
13. Rück-Braun, K., Stamm, A., Engel, S., and Kunz, H. (1997) *J. Org. Chem.*, **62**, 967–975.
14. Kunz, H., Pfrengle, W., and Sager, W. (1989) *Tetrahedron Lett.*, **30**, 4109–4110.
15. Knauer, S., Kranke, B., Krause, L., and Kunz, H. (2004) *Curr. Org. Chem.*, **8**, 1739–1761.
16. Hon, Y.-S., Cheng, F.-L., Huang, Y.-P., and Lu, T.-J. (1991) *Tetrahedron: Asymmetry*, **2**, 879–882.
17. Totani, K., Nagatsuka, T., Takao, K., Ohba, S., and Tadano, K. (1999) *Org. Lett.*, **1**, 1447–1450.
18. Totani, K., Nagatsuka, T., Yamaguchi, S., Tadano, K., Ohba, S., and Tadano, K. (2001) *J. Org. Chem.*, **66**, 5965–5975.
19. Munakata, R., Totani, K., Takao, K., and Tadano, K. (2000) *Synlett*, 979–982.
20. Totani, K., Asano, S., Takao, K., and Tadano, K. (2001) *Synlett*, 1772–1776.
21. Asano, S., Tamai, T., Totani, K., Takao, K., and Tadano, K. (2003) *Synlett*, 2252–2254.
22. Enders, D., Berner, O.M., Vignola, N., and Bats, J.W. (2001) *Chem. Commun.*, 2498–2499.
23. Enders, D., Berner, O.M., Vignola, N., and Harnying, W. (2002) *Synthesis*, 1945–1952.
24. Kunz, H., Sager, W., Schanzenbach, D., and Decker, M. (1991) *Liebigs Ann. Chem.*, 649–654.
25. Kunz, H. and Pfrengle, W. (1989) *Angew. Chem. Int. Ed. Engl.*, **28**, 1067–1068.
26. Weymann, M., Schultz-Kukula, M., and Kunz, H. (1998) *Tetrahedron Lett.*, **39**, 7835–7838.
27. Zech, G. and Kunz, H. (2003) *Angew. Chem. Int. Ed.*, **42**, 787–790.
28. Zech, G. and Kunz, H. (2004) *Chem. Eur. J.*, **10**, 4136–4149.
29. Kranke, B., Hebrault, D., Schultz-Kukula, M., and Kunz, H. (2004) *Synlett*, 671–674.
30. Follmann, M. and Kunz, H. (1998) *Synlett*, 989–990.
31. Klegraf, E., Follmann, M., Schollmeyer, D., and Kunz, H. (2004) *Eur. J. Org. Chem.*, 3346–3360.
32. Klegraf, E., Knauer, S., and Kunz, H. (2006) *Angew. Chem. Int. Ed.*, **45**, 2623–2626.

3
Reaction of Enolates

Noureddine Khiar, Inmaculada Fernández, Ana Alcudia, Maria Victoria García and Rocío Recio

3.1
Introduction

The aldol reaction is one the best known and widely employed methods for the stereoselective buildup of carbon–carbon bonds [1]. Within the different approaches developed so far, the chiral auxiliary-based aldol reaction is the best-studied transformation and is well understood in terms of the reaction mechanism. While, in principle, a stoichiometric auxiliary could be used in various ways in the aldol reaction, the consensus is that the chiral-directing group should be incorporated into the (enolate) nucleophile. Although uncommon, there are examples where the chiral auxiliary is in the electrophilic component [2]. The chiral auxiliary-based aldol reaction has reached such efficiency and reliability that it is employed in innumerable syntheses of active pharmaceutical ingredients (APIs) needed for preclinical and early clinical studies on a multi-kilogram scale [3]. The main reasons for this are that diastereoselective aldol reactions are generally more robust in terms of scale than catalytic processes, and isomeric byproducts are diastereomer of the desired product, which can easily be removed from the mixture.

In an aldol reaction, one C–C bond and up to two stereogenic centers can be formed in a single chemical step. Aldol reactions can be categorized in two major families based on the structure of the enolate component (Scheme 3.1).

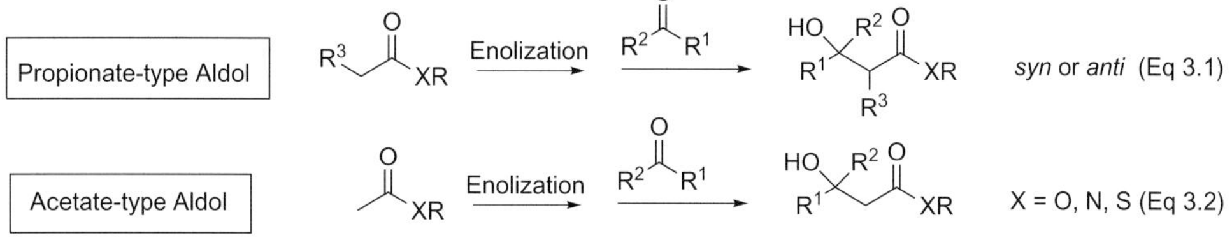

Scheme 3.1 Types of aldol reaction.

Carbohydrates – Tools for Stereoselective Synthesis, First Edition. Edited by Mike Martin Kwabena Boysen.
© 2013 Wiley-VCH Verlag GmbH & Co. KGaA. Published 2013 by Wiley-VCH Verlag GmbH & Co. KGaA.

The first type involves the use of enolates derived from acetates bearing a substituent $R^3 \neq H$ [Eq. (3.1)]. Most often, this substituent is simply a methyl group, and this class is termed "propionate-type" aldol reactions. Aldol reactions using simple acetate enolates ($R^3 = H$) are, naturally, called "acetate-type" aldols [Eq. (3.2)]. This classification is used because propionate-type enolates and simple acetate enolates behave quite differently in chiral auxiliary-controlled aldol reactions. While many chiral auxiliaries have been shown to effectively control the stereochemical course of propionate-type aldol reactions, the stereocontrol of an acetate-type aldol reaction turns out to be far more challenging. As a consequence, asymmetric acetate aldol reactions have been utilized much less frequently in the total syntheses of natural products as well as novel biologically active compounds.

3.2
Aldol Alkylation

Virtually all chiral auxiliaries developed so far have been used in the aldol reaction with more or less success. The use of carbohydrates as chiral auxiliaries has been scarce in this important transformation and the results obtained cannot compete with the gold standard in the literature [4]. Carbohydrates suffer historically from the perception of being highly complex molecules, with too many functional groups and too many chiral centers, to be of practical utility in stereodifferentiating processes. Indeed, the first results obtained by the group of Heathcock [5], at the beginning of the 1980s, seem to reinforce this negative opinion[1]. Condensation of the lithium enolate of propionate **1** derived from D-fructopyranose with benzaldehyde afforded the four diastereoisomers of the aldol adducts **2** with practically no stereodifferentiation, apart from a small *anti/syn* preference (Scheme 3.2).

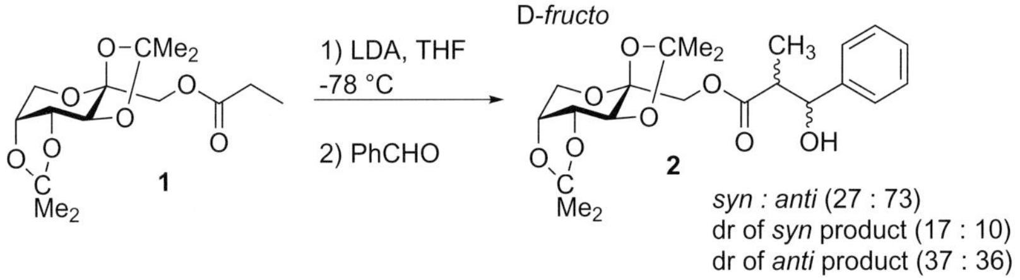

Scheme 3.2 Propionate-type aldol reaction using diacetone-D-fructopyranose as chiral auxiliary.

A detailed study on the same reaction was carried out 15 years later by Costa et al., who demonstrated that the low diastereoselectivity obtained is due to the loss of integrity of the enolate geometry during the reaction, as a consequence of the various sites to which the lithium cation of the enolate may coordinate [6]. The

---

1) Even though it is an aldol condensation, we include this example here because of its historical importance in the development of carbohydrates as chiral auxiliaries in asymmetric synthesis.

reaction of LDA (lithium diisopropylamide) with ester **1** followed by addition of benzyl bromide leads to the alkylated product with modest yield and 2:1 diastereoselectivity, which does not reflect the 3.35:1 (E)-to-(Z) enolate geometry. The authors rationalized this result by invoking an interchange of the favored chelate of the (E) enolate **3** with another chelate **4**, which is stabilized by coordination of the lithium to O3 of the fructopyranose scaffold (Scheme 3.3).

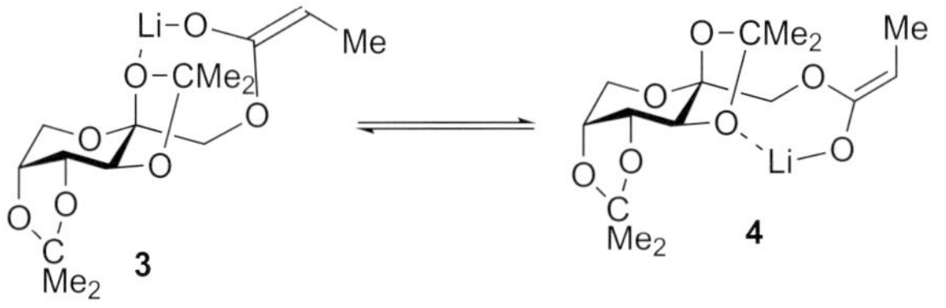

Scheme 3.3 Possible geometries of the (E)-enolate derived from **1**.

A further example of the negative influence of the multiple coordinative sites of the carbohydrate framework for metal ions has been reported by Kunz in the alkylation of the prochiral enolate of 3-O-propionyl diacetone-D-glucose **5**. Deprotonation of **5** with LDA, followed by addition of ethyl iodide gave alkylated product **6** in 30% yield and with low (2:1) diastereoselectivity in favor of the (2R)-diastereoisomer, together with compound **7** (50% yield) (Scheme 3.4) [7].

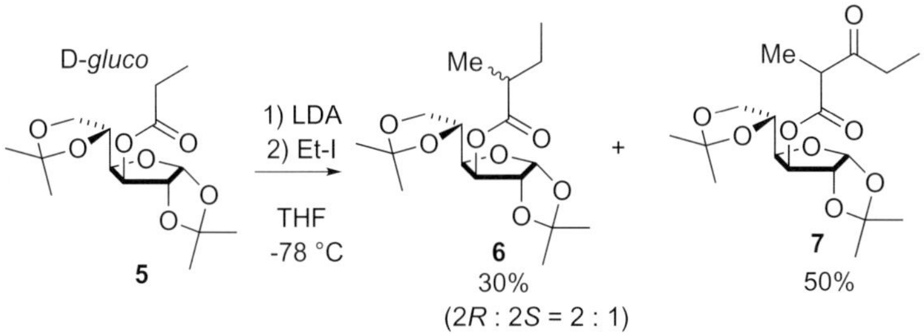

Scheme 3.4 Diastereoselective alkylation of a lithium enolate using diacetone-D-glucose as chiral auxiliary.

The authors demonstrated that this is a consequence of the intermediate enolate **8**, which decomposes reversibly even at −70 °C into ketene **10** and carbohydrate alkoxide **9** (Scheme 3.5) [7]. The reaction of the intact enolate **8** with ketene **10** explains the formation of the ester condensation product **7**.

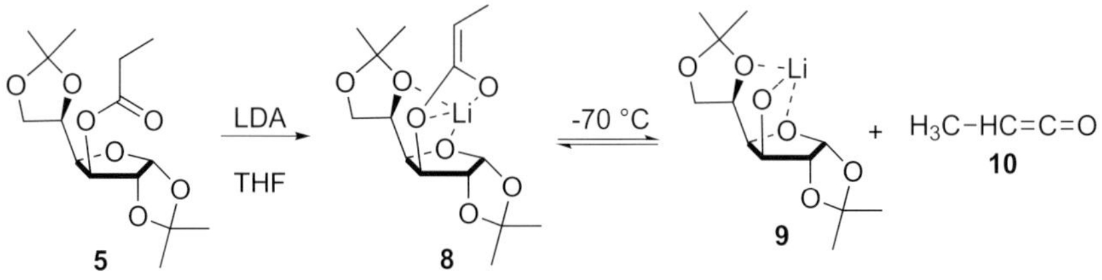

Scheme 3.5 Rationalization of the low yield obtained in the alkylation of ester **5**.

3 Reaction of Enolates

Interestingly, and in contrast to enolate **8**, the enolates of esters of 3-epimeric D-allofuranose **11** are more robust and considerably more reactive. Deprotonation at −90 °C led to *(E)* enolates **12**, which react with methyl iodide to give the alkylated products **13** with high diastereoselectivity (Scheme 3.6), indicating the great influence of the sugar structure on yield and diastereoselectivity (Scheme 3.6).

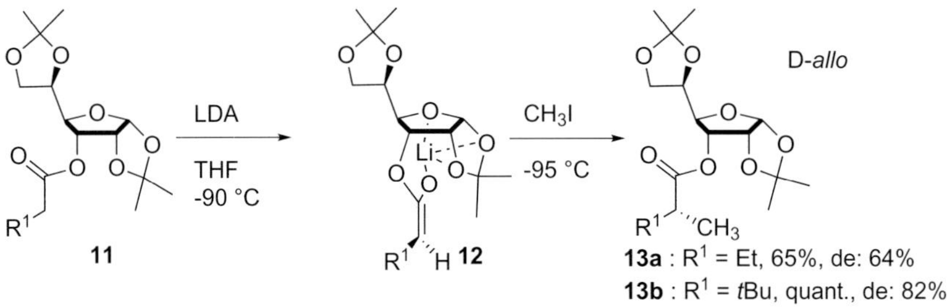

Scheme 3.6 Diastereoselective alkylation of a lithium enolate using diacetone-D-allofuranose as chiral auxiliary.

Along these lines, results from the Mulzer group corroborate the difficulty in predicting the reactivity of carbohydrate-derived enolates [8]. The enolates of 3-O-acyl derivatives of D-gulofuranose **14** (Scheme 3.7) – predicted to be strongly chelated by the "basket"-like disposition of the oxygens on the α-face of the sugar and therefore to readily eliminate the acyl moiety as ketene – afforded different results. Reaction with LDA or LTMP was mostly *(E)*-selective for all the esters, and methylation occurred on the *Si* face, affording the alkylated products with 4 : 1 to 10 : 1 selectivity (Scheme 3.7). Additives that could interfere with lithium chelation (TMEDA, HMPT) or that might enhance complex formation (MgBr$_2$, ZnCl$_2$) did not affect the observed selectivity.

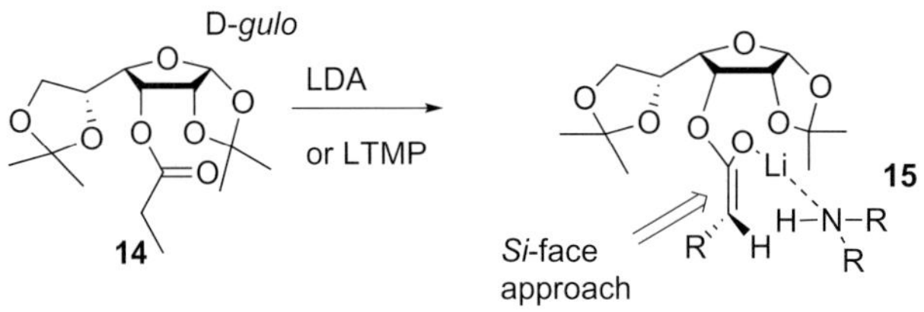

Scheme 3.7 *Si*-face preference in the alkylation of lithium enolate **15**.

The problem of intramolecular chelation, and the subsequent formation of ketenes, could possibly be circumvented by the formation of enol silyl ethers as reactive intermediates. Indeed, the reaction of *N*-chlorosuccinimide (NCS) with silylketene acetals **16** afforded the α-chlorocarboxylic acid esters **17** with excellent diastereoselectivity (9 : 1 to 49 : 1), as a consequence of the more favorable frontal attack of NCS on **16** (Scheme 3.8) [9].

3.2 Aldol Alkylation

Scheme 3.8 Diastereoselective chlorination of silylketene acetals **16**.

The halo-acid products **18** were cleaved from the diacetone-D-glucose auxiliary by treatment with cold lithium hydroperoxide, without racemization of the newly created chiral center. The two steps typically gave the halo-acids in good overall yields (about 70%), and the auxiliary was also recovered in good yield.

One of the first successful examples of enolate alkylation on a furanoside-type framework was developed by the Kakinuma group [10]. They showed that the introduction of a hydroxymethyl group at the 3-position of dicyclohexylidene-D-glucose provided a more controlled environment for enolization and alkylation. Treatment of lithium enolate of **19** (R^1 = H) with benzyl bromide and an excess of LiCl led to the benzylated compound in **20a** 60% yield and an excellent 92% de. In contrast, on using the silylated derivative of **19** the diastereoselectivity dropped to 2:1. However, in the presence of HMPA the reaction became highly selective, affording the complementary *(S)*-benzylated product in **20b** 91% de (Scheme 3.9).

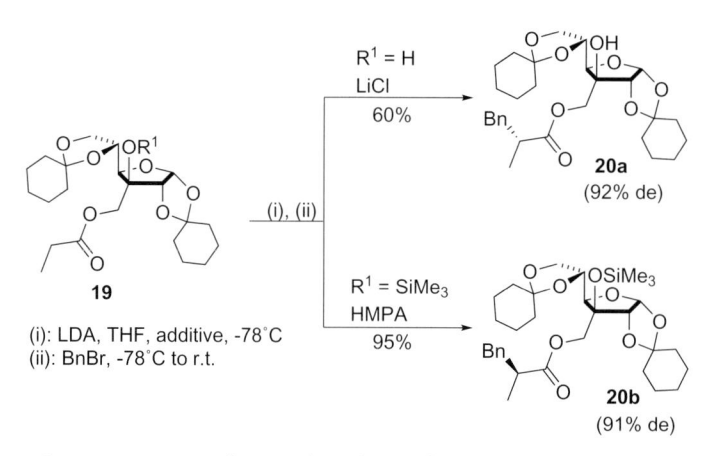

Scheme 3.9 Diastereodivergent benzylation of the lithium enolate derived from **19**.

In clear analogy with the work of Evans, Köll and Lützen synthesized in two steps a xylose-derived oxazolidinone auxiliary, and explored the behavior of the corresponding imide **21** in several alkylation, acylation, and halogenation reactions [11]. Alkylation of the corresponding lithium enolate afforded the products with moderate yields even with active alkyl halides. The selectivity of the process was around 5–10 : 1, with an intriguing reversal of the asymmetric induction by changing the R group from an aliphatic to an aromatic group. The authors proposed that their aliphatic imides formed lithium-chelated *(Z)*-enolates **22**, which were alkylated on the exposed *Si*-face to yield adducts **24**. By an undetermined stereoelectronic effect between the carbohydrate auxiliary's oxygen and the aromatic ring of the imide, aryl substituted imides seem to form preferentially the *(E)* enolates **23**, leading to the alkylated products **25** through *Re*-face attack (Scheme 3.10).

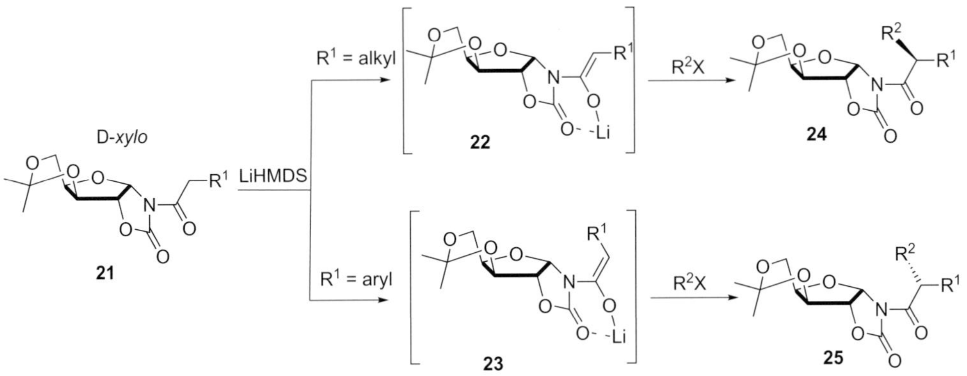

Scheme 3.10 Aldol reaction using a xylose-derived oxazolidinone as chiral auxiliary.

More recently, Tadano's group has successfully used carbohydrate auxiliaries in several stereoselective C–C bond formations, such as 1,4-additions of organocopper reagents, conjugate additions of alkyl radicals, and Diels–Alder reactions [12]. In the case of enolate alkylation, they have found that D-glucopyranoside **26** gave the best results. Alkylation of the sodium enolate derived from 4-*O*-acyl derivative **27** with alkyl halides provided the desired compounds **29** with excellent yield and 90–96% de (Scheme 3.11) [13].

The high stereoselectivity observed was explained by invoking the formation of *(Z)*-enolate **28**, as a consequence of an unfavorable steric interaction between the R^1 group and the carbohydrate framework and an effective blockade of the front side of the enolate by the TBS (*t*-butyldimethylsilyl) group on O3.

The same explanation was invoked for the high diastereoselectivity obtained during the asymmetric synthesis of compounds with an all-carbon stereogenic center through a double alkylation of the α-carbon of 4-*O*-acetoacetyl derivative **30**, which was also prepared by acylation of **26** (Scheme 3.12) [14]. The acetoacetate **30** exists as a 3 : 1 mixture of the diketo and keto-enol forms. The C-methylation at the α-carbon of this mixture occurred on using K_2CO_3 with methyl iodide at 40 °C. NMR analysis of the mixture indicated that the mono C-methylated product **31** exists as a 5 : 2 : 2 mixture of three tautomeric forms. The second alkylation reaction of this mixture using sodium methoxide as base and benzyl bromide or

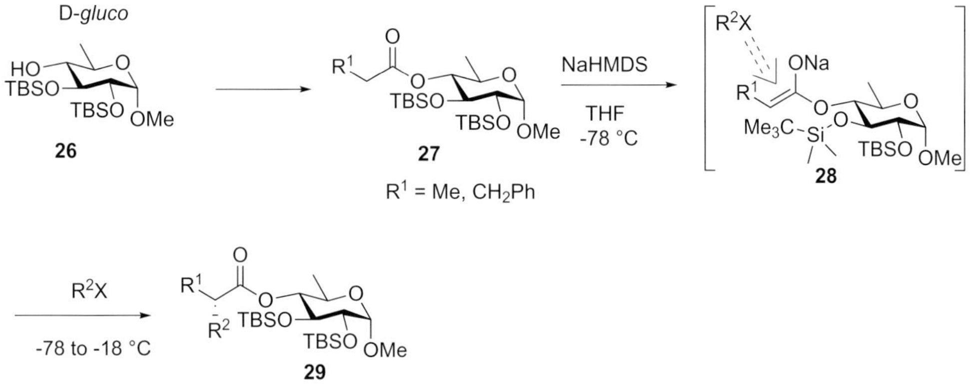

Scheme 3.11 Sodium enolate alkylation using methyl 6-deoxy-2,3-di-O-(t-butyldimethylsilyl)-α-D-glucopyranoside **26** as chiral auxiliary.

For compound **29**:
- R^1 = Me, R^2 = CH$_2$Ph, 97% yield, 90% de
- R^1 = CH$_2$Ph, R^2 = Me, 80% yield, 96% de

allyl bromide as electrophile afforded the dialkylated products in (R)-configuration as a single diastereoisomer. The absolute configuration has been determined by chemical correlation by converting compounds **32** into the known stereochemically defined pyrazoline derivatives **33** (Scheme 3.12).

For the synthesis of both enantiomers of the β keto ester product with the quaternary chiral carbon, the same group developed recently a new chiral auxiliary, namely, the D-glucopyranoside **34** (Scheme 3.13). The use of this compound in the same synthetic route afforded the final compounds with the opposite absolute configuration at the newly created quaternary center (Scheme 3.13). The diastereoselectivity observed has been rationalized by the transition state **36** in which the (Z)-enolate was attacked at the Si-face as the rear side was efficiently shielded by the bulky TBSO group on O3 [15].

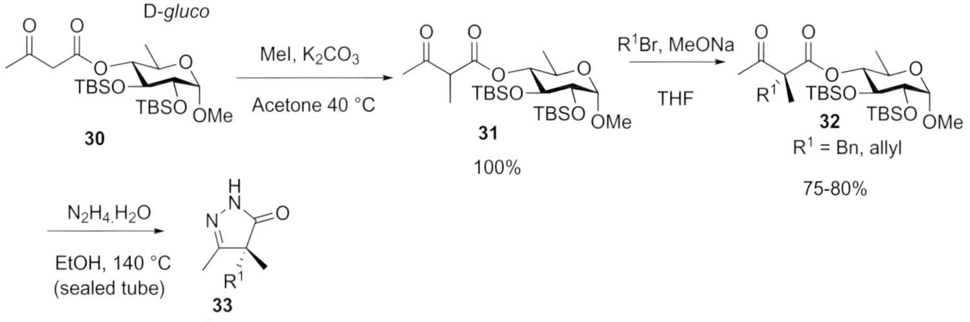

Scheme 3.12 Diastereoselective synthesis of pyrazoline derivatives **33**.

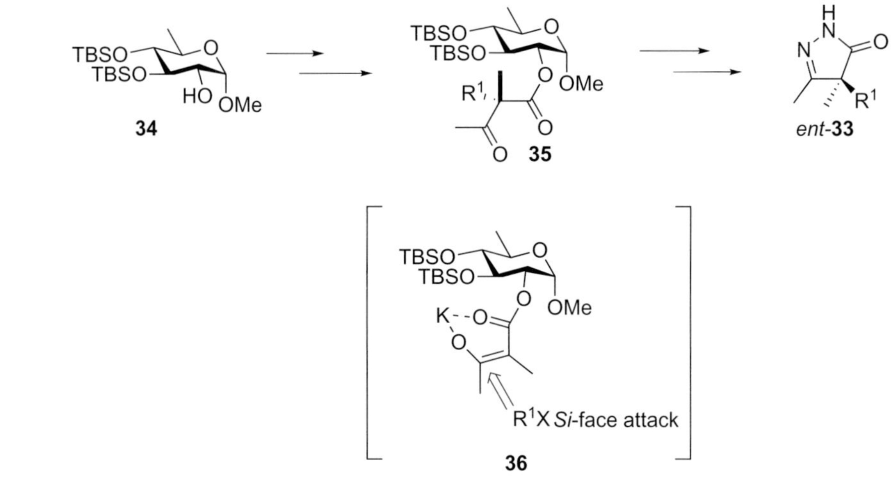

Scheme 3.13 Diastereoselective synthesis of pyrazoline derivatives *ent*-33.

An exhaustive study directed towards the asymmetric synthesis of α-hydroxy-carboxylic acids has been carried out by Wang's group using diacetone-D-fructose **37** as chiral auxiliary [16]. Esterification with various glycolic acid derivatives carrying 11 different protective groups afforded the glycolate derivatives **39** in good yields. A study of the enolization reaction shows that the use of LiHMDS in the presence or absence of HMPA gave the best results depending on the reactivity of the electrophile (Scheme 3.14).

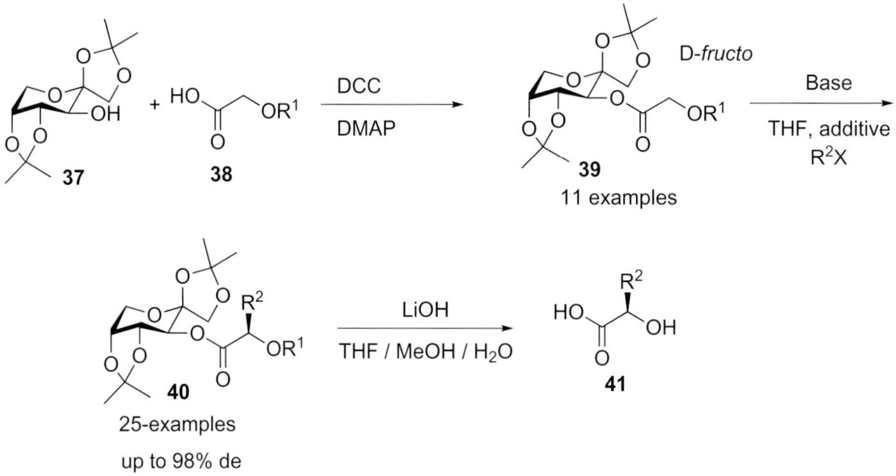

Scheme 3.14 Asymmetric synthesis of α-hydroxy carboxylic acids through aldol alkylation using diacetone-D-fructose as chiral auxiliary.

3.2 Aldol Alkylation

The hydroxy protective group has little effect on the diastereoselectivity of the process, but it significantly influences the yield, probably by affecting the stability of enolates, with silyl groups being most successful (Table 3.1). The alkylated products **40** were generally obtained good yields and excellent diastereoselectivity (up to 98%). The absolute configuration of the newly created chiral center *(R)* product was determined X-ray analysis and by chemical correlation of the free hydroxy acids **41**, which was obtained without racemization in the saponification step.

Compared to other approaches developed for the asymmetric synthesis of α-hydroxy acids, the fructose auxiliary gave slightly lower yield and stereoselectivity than Katsuki's pyrrolidine [17] and Wang's [18] and Pearson's [19] dioxolane auxiliaries. It is not quite as selective as the Evans auxiliary [20], but it gave good yields with various electrophiles, including many less reactive iodides (Table 3.2).

Table 3.1 Diastereoselective methylation of the enolate derived from **39**.

Entry	Substrate	R[a]	Product	Yield (%)	de (%)[b]
1	39a	Bn	40a	73	86
2	39b	$CH_2C_6H_4NO_2$ (*p*)	40b	19	89
3	39c	PMB	40c	66	91
4	39d	$CH_2C_6H_4F$ (*p*)	40d	75	93
5	39e	Me	40e	66	92
6	39f	MOM	40f	65	91
7	39g	BOM	40g	68	87
8	39h	TBS	40h	76	84[c]
9	39i	TBDPS	40i	83	89[c]
10	39j	TES	40j	77	88[c]
11	39k	TIPS	40k	70	84

a) PMB = *p*-methoxybenzyl; MOM = methoxymethyl; BOM = benzyloxymethyl; TBS = *t*-butyldimethylsilyl; TBDPS = *t*-butyldiphenylsilyl; TES = triethylsilyl; TIPS = tris(isopropyl) silyl.
b) Determined by 1H NMR.
c) Reaction conducted at −78 °C.

Table 3.2 Addition of different electrophiles to the TES (triethylsilyl) glycolate **39j**.

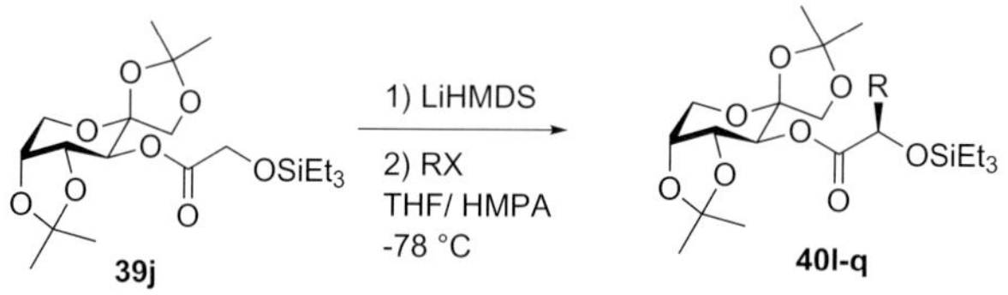

Entry	Electrophile (RX)	Product	Yield (%)	de (%)
1	CH$_2$=CHCH$_2$I	40l	71	91
2	BnBr	40m	75	96
3	CH$_3$CH$_2$I	40n	83	88
4	PhCH$_2$CH$_2$I	40o	61	60
5	CH$_3$(CH$_2$)$_4$I	40p	58	83
6	2-NphCH$_2$Br	40q	71	91

3.3
Aldol Addition

Along with the work of Heathcock cited earlier, pioneering work on the use of carbohydrates as chiral auxiliaries was carried out by the group of Brandänge on the first, and at present sole, example of acetate-type aldolization. The aldol condensation of the 3-O-acetate of diacetone-D-glucose with acetophenone was performed under various conditions [21]. The best results were obtained by deprotonation of **42** with lithium N-isopropyl-(–)-menthyl amide (Li-IMA) leading to tertiary carbinol **43** with 50% de in favor of the (S)-isomer (Scheme 3.15). The influence of the complexation of Li$^+$ in these processes was pivotal, since addition of MgCl$_2$ to the enolate of **42** favored the (R)-diastereoisomer of **43**.

Köll has also used his D-xylofuranosyl oxazolidinone derivatives in propionate-type aldol condensation [22]. Deprotonation of imides **21** with LiHMDS and then quenching the enolates with simple aliphatic aldehydes gave syn (2′R,3′S)-aldols

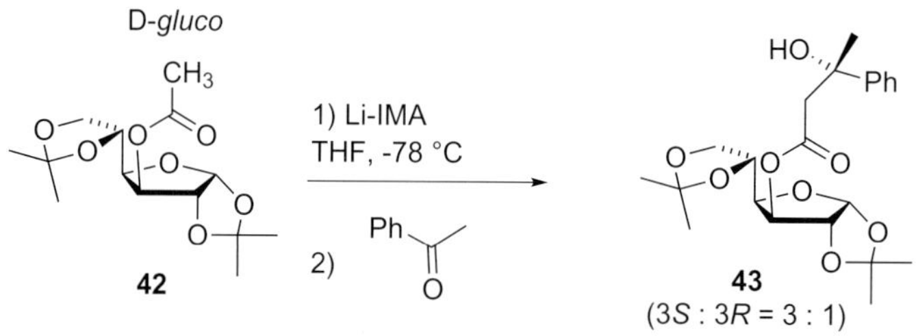

Scheme 3.15 Acetate-type aldolization using diacetone-D-glucose as chiral auxiliary.

44 via (Z)-lithium enolates. Interestingly, the diastereomeric syn (2'S,3'R)-aldols 45 were obtained via the (E)-enolate with aromatic aldehydes (Table 3.3).

The best results were obtained with aliphatic aldehydes and in general the major products were obtained with 5–15:1 diastereoselectivity. The reactions with aromatic aldehydes showed poor diastereoselectivity and were erratic, with the major product depending on the employed aldehyde. The authors rationalized these results by invoking a chair-like transition state in the case of aliphatic aldehydes and competitive boat and twist-like transition states for aromatic aldehydes. The method, which has the advantages of leading to "non-Evans" syn-aldol compounds, suffers from the poor yields for most of the cases studied.

The spiro-oxazolidinone **46**, obtained from D-galactose in three high-yielding steps, featuring an interesting intramolecular nitrene insertion, has also been used in aldol condensation by Banks et al. (Scheme 3.16) [23]. Lithium enolate of propionyl imide **47** was reacted with benzaldehyde to afford at −78 °C a mixture of three diastereomeric aldol products in a 89:6:5 ratio (78% de), from which the (2'S,3'S) erythro product **48** was obtained in 83% yield. The absolute configuration of **48** was confirmed by X-ray analysis as well as by cleavage of the amide **48** with lithium borohydride to afford (1S,2R)-1-phenyl-2-methylpropane-1,3-diol **49**.

Table 3.3 Aldol reactions of N-acyl derivatives with aldehydes [23].

Entry	Substrate 21 (R¹)	Aldehyde	Diastereomeric ratio	Major product	Configuration of major product	Yield (%)
1	21a (Me)	Acetaldehyde	10:1	44a	syn-(R,S)	37
2	21a (Me)	Acrolein	5:1	44b	syn-(R,S)	39
3	21a (Me)	Butyraldehyde	6:1	44c	syn-(R,S)	29
4	21a (Me)	iso-Butyraldehyde	15:1	44d	syn-(R,S)	39
5	21a (Me)	Cinnamaldehyde	2:1	44e	syn-(R,S)	46
6	21b (Et)	Acrolein	4:1	44f	syn-(R,S)	34
7	21b (Et)	iso-Butyraldehyde	6:1	44g	syn-(R,S)	38
8	21c (Ph)	Acetaldehyde	12:1	45a	syn-(S,R)	22
9	21d (Bn)	Acetaldehyde	15:1	45b	syn-(S,R)	26

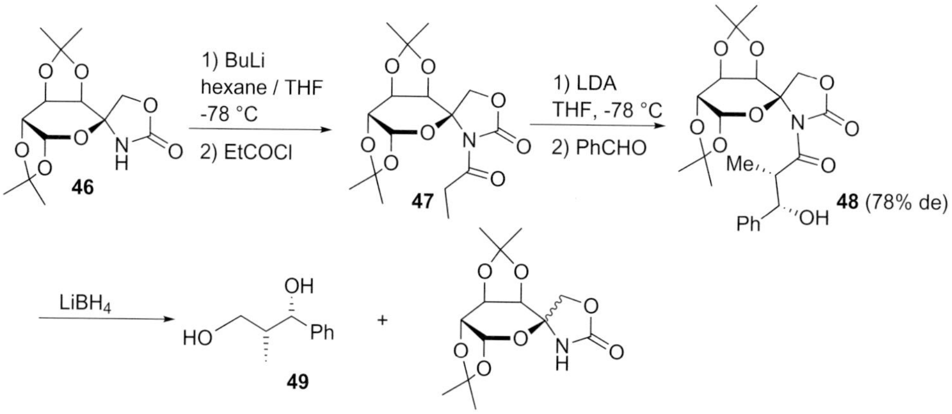

Scheme 3.16 Propionate-type aldolization using a spiro-oxazolidinone derived from galactose as chiral auxiliary.

However, subsequent studies have shown that the recovered spiro-oxazolidinone auxiliary **46** undergoes epimerization about the spiro-carbon atom upon cleavage with lithium borohydride.

To avoid the epimerization process, the same group has reported the synthesis of oxazinone **50** in four steps starting from gulonic acid [24]. Lithium enolate mediated aldol condensation of **51** with benzaldehyde gave only the *syn* adducts in 88% yield and a 91:9 ratio in favor of the diastereoisomer **52** with (2′S,3′S) configurations. Reductive cleavage of **52** using LiBH$_4$ led to (1S,2R)-1-phenyl-2-methylpropane-1,3 diol-**49**, together with recovery of the chiral auxiliary **50** with no appreciable epimerization (Scheme 3.17) [20].

Solladié's group has studied the reaction of an imino glycinate derivative of a carbohydrate with an aldehyde *en route* to chiral α-amino acids derivatives (Scheme 3.18). The aldol condensation of diacetone-D-glucose iminoglycinate **53**, obtained in three steps from diacetone-D-glucose, with tetradecanal under phase-transfer

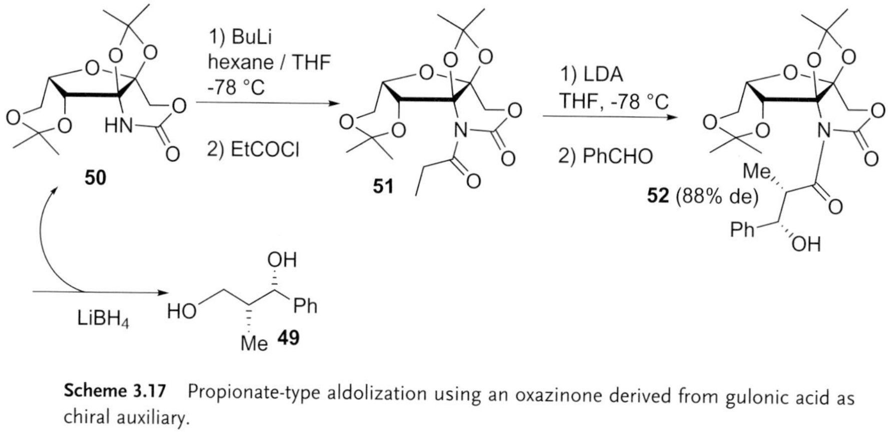

Scheme 3.17 Propionate-type aldolization using an oxazinone derived from gulonic acid as chiral auxiliary.

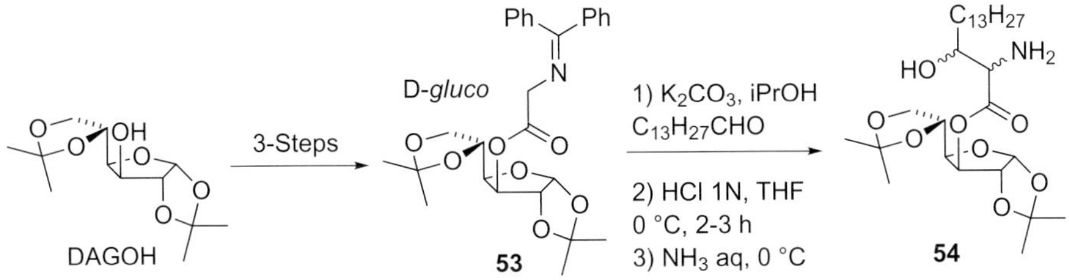

Scheme 3.18 Aldolization of an imino glycinate derived from diacetone-D-glucose.

conditions [K_2CO_3, iPrOH] leads to the amino alcohol derivative **54** in a disappointingly low diastereoselectivity and enantioselectivity [25]. The use of the corresponding lithium enolate of **53** (generated by LDA in THF) to obtain a well-defined chelated transition state, did not improve the diastereoselectivity (16%), but did improve the enantioselectivity (45%).

A carbohydrate auxiliary can also be linked to the electrophilic component of an aldol condensation. Highly selective Mukaiyama aldol reactions have been obtained using cyclitol pyruvate or phenylglyoxylate esters **55** (Table 3.4). This approach

Table 3.4 Results of the $SnCl_4$-promoted aldol reaction of **55a,b** with various nucleophiles.

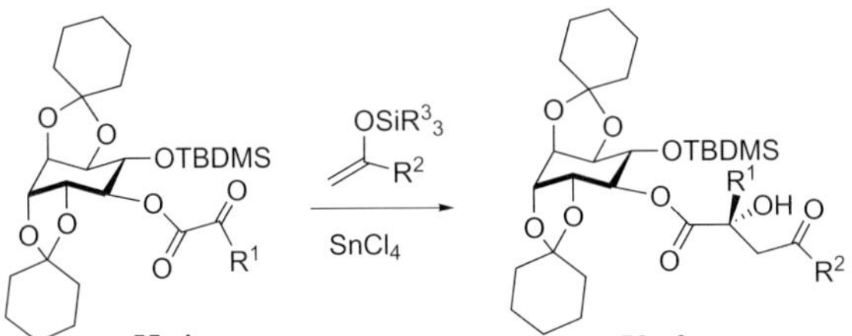

Entry	Substrate 55 (R¹)	OSiR³ / R²	Reaction conditions	Product 56 (R²)	Yield (%)	de (%)
1	55a (Ph)	OSiMe₃ / tBu	−78 °C, 1.5 h	56a (t-Bu)	89	>98
2	55a (Ph)	OSiMe₃ / Ph	−78 °C, 1.5 h	56b (Ph)	70	>98
3	55b (Me)	OSiMe₃ / tBu	−78 °C, 1.5 h	56c (t-Bu)	98	>98
4	55b (Me)	OSiMe₃ / Ph	−78 °C, 1.5 h	56d (Ph)	97	94
5	55a (Ph)	OSiMe₂tBu / OEt	−20 °C, 45 min	56e (OEt)	92	>98
6	55b (Me)	OSiMe₂tBu / OEt	−20 °C, 20 min	56f (OEt)	80	>98

provides an efficient enantioselective synthesis of chiral tertiary aldols, which are often difficult to prepare. Essentially, a single diastereoisomer was obtained in 70–97% yield by the reaction of **55** with several silyl enol ethers or silyl ketene acetals in the presence of SnCl$_4$. The chiral auxiliary was easily removed from the aldol adduct by simple base-promoted hydrolysis of **56** [26].

Tadano et al. have used propionyl ester **27** in the 1,4-addition to α,β-unsaturated methyl crotonate (Scheme 3.19). The *anti*-Michael adduct **57** was obtained as a single product in 75% yield and 90% diastereoselectivity. The synthetic usefulness of the approach was demonstrated by a three-step asymmetric synthesis of the acyclic monoterpene (–)-lasiol **58**, a sex attracting pheromone of the ant *Lasius meridionalis*. As before the high diastereoselectivity was accounted for by invoking the formation of *(Z)*-enolate **28** and the subsequent approach of the methyl crotonate from the *Si* face (Scheme 3.19) [27].

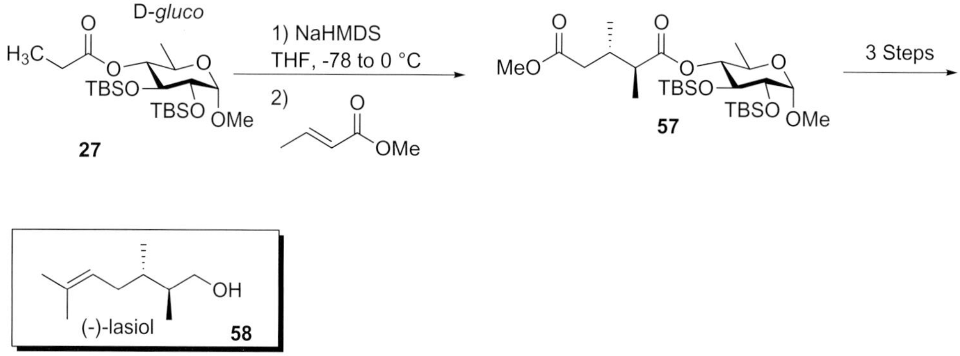

Scheme 3.19 Asymmetric synthesis of (–)-lasiol using methyl 6-deoxy-2,3-di-O-(*t*-butyldimethylsilyl)-α-D-glucopyranoside as chiral auxiliary.

3.4
Concluding Remarks

From the examples presented in this section some concluding remarks can be drawn. As stated in the introduction, auxiliary-based aldol condensation has reached great sophistication, efficiency, and reliability in predicting the stereochemical outcome of the reaction. As a result, this approach is commonly employed by pharmaceutical companies at the preclinical and clinical stage to efficiently and rapidly produce enantiopure APIs on a kilogram scale. An illustrative example of such applications is the impressive large-scale synthesis of the anticancer natural compound discodermolide by Novartis [28]. The approach is based on an Evans aldol reaction using an oxazilidinone as chiral auxiliary, which was, importantly, conducted on a production scale with little or no modification from the originally reported standard conditions (n-Bu$_2$BOTf, NEt$_3$, CH$_2$Cl$_2$ or toluene, −78 °C to room temperature) to produce, in almost all cases, the *syn*-aldol adduct with extremely high diastereoselectivity [29]. The large-scale synthesis of discodermolide employs a hybrid route that combines the best reactions of total syntheses developed by the groups of Smith [30] and Paterson [31]. The *(R)*-3-propionyl-4-benzyloxazolidinone **59** was used in two aldol reactions to install 8 out of the 13 stereocenters embedded in discodermolide at C2-C3, C11-C12, C16-C17, and C18-C19 (Scheme 3.20).

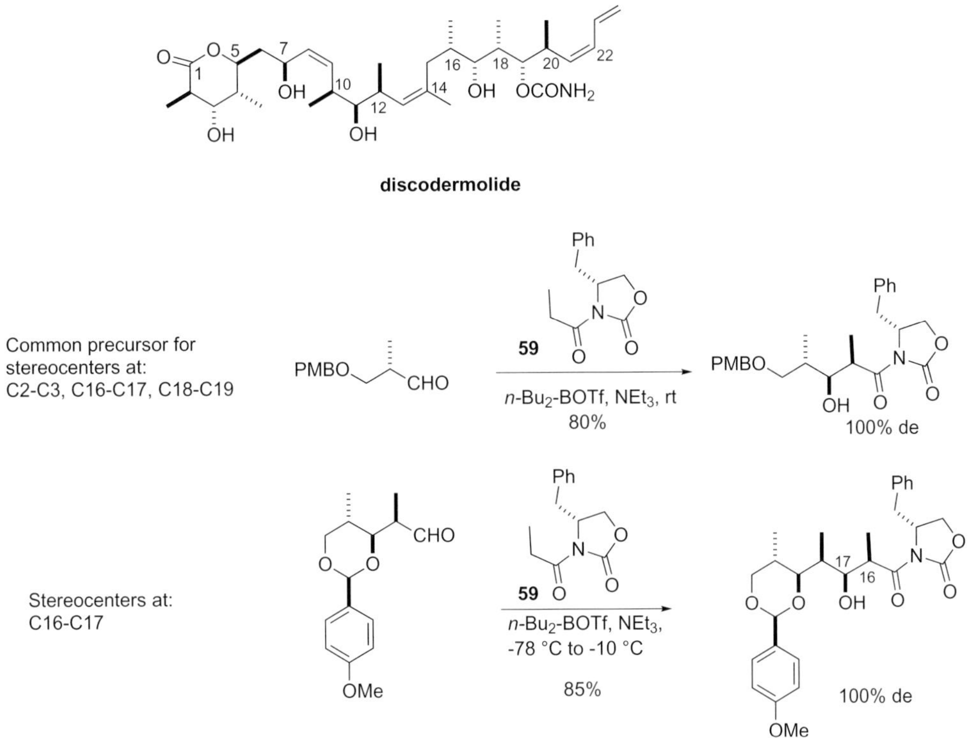

Scheme 3.20 Large-scale synthesis discodermolide based on the Evans aldol reaction.

Obviously, the carbohydrate-based aldol reactions developed so far cannot compete yet with the gold standard developed in the literature. One of the main reasons for this state of affair is paradoxically the rich stereochemical information encoded by the polyfunctional sugar ring. In substrate-controlled chiral transformations, a decisive requirement for efficient transfer of chirality from the chiral auxiliary to the product is the formation of a well-organized transition state, generally granted by appropriate coordination between the reagent and the chiral template. However, in the case of polyhydroxylated carbohydrates, the multiple sites for coordination of metal or Lewis acids can be a problematic issue, and in some cases constitute a real drawback. This limits the use of carbohydrates "as such" in that it necessitates designing and tailoring the sugar template for a given transformation. Additionally, given the importance of metal enolates for the stereochemical outcome of the aldol reactions, it is surprising that most carbohydrate-based aldol reactions depend strongly on the coordination of lithium and sodium enolates. Thus, tailoring the carbohydrate template for an "effective" metal coordination, and using other Lewis acids such as boron, titanium, or tin, may permit the development of efficient carbohydrate-based aldol reactions in the future.

References

1 (a) Trost, B.M. and Brindle, C.S. (2010) *Chem. Soc. Rev.*, **39**, 1600; (b) Palomo, C., Oiarbide, M., and Garcia, J.M. (2002) *Chem.-Eur. J.*, **8**, 37; (c) Geary, L.M. and Hutin, P.G. (2009) *Tetrahedron: Asymmetry*, **20**, 131.

2 Garcia Ruano, J.L., Fernandez-Ibanez, M.A., and Maestro, M.C. (2006) *Tetrahedron*, **62**, 12297.

3 Farina, V., Reeves, J.T., Sennayake, C.H., and Song, J.J. (2006) *Chem. Rev.*, **106**, 2734.

4 (a) Kunz, H. and Rük, K. (1993) *Angew. Chem. Int. Ed. Engl.*, **32**, 336; (b) Hultin, P.G., Earle, M.A., and Sudharshan, M. (1997) *Tetrahedron*, **53**, 14823.

5 Heathcock, C.H., White, C.T., Morrison, J.J., and van Derver, D. (1981) *J. Org. Chem.*, **46**, 1296.

6 Costa, P.R.R., Ferreira, V.F., Alencar, K.G., Filho, H.C.A., Ferreira, C.M., and Pinheiro, S. (1996) *J. Carbohydr. Chem.*, **15**, 691.

7 Kunz, H. and Mohr, J. (1988) *J. Chem. Soc. Chem. Commun.*, 1315.

8 Mulzer, J., Hiersemann, M., Bushmann, J., and Luger, P. (1996) *Liebigs Ann. Chem.*, 649.

9 Duhamel, L., Angiland, P., Desmurs, J.R., and Valuot, J.Y. (1991) *Synlett*, 807.

10 Kishida, M., Eguchi, T., and Kakinuma, K. (1996) *Tetrahedron Lett.*, **37**, 2061.

11 (a) Köll, P. and Lützen, A. (1995) *Tetrahedron: Asymmetry*, **6**, 43; (b) Köll, P. and Lützen, A. (1996) *Tetrahedron: Asymmetry*, **7**, 637; (c) Lützen, A. and Köll, P. (1997) *Tetrahedron: Asymmetry*, **8**, 29.

12 Totani, K., Takao, K.-I., and Tadano, K.-I. (2004) *Synlett*, 2066.

13 Totani, K., Asano, S., Takao, K.-I., and Tadano, K.-I. (2001) *Synlett*, 1772.

14 Kozawa, I., Akashi, Y., Takiguchi, K., Sasaki, D., Sawamoto, D., Takao, K.-I., and Tadano, K.-I. (2007) *Synlett*, 399.

15 Akashi, Y., Takiguchi, K., Sasaki, D., Sawamoto, D., Takao, K.-I., and Tadano, K.-I. (2009) *Tetrahedron Lett.*, **50**, 1139.

16 Yu, H., Ballard, C.E., Boyle, P.D., and Wang, B. (2002) *Tetrahedron*, **58**, 7663.

17 Enomoto, M., Ito, Y., Katsuki, T., and Yamaguchi, M. (1985) *Tetrahedron Lett.*, **26**, 1343.

18 Chang, J.W., Jang, D.P., Uang, B.J., Liao, F.L., and Wang, S.L. (1999) *Org. Lett.*, **1**, 2061.

19 Pearson, W.H. and Cheng, M.C. (1986) *J. Org. Chem.*, **51**, 3746.

20 Crimmins, M.T., Emmite, K.A., and Katz, J.D. (2000) *Org. Lett.*, **2**, 2165.

21 Brandänge, S., Josephson, S., Mörch, L., and Vallèn, S. (1981) *Acta Chem. Scand. Ser. B*, **35**, 273.

22 Lützen, A. and Köll, P. (1997) *Tetrahedron: Asymmetry*, **8**, 1193.

23 Banks, M.R., Blake, A.J., Cadogan, J.I.G., Dawson, M.I., Gaur, S., Gosney, I., Gould, R.O., Grant, K.J., and Hodgson, P.K.G. (1993) *J. Chem. Soc., Chem. Commun.*, 1146.

24 Banks, M.R., Cadogan, J.I.G., Gosney, I., Gaur, S., and Hodgson, P.K.G. (1994) *Tetrahedron: Asymmetry*, **5**, 2447.

25 Solladié, G., Saint Clair, J.-F., Philippe, M., Semeria, D., and Maignan, J. (1996) *Tetrahedron: Asymmetry*, **7**, 2358.

26 Akiyama, T., Ishikawa, K., and Ozaki, S. (1994) *Synlett*, 275.

27 Asano, S., Tamai, T., Totani, K., Takao, K.-I., and Tadano, K.-I. (2003) *Synlett*, 2252.

28 (a) Loiseleur, O., Koch, G., and Wagner, T. (2004) *Org. Process Res. Dev.*, **8**, 597; (b) Mickel, S.J., Sedelmeier, G.H., Niederer, D., Daeffler, R., Osmani, A., Schreiner, K., Seeger-Weibel, M., Berod, B., Schaer, K., Gamboni, R., Chen, S., Chen, W., Jagoe, C.T., Kinder, F.R., Jr., Loo, M., Prasad, K., Repic, O., Shieh, W.-C., Wang, R.-M., Waykole, L., Xu, D.D., and Xue, S. (2004) *Org. Process Res. Dev.*, **8**, 92; (c) Mickel, S.J., Sedelmeier, G.H., Niederer, D., Schuerch, F., Koch, G., Kuesters, E., Daeffler, R., Osmani, A., Seeger-Weibel, M., Schmid, E., Hirni, A., Schaer, K., Gamboni, R., Bach, A., Chen, S., Chen, W., Geng, P., Jagoe, C.T., Kinder, F.R., Jr., Lee, G.T., McKenna, J., Ramsey, T.M., Repic, O., Rogers, L., Shieh, W.-C., Wang, R.-M., and Waykole, L. (2004) *Org. Process Res. Dev.*, **8**, 107.

29 Evans, D.A., Bartroli, J., and Shih, T.L. (1981) *J. Am. Chem. Soc.*, **103**, 2127.

30 (a) Smith, A.B., III, Beauchamp, T.J., LaMarche, M.J., Kaufman, M.D., Qui, Y., Arimoto, H., Jones, D.R., and Kobayashi, K. (2000) *J. Am. Chem. Soc.*, **122**, 8654; (b) Smith, A.B., III, Kaufman, M.D., Beauchamp, T.J., LaMarche, M.J., and Arimoto, H. (1999) *Org. Lett.*, **1**, 1823; (c) Smith, A.B., III, Qui, Y., Jones, D.R., and Kobayashi, K. (1995) *J. Am. Chem. Soc.*, **117**, 12011.

31 (a) Paterson, I., Delgado, O., Florence, G.L., Lyothier, I., Scott, J.P., and Sereinig, N. (2003) *Org. Lett.*, **5**, 35; (b) Paterson, I., Florence, G.J., Gerlach, K., Scott, J.P., and Sereinig, N. (2001) *J. Am. Chem. Soc.*, **123**, 9535; (c) Paterson, I. and Florence, G.J. (2000) *Tetrahedron Lett.*, **41**, 6935; (d) Paterson, I., Florence, G.J., Gerlach, K., and Scott, J.P. (2000) *Angew. Chem. Int. Ed.*, **39**, 377.

4
Cycloadditions

Mike M.K. Boysen

4.1
Diels–Alder Reactions

Asymmetric Diels–Alder reactions are one of the most important methods for the buildup of six-membered carbocycles while the hetero-Diels–Alder variant grants access to several valuable heterocyclic compounds. The application of chiral auxiliaries to these reactions constitutes a powerful approach to access these important products in a stereoselective manner. In principle, the chiral auxiliary may be attached to either the diene or the dienophile; however, for practical reasons, the modification of dienophiles with chiral groups is more common.

4.1.1
Dienophiles Attached to the Auxiliary

Most important dienophiles for Diels–Alder reactions contain an acrylic acid moiety, which can be conveniently attached to chiral scaffolds via ester or imide formation. Therefore, incorporation of a chiral auxiliary into the dienophile is a very popular approach to asymmetric Diels–Alder reactions. As carbohydrate scaffolds offer a whole choice of hydroxyl groups, it is unsurprising that mainly carbohydrate-derived alcohols have been explored as chiral auxiliaries. Scheme 4.1 shows the general reactions of auxiliary-modified acrylic esters (X = O) and imides [X = N-(CO)R] **1** as dienophile with various cyclic and acyclic dienes **2–4** in the presence of a Lewis acid. Conducted at low temperatures these reactions lead predominantly to the *endo*-diastereomers of the Diels–Alder products **7–11**.

Since the late 1980s several carbohydrate acrylates have been employed as dienophiles in asymmetric Diels–Alder reactions. The group of Kunz first examined acrylic esters of furanosides derived from D-glucose and D-xylose respectively [1]; later they introduced acrylic esters of pyranosidic 1,5-anhydro-2-deoxy derivatives of D-glucose and L-rhamnose as pseudo-enantiomers [2].

Scheme 4.2 shows the syntheses of chirally modified dienophiles **16** and **20**, which were prepared from acetylated glucal (**13**) and rhamnal (**18**) respectively. While the synthesis of **16** proceeded in high yields for all eight steps, the shorter

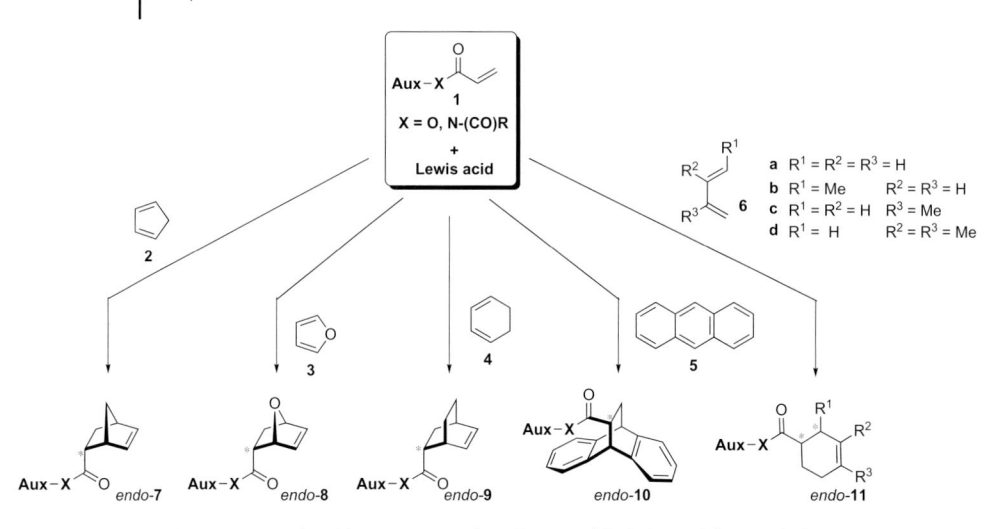

Scheme 4.1 Diels–Alder reactions of auxiliary-modified dienophile **1** with dienes **2–5**.

Scheme 4.2 Preparation of D-*gluco*- and L-*rhamno*-configured acrylic esters **16** and **20** as dienophiles for asymmetric Diels–Alder reactions.

synthesis of **20**, consisting of only four steps, suffered from moderate yields for the introduction of the acryloyl moiety.

Furanosidic D-*gluco* and D-*xylo* acrylates **21** and **23** as well as pyranosidic D-*gluco* and L-*rhamno* acrylates **16** and **20** were employed in Diels–Alder reactions with cyclopentadiene (**2**) in the presence of titanium-based Lewis acids (Scheme 4.3). The furanosidic dienophiles **21** and **23** were first per-silylated using TMSCl, then TiCl$_4$ was added, resulting in de-silylation and concomitant chelation of Ti(IV) by the acrylic ester and hydroxy groups of the carbohydrate scaffold. Reaction of these complexes with cyclopentadiene at −78 °C resulted in high selectivity for the *endo*-product with values of dr above 90 : 10 for both dienophiles. The authors rationalized the stereochemical outcome by invoking transition states **TS-1** and **TS-2**, in

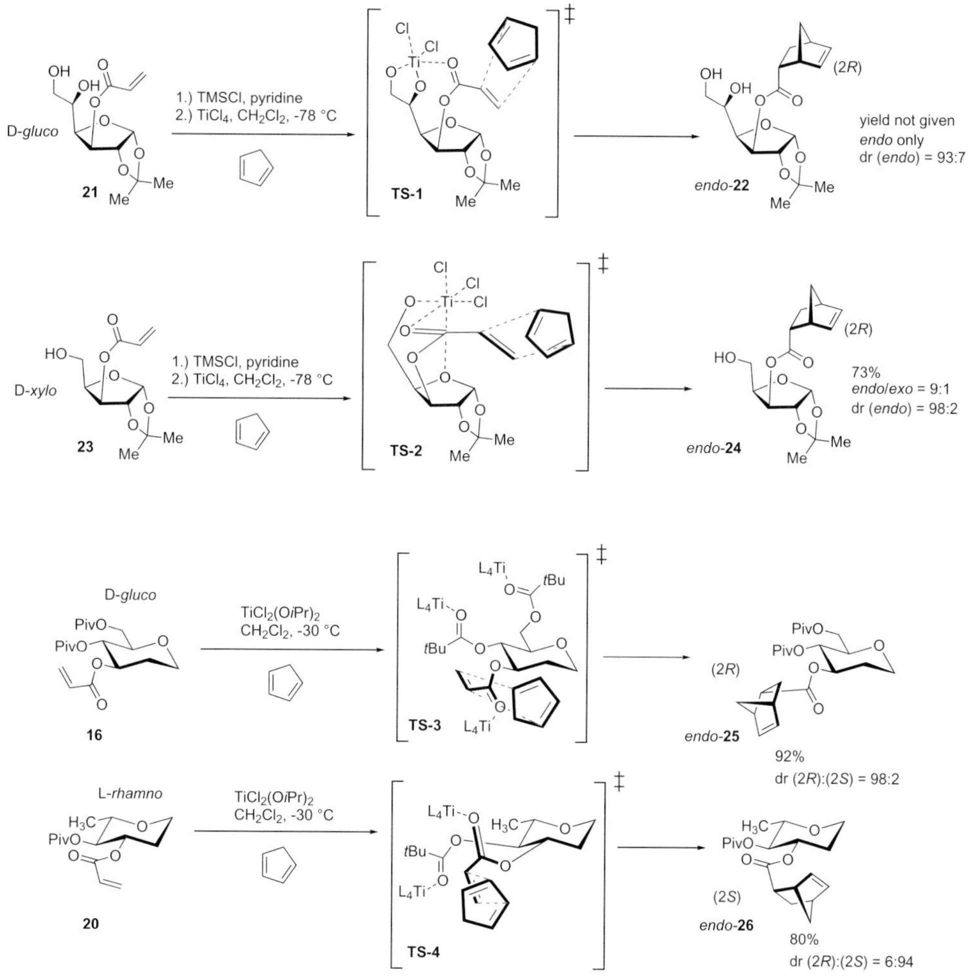

Scheme 4.3 Asymmetric Diels–Alder reactions using carbohydrate-modified dienophiles **21**, **23**, **16**, and **20** and plausible transition states explaining the observed selectivities.

which the acryloyl moiety adopts an s-trans conformation and the attack of the diene is directed to the upper face of the acrylic double bond. Per-silylation prior to the addition of the Lewis acid proved to be essential as otherwise the formation of HCl and chlorotitanate salts severely hampered the Diels–Alder reaction. For fully protected dienophiles **16** and **20**, the milder Lewis acid $TiCl_2(OiPr)_2$ was sufficient and the reaction was conducted at −30 to 0 °C. Both auxiliary-modified dienophiles offered high yields of the *endo*-products in excellent values of dr but even more importantly they led to a complete reversal of the asymmetric induction of the reaction. The Diels–Alder products attached to the carbohydrate backbones of *endo*-**25** and *endo*-**26** were obtained in opposite configuration and thus dienophiles **16** and **20** act as efficient *pseudo-enantiomers*. Again the authors offer a plausible explanation of the stereochemical outcome for both dienophiles. Unlike in the case of **21** and **23**, which act as chelate ligands for one Ti(IV) center, **16** and **20** presumably do not form chelates. Instead, the pivalates and the acrylate coordinate to one Ti(IV) center each, making the activated dienophiles very bulky species. In the favored transition states **TS-3** and **TS-4** the acrylic ester moiety exists in an *s-trans* conformation and is preferentially attacked by the diene from the face opposite to the bulky pivalate at C4.

Using **16** and **20** Diels–Alder reactions of an impressive range of cyclic and acyclic dienes were performed (Scheme 4.4), giving products in moderate to

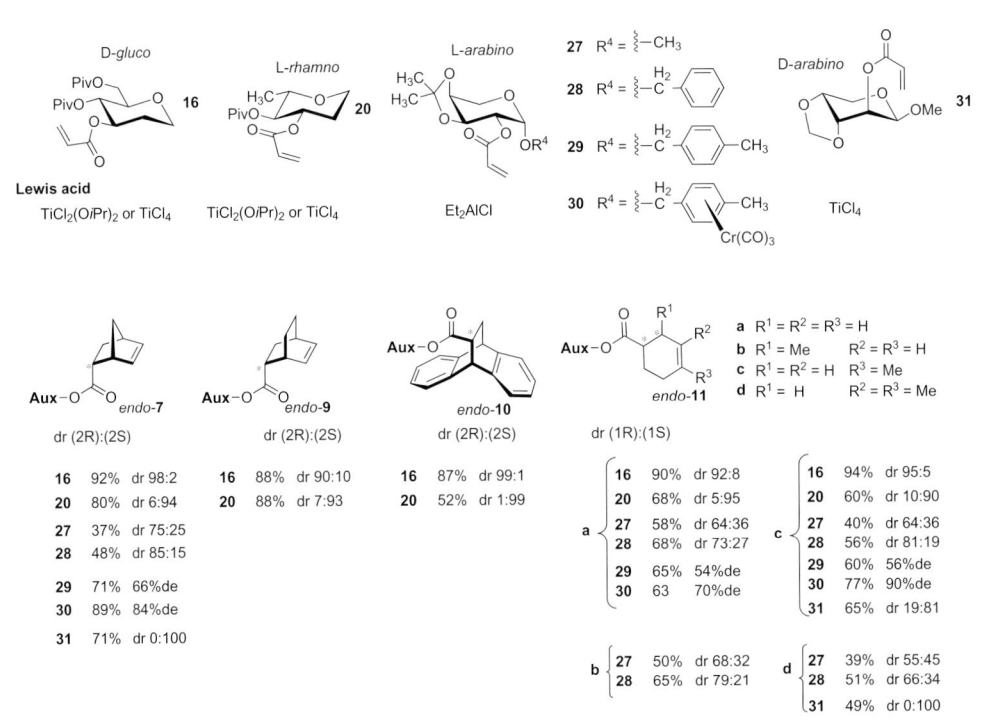

Scheme 4.4 Diels–Alder reactions of various dienes with auxiliary-modified dienophiles by Kunz, (**16, 20**), Shing (**27–30**), and Nouguier (**31**).

excellent yields and high drs. These good results, even in the case of the conformationally flexible acyclic dienes, highlight the generality of the pseudo-enantiomeric dienophiles **16** and **20**.

The group of Shing has published a series of dienophiles derived from D-arabinose (**27–30**) for Diels–Alder reactions catalyzed by Et$_2$AlCl [3] (Scheme 4.4). Employing arabinose-based auxiliaries is interesting as both enantiomers of this monosaccharide are commercially available at low price. Shing studied the influence of the anomeric residue on stereoselectivity for the reactions of cyclopentadiene (**2**), butadiene (**6a**), and isoprene (**6c**). While dienophile **27** with a small methyl residue gave hardly any diastereoselectivity (dr around 60:40), the situation improved for dienophile **28** with a larger anomeric benzyl group, but the drs were still only moderate [3a]. In further studies 4-methylbenzyl-modified dienophile (**29**) and derivative **30** containing a η^6-coordinated chromium(0) center on the benzylic aglycon moiety were explored. While the methylbenzyl derivative **29** gave only low to moderate drs, dienophile **30** with its organometallic aglycon led to significantly improved selectivities of up to 90% de in the case of isoprene (**6c**). Acrylate **31**, containing a D-arabinose scaffold, which is closely related to **27**, was reported by Nouguier [4]. The dienophiles **31** and **27** of Shing and Nouguier differ (apart from the absolute configuration of the arabinose scaffold) only in the cyclic 3,4-O-acetal protective group: Dienophile **27** contains a 3,4-O isopropylidene acetal, which is exchanged for a 3:4-O methylidene acetal in **31**. With TiCl$_4$ as Lewis acid, **31** gave the Diels–Alder products of cyclopentadiene and isoprene in opposite configuration compared to **27**; however, the diastereoselectivity was much higher. In case of cyclopentadiene as substrate, **31** gave a dr of almost 0:100, while **27** only yielded a dr of 75:25; a similar observation was made in the case of isoprene (**6c**). The substantial improvement appears to be mainly due to the different choice of the Lewis acid, as the reaction of **31** with cyclopentadiene and Et$_2$AlCl as Lewis acid led to a dr of only 27:73 [4b], a result comparable to the one reported by Shing for **27** [3a].

Tadano and coworkers prepared acrylic esters **36** and **37** based on D-glucose and D-mannose in which the acryl ester moiety is attached to C4 or C2 of the pyranose units, respectively [5, 6]. To optimize the carbohydrate scaffolds for the Diels–Alder reaction with cyclopentadiene (**2**), protective groups with varying steric demand were attached to the remaining pyranose hydroxy groups (Bn, Piv, and TBS at the hydroxy groups at C2 and C3 for *gluco*-acrylate **36**; Bn, Piv, Bz, and Ac at the hydroxy group at C3 for *manno*-acrylate **37**) (Table 4.1). Under thermal conditions, dienophiles **36a–c** gave high yields but unsatisfactory *endo:exo* selectivities. In all cases the (2S)-products were predominantly obtained but the dr of the reaction depended strongly on the bulk of the pyranose O-substituents and increased from **36a** (R^1 = Bn) to **36c** (R^1 = TBS) (entries 1–3). Under Lewis acid catalysis, the *endo:exo* selectivities improved considerably but at the price of diminished yields and complete loss of diastereoselectivity for **36a** (R^1 = Bn) while the TBS-modified dienophile **36c** did not promote the reaction (entries 3 and 6). Dienophile **36b** (R^1 = Piv) gave the best result in this series with still moderate yield but a high dr. Interestingly, the direction of the

Table 4.1 Diels–Alder reactions of dienophiles **36–38** with cyclopentadiene (**2**).

2 X = CH$_2$
3 X = O
36–38
endo-**7** or endo-**8**

Entry	Dienophile				Lewis acid, temperature	Substrate	Yield (%)	endo:exo	dr (2R):(2S)
			R^1						
1			Bn	36a	None, room temp.		96	79:21	29:71
2	D-gluco		Piv	36b	None, room temp.		96	89:11	13:87
3			TBS	36c	None, room temp.		97	80:20	5:95
4			Bn	36a	EtAlCl$_2$, –78 °C		63	95:5	50:50
5			Piv	36b	EtAlCl$_2$, –78 °C		79	95:5	90:10
6			TBS	36c	EtAlCl$_2$, –78 °C		No reaction	—	—
7			Bn	37a	EtAlCl$_2$, –78 °C		83	95:5	67:33
8	D-manno		Piv	37b	EtAlCl$_2$, –78 °C	**2**	89	95:5	86:14
9			Bz	37c	EtAlCl$_2$, –78 °C		84	95:5	91:9
10			Ac	37d	EtAlCl$_2$, –78 °C		98	95:5	89:11
11	D-gluco				None, room temp.		96	74:26	91:9
12			—	38	EtAlCl$_2$, –78 °C		97	94:6	8:92
13	D-manno		Piv	37b	EtAlCl$_2$, –78 °C		89	95:5	99:1
14	D-gluco		—	38	EtAlCl$_2$, –78 °C	**3**	88	91:9	2:98

asymmetric induction in the presence of a Lewis acid was opposite to that observed under thermal conditions and a dr of 90:10 in favor of the (2R)-product was obtained (cf. entries 2 and 5). The authors explain this switch in the stereoselectivity by a conformational change within the acrylic ester moiety, which is triggered by complexation by the Lewis acid (cf. Scheme 4.6 below). The D-*manno*-configured dienophiles **37a–d** were tested under thermal conditions (not shown) and in the presence of EtAlCl$_2$ (entries 7–10). While thermal conditions did not give any stereoselectivity, good to high yields and *endo:exo* selectivities were observed for **37a–d** in the presence of EtAlCl$_2$ but again the steric bulk of the C3 substituent had strong impact on stereoselectivity. A dr of 91:9 in favor of the (2R)-product was obtained for R^1 = Bz (**37c**), while smaller and larger groups led to reduced stereoselectivity (**37a,b** and **37d**).

The alternative D-gluco configured dienophile **38** carrying the acrylic ester on C2 was also explored under thermal conditions and Lewis acid catalysis (Table 4.1, entries 11, 12) [5b]. High yields were obtained in both cases, but the *endo:exo* selectivity was rather low under thermal conditions. As in case of dienophile **36b**, thermal conditions and Lewis acid catalysis led to opposite stereochemistry in the Diels–Alder addition products, which were obtained in good drs in both cases. Thus, the change in reaction conditions can be used to obtain both product enantiomers with the same carbohydrate dienophile.

Finally, Tadano used dienophiles **37b** and **38** in the asymmetric Diels–Alder reaction with furan (**3**) [6]. In the presence of EtAlCl$_2$ good yields and high *endo:exo* selectivities were observed; **37b** and **38** yielded the product in opposite configuration and are therefore efficient pseudo-enantiomers in this reaction (Table 4.1 entries 13, 14).

After chromatographic purification, which removed the *exo*-diastereomers, the Diels–Alder products were cleaved off the carbohydrate auxiliaries by DIBALH reduction, as exemplified for the product of dienophile **36c** (Scheme 4.5). Upon reduction (2S)-**39c** yielded norbornylmethanol, (2S)-**40**, in 94% ee, while the carbohydrate scaffold **41c** was isolated in 76% yield and may be re-used [5b].

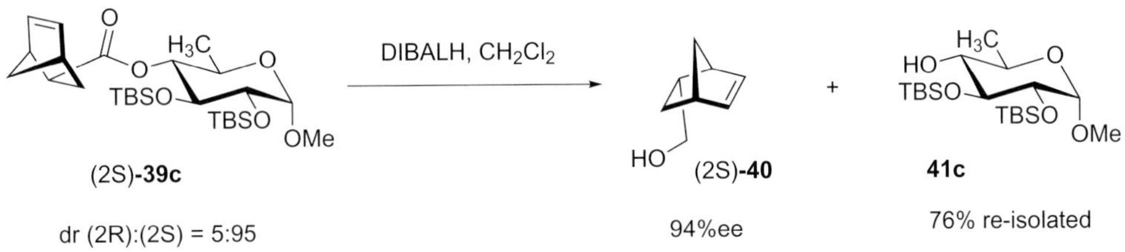

Scheme 4.5 Cleavage of the product from the carbohydrate scaffold and re-isolation of the auxiliary.

Scheme 4.6 presents a rationale explaining the diastereoselectivities observed with **36b** under thermal conditions and Lewis acid catalysis. The authors suggest that the acrylic ester moiety predominantly adopts an *s-cis* conformation in the absence of a Lewis acid, while coordination of a Lewis acid triggers a

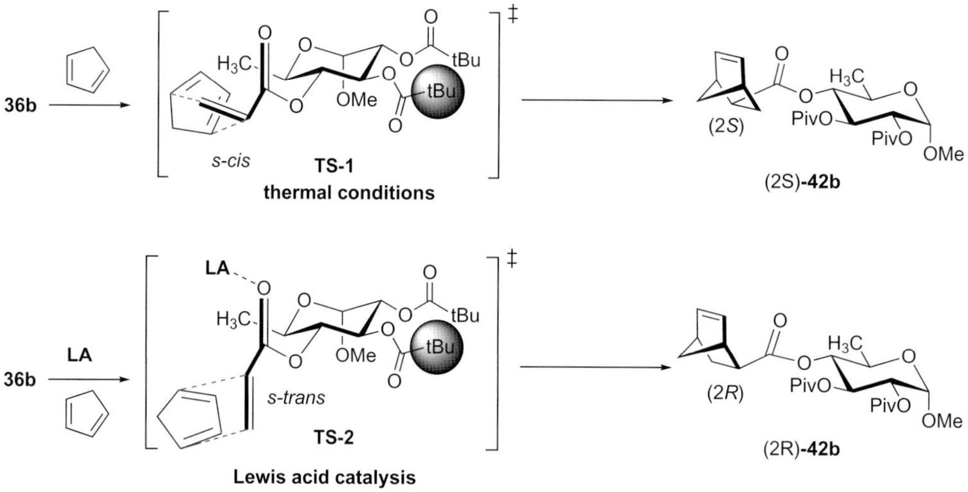

Scheme 4.6 Plausible transition states rationalizing the stereoselectivities observed with dienophile **36b** in the absence and presence of a Lewis acid catalyst.

conformational change in favor of the *s-trans* conformer. In both cases the diene attacks the dienophile preferentially from the less hindered face opposite to the bulky pivalate group at C3, but due to the different conformations of the acryloyl moiety opposite faces of its double bond are exposed in transition states **TS-1** and **TS-2**, leading to products (2*S*)-**42b** and (2*R*)-**42b**, respectively, with opposite stereochemistry.

A selection of carbohydrate-derived acrylic esters with highly diverse architectures have been reported by Ferreira [7], who tested diisopropylidene derivatives of D-glucose, D-allose, D-galactose, and D-fructose (**43–46**) as scaffolds. These dienophiles were tried under thermal conditions as well as in the presence of various Lewis acid catalysts. Yields as well as drs were generally moderate to low. The best results of these studies were obtained with aluminum-based Lewis acids (Scheme 4.7). In contrast, fructose-based dienophile **47**, which was reported by Nouguier [8], gave excellent yield and dr using $SnCl_4$ as the Lewis acid; a dr of 10:90 was obtained using $EtAlCl_2$. Fructose acrylate **46** was also employed by Enholm and coworkers [9] under conditions identical to those reported by Ferreira [7], but contrary to Ferreira *et al.*, who did not observe any stereoselectivity, Enholm and coworkers isolated optically active material after reductive cleavage of the carbohydrate auxiliary – unfortunately, no dr was reported. Enholm also successfully performed an interesting intramolecular Diels–Alder reaction using the fructose scaffold of **46** but again no exact dr was given.

More recently, Suárez [10] published conceptually new chiral auxiliaries derived from levoglucosenone (**48**), which is accessible by pyrolysis of cheap cellulose sources as, for example, waste paper. Diels–Alder reaction of **48** with anthracene (**5**) or 9-methoxymethylanthracene (**49**) yielded bulky ketones **50a,b**, which were subsequently converted into dienophiles **52a** [10a, b, d] and **52b** [10c] by a sequence

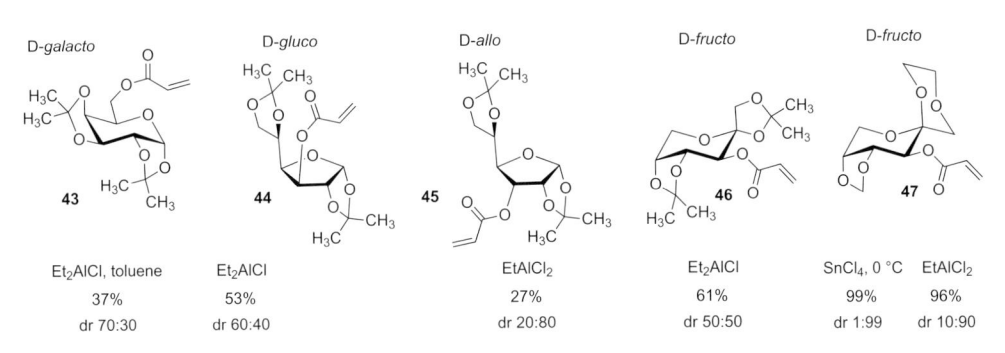

Scheme 4.7 Diels–Alder reactions of dienophiles by Ferreira (**43–46**) and Nouguier (**47**) with cyclopentadiene (**2**).

of stereoselective reduction and acryloylation (Scheme 4.8). Structures related to **52a,b** were prepared by cycloaddition of levoglucosenone (**48**) with cyclopentadiene [10d], and simplified dienophiles were obtained from **48** via hydrogenation of the C=C double bond, stereoselective reduction of the keto group, and subsequent acryloylation of the resulting alcohol [10e].

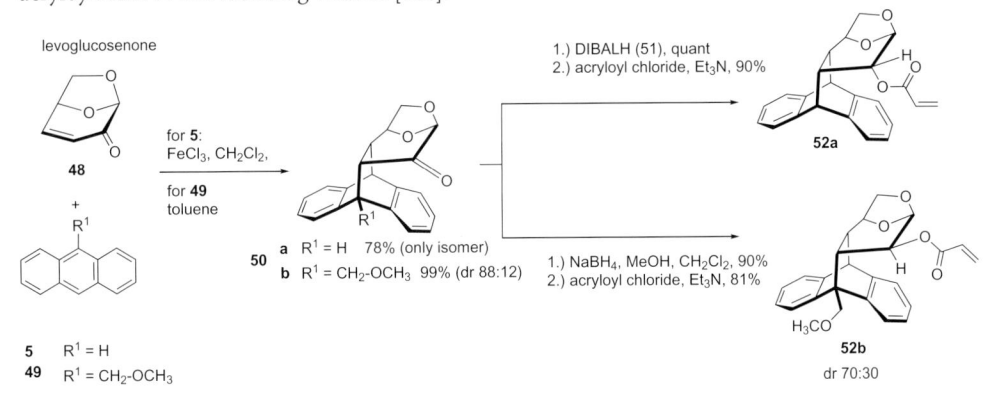

Scheme 4.8 Preparation bulky dienophiles **52a,b** from levoglucosenone (**48**) and their application in Diels–Alder reactions.

The authors explored various conditions and Lewis acids for the Diels–Alder reaction of dienophiles **52a,b** and related structures but the best results were again obtained with Et$_2$AlCl. Scheme 4.9 summarizes the most important findings of these studies. Under optimized conditions dienophiles **52a** and **52b** gave high yield and excellent *endo:exo* selectivity; regarding the dr of the cycloaddition, compound **52a** gave the diastereomeric products in a ratio of 90 : 10 while **52b** offered nearly perfect stereoselectivity [10c].

Scheme 4.9 Diels–Alder reactions of dienophiles **52a,b** with cyclopentadiene (**2**).

Compared to carbohydrate acrylates, imine-dienophiles derived from carbohydrates have rarely been reported. The reason for this may be that the choice of carbohydrate scaffolds containing amino groups is rather limited and formation of acrylic esters is much more straightforward. Banks et al. have introduced carbohydrate acryloyl imines **53-55** (Scheme 4.10) [11]; interestingly, none of these derivatives was prepared from natural nitrogen-containing carbohydrates. Instead, D-galactose, D-fructose, and the diisopropylidene derivative of 2-keto-L-gulonic acid

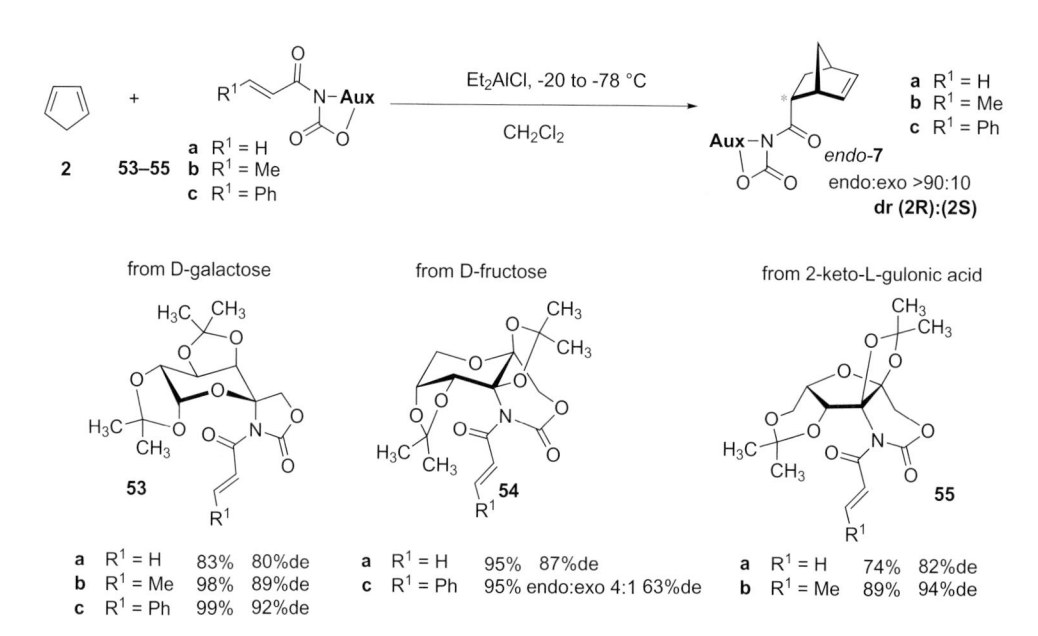

Scheme 4.10 Diels–Alder reactions of carbohydrate imine dienophiles **53–55** with cyclopentadiene (**2**).

were used as starting materials. The latter compound is an intermediate from the synthesis of L-ascorbic acid from D-glucose by the Reichenstein process [12]. All three dienophiles were obtained using an intramolecular nitrene insertion reaction as the key step for construction of the oxazolidinone and oxazinone ring systems of **53-55** [11]. In Lewis acid promoted Diels–Alder reactions with cyclopentadiene (**2**), imines **53-55** led to high yields and good to excellent des in almost all cases. Notably, imines **53-55** are one of the rare examples in which carbohydrate-derived crotonic and cinnamic esters were employed as dienophiles.

Hetero-Diels–Alder reactions were among the earliest asymmetric transformations conducted with carbohydrate-derived auxiliaries. As in the case of Diels–Alder reactions, carbohydrate-modified dienophiles are most frequently employed. In 1984 Vasella *et al.* introduced α-chloronitroso compound **60** as a chirally modified heterodienophile [13]. The synthesis of **60** started from D-mannose (**56**), which was transformed into diisopropylidene derivative **57**, which in turn was reacted with hydroxyl amine to furnish acyclic carbohydrate oxime **58** [14]. Oxidation of **58** with manganese(IV) oxide [15] followed by treatment with *tert*-butyl hypochloride yielded heterodienophile **60** as virtually one anomer (Scheme 4.11).

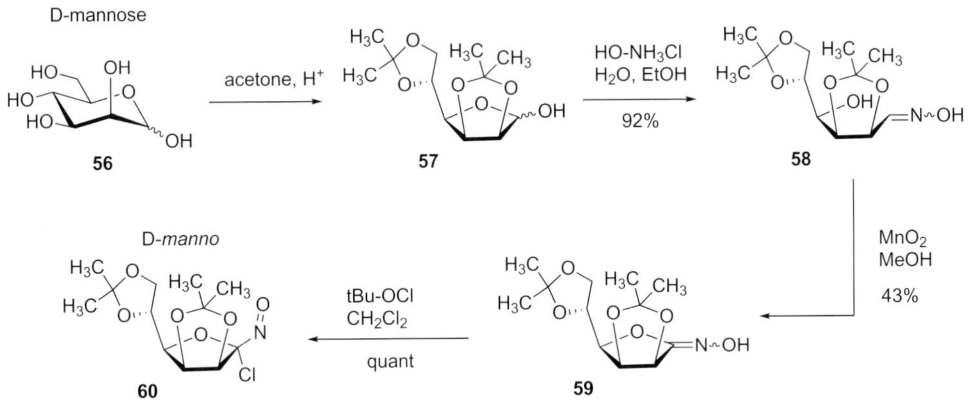

Scheme 4.11 Preparation of D-*manno* configured α-chloronitroso compound **60** from D-mannose (**56**).

In contrast to simple alkyl-α-chloronitroso compounds **60** exhibited enhanced reactivity towards diene substrates. This is due to the location of the α-chloronitroso moiety at the anomeric center of the carbohydrate scaffold, which exerts an electron-withdrawing effect on the nitroso group via the endocyclic oxygen atom. Heterodienophile **60** was employed in cycloadditions with cyclohexadiene (**4**), racemic cyclohexadiene derivative *rac*-**67**, and acyclic dienes **69**, **71**, and **6b** [13, 16, 17]. To access both enantiomers of the cycloaddition products, ribose-derived heterodienophile **61** was designed as a pseudo-enantiomer to **60** [17]. The reactions with heterodienophiles **60** and **61** were conducted in EtOH or CH$_2$Cl$_2$ and subsequent workup under acidic conditions was accompanied by cleavage of the products from the carbohydrate scaffolds. These were re-isolated as lactones **64** and **65** and could be re-used for further hetero-Diels–Alder reaction after reaction with hydroxyl amine. Table 4.2 summarizes the results of all reactions. Irrespective of

Table 4.2 Hetero-Diels–Alder reactions of carbohydrate-derived α-chloronitroso compound **60** with cyclic and acyclic diene substrates.

Entry	Heterodienophile	Diene	Product	Yield (%)	ee (%)
1	60 D-manno	4		70–89	>96
2	61 D-ribo	4	66 ent-66	96	>96
3	60 D-manno	rac-67	68	72–84	>96
4	61 D-ribo	rac-67	ent-68	82	>96
5	60 D-manno	69	70	69–79	>96
6	61 D-ribo	69	ent-70	74	>96
7	60 D-manno	71	72	63–68	>96
8	61 D-ribo	71	ent-72	92	>96

Table 4.2 (Continued)

Entry	Heterodienophile	Diene	Product	Yield (%)	ee (%)
9	60 D-*manno*	6b	73a + 73b (2:1) *ent*-73a + *ent*-73b (2:1)	63–82 (73a + 73b)	>96 (73a)
10	61 D-*ribo*	6b		77	>96 (*ent*-73a)
11	60 D-*manno*	74	75	89	>99
12	60 D-*manno*	76	77	82	>97

the diene substrate, all hetero-Diels–Alder reactions with D-manno compound **60** proceeded in fair to excellent yield and gave products **66, 68, 70, 72, 73** in excellent ee (entries 1, 3, 5, 7, 9) [17]. The D-ribo derivative **61** proved to be a highly efficient pseudo-enantiomer, giving products *ent*-**66**, *ent*-**68**, *ent*-**70**, *ent*-**72**, and *ent*-**73** in almost identically good yields and ees (entries 2, 4, 6, 8, 10) [17]. Notably, in the case of racemic substrate *rac*-**67** an efficient kinetic racemic resolution took place. Unsymmetrically substituted substrate **6b** gave rise to a regioisomeric mixture of products **73a** + **73b** and *ent*-**73a** + *ent*-**73b**, respectively. While the regioselectivity was modest, **73a** and *ent*-**73a** were obtained in high ee.

Reactions of *meso*-dienes **74** and **76** with D-manno heterodienophile **60** were reported by Werbitzky [18] and Piepersberg [19]. The corresponding products **75** and **77** were formed in high yields and ees (Table 4.2, entries 11, 12) and were employed in the preparation of aminocyclitols **79** [18] and **81** [19] (Scheme 4.12).

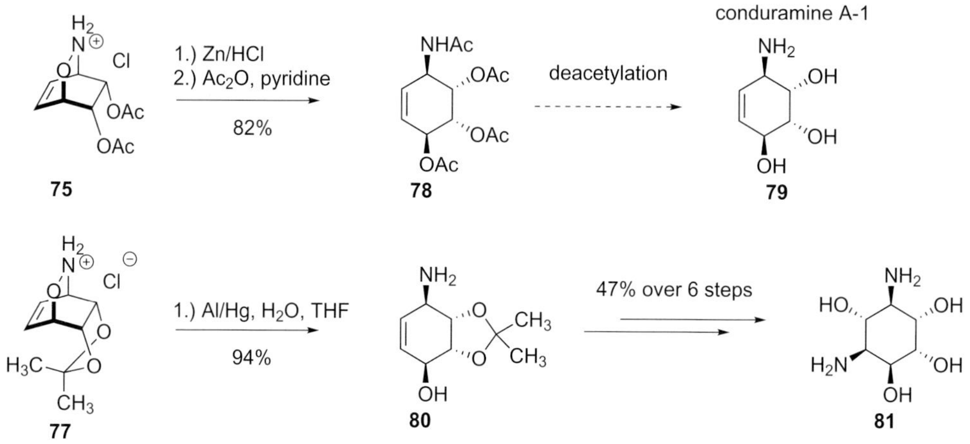

Scheme 4.12 Synthesis of aminocyclitols **79** and **81** from hetero-Diels–Alder products **74** and **76**.

The group of Defoin used heterodienophile **60** in the preparation of some nojirimycin derivatives [20]. Scheme 4.13 shows the key steps towards 6-deoxy-L-allo-nojirimycin (**85**) from diene **82** and heterodienophile **60** [20c]. Even though **82** was employed as a diastereomeric mixture of the (4*E*/*Z*) isomers, the cycloaddition step gave **83** as the only stereoisomer. Probably, de-acetalization of **82** takes place under the reaction conditions and the resulting aldehyde, which is prone to (4*E*/*Z*) isomerization, reacts with **60**. The resulting addition product is then re-acetalized by methyl orthoformate [20c]. N-Protection and diastereoselective dihydroxylation yielded **84** in 35% yield over three steps in excellent ee. This product was subsequently transformed into **85**.

As an alternative to **60**, Wightman and coworkers designed α-chloronitroso compound **88**, which is accessible from D-xylose (**86**) in 69% overall yield (Scheme 4.14) [21]. In **88** the chloronitroso moiety resides on C3 of the xylofuranose scaffold

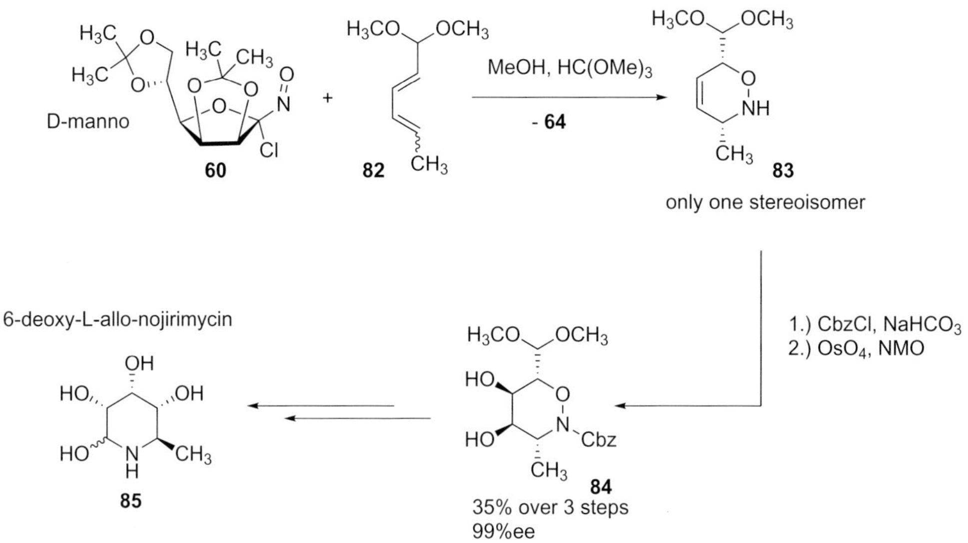

4.1 Diels–Alder Reactions | 79

Scheme 4.13 Preparation of nojirimycin derivatives using hetero-Diels–Alder reactions with **60** as key step.

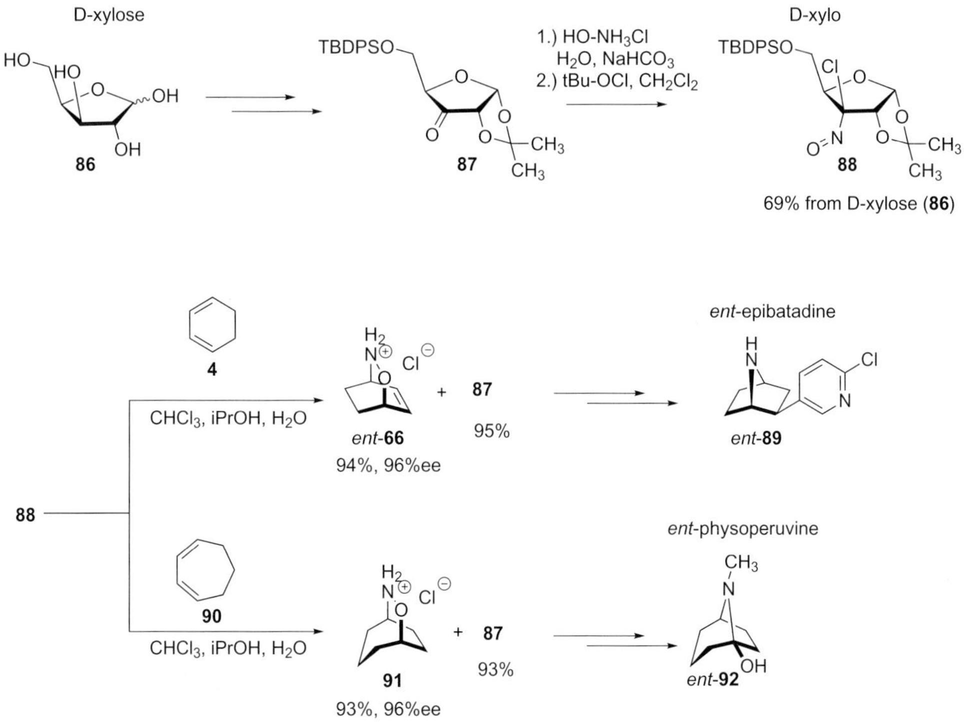

Scheme 4.14 Heterodienophile **88** in the preparation of *ent*-epibatadine (*ent*-**89**) and *ent*-physoperuvine (*ent*-**92**).

rather than at the anomeric center. Reaction of **88** with cyclohexadiene (**4**) yielded *ent*-**66** and thus acts as another pseudo-enantiomer to **60**. With cycloheptadiene (**16**) as substrate **91** was formed, while cyclopentadiene did not yield any of the expected products. Products *ent*-**66** and **91** were isolated in high yields and ees and the xylose auxiliary **87**, which was cleaved from the products during workup, could be re-isolated almost quantitatively. The products *ent*-**66** and **91** were used as starting materials for stereoselective syntheses of the natural products epibatadine and physoperuvine, respectively. Owing to the configurations of the hetero-Diels–Alder adducts, the enantiomers of both natural products (*ent*-**89** and *ent*-**92**) were obtained. Wightman also reported the preparation of a pseudo-enantiomer to **88**, which was prepared from L-sorbose and gives product **66** in the reaction with cyclohexadiene in high yield and ee [21].

In conclusion, heterodienophiles **60**, **61**, and **88** are valuable tools for asymmetric synthesis. This is impressively highlighted by the excellent stereoselectivities, broad spectrum of dienes that have been successfully converted, and the numerous applications in the synthesis of natural products.

Carbohydrate-derived enol and allenyl ethers in cycloadditions with heterodienes have been reported by the Reissig group [22]. As an example for the synthesis of such dienophiles, Scheme 4.15 shows the preparation of enol ethers **95a,b** and allenyl ether **95c** from diisopropylidene glucose [22a].

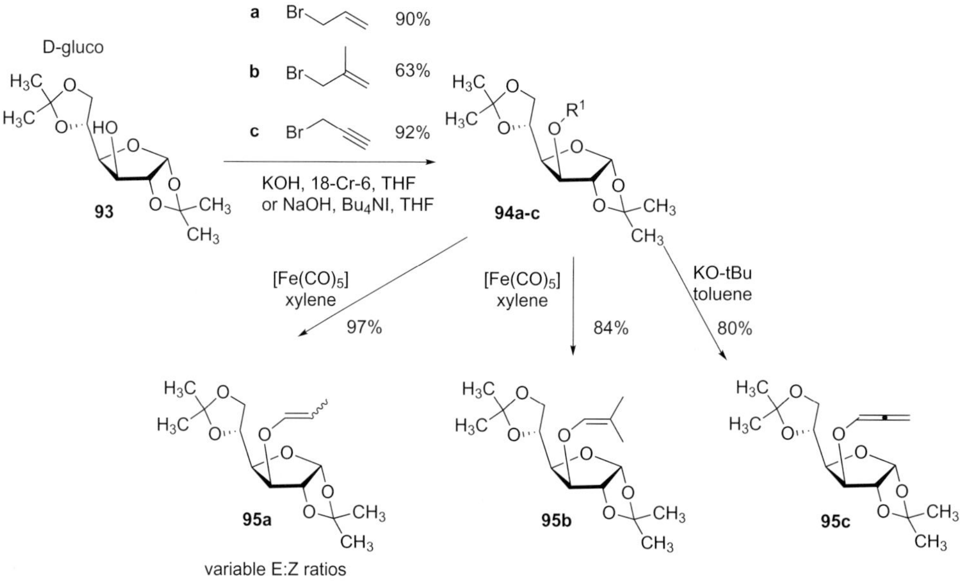

Scheme 4.15 Preparation D-glucose derived enol ethers **95a,b** and allenyl ether **95c** as dienophiles in hetero-Diels–Alder reactions.

Dienophiles **95a–c** were reacted with nitrosoalkenes prepared *in situ* from α-halo-oximes. Table 4.3 presents the results for enol ethers **95a** and **95b**. Enol ether **95a** was employed as an isomeric mixture with variable (*Z*/*E*)-ratio and, therefore, the cycloadditions with this substrate yielded **97-99** as a cis/trans mixture

Table 4.3 Diels–Alder reactions of enol ethers **95a,b** with heterodienes **96a–c**.

Entry	Dienophile	Heterodiene	Product	Yield (%)	dr
1	(Z)-95a + (E)-95a, Z:E = 94:6	96a (R² = Ph)	cis-97 + trans-97 (92:8)	54	cis-97 79:21; trans-97 95:5
2	(Z)-95a + (E)-95a, Z:E = 92:8	96b (R² = CO₂Et)	cis-98 + trans-98 (75:25)	33	cis-98 81:19; trans-98 95:5
3	(Z)-95a + (E)-95a, Z:E = 70:30	96c (R² = CF₃)	cis-99 + trans-99 (82:18)	65	cis-99 93:7; trans-99 90:10
4	95b (FCb)	96a	100	60	88:12

(entries 1–3). The *(E)*-configured isomer of dienophile **95a** is more reactive than its *(Z)*-counterpart; therefore, the cis/trans ratio of products **97–99** does not reflect the *(Z/E)* ratio of the starting material. With dienes **96a** and **96b** the drs were significantly higher for the *trans*-configured products (entries 1, 2), while similarly high drs were obtained for both isomers of **99** (entry 3). Dienophile **95b** gave a dr of 88:12 and 60% yield (entry 4).

Allenyl ether dienophile **95c** was reacted with dienes **96a–c**, and the resulting products were subjected to basic isomerization, yielding **101–103** (Table 4.4).

Table 4.4 Diels–Alder reactions of allenyl ethers **95c** with heterodienes **96a–c**.

Entry	Dienophile	Heterodiene	Product	Yield (%)	Dr
1	D-*gluco* **95c**	**96a** (R^1 = Ph)	**101**	72	5:95
2		**96b** (R^1 = CO_2Et)	**102**	82	21:79
3		**96c** (R^1 = CF_3)	**103**	69	5:95
4	D-*allo* **109**	**96a**	**104**	83	50:50
5	D-*manno* **110**	**96a**	**105**	56	13:81
6	D-*fructo* **111**	**96a**	**106**	56	78:22
7		**96b**	**107**	60	81:19
8		**96c**	**108**	63	85:15

R^2 = Ph, CO_2, CF_3 (for **96** and products **101–108**)

dr (6R):(6S)

High drs were only observed for the reactions with heterodienes **96a** and **96c** [22b, c] (entries 1, 3). In a further study Reissig and coworkers explored carbohydrate allenyl ether dienophiles **109–111** [22b, c], which were prepared from corresponding propargylated precursors [23]. The drs of the cycloaddition reactions with **109–111** strongly depended on the carbohydrate scaffold [22b, c]. While allo-configured dienophile **109** did not induce any stereoselectivity (entry 4), fructose-based compound **111** led to a reversal in the stereoinduction (entries 6–8).

Adducts **101** and **106** derived from gluco-compound **95c** and fructo-dienophile **111**, respectively, were subjected to catalytic hydrogenation with palladium on charcoal. This resulted in opening of the 1,2-oxazine ring system, stereoselective hydrogenation of the C4=C5 double bond, and concomitant cleavage of the auxiliaries, yielding chiral amine **112** (Scheme 4.16). Compounds **101** and **106** gave the enantiomeric amine products *(R)-* and *(S)-***112**, respectively, in 80 and 60% ee. The ees observed for *(R)-* and *(S)-***112** are lower than expected from the drs of the starting materials, indicating that the stereoselectivity of the hydrogenation step is not complete.

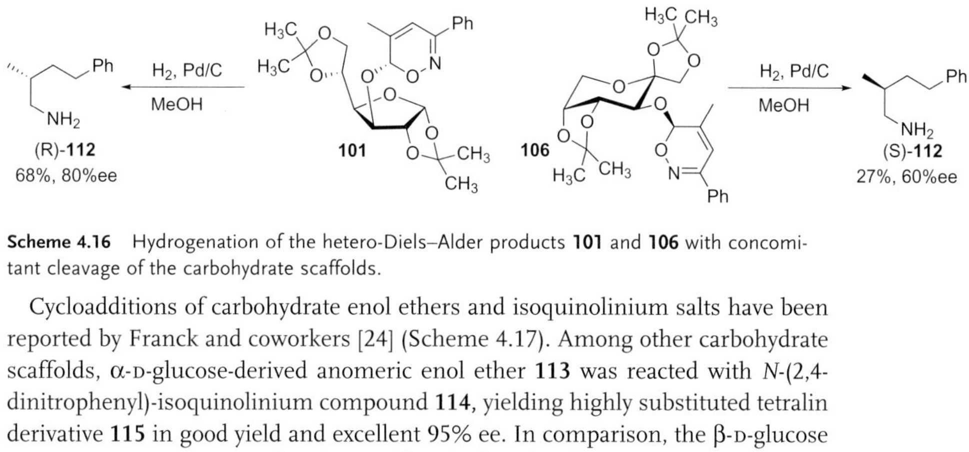

Scheme 4.16 Hydrogenation of the hetero-Diels–Alder products **101** and **106** with concomitant cleavage of the carbohydrate scaffolds.

Cycloadditions of carbohydrate enol ethers and isoquinolinium salts have been reported by Franck and coworkers [24] (Scheme 4.17). Among other carbohydrate scaffolds, α-D-glucose-derived anomeric enol ether **113** was reacted with N-(2,4-dinitrophenyl)-isoquinolinium compound **114**, yielding highly substituted tetralin derivative **115** in good yield and excellent 95% ee. In comparison, the β-D-glucose counterpart to **113** gave only a low yield and a low de of 50%.

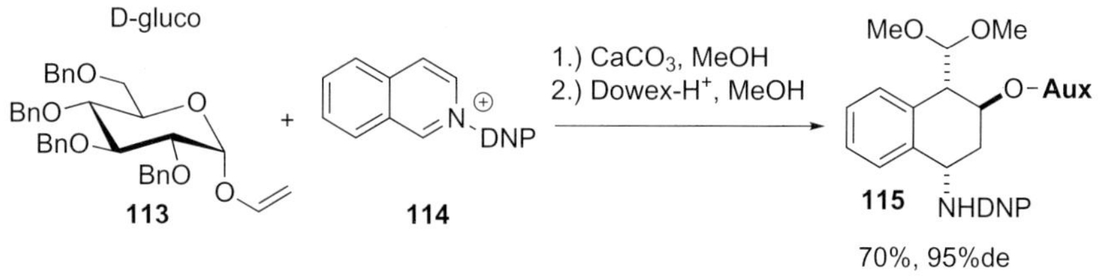

Scheme 4.17 Hetero-Diels–Alder reaction of an isoquinolinium salt.

4.1.2
Dienes Attached to the Auxiliary

Even though the attachment of a diene component to a carbohydrate scaffold is not nearly as straightforward as that of a dienophile, carbohydrate-modified dienes were reported far earlier. A very comprehensive study on glycosylated dienes was

4 Cycloadditions

performed by Stoodley and coworkers. At the outset, they prepared a carbohydrate analog (β-**119**) of Danishefsky's diene [25] (Scheme 4.18): reaction of acetobromoglucose (**116**) and sodium enolate (**117**) gave β-glucoside **118**, which was converted into the desired diene (β-**119**) [26].

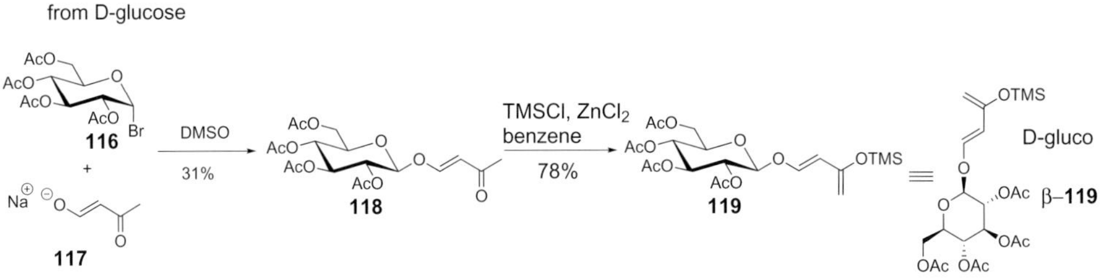

Scheme 4.18 Preparation of anomeric β-gluco diene β-**119** from acetobromoglucose **116**.

This diene was employed in diastereoselective Diels–Alder reactions with several cyclic dienophiles (Table 4.5). Out of the possible stereoisomers, reaction of β-**119**

Table 4.5 Diels–Alder reactions of carbohydrate-derived diene **119** with cyclic dienophiles.

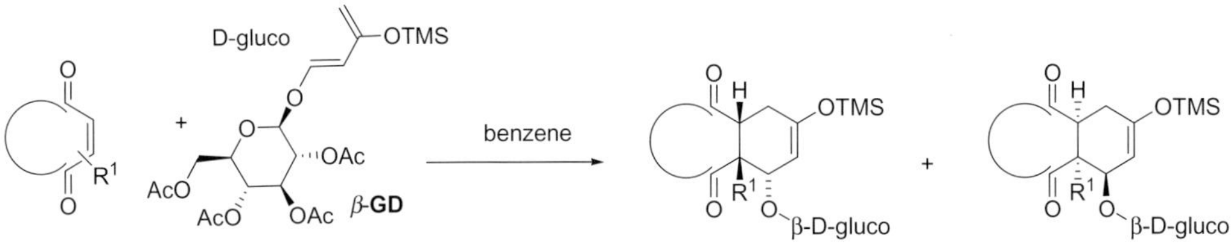

Entry	Dienophile	Major product	dr (major/minor) raw product	Isolated yield (major product) (%)
1	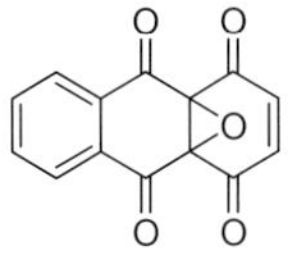 **120**	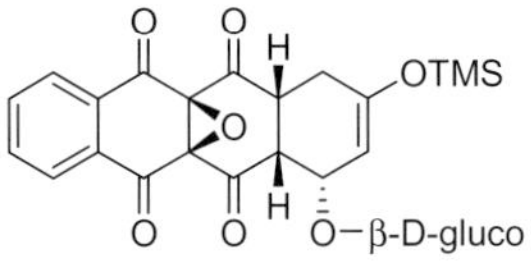 **121**	75:25	43
2	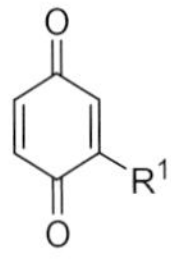 **122** a, $R^1 = H$	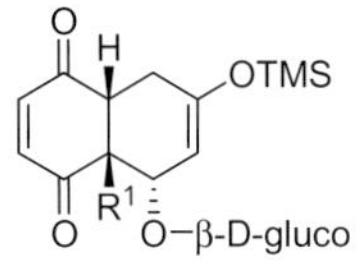 **123**	89:11	46
3	**122** b, $R^1 = CO_2Et$	**123**	88:12	75
4	**122** c, $R^1 = Ac$	**123**	75:25	43
5	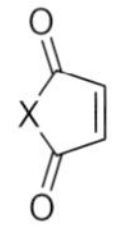 **124** a, X = N-Ph	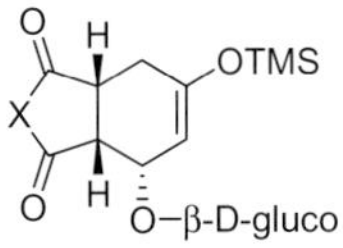 **125**	86:14	58
6	**124** b, X = NH	**125**	85:15	37
7	**124** c, X = O	**125**	Approx. 6:1	52

with *meso*-tetrone **120** yielded only two products in a dr of 75:25 and the major diastereomer **121** was isolated in pure form and 43% isolated yield [26] (entry 1). Regarding the complexity of product **121** and the number of stereocenters generated in only one step this is an excellent result. Diels–Alder reactions with *para*-benzoquinones **122a,b**, maleimides **124a,b**, and maleic anhydride **124c** gave even better drs and higher yields in most cases [27].

For cycloadditions with diene β-**119** the authors proposed transition state models **TS-1a** and **TS-2a** (Scheme 4.19) [27, 28]. The relative position of the anomeric diene moiety is supposedly governed by the exo-anomeric effect, giving rise to two possible spatial orientations of the diene. If these opposite faces are exposed to a dienophile, attack will commence from the unhindered top face. Thus **TS-1a** and **TS-2a** lead to products with opposite absolute configurations but, due to unfavorable steric interactions between the anomeric hydrogen and H2 of the diene

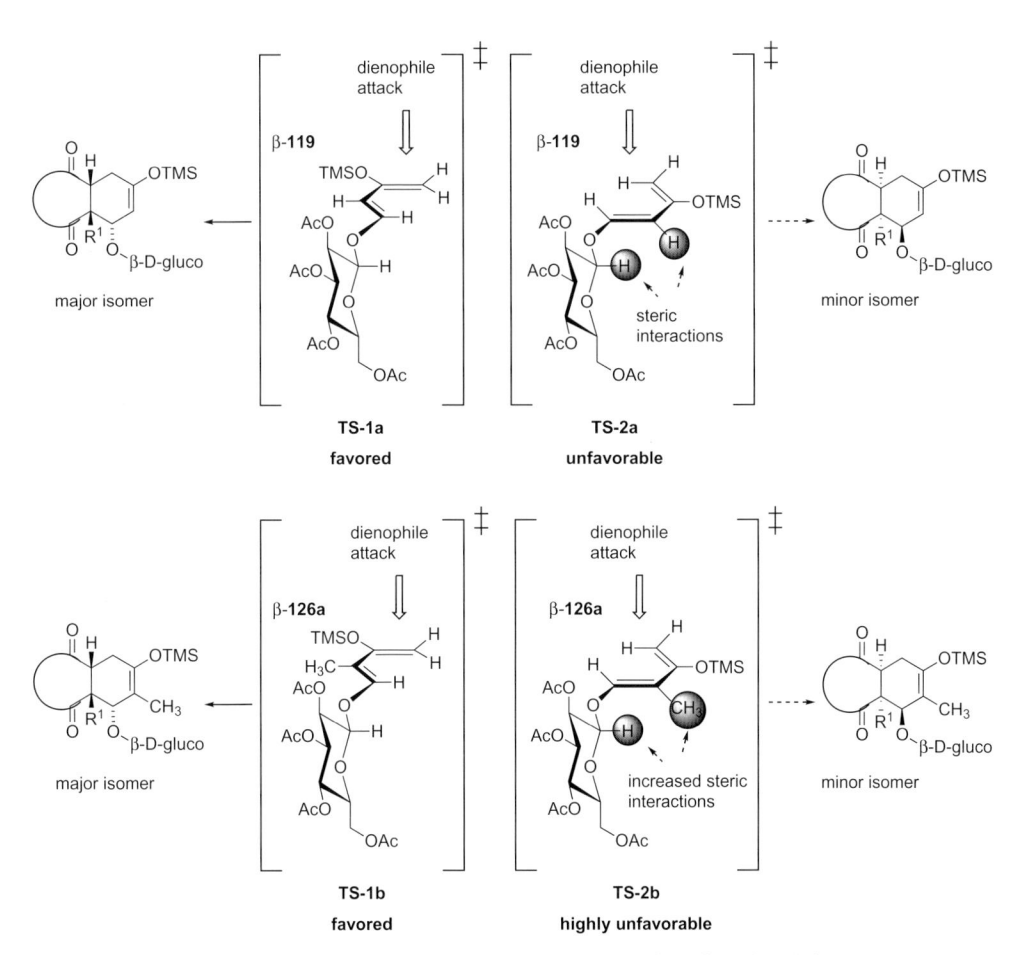

Scheme 4.19 Plausible transition states for Diels–Alder reactions with β-configured carbohydrate dienes β-**119** and β-**126a**.

moiety in **TS-2a**, the reaction should proceed mainly via **TS-1a**, giving a viable explanation for the drs detailed in Table 4.5. To confirm these considerations, the authors explored diene β-**126a**, containing a 2-methyl group on the butadiene moiety. Owing to enhanced unfavorable steric interactions between the anomeric hydrogen and the 2-methyl group of the diene in **TS-2b** a reaction via **TS-1b** should be even more favored and an increase in the drs was expected.

Indeed, reactions with diene β-**126a** proceeded with significantly improved stereoselectivities, often giving exclusively the diastereomer expected to result from **TS-1b** (Table 4.6, entries 1–3) [28, 29]. The influence of diene structure on stereoselectivity was explored further using a whole array of differently substituted dienes (β-**126b–g**) [28, 29] From the results of this study (Table 4.6) a clear trend emerges: all dienes with a 2-methyl group gave higher drs (entries 4, 7, 10) than those without this group. The influence of substituents at other positions of the diene moiety has no significant influence on the dr, which was in the same range as the for β-**119** [28, 29]. Diene β-**126c** was also reacted with dienophile **135**, yielding adduct **136** along with three minor stereoisomers (entry 11) [30].

Diels–Alder products **121** and **136** were used in the asymmetric synthesis of demethoxydaunomycinone **137** [26], an anthracycline derivative, and the preparation of **138**, the enantiomer of the natural product (−)-bostrycin [30], respectively (Scheme 4.20).

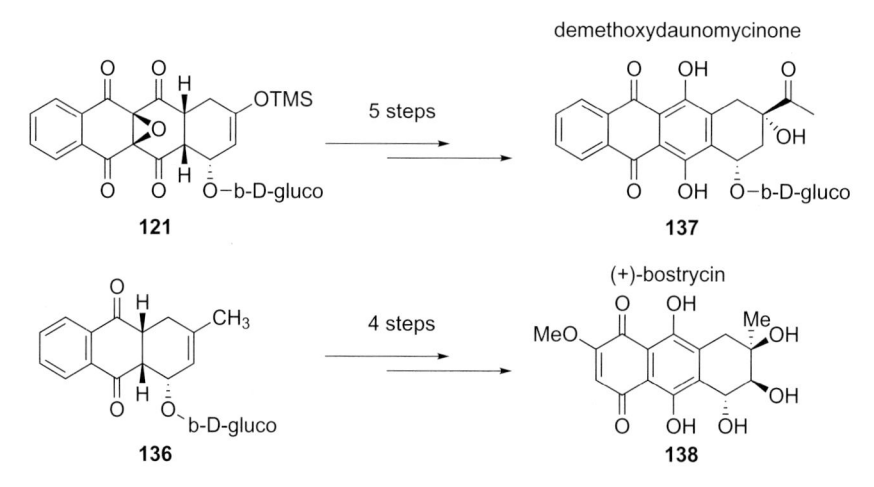

Scheme 4.20 Preparation of demethoxydaunomycinone **137** and (+)-bostrycin (**138**) from Diels–Alder products **121** and **136**, respectively.

To explore the influence of the anomeric configuration, Stoodley and coworkers prepared α-configured carbohydrate diene α-**139** [31]. The authors reasoned that the α-configuration of the carbohydrate dienes should cause the reaction to proceed via **TS-2a**, as now **TS-1a** is rendered unfavorable by steric interactions (Scheme 4.21). Consequently, the products should be obtained with opposite stereochemistry to those of reactions with diene β-**119**.

Table 4.6 Diels–Alder reactions of modified carbohydrate dienes β-**126** with cyclic dienophiles.

Entry	Dienophile	Diene	Major product	dr (major/minor) raw product	Isolated yield (major product) (%)
1	**122**	R^1 = H **a**	**127**	>99:1	69
2		R^1 = CO$_2$Et **b**		>99:1	72
3		R^1 = Ac **c**		Not given	65
4	**124a**	β-**126a**	**128**	95:5	87
5		β-**126b**	**129**	86:14	59
6		β-**126c**	**130**	88:12	60
7		β-**126d**	**131**	95:5	68
8		β-**126e**	**132**	89:11	59

Table 4.6 (Continued)

Entry	Dienophile	Diene	Major product	dr (major/minor) raw product	Isolated yield (major product) (%)
9		OTBS diene β-126f	133	85:5	70
10		diene β-126g	134	95:5	Not given
11	135	β-126c	136	63:21:13:3 more than two isomers observed	54

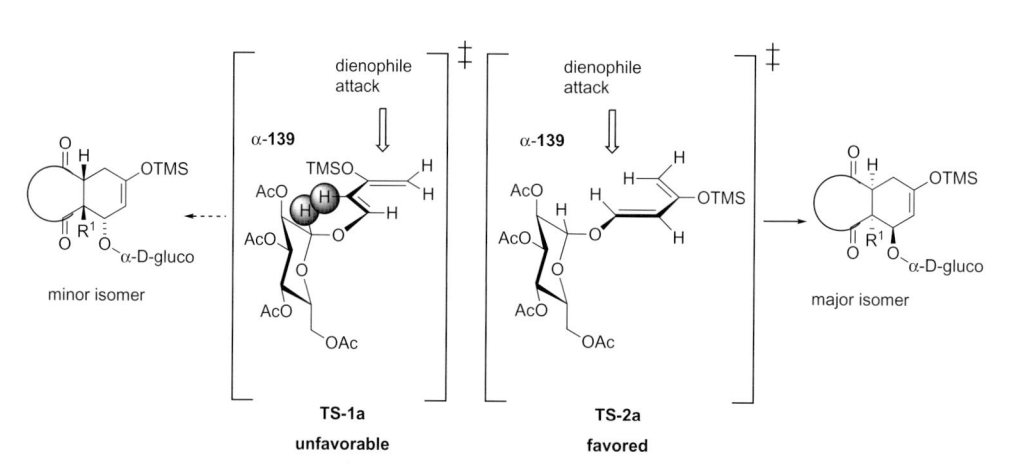

Scheme 4.21 Plausible transition states for Diels–Alder reactions with α-configured carbohydrate diene α-**139**.

With diene α-**139** a reversal of the asymmetric induction was indeed observed; however, the dr of the reaction product with phenylmaleimide (**124a**) was only low [31]. Further, the influence of the substituents at C2 and C6 of the pyranose scaffolds were explored for both anomeric series, revealing the largest effect for α-configured dienes [32].

The group of Lubineau studied dienes with unprotected carbohydrate moieties for Diels–Alder reactions in water [33]. Table 4.7 summarizes the results obtained with β-glucose-derived diene β-**140** and three dienophiles. All reactions yielded only two out of four possible stereoisomers, which were formed via an *endo*-transition state. Acrolein (**141**) and methacrolein (**145**) produced only modest drs with the carbohydrate diene, while acrylic ester **143** gave a dr of 70 : 30, the highest selectivity in this study. When the corresponding α-glucosyl diene was employed in the reaction with methacrolein, the *endo*-selectivity was severely reduced and a mixture of all four possible Diels–Alder adducts resulted. The major product of

Table 4.7 Diels–Alder reactions in water using diene β-**140** with a deprotected glucose.

Entry	Dienophile	Major product	dr (major/minor) raw product	Isolated yield (major + minor) (%)
1	**141** (acrolein)	**142** (β-D-gluco)	60 : 40	Not given
2	**143** (MeO-acrylate)	**144** (β-D-gluco)	70 : 30	66
3	**145** (methacrolein)	**146** (β-D-gluco)	60 : 40	90

this reaction had opposite stereochemistry compared to the one obtained from the reaction with the β-configured diene [33b]. The Diels–Alder products could be cleaved from the carbohydrate scaffold either enzymatically or, more traditionally, under acidic conditions.

In further studies the same authors made systematic modifications of the hydroxy groups at C2 and C6 in β- and α-configured glucose dienes [34]. The results are similar to those obtained by Stoodley et al. [32] (vide supra) who conducted their experiments independently.

Cyclic dienes attached to a carbohydrate scaffold are a rare occurrence. One interesting example was reported by Bird and Lewis, who used cyclohexadienyl ethers of isopropylidene glucose **147a,b** in the cycloaddition with naphthoquinone (**148**) [35]. Acidic treatment of the Diels–Alder products induced a rearrangement to the benzonaphthofuranones **150a,b** with concomitant loss of the carbohydrate auxiliary. In this manner, excellent yields of the (+)-isomers of **150a** and **150b** were obtained (Scheme 4.22).

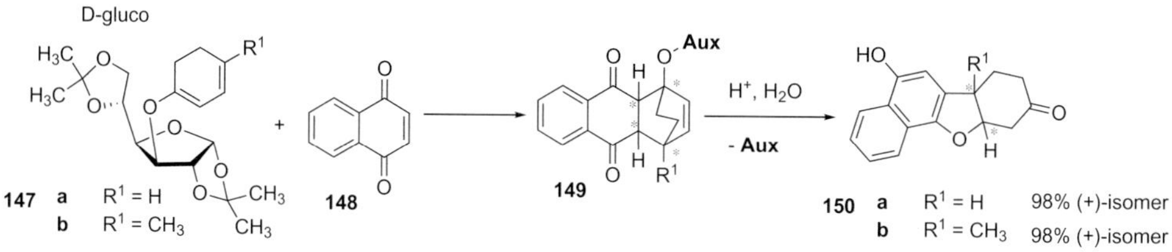

Scheme 4.22 Cycloadditions with carbohydrate-modified cyclic dienes **147** derived from diisopropylidene glucose.

Hetero-Diels–Alder reactions with carbohydrate-bound dienes were among the very first examples of the application of carbohydrate tools in stereoselective synthesis. Pioneering work was performed by David et al. who reported examined reactions of a diene derived from diisopropylidene glucose as early as 1974 [36]. Diene **153** was obtained in two steps from diisopropylidene glucose **93**. Interestingly, in contrast to the carbohydrate dienes that were later extensively employed by Stoodley and Lubineau, **153** is not an anomeric diene (Scheme 4.23).

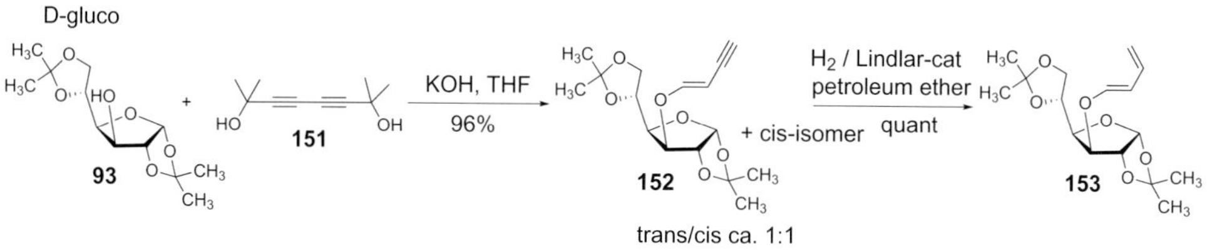

Scheme 4.23 Preparation of carbohydrate diene **153** from diisopropylidene glucose **93**.

Cycloaddition **153** with of butyl glyoxylate gave a mixture of the diastereomeric products **154a–d**, which could be separated into two fractions containing **154a** +

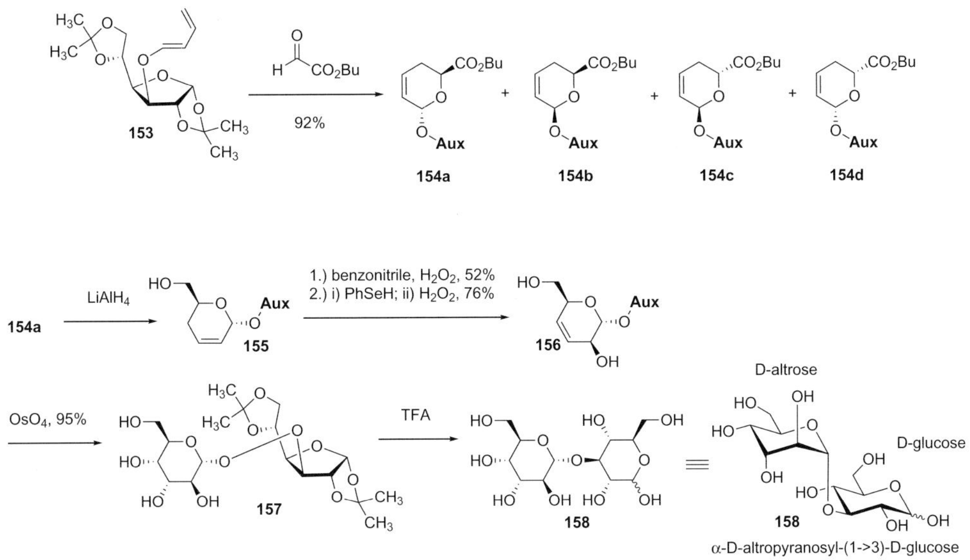

Scheme 4.24 Hetero-Diels–Alder reaction of carbohydrate diene **153** with butyl glyoxylate and application in an alternative synthesis of disaccharides.

154d and **154b** + **154c**, respectively (Scheme 4.24). These compounds may be viewed as 2,3-unsaturated carbohydrate derivatives.

Product **154a** was subsequently used for the construction of disaccharide derivative **158** via a sequence of diastereoselective functionalization reactions [37]. Employing this strategy, the same authors also prepared derivatives of the blood group A carbohydrate antigen [38]. Further studies were conducted on reactions of cis-configured carbohydrate dienes [39] and dienes carrying an additional 4-benzyloxy residue [40]. Even though these early examples of hetero-Diels–Alder reactions on carbohydrate dienes were fraught with rather modest diastereoselectivities and the preparation of di- and trisaccharides from the cycloaddition products required functionalization via multistep protocols, they impressively demonstrate the potential of carbohydrates as asymmetric inductors.

Heterodienes fixed to carbohydrate scaffolds have been described by Marazano [41]. Using the 1,2-dihydropyridine *N*-glycosides of D-glucose, D-xylose, and D-arabinose carrying acetyl or pivaloyl groups, cycloadditions with methyl acrylate were explored (Scheme 4.25). The isoquinuclidine derivatives **162**, which were obtained after treatment of the initial hetero-Diels–Alder products with methyl chloroformate, are interesting intermediates for alkaloid syntheses. To determine the stereoselectivities of the reactions, products **162** were converted into the diastereomeric compounds **163** and **164** by hydride reduction and treatment with camphorsulfonyl chloride. The enantiomeric excesses were strongly dependent on the carbohydrate scaffold and its protective groups. The bulkier pivalate derivatives **159b–161b** gave significantly better ees for than the acetates, and the D-gluco architecture of dienes **159** was clearly superior to the D-xylo compounds **160**. While

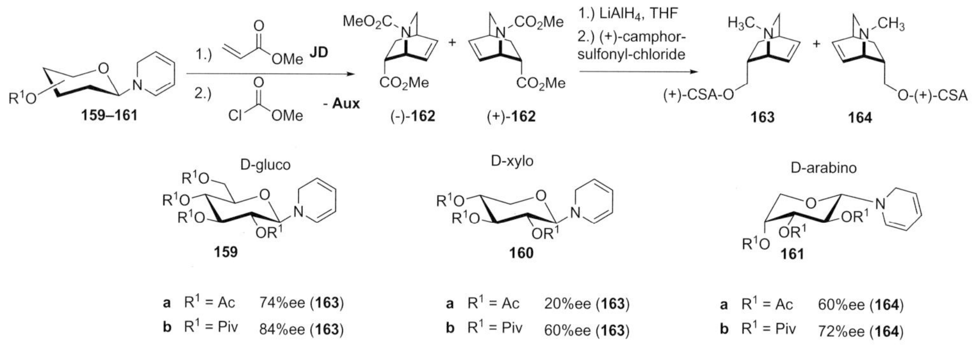

Scheme 4.25 Hetero-Diels–Alder reactions of N-glycosyl 1,2-dihydropyridines **159-161** with methyl acrylate.

159 and **160** predominantly produced **163**, D-arabino derivative **161** acts as a pseudo-enantiomer and gave rise to product **164**.

Stoodley and coworkers used glycosylated diene β-**165** in reactions with electron-poor azo compounds as heterodienophiles (Scheme 4.26) [42]. Hetero-Diels–Alder reaction of β-**165** with *tert*-butyl azodicarboxylate **166** gave adduct **167** in good yield as a single diastereomer, which was elaborated into *(S)*-tetrahydropyridazine

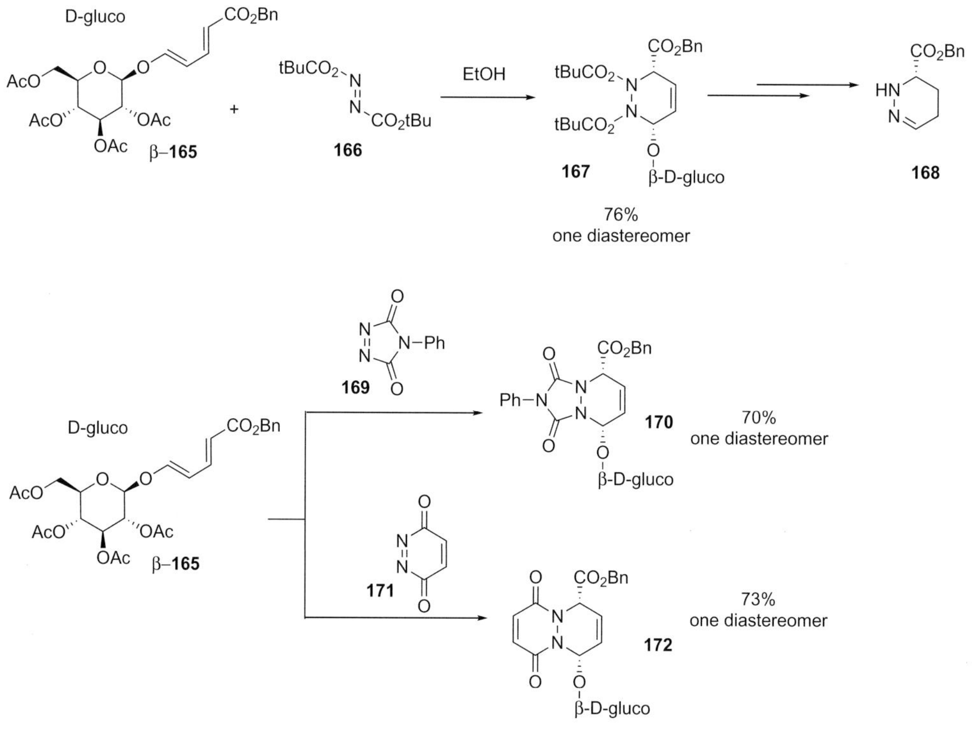

Scheme 4.26 Hetero-Diels–Alder reactions of glycosylated diene β-**165** with electron-poor azo compounds **166, 169,** and **171** and preparation of *(S)*-tetrahydropyridazine carboxylic acid (**168**).

carboxylic acid, a constituent of the antibiotically active heptapeptide antrimycin Dv [42a]. In further studies 1,2,4-triazoline-3,5-dione **169** and pyridazine-3,6-dione **171**, both generated *in situ*, were employed as heterodienophiles [42b]. The respective adducts **170** and **172** were isolated in high yield and as single diastereomers (Scheme 4.26).

4.2
1,3-Diploar Cycloadditions

This type of cycloaddition is not only attractive for the preparation of heterocycles but also a powerful approach for the construction of acyclic molecules containing heteroatom substituents. To effect asymmetric induction, either the 1,3-dipole or the dipolarophile may be attached to a chiral auxiliary. In the case of carbohydrates as tools for asymmetric induction, carbohydrate-bound 1,3-dipoles clearly dominate, while only a few reports of dipolarophiles on carbohydrate scaffolds have appeared.

4.2.1
1,3-Dipoles Attached to the Auxiliary

Virtually all examples of preparative interest reported so far concern N-glycosyl nitrones as 1,3-dipoles. Again, this field was pioneered by Vasella, who first reported D-ribose-based nitrones and their application as early as 1977 [43]. For the preparation of these nitrones ribose derivative **174** was reacted with hydroxylamine to give acyclic ribose oxime **175** as a mixture of *(E)*- and *(Z)*-isomers. In condensation reactions with aldehydes or ketones **175** formed N-glycosyl nitrones **176** [43a]. Starting from mannose-derived acyclic oxime **58**, which is also useful in asymmetric hetero-Diels–Alder reactions (cf. Scheme 4.11), manno-configured nitrones **177** were obtained in a similar manner (Scheme 4.27) [43b].

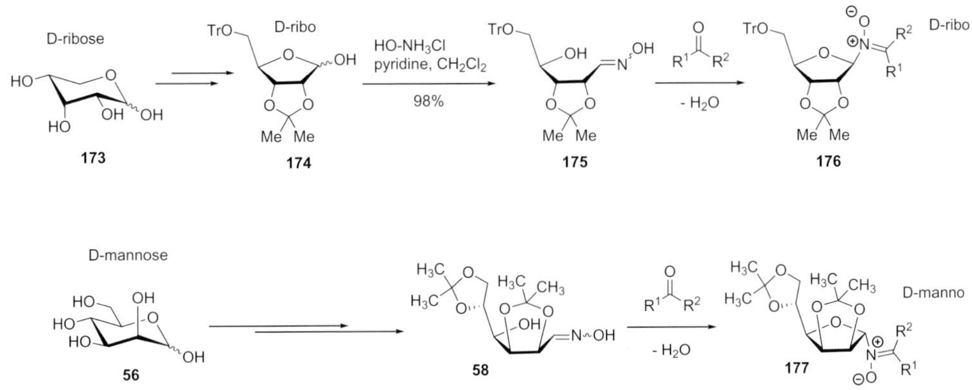

Scheme 4.27 N-Glycosyl nitrones **176** and **177** derived from D-ribose and D-mannose, respectively.

With ribo-nitrones **176a** and **176b** derived from formaldehyde (**178a**) and acetone (**178b**), respectively, 1,3-dipolar cycloaddition reactions with methyl methacrylate (**179**) gave the isoxazolidines *(R)*- and *(S)*-**180** in excellent yields (Scheme 4.28). In the case of formaldehyde nitrone **176a** the de was 67% while the acetone-derived nitrone **176b** gave more than 95% de; in both cases the new stereocenter was formed in the *(R)*-configuration. Cleavage of the products from the carbohydrate scaffolds was possible under acidic conditions [43a, b]. The mannose-derived nitrones **177a** and **177b** from formaldehyde and acetone, respectively, also reacted smoothly with methyl methacrylate. The diastereoselectivities for the products **182** were on a similar level as in case of ribo-dipolarophiles **176** but the direction of the asymmetric induction was reversed. Thus, products **181** were obtained in enantiomeric configuration after cleavage from the manno-scaffold [43b].

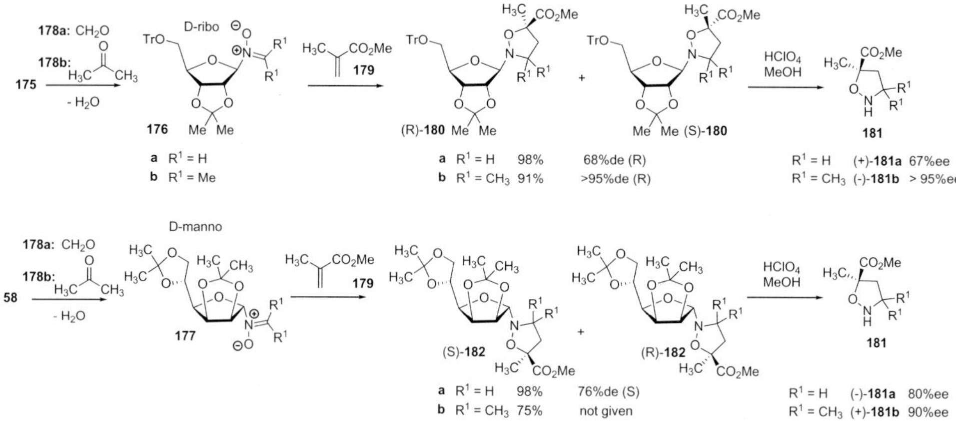

Scheme 4.28 Asymmetric 1,3-dipolar cycloadditions using *N*-glycosyl nitrones **176a,b** and **177a,b** derived from formaldehyde and acetone with methyl methacrylate (**179**) as dipolarophile.

The reaction of ribo-configured oxime **175** with acetaldehyde (**178c**) yielded an *(E/Z)*-mixture of the nitrones **176c**, which upon 1,3-dipolar cycloaddition with methyl methacrylate gave a mixture of four diastereomeric isoxazolidines (**180c**). The two major diastereomers of this mixture had the same configuration at the quaternary stereocenter of the isoxazolidine moiety and differed only in the configuration at the stereocenter carrying the methyl group (Scheme 4.29) [43b]. Therefore, with respect to the quaternary center, the asymmetric induction is independent of the *(E)*- or *(Z)*-configuration of the nitrone **176c**, which, however, determines the stereochemistry at the chiral carbon bearing the methyl group. The author also described the reaction of non-activated alkenes with formaldehyde-derived nitrone **176a** (Scheme 4.29). Reactions of **176a** with 1-octene (**183**) and cyclohexene (**185**)

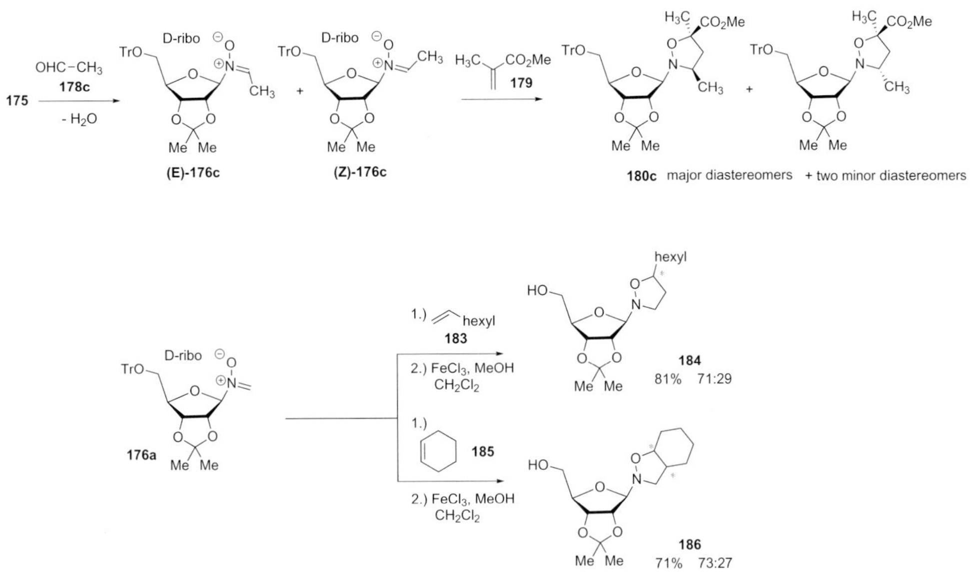

Scheme 4.29 Asymmetric 1,3-dipolar cycloadditions using N-glycosyl nitrone **176c**, from acetaldehyde (**178c**) with methyl methacrylate (**179**), and N-glycosyl nitrone **176a**, derived from formaldehyde, with non-activated alkenes 1-octene (**183**) and cyclohexene (**185**).

yielded the expected adducts, which were detritylated for purification. The resulting N-ribosyl isoxazolidines **184** and **186** were obtained in good yield but only in drs around 70:30. Nevertheless these two examples are remarkable as 1,3-dipolar cycloadditions with simple alkyl-substituted olefins are challenging.

Based on the stereochemical outcome observed for the 1,3-dipolar cycloaddition of manno-configured nitrone **177a** with methyl methacrylate and additional mechanistic studies, the Vasella group suggested two plausible transition states for the reaction (Scheme 4.30) [43, 44]. Nitrone **177a** is assumed to adopt either the O-*exo*- or the O-*endo*-conformation, which differ in the relative orientation of the nitrone oxygen to the carbohydrate scaffold. Of these two, the O-*endo* conformer is supposedly favored due to steric reasons. To form product *(S)*-**182** the dipolarophile **179** can approach the nitrone moiety in the O-*endo* conformer either from the top face (*syn* attack) or from the bottom face (*anti* attack), adopting a methyl *endo* or an ester *endo* orientation with respect to the nitrone. These two attacks lead to transition states **TS-1** and **TS-2**, respectively, of which **TS-1** is assumed to be more favorable for stereoelectronic reasons: formation of the isoxazoline moiety during the cycloaddition is accompanied by rehybridization of the nitrone nitrogen atom from sp^2 to sp^3 and generation of a free electron pair. In **TS-1** the orbital containing this nonbonding electron pair is oriented parallel to the σ* orbital of the endocyclic C–O bond, which should result in an overall stabilization of the newly formed molecule.

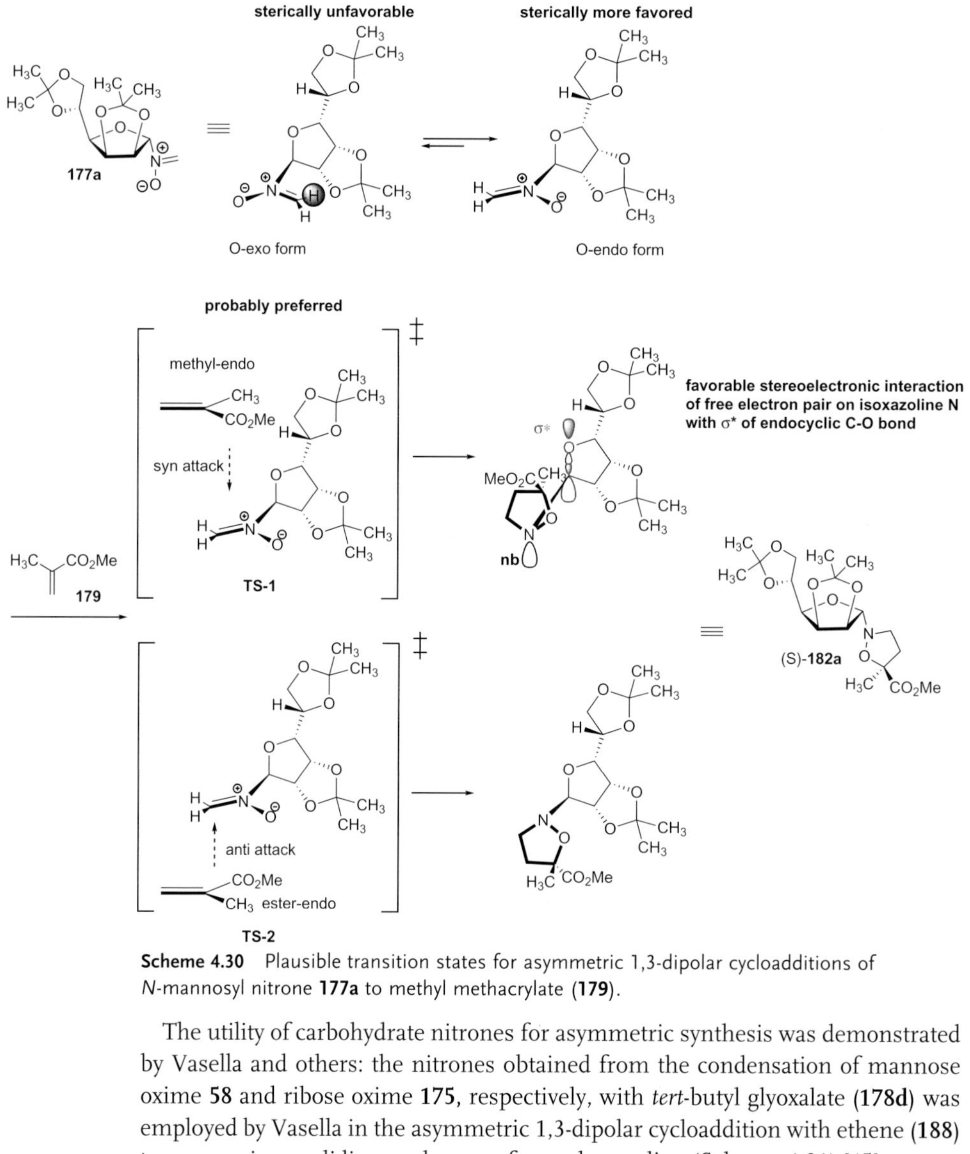

Scheme 4.30 Plausible transition states for asymmetric 1,3-dipolar cycloadditions of N-mannosyl nitrone **177a** to methyl methacrylate (**179**).

The utility of carbohydrate nitrones for asymmetric synthesis was demonstrated by Vasella and others: the nitrones obtained from the condensation of mannose oxime **58** and ribose oxime **175**, respectively, with *tert*-butyl glyoxalate (**178d**) was employed by Vasella in the asymmetric 1,3-dipolar cycloaddition with ethene (**188**) to prepare isoxazolidine analogues of L- and D-proline (Scheme 4.31) [45].

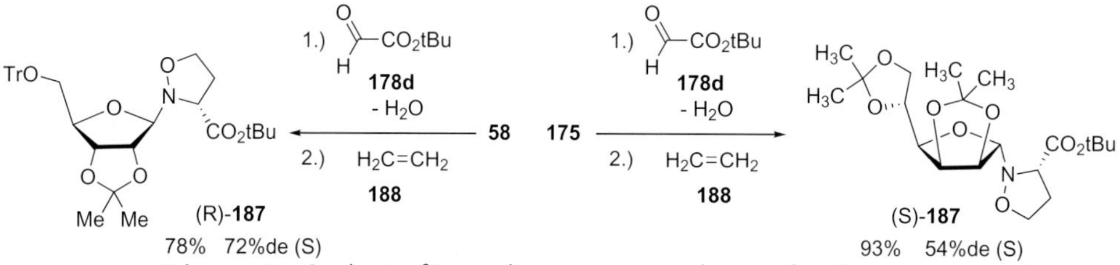

Scheme 4.31 Synthesis of isoxazoline-containing analogues of proline using carbohydrate oximes **58** and **175**.

1,3-Dipolar cycloadditions using carbohydrate oximes **175** and **58** or related structures have also been applied as key steps in the syntheses of complex natural products. As the isoxazolidine N–O bond is readily cleaved under reductive conditions, these cycloadditions are also useful for the preparation of compounds not containing isoxazolidine motifs. In an elegant approach Vasella used mannofuranose hydroxylamine in the stereoselective total synthesis of nojirimycin (Scheme 4.32) [46].

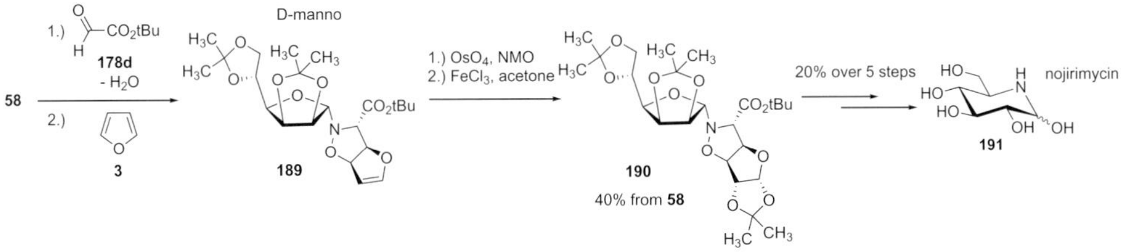

Scheme 4.32 Asymmetric synthesis of nojirimycin via 1,3-dipolar cycloaddition with carbohydrate-derived nitrones.

The group of Kibayashi reported an approach for the synthesis of the peptide-analog natural products (+)-negamycin (**198**) and (−)-epinegamycin (**199**) using cyclohexylidene-modified carbohydrate nitrone **194** derived from D-gulonic-γ-lactone **192** (Scheme 4.33) [47]. The 1,3-dipolar cycloaddition of **194** to N-Cbz allyl

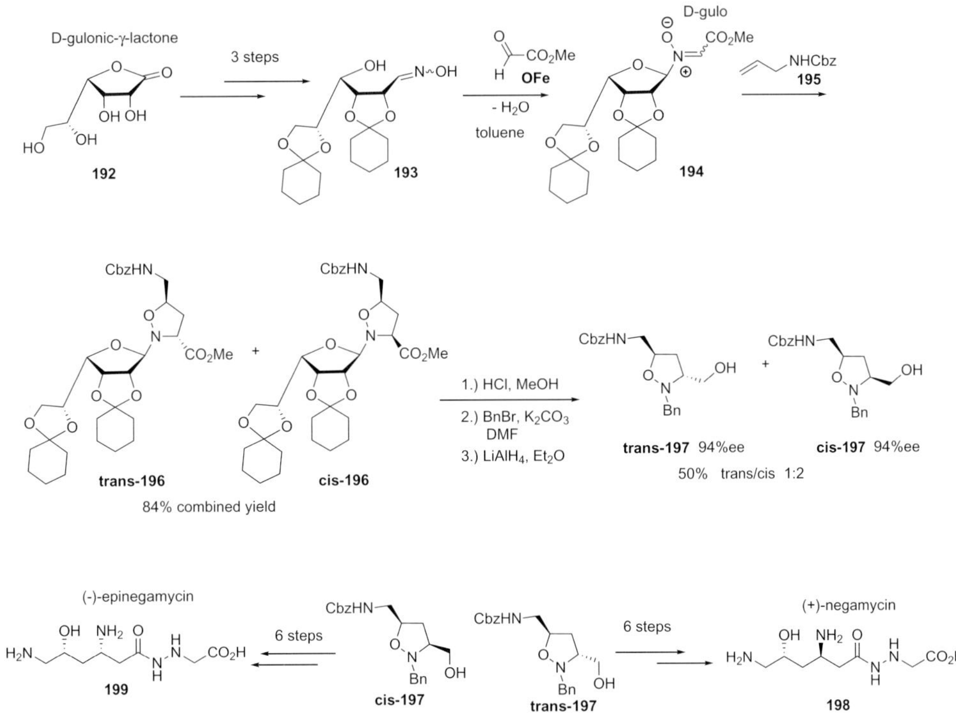

Scheme 4.33 Asymmetric synthesis of (+)-negamycin (**198**) and (−)-epinegamycin (**199**) with carbohydrate nitrone **194** derived from D-gulono-γ-lactone (**192**).

amine (**195**) yielded products *trans*- and *cis*-**196** in good combined yield. Cleavage of the isoxazoline products from the carbohydrate scaffold followed by N-benzylation and reduction of the ester to a primary alcohol yielded *trans*- and *cis*-**197** in 50% yield but only in a very modest ratio of 1:2; however, the enantiomeric purity of the diastereomeric products was excellent. Attempts to use cyclohexylidene-modified nitrones of D-ribose and D-mannose in an analogous fashion led to significantly lower ees for *trans*- and *cis*-**197** [47b]. Both diastereomers were used in the synthesis of peptide-like target compounds in a six-step sequence, including reductive cleavage of the isoxazolidine N–O bond, that yielded (+)-negamycin (**198**) from *trans*-**197** and (−)-epinegamycin (**199**) from *cis*-**197** [47a, b].

Whitney and coworkers used the successful ribose-derived nitrone **176a** in the stereoselective preparation of acivicin (**203**) using methylidene-protected vinylglycine derivative **200** as dipolarophile (Scheme 4.34) [48]. The cycloaddition gave the adducts *(S)*- and *(R)*-**201** and acidic cleavage of the carbohydrate scaffolds with subsequent oxidation using NCS yielded isoxazolines *(S)*- and *(R)*-**202** in good overall yield and excellent dr. After chlorination of the isoxazoline moiety and removal of all protective groups acivicin (**203**) was isolated in 39% overall yield.

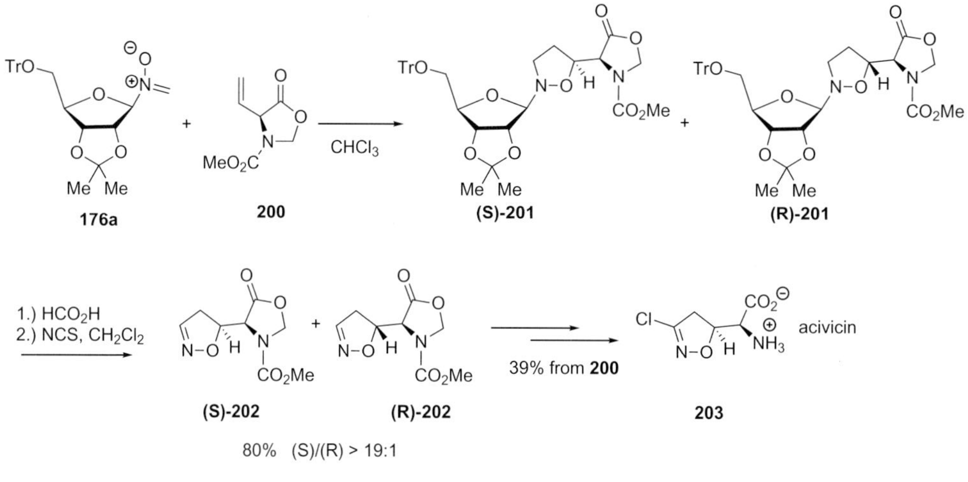

Scheme 4.34 Stereoselective synthesis of acivicin (**203**) via 1,3-dipolar cycloaddition of ribo-configured nitrone **176a** with vinyl glycine derivative **200**.

Sakamoto has published an elegant stereoselective synthesis of the N-terminal amino acid part of nikkomycin Bz [49], an unusual dipeptide natural product containing a nucleosidic motif (Figure 4.1). The synthetic approach used cyclohexylidene-protected oxime *ent*-**193a**, which was obtained from L-gulonic-γ-lactone (*ent*-**192**). The corresponding D-gulo configured oxime **193** had previously been designed by Kibayashi [47] (Scheme 4.33). Condensation of *ent*-**193a** with methylglyoxalate hemiacetal **178f** yielded nitrone *ent*-**194a**, which was subjected to Lewis-acid-catalyzed transesterification with methoxy cinnamyl alcohol (**205**). The resulting nitrone **206** underwent a highly diastereoselective intramolecular 1,3-dipolar cycloaddition, affording adduct **207** in 75% yield as a single diastere-

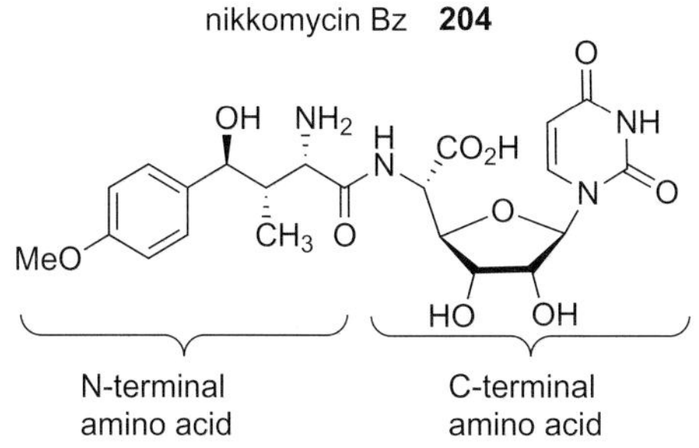

Figure 4.1 Unusual dipeptide natural product nikkomycin Bz (**204**).

omer. The N-terminal amino acid of nikkomycin Bz was prepared from **207** in four steps, including reductive cleavage of the isoxazolidine N–O bond [49]. The reaction sequence is summarized in Scheme 4.35.

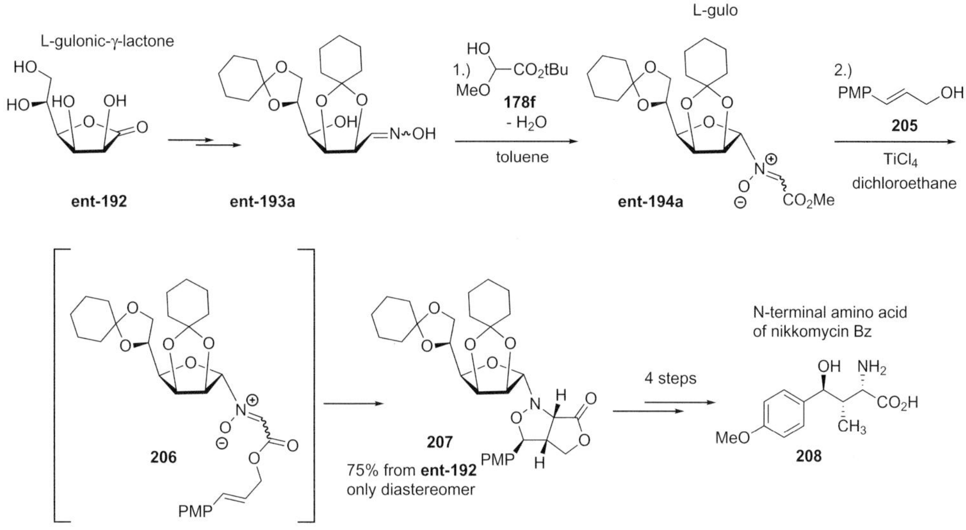

Scheme 4.35 Asymmetric synthesis of the N-terminal amino acid of nikkomycin via intramolecular 1,3-dipolar cycloaddition with L-gulosyl nitrone **206** as key step.

Apart from these examples, there have been miscellaneous reports of glycosyl nitrones used as tools for asymmetric 1,3-dipolar cycloadditions. Brandi and coworkers used nitrones from manno-configured oxime **175** and glyoxalic esters for the preparation of 4-oxopipecolic acid using methylenecyclopropane as the dipolarophile. The reaction was successful, albeit the diastereoselectivity remained modest [50]. The group of Fišera reported 1,3-dipolar cycloadditions of N-aryl-maleimides to nitrones derived from D-glucopyranose, and while for most examples the dr was around 70:30 in one case a dr of 95:5 was obtained [51]. Chiacchio has prepared isoxazolidinyl-thymidines as analogues of regular thymine nucleosides by employing a modified ribose oxime for nitrone formation with ethyl glyoxylate [52].

4 Cycloadditions

A new approach towards *N*-glycosyl nitrones has been reported by Goti *et al.* [53] Reaction of various anomerically unprotected monosaccharide derivatives with *N*-alkyl hydroxylamines yielded *N*-glycosyl *N*-alkyl hydroxylamines, of which only the cyclic form was detectable. Scheme 4.36 shows the preparation of

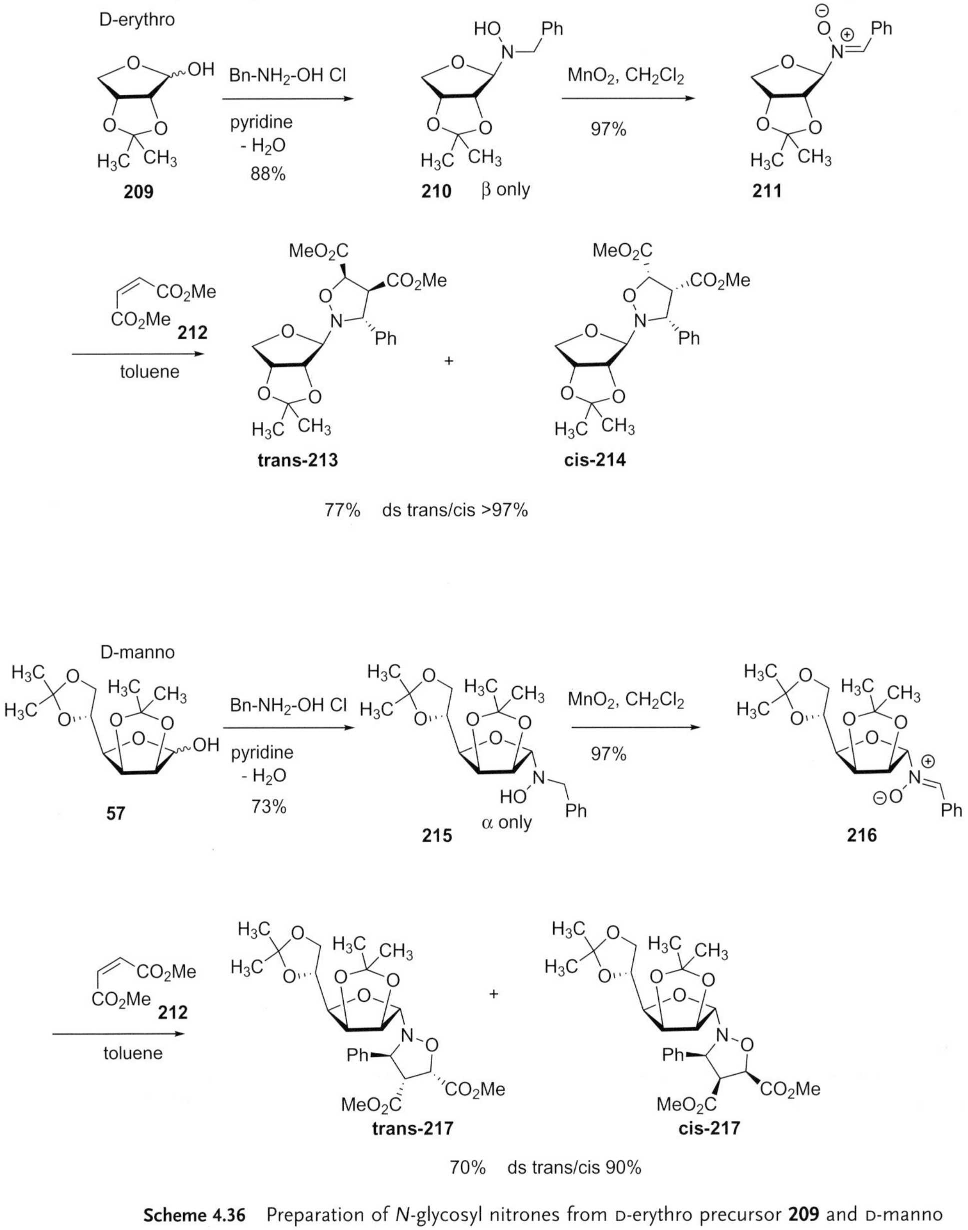

Scheme 4.36 Preparation of *N*-glycosyl nitrones from D-erythro precursor **209** and D-manno precursor **57** by oxidation of the corresponding *N*-glycosyl *N*-alkyl hydroxylamines **210** and **215**, including their application in stereoselective 1,3-dipolar cycloadditions with dimethyl maleate (**212**).

D-N-erythrosyl and N-mannosyl N-alkyl hydroxylamines **210** and **215**, which were oxidized to the corresponding N-erythrosyl and N-mannosyl nitrones **211** and **216** with manganese(IV) oxide. Both nitrones were subjected to 1,3-dipolar cycloadditions with diethyl maleate (**212**) as dipolarophile. Both nitrones **211** and **216** gave the corresponding adducts, each with three new stereocenters, in high yield and stereoselectivity.

4.2.2
Dipolarophiles Attached to the Auxiliary

Examples of dipolarophiles attached to carbohydrate auxiliaries are scarce. One example was reported by Tadano and coworkers, who deployed their D-glucose-derived acrylic ester **36c** in the asymmetric 1,3-dipolar cycloadditions of nitriloxides **218** [54]. Ester **36c**, which had already successfully been employed in asymmetric Diels–Alder reactions (Table 4.1), gave the adduct *(R)*-**219** almost diastereomerically pure and in excellent yield (Scheme 4.37).

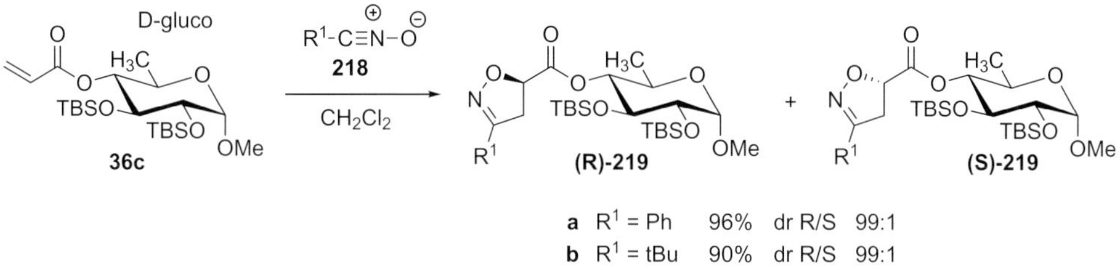

a	R^1 = Ph	96% dr R/S 99:1
b	R^1 = tBu	90% dr R/S 99:1

Scheme 4.37 Asymmetric 1,3-dipolar cycloadditions of nitrile oxides **218** to carbohydrate-derived dipolarophile **36c**.

4.3
[2 + 2] Cycloadditions

Only a few stereoselective [2 + 2] cycloadditions using carbohydrate auxiliaries has been reported. Early examples were published by the group of Chmielewski who examined reactions of carbohydrate enol ethers with tosyl isocyanate [55] and chlorosulfonyl isocyanate [56] leading to β-lactams. The yields of the reactions were moderate and the stereoselectivity turned out to be strongly dependent on the nature of the carbohydrate scaffold and the relative positioning of the enol ether moiety (Scheme 4.38). Only non-anomeric enol ethers were employed for these studies and the stereocenters generated in the reactions are all part of hemiaminal motifs; unsurprisingly, therefore, no attempts were made to cleave the chiral products off the carbohydrate scaffold.

Another approach towards chiral β-lactams is the Staudinger ketene-imine cycloaddition, which was explored by Balogh and by Georg using glucose- and galactose-derived auxiliaries. Balogh used 2,3-deoxy-glycoside **225**, which was prepared via Ferrier rearrangement of glucal **13**. Treatment of **225** with oxalyl chloride and triethylamine led to glycosylated ketene **226**, which reacted with cinnamic

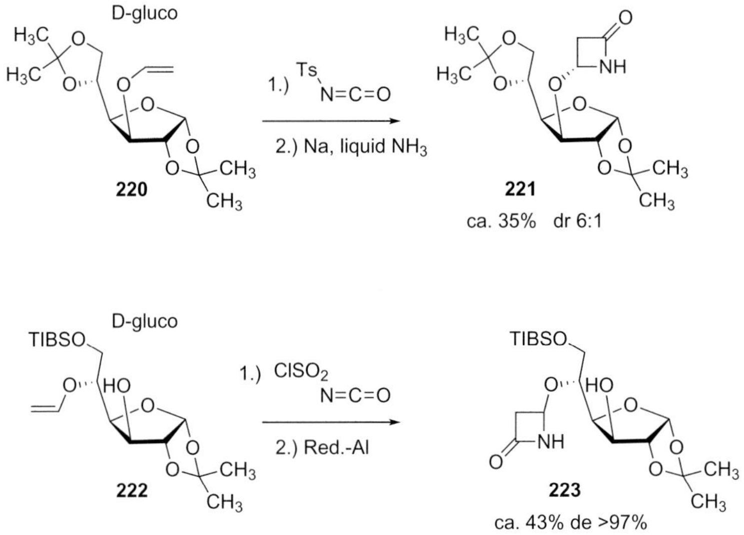

Scheme 4.38 [2 + 2] Cycloadditions of non-anomeric carbohydrate enol ether with tosyl isocyanate and chlorosulfonyl isocyanate.

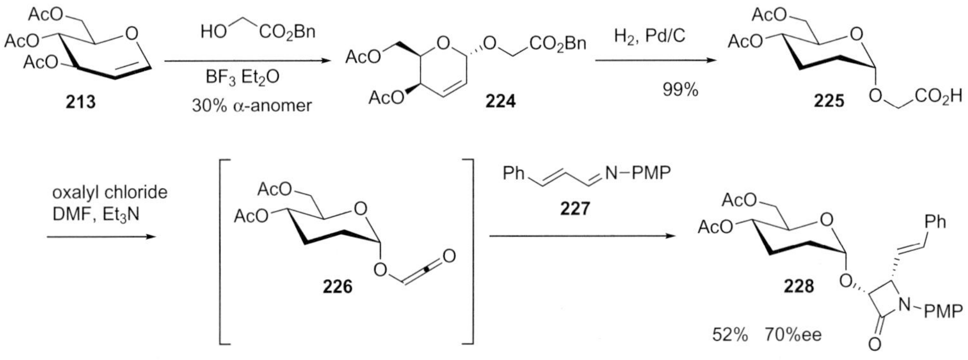

Scheme 4.39 Preparation 2,3-deoxy glucoside **225** as precursor for glycosyl ketene and its [2 + 2] Staudinger cycloaddition to imine **227**.

imine **227** to yield cis-configured β-lactam **228** in 52% yield and 70% ee (Scheme 4.39) [57].

The group of Georg used N-galactosyl imine **229** for a Staudinger cycloaddition with the ketene prepared from acid chloride **230** (Scheme 4.40) [58]. The diastereomeric products **231** were obtained in good yield, albeit with only low diastereoselectivity.

The Kunz group used anomeric enol ether of D-galactose **232** in a [2 + 2] cycloaddition with dichloroketene, prepared in situ from trichloroacetic acid chloride (**233**) [59]. The products (R)- and (S)-**234** were isolated after hydride reduction of the

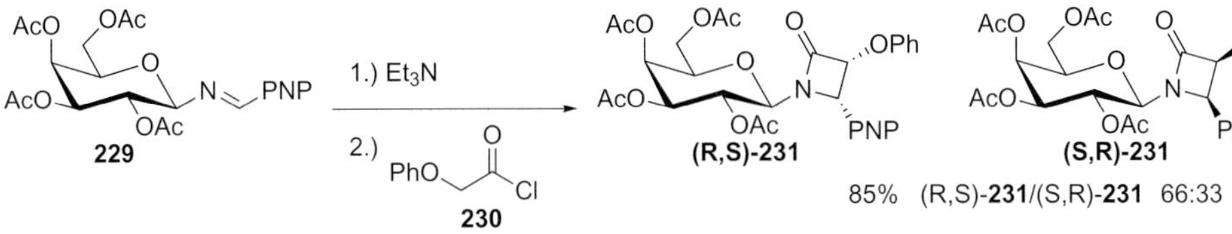

Scheme 4.40 Staudinger cycloaddition of galactose-modified imine **229** and a ketene prepared from acid chloride **230**.

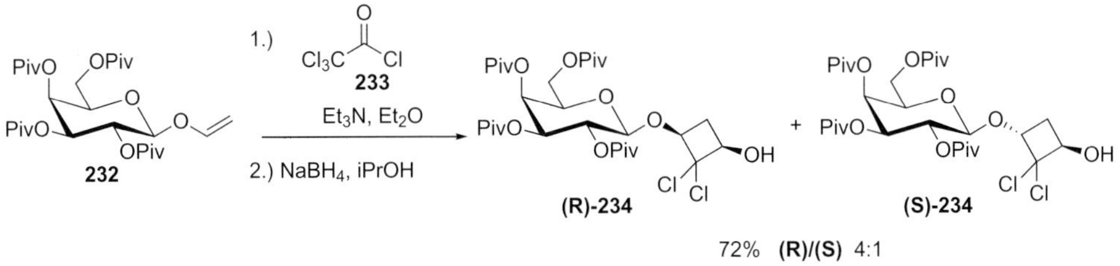

Scheme 4.41 [2 + 2] Cycloaddition of anomeric enol ether **232** with dichloroketene prepared from trichloroacetic acid chloride (**233**).

initially formed cyclobutenones in 72% yield and good diastereoselectivity (Scheme 4.41).

Scharf and coworkers reported Paterno–Büchi reactions of carbohydrate glyoxalates **235a–d** and furan (**3**) [60]. Several carbohydrate scaffolds were tested, which differed considerably regarding their efficiency for asymmetric induction (Scheme 4.42). All derivatives gave the expected cycloaddition products in about 80% yield, but while D-glucofuranose glyoxalate **235a** led to moderate de,

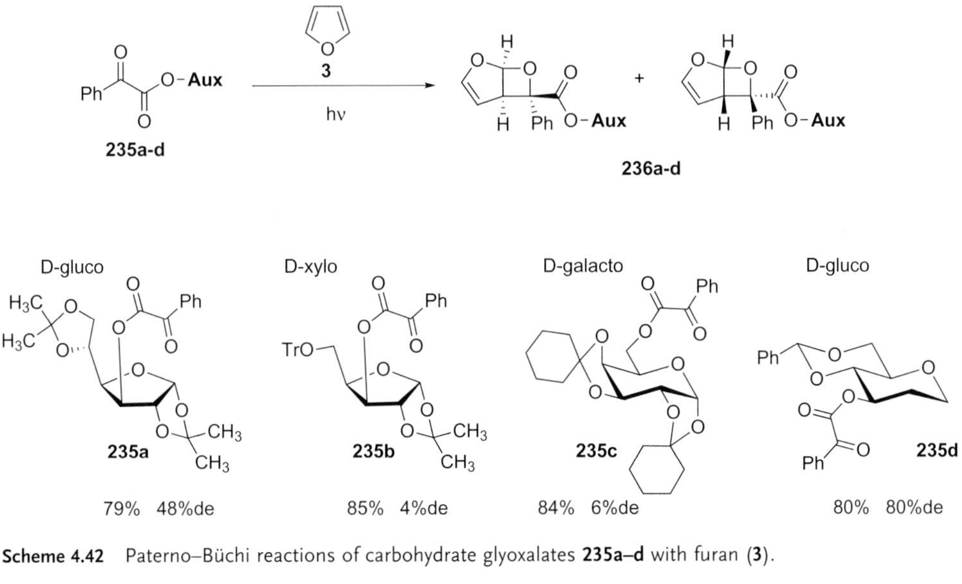

Scheme 4.42 Paterno–Büchi reactions of carbohydrate glyoxalates **235a–d** with furan (**3**).

D-xylofuranose **235b** and D-galactopyranose **235c** gave virtually no diastereoselectivity. The best result was achieved using 1,2-deoxy D-glucose derivative **235d**, which offered up to 80% de. This study highlights, once again, how strong an impact the carbohydrate architecture may have on stereoselectivity.

References

1 Kunz, H., Müller, B., and Schanzenbach, D. (1987) *Angew. Chem. Int. Ed. Engl.*, **26**, 267.
2 Stähle, W. and Kunz, H. (1991) *Synlett*, 260.
3 (a) Shing, T.K.M. and Lloyd-Williams, P. (1987) *J. Chem. Soc., Chem. Commun.*, 423; (b) Shing, T.K.M., Chow, H.-F., and Chung, I.H.F. (1996) *Tetrahedron Lett.*, **37**, 3713.
4 (a) Nouguier, R., Gras, J.-L., Giraud, B., and Virgili, A. (1991) *Tetrahedron Lett.*, **32**, 5529; (b) Nouguier, R., Gras, J.-L., Giraud, B., and Virgili, A. (1992) *Tetrahedron*, **48**, 6245.
5 (a) Nagatsuka, T., Yamaguchi, S., Totani, K., Takao, K., and Tadano, K.-I. (2001) *Synlett*, 481; (b) Nagatsuka, T., Yamaguchi, S., Totani, K., Takao, K., and Tadano, K. (2001) *J. Carbohydr. Chem.*, **20**, 519.
6 Totani, K., Takao, K., and Tadano, K. (2004) *Synlett*, 2066.
7 Ferreira, M.L.G., Pinheiro, S., Perrone, C.C., Costa, P.R.R., and Ferreira, V.F. (1998) *Tetrahedron: Asymmetry*, **9**, 2671.
8 Nouguier, R., Mignon, V., and Gras, J.-L. (1999) *J. Org. Chem.*, **64**, 1412.
9 Enholm, E.J. and Jiang, S. (2000) *J. Org. Chem.*, **65**, 4756.
10 (a) Sarotti, A.M., Spanevello, R.A., and Suárez, A.G. (2004) *Tetrahedron Lett.*, **45**, 8203; (b) Sarotti, A.M., Spanevello, R.A., and Suárez, A.G. (2005) *Tetrahedron Lett.*, **46**, 6987; (c) Sarotti, A.M., Spanevello, R.A., and Suárez, A.G. (2006) *Org. Lett.*, **8**, 1487; (d) Sarotti, A.M., Spanevello, R.A., Duhayon, C., Tuchagues, J.-P., and Suárez, A.G. (2007) *Tetrahedron*, **63**, 241; (e) Zanardi, M.M. and Suárez, A.G.A.G. (2009) *Tetrahedron Lett.*, **50**, 999.
11 (a) Banks, M.R., Blake, A.J., Cadogan, J.I.G., Dawson, I.M., Gaur, S., Gosney, I., Gould, R.O., Grant, K.J., and Hogson, P.K.G. (1993) *J. Chem. Soc., Chem. Commun.*, 1146; (b) Banks, M.R., Cadogan, J.I.G., Gosney, I., Gaur, S., and Hogson, P.K.G. (1994) *Tetrahedron: Asymmetry*, **5**, 2447; (c) Banks, M.R., Cadogan, J.I.G., Gosney, I., Gould, R.O., Hogson, P.K.G., and McDougall, D. (1998) *Tetrahedron*, **54**, 9765.
12 Reichenstein, T. and Grüssner, A. (1934) *Helv. Chim. Acta*, **17**, 311.
13 (a) Felber, H., Kresze, G., Braun, H., and Vasella, A. (1984) *Tetrahedron Lett.*, **25**, 5381; (b) Felber, H., Kresze, G., Prewo, R., and Vasella, A. (1986) *Helv. Chim. Acta*, **69**, 1137.
14 Vasella, A. (1977) *Helv. Chim. Acta*, **60**, 1273.
15 Beer, D. and Vasella, A. (1985) *Helv. Chim. Acta*, **68**, 2254.
16 The absolute configuration of the hetero-Diels–Alder product of **60** with cyclohexadiene was at first assigned incorrectly; the correct stereochemistry was published later by these authors: Braun, H., Charles, R., Kresze, G., Sabuni, M., and Winkler, J. (1987) *Liebigs Ann. Chem.*, 1129.
17 Braun, H., Felber, H., Kresze, G., Schmidtchen, F., Prewo, R., and Vasella, A. (1993) *Liebigs Ann. Chem.*, 261.
18 Werbitzky, O., Klier, K., and Felber, H. (1990) *Liebigs Ann. Chem.*, 267.
19 Schurrle, K., Beier, B., and Piepersberg, W. (1991) *J. Chem. Soc., Perkin Trans. 1*, 2407.
20 (a) Defoin, A., Sarazin, H., Strehler, C., and Streith, J. (1994) *Tetrahedron Lett.*, **35**, 5653; (b) Defoin, A., Sarazin, H., and Streith, J. (1996) *Helv. Chim. Acta*, **79**, 560; (c) Defoin, A., Sarazin, H., and Streith, J. (1997) *Tetrahedron*, **53**, 13769; (d) Defoin, A., Sarazin, H., Sifferlen, T., Strehler, C., and Streith, J. (1998) *Helv. Chim. Acta*, **81**, 1417.

21 Hall, A., Bailey, P.D., Rees, D.C., and Wightman, R.H. (1998) *Chem. Commun.*, 2251.
22 (a) Reissig, H.-U. (1992) *Angew. Chem. Int. Ed. Engl.*, **31**, 288; (b) Arnonld, T., Orschel, B., and Reissig, H.-U. (1992) *Angew. Chem. Int. Ed. Engl.*, **31**, 1033; (c) Zimmer, R., Orschel, B., Scherer, S., and Reissig, H.-U. (2002) *Synthesis*, 1553.
23 Hausherr, A., Orschel, B., Scherer, S., and Reissig, H.-U. (2001) *Synthesis*, 1377.
24 Choudhury, A., Franck, R.W., and Gupta, R.B. (1989) *Tetrahedron Lett.*, **30**, 4921.
25 Danishefsky, S.J. and Kitahara, T. (1974) *J. Am. Chem. Soc.*, **96**, 7807.
26 (a) Gupta, R.C., Harland, P.A., and Stoodley, R.J. (1983) *J. Chem. Soc., Chem. Commun.*, 754; (b) Gupta, R.C., Harland, P.A., and Stoodley, R.J. (1984) *Tetrahedron*, **40**, 4657.
27 (a) Gupta, R.C., Slawin, A.M.Z., Stoodley, R.J., and Williams, D.J. (1986) *J. Chem. Soc., Chem. Commun.*, 668; (b) Gupta, R.C., Slawin, A.M.Z., Stoodley, R.J., and Williams, D.J. (1986) *J. Chem. Soc., Chem. Commun.*, 1116; (c) Gupta, R.C., Raynor, C.M., and Stoodley, R.J. (1988) *J. Chem. Soc., Perkin Trans. 1*, 1773.
28 Gupta, R.C., Larsen, D.S., and Stoodley, R.J. (1989) *J. Chem. Soc., Perkin Trans. 1*, 739.
29 Larsen, D.S. and Stoodley, R.J. (1989) *J. Chem. Soc., Perkin Trans. 1*, 1841.
30 Beagley, B., Larsen, D.S., Pritchard, R.G., and Stoodley, R.J. (1989) *J. Chem. Soc., Chem. Commun.*, 17.
31 Larsen, D.S. and Stoodley, R.J. (1989) *J. Chem. Soc., Perkin Trans. 1*, 1339.
32 Beagley, B., Larsen, D.S., Pritchard, R.G., and Stoodley, R.J. (1990) *J. Chem. Soc., Perkin Trans. 1*, 3113.
33 (a) Lubineau, A. and Queneau, Y. (1985) *Tetrahedron Lett.*, **26**, 2653; (b) Lubineau, A. and Queneau, Y. (1987) *J. Org. Chem.*, **52**, 1001.
34 Lubineau, A. and Queneau, Y. (1989) *Tetrahedron*, **45**, 6697.
35 Bird, C.W. and Lewis, A. (1989) *Tetrahedron Lett.*, **30**, 6227.
36 David, S., Eustache, J., and Lubnieau, A. (1974) *J. Chem. Soc., Perkin Trans. 1*, 2274.
37 (a) David, S., Lubnineau, A., and Vatèle, J.-M. (1975) *J. Chem. Soc., Chem. Commun.*, 701; (b) David, S., Lubineau, A., and Vatèle, J.-M. (1976) *J. Chem. Soc., Perkin Trans. 1*, 1831.
38 David, S., Lubnineau, A., and Vatèle, J.-M. (1978) *J. Chem. Soc., Chem. Commun.*, 535.
39 Davide, S., Eustache, J., and Lubineau, A. (1979) *J. Chem. Soc., Perkin Trans. 1*, 1795.
40 David, S. and Eustache, J. (1979) *J. Chem. Soc., Perkin Trans. 1*, 2230.
41 Marazano, C., Yannic, S., Genisson, Y., Mehmadndoust, M., and Das, B.C. (1990) *Tetrahedron Lett.*, **31**, 1995.
42 (a) Aspinall, I.H., Cowley, P.M., Mitchell, G., and Stoodley, R.J. (1993) *J. Chem. Soc., Chem. Commun.*, 1179; (b) Aspinall, I.H., Cowley, P.M., and Stoodley, R.J. (1994) *Tetrahedron Lett.*, **35**, 3397.
43 (a) Vasella, A. (1977) *Helv. Chim. Acta*, **60**, 426; (b) Vasella, A. (1977) *Helv. Chim. Acta*, **60**, 1273.
44 (a) Bernet, B., Krawcyk, E., and Vasella, A. (1985) *Helv. Chim. Acta*, **68**, 2299; (b) Huber, R. and Vasella, A. (1990) *Tetrahedron*, **46**, 33.
45 Vasella, A. and Voeffray, R. (1981) *J. Chem. Soc., Chem. Commun.*, 97.
46 Vasella, A. and Voeffray, R. (1982) *Helv. Chim. Acta*, **65**, 1134.
47 (a) Iida, H., Kasahara, K., and Kibayashi, C. (1986) *J. Am. Chem. Soc.*, **108**, 4647; (b) Kasahara, K., Iida, H., and Kibayashi, C. (1989) *J. Org. Chem.*, **54**, 2225.
48 (a) Mzengeza, S., Yang, C.M., and Whitney, R.A. (1987) *J. Am. Chem. Soc.*, **109**, 276; (b) Mzengeza, S. and Whitney, R.A. (1988) *J. Org. Chem.*, **53**, 4074.
49 Tamura, O., Mita, N., Kusaka, N., Suzuki, H., and Sakamoto, M. (1997) *Tetrahedron Lett.*, **38**, 429.
50 Machetti, F., Cordero, F.M., Sarlo, F.D., Guarna, A., and Brandi, A. (1996) *Tetrahedron Lett.*, **37**, 4205.
51 Fišera, L., Al-Tiari, U.A.R., Ertl, P., and Prónayová, N. (1993) *Monatsh. Chem.*, **124**, 1019.
52 (a) Chiaccio, U., Corsaro, A., Gumina, G., Rescifina, A., Iannazzo, D., Piperno, A., Romeo, G., and Romeo, R. (1999) *J. Org. Chem.*, **64**, 9321; (b) Chiacchio, U., Rescifina, A., Corsaro, A., Pistarà, V.,

Romeo, G., and Romeo, R. (2000) *Tetrahedron: Asymmetry*, **11**, 2045.
53 Cicchi, S., Marradi, M., Corsi, M., Faggi, C., and Goti, A. (2003) *Eur. J. Org. Chem.*, 4152.
54 Tamai, T., Asano, S., Totani, K., Takao, K., and Tadano, K. (2003) *Synlett*, 1865.
55 Kaluza, Z., Fudong, W., Belzeki, C., and Chmielewski, M. (1989) *Tetrahedron Lett.*, **30**, 5171.
56 (a) Kaluza, Z., Furman, B., Patel, M., and Chmielewski, M. (1994) *Tetrahedron: Asymmetry*, **5**, 2179; (b) Kaluza, Z., Furman, B., and Chmielewski, M. (1995) *Tetrahedron: Asymmetry*, **6**, 1719; (c) Furman, B., Kaluza, Z., and Chmielewski, M. (1996) *Tetrahedron*, **52**, 6019.
57 Borer, B.C. and Balogh, D.W. (1991) *Tetrahedron Lett.*, **32**, 1039.
58 (a) Georg, G.I., Mashava, P.M., Akgün, E., and Milstead, M.W. (1991) *Tetrahedron Lett.*, **32**, 3151. (1992) Corrigendum. *Tetrahedron Lett.*, **33**, 1524; (b) Georg, G.I., Akgün, E., Mashava, P.M., Milstead, M., Ping, H., Wu, Z., and Vander Velde, D. (1992) *Tetrahedron Lett.*, **33**, 2111.
59 Ganz, I. and Kunz, H. (1994) *Synthesis*, 1353.
60 Pelzer, R., Jütten, P., and Scharf, H.-D. (1989) *Chem. Ber.*, **122**, 487.

5
Cyclopropanation

Noureddine Khiar, Inmaculada Fernández, Ana Alcudia, Maria Victoria García and Rocío Recio

5.1
Introduction

The asymmetric synthesis of chiral non-racemic cyclopropane subunits has been of great interest in recent decades as several natural [1] and synthetic [2] products contain a cyclopropane ring in their structure. Analysis of the literature shows that three main approaches have been used in the catalytic and stoichiometric stereoselective construction of the cyclopropane framework: (i) halomethyl-metal-mediated cyclopropanation reactions, (ii) transition-metal-catalyzed decomposition of diazo compounds, and (iii) nucleophilic addition–ring closure sequence [3]. Of these approaches the Simmons–Smith reaction [4], the paradigmatic example of halomethyl-metal-mediated cyclopropanations, is one of the successful and general approaches for the synthesis of enantiopure cyclopropanes [3]. Particularly powerful are diastereoselective cyclopropanation reactions of substituted allylic alcohol or α,β-unsaturated carbonyl compounds linked to a chiral auxiliary. In this regard Charette *et al.* has developed one of the most efficient diastereoselective cyclopropanation approaches using glucose as chiral platform (Table 5.1).

The Simmons–Smith cyclopropanation of allylic D-glucopyranose **1** affords the products **2**, usually with a diastereomeric ratio larger than 50:1 [5]. The different 2-hydroxyglucopyranisides **1a–g** were readily accessible using Danishefsky's glycosylation method with commercially available tri-*O*-benzyl-D-glucal and several allylic alcohols. Structure–diastereoselectivity relationships of the starting allylglycosides show that the free hydroxyl group at C2 plays a key role in the diastereofacial differentiation in **1**. This group reacts with diethylzinc and forms a coordinate anchor for the Simmons–Smith intermediate with the iodomethyl-zinc unit. Indeed, allyl glycopyranosides in which the 2-position is protected showed only a modest asymmetric induction in cyclopropanation under standard conditions. The scope of the reaction is very broad, as illustrated by the high level of asymmetric induction observed with several differently substituted allylic alcohols (Table 5.1).

Carbohydrates – Tools for Stereoselective Synthesis, First Edition. Edited by Mike Martin Kwabena Boysen.
© 2013 Wiley-VCH Verlag GmbH & Co. KGaA. Published 2013 by Wiley-VCH Verlag GmbH & Co. KGaA.

Table 5.1 Cyclopropanation of substituted allylic D-glucopyranosides **1**.

Entry	Allyl glucoside	(structure with R¹, R², R³)	Temperature (°C)	Diastereoselectivity[b]
1	1a	O-CH₂-CH=CH-Pr	−35 to 0	>50 : 1 (124 : 1)
2	1b	O-CH₂-CH=CH-Me	−35 to 0	>50 : 1
3	1c	O-CH₂-CH=CH-Ph	−35 to 0	>50 : 1 (130 : 1)
4	1d	O-CH₂-C(Pr)=CH₂ (cis)	−35 to 0	>50 : 1 (114 : 1)
5	1e	O-CH₂-CH=CH-CH₂-OTBDPS	−20 to 0	>50 : 1
6	1f	O-CH₂-CH=C(Me)₂	−50 to −20	>50 : 1 (111 : 1)
7	1g	O-CH₂-cyclohexenyl	−35 to 0	>50 : 1 (100 : 1)

a) Isolated yields of purified products.
b) The diastereoselectivity was determined by ^{13}C NMR. The values in parenthesis were obtained by capillary GC.

To enhance the synthetic utility of this process, Charette's group has solved the two main problems associated with the use of carbohydrates as chiral auxiliaries, namely, the synthesis of a pseudo-enantiomeric chiral inducer and an efficient method for removing the chiral auxiliary. With regard to the first issue, when the reaction was conducted with the auxiliary **3** derived from L-rhamnal, cyclopropanes **4** with the opposite configuration in the cyclopropane units were obtained (Scheme 5.1). Similarly, in the reaction with the α-allyl glucosides corresponding to **1**, preferential formation of the cyclopropylmethylglucosides with the opposite configuration to that of **2** were obtained [6].

To liberate the desired cyclopropyl compounds, Charette *et al.* developed an approach based on a ring contraction reaction of the pyranosidic auxiliary. After activating the 2-OH group of **1g** as a triflate, the ring contraction occurs by heating at 160 °C in a mixture of DMF, pyridine, and water (Scheme 5.2). Subsequent reduction with NaBH$_4$ leads to the desired cyclopropylmethanol **5** in 90% yield

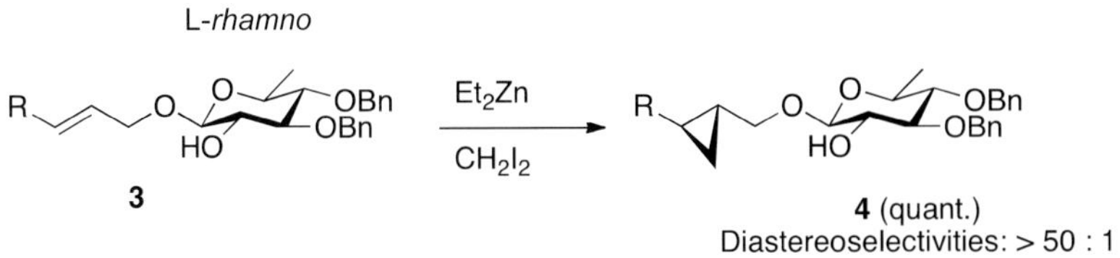

Scheme 5.1 Simmons–Smith cyclopropanation of allylic glycosides using L-rhamnal as chiral auxiliary.

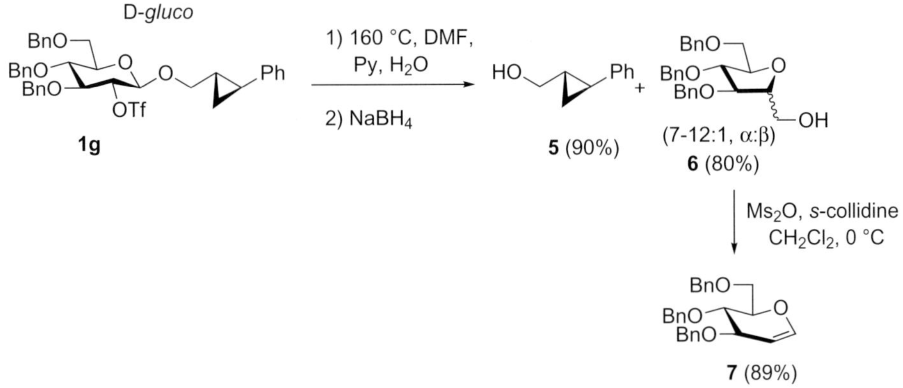

Scheme 5.2 Cleavage of the chiral auxiliary in the synthesis enantiopure cyclopropyl alcohols.

together with ring-contracted chiral auxiliary **6** in 80% yield. Subsequent dehydration via a mesylate derivative of **6** regenerated tri-O-benzyl-D-glucal **7** [7].

The synthetic utility of the approach has been demonstrated by the asymmetric synthesis of all four isomers of coronamic acid, a cyclopropyl amino acid with important agrochemical applications [8]. With these results it is clear that Charette's chiral auxiliary mediated Simmons–Smith cyclopropanation can compete satisfactorily with the best methods in the literature for the synthesis of chiral non racemic cyclopropanes [3]. Following the pioneering work of Charette, other Simmons–Smiths reactions using carbohydrates as chiral templates have been developed, directed mainly at simplifying the rather long route for regeneration of the chiral auxiliary. To do so, Kang et al. coupled the alkene moiety via an acetal to fructopyranose **8** as chiral template (Scheme 5.3). The acetals **9** were formed as a mixture

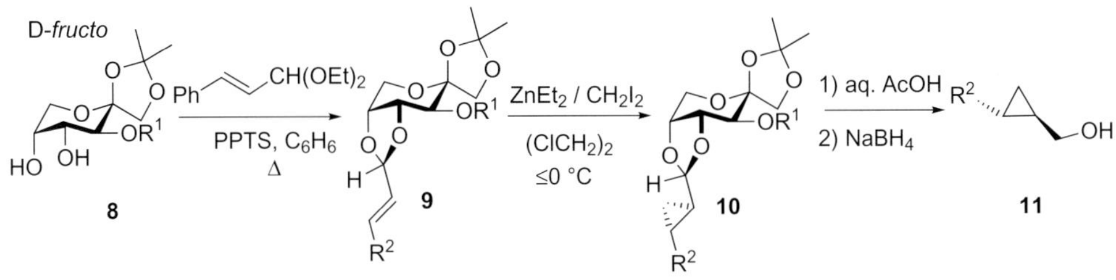

Scheme 5.3 Cyclopropanation of alkenylidene acetals derived from fructopyranose.

of *exo-* and *endo-*diastereoisomers. The *endo* isomer was a major product in all cases, and the mixtures were easily separated by chromatography.

In the case of bulky residues R^1 cyclopropanation of **9** occurred predominantly on the back face of the alkene, affording cyclopropane derivative **10** (Table 5.2). Mild acid hydrolysis of **10** followed by reduction provided the hydroxymethyl-cyclopropane **11** in a modest 65–85 ee in the case of *endo-*acetals and in lower ee for *exo-*acetals [9].

Table 5.2 Effect of the alkyl chain R^1 at C3 of **9** (R^2 = Ph, in Scheme 5.3) in cyclopropanation.

Entry	Starting material 9 (R^2 = Ph)	R^1	Cyclopropanated product, 10	Yield (%)	Hydrolysis–reduction 11 (R^2 = Ph) Yield (%)	ee (%)
1	9a	Bn	10a	85	88	71
2	9b	CH_3	10b	88	88	46
3	9c	o-MeC$_6$H$_4$CH$_2$	10d	72	86	68
4	9d	m-MeC$_6$H$_4$CH$_2$	10d	87	89	72
5	9e	p-MeC$_6$H$_4$CH$_2$	10e	43	90	82
6	9f	p-MeOC$_6$H$_4$CH$_2$	10f	87	85	66
7	9g	p-tBuC$_6$H$_4$CH$_2$	10g	89	91	69
8	9h	p-PhC$_6$H$_4$CH$_2$	10h	85	93	77
9	9i	2-NpCH$_2$	10i	85	90	64

An exhaustive study of the Simmons–Smith cyclopropanation of five- and six-membered alkenylidene acetals derived from carbohydrates has recently been carried out by Vega-Pérez and Iglesias-Guerra (Scheme 5.4). The stereoselective cyclopropanation of unsaturated acetals **12a–f** derived from methyl L-rhamnopyranoside, 1,2-*O*-isopropylidene-D-glucofuranose, methyl D-glucopyranoside, and dodecyl *N*-acetyl-2-amino-2-deoxy-D-glucopyranoside have been studied using Charette's initial conditions. Even when the problem of *endo-*, *exo-*adducts was solved, the cyclopropanated products **13a–f** were obtained only with modest diastereoselectivities (Scheme 5.4) [10].

Utilization of 1,2-*O*-isopropylidene-D-xylofuranoside as chiral auxiliary afforded better results (Table 5.3). Indeed, the cyclopropanation of alkenylidene acetals derived from 1,2-*O*-isopropylidene-D-xylofuranoside **14a–h** in 1,2-dicloroethane leads to cyclopropanated products **15a–h** in good yields and good diastereoselectivities. The absolute configuration has been determined by chemical correlation with known hydroxymethyl cyclopropanes **16** obtained by acid hydrolysis of the acetals followed by reduction of the obtained aldehyde [11].

Beside Simmons–Smith cyclopropanation reactions, transition-metal-catalyzed decomposition of diazo compounds using carbohydrates as chiral auxiliaries has also been examined for the diastereoselective synthesis of cyclopropanes. In this regard, the Ferreira group studied the stereoselectivity in the cyclopropanation with several sugars tethered diazoacetates **17a-e**, using achiral Rh$_2$(OAc)$_4$ as

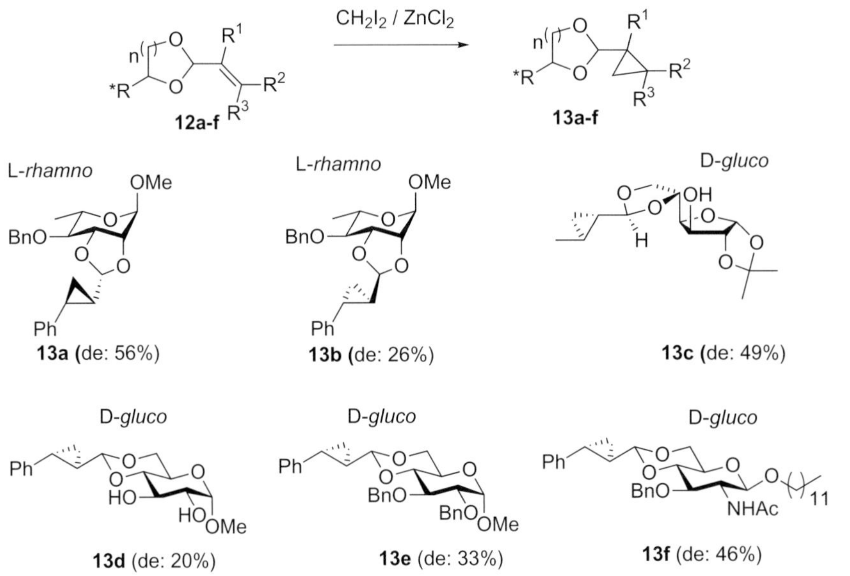

Scheme 5.4 Cyclopropanation of alkenylidene acetals derived from different sugar-based chiral auxiliaries.

catalyst or a chiral Cu(I)-catalyst derived from the fluorinated bis oxazoline **20** (Scheme 5.5) [12].

The cyclopropanes *trans*-**18a–e** and *cis*-**19a–e** were obtained as a mixture in low yield and low to moderate stereoselectivities (Table 5.4). The best diastereoselectivities were observed in the case of α-diazoacetate **17a** derived from the

Table 5.3 Cyclopropanation of alkenylidene acetals from 1,2-O-isopropylidene-D-xylofuranoside **14a–h**.

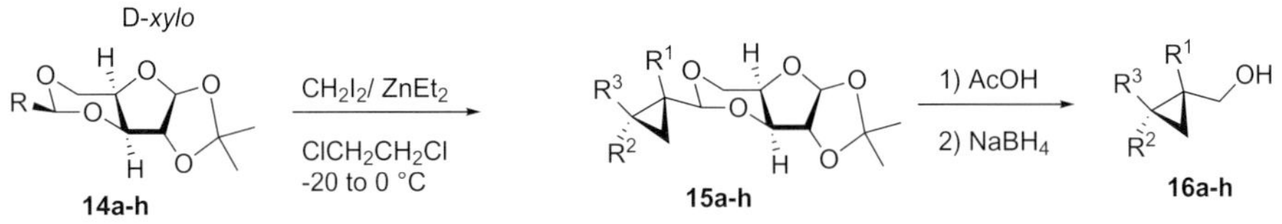

Entry	R^1	R^2	R^3	Cyclopropanated product, 15	Yield (%)	de (%)
1	H	H	Ph	15a	75	82
2	H	H	o-NO$_2$C$_6$H$_5$	15b	84	78
3	H	H	o-MeOC$_6$H$_5$	15c	86	77
4	H	H	p-MeOC$_6$H$_5$	15d	77	46
5	H	H	n-C$_7$H$_{15}$	15e	74	100
6	H	Ph	Ph	15f	69	100
7	Me	H	Ph	15g	75	76
8	n-C$_8$H$_{13}$	H	Ph	15h	72	66

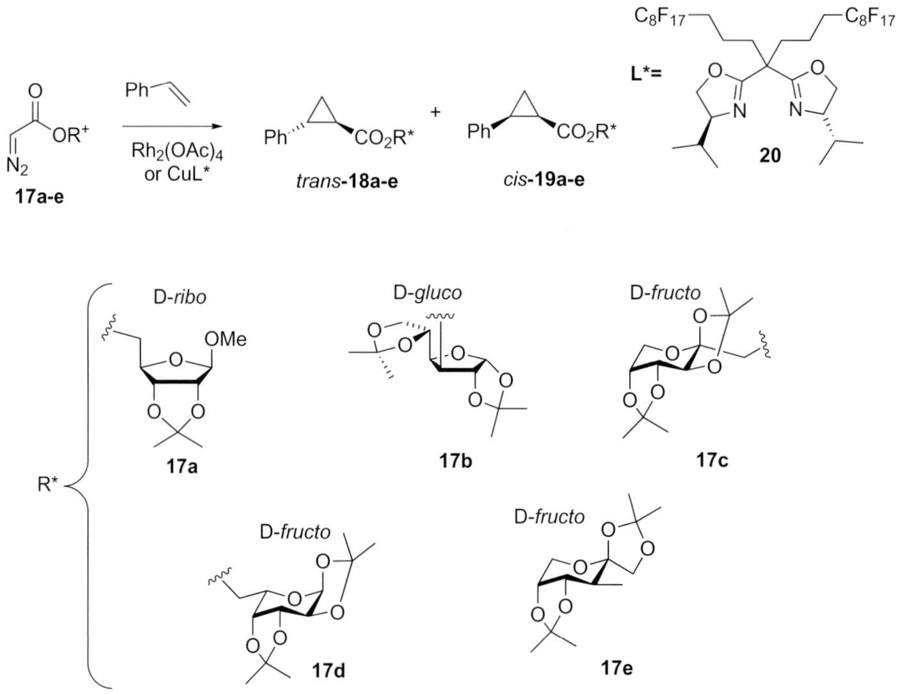

Scheme 5.5 Diastereoselective metal-catalyzed cyclopropanation of sugar tethered diazoacetates.

Table 5.4 Stereoselectivities in the cyclopropanation of **17a–e**.

Entry	Diazoacetate	Catalyst	Yield (%)	Trans : cis ratio 22 : 23	18 (% de)	19 (% de)
1	17a	Rh$_2$(OAc)$_4$	20	86 : 14	60	46
2	17a	20, CuOTf	30	95 : 05	60	53
3	17b	Rh$_2$(OAc)$_4$	21	72 : 28	20	26
4	17b	20, CuOTf	25	80 : 20	19	26
5	17c	Rh$_2$(OAc)$_4$	35	68 : 32	<10	<10
6	17c	20, CuOTf	62	75 : 25	<10	<10
7	17d	Rh$_2$(OAc)$_4$	62	85 : 15	33	12
8	17d	20, CuOTf	35	96 : 04	34	02
9	17e	Rh$_2$(OAc)$_4$	37	80 : 20	17	68
10	17e	20, CuOTf	15	85 : 15	2	92

D-ribose. In all cyclopropanations of α-diazoacetates **17a–e**, the chiral catalysts **20**-Cu(I) gave a better *trans/cis* ratio than reactions performed with the achiral Rh(II) catalyst, showing the prominent role of the bis-oxazoline on this selectivity. Nevertheless, the similar diastereomeric excess obtained for each of the main adducts **18a–d** (Table 5.4) with both systems demonstrates the effective and unexpected role of the carbohydrate auxiliary in control of the diastereoselectivity of *trans*-cyclopropanes.

5.2
Epoxidation

The epoxidation of C=C double bonds is surely one of the most useful transformations in organic chemistry [13]. The resulting epoxy functionality can be easily transformed either by base-catalyzed and/or acid-catalyzed diastereoselective ring-opening into a wide number of synthetic and biologically relevant molecules [14]. The powerfulness of the catalytic Sharpless–Katsuki [15] epoxidation of allylic alcohols and the catalytic Jacobsen–Katsuki epoxidation of unfunctionalized olefins [16] has completely supplanted other methodologies; thus, unsurprisingly, investigations of the application of chiral auxiliaries in asymmetric epoxidation have been very scarce [17]. As in the case of the Simmons–Smith cyclopropanation, the use of carbohydrates in this area was dictated by the facility of controlling their conformational behavior and the possibility of controlling electronic, steric, and stereoelectronic interactions between the carbohydrate and the oxidants. The alkene moiety has been linked to the carbohydrate auxiliary either as allyl glycosides or through the nitrogen atom of amino sugars, that is, as amides. Following their excellent results in the cyclopropanation reaction of allyl glycosides, Charette's group studied the epoxidation of the same substrates as for the cyclopropanation under various conditions. Surprisingly, the vanadium-, molybdenum-, or titanium-catalyzed epoxidation of *trans*-2'-butenyl 3,4,6-tri-*O*-benzyl-β-D-glucopyranoside **1b** produced the desired epoxide **21** in very low yields (0–40%) and no diastereoselectivity [18]. The use of *m*-CPBA (*meta*-chloroperoxybenzoic acid) as oxidizing agent afforded the diastereomeric epoxides in 80% diastereoselectivity when the C2 hydroxyl was free, but with no selectivity when it was protected (Scheme 5.6). This fact combined with the absolute configuration of the major isomer suggests preferential electrophilic attack of the peracid from the *Re* face (relative to the C2 position of the C=C double bond) of the allylic substrate, assisted through hydrogen bonding between the OH group at the C2 center of the sugar and the peracid. Surprisingly, it is interesting to stress that the major epoxide obtained (**21**) has the opposite configuration of that obtained in cyclopropanation of the same substrate, indicating that oxygen delivery from the

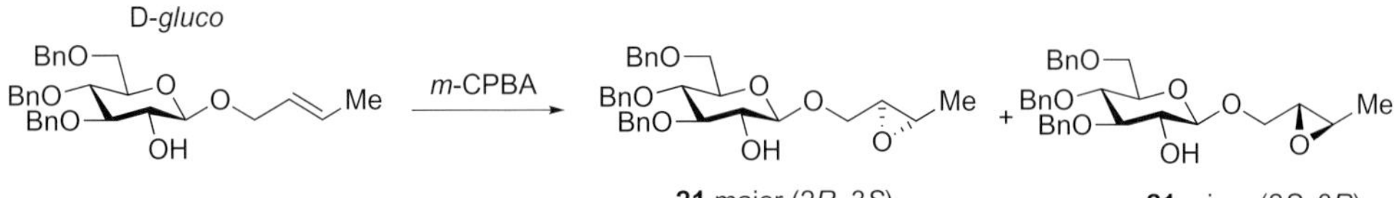

Scheme 5.6 Diastereoselective epoxidation of allylic glucopyranoside **1b**.

Table 5.5 Epoxidation of propenyl-linked peracetylated N-acetyl glucosamine **22a–d**.

Entry	R^1	R^2	R^3	Epoxide	Yield (%)	de (%)
1	H	H	H	23a	41	80
2	Me	H	H	23b	57	95
3	H	Ph	H	23c	67	88
4	H	Me	Me	23d	52	43

peroxide in the epoxidation reaction takes place at the opposite side of the carbenoid species compared to the cyclopropanation reaction.

The epoxidation of propenyl-linked peracetylated N-acetyl glucosamine **22a–d** has been studied by the groups of Vega-Pérez and Takasu [19]. The use of m-CPBA as oxidizing agent in CHCl$_3$ at −15 °C afforded the epoxides **23a–d** with the (R) absolute configuration in acceptable yields (42–71%) and moderate to good diastereoselectivities (Table 5.5).

Better yields and diastereoselectivities were obtained by the use of conformationally rigid, bicyclic, benzylidene acetals of glucosamine with a free or protected hydroxyl group at C3 [20]. Oxidation of the allyl acetals **24**, with a large amide functionality at C2 of the sugar, with m-CPBA leads to the corresponding epoxides **25** in very high diastereoselectivity (up to 95% de) and good yields (Table 5.6). This high facial control, and the fact that the major epoxide obtained has (R)-configuration, implies that the hydrogen atom of the amide stabilizes the transition state of the oxygen transfer through hydrogen bonding with the attacking peracid. Interestingly, even though the hydroxyl group at C3 is far from the alkene function its protection decreases the diastereoselectivity of the oxidation. This implies that besides the hydrogen bonding with amide function, the hydroxyl group at C3 of the sugar helps in stabilizing the transition state for an efficient oxygen transfer.

Alkenylidene acetals with five- and six-membered rings (**26a–m**) related to those already used in the Simmons–Smith cyclopropanation (Table 5.3) have also been used as substrates in the diastereoselective epoxidation with m-CPBA (Scheme 5.7) [21]. The final epoxides derived from N-acetyl-2-amino-2-deoxy-D-allopyranose, D-altropyranose, D-galactopyranose, N-acetyl-D-glucosamine, D-glucofuranose, and xylofuranose have been obtained with acceptable yields (62–90%) and low to moderate diastereoselectivities (0–74% de).

Table 5.6 Diastereoselective epoxidation of the amide-functionalized allyl glycosides **24** with m-CPBA [20].

Entry	Alkene	R^1	R^2	R^3, R^4, R^5 group	Yield (%)	de (%)[a]
1	24a	H	CH$_3$	O-CH$_2$-CH=CH$_2$	48	100
2	24b	H	CH$_3$	O-CH$_2$-C(CH$_3$)=CH$_2$	49	97
3	24c	H	CH$_3$	O-CH$_2$-CH=CH-CH$_3$	66	82
4	24d	H	CH$_3$	O-CH$_2$-CH=CH-Ph	65	87
5	24e	H	CH$_3$	O-CH$_2$-CH=C(CH$_3$)$_2$	58	61
6	24f	H	n-C$_7$H$_{15}$	O-CH$_2$-CH=CH$_2$	80[b]	>95
7	24g	H	n-C$_5$H$_{11}$	O-CH$_2$-CH=CH$_2$	76[b]	>95
8	24h	Ac	CH$_3$	O-CH$_2$-CH=CH$_2$	82	70
9	24i	Me	CH$_3$	O-CH$_2$-CH=CH$_2$	47	74
10	24j	Bn	CH$_3$	O-CH$_2$-CH=CH$_2$	85	68
11	24k	TBDMS	CH$_3$	O-CH$_2$-C(CH$_3$)=CH$_2$	83	64
12	24l	TBDMS	CH$_3$	O-CH$_2$-C(CH$_3$)=CH$_2$	76	74
13	24m	n-C$_5$H$_{11}$CO	n-C$_5$H$_{11}$	O-CH$_2$-CH=CH$_2$	74[b]	>95

a) Determined by ^{1}H NMR spectroscopy.
b) Reactions performed at room temperature.

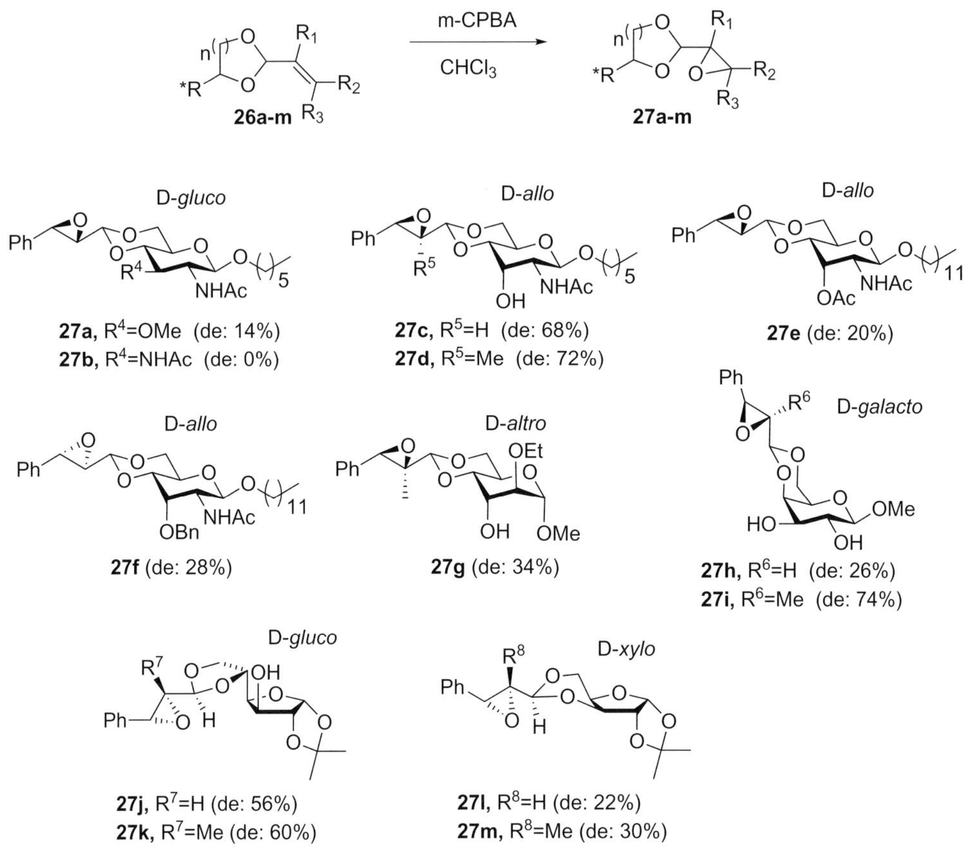

Scheme 5.7 Diatereoselective epoxidation of alkenylidene acetals derived from various sugar auxiliaries.

α,β-Unsaturated amides **29** and **30**, derived from carbohydrates, have been obtained by condensation of sugar amines with the appropriate activated unsaturated acid. These were subsequently treated with *m*-CPBA in dichloromethane (Scheme 5.8).

The diastereoselectivity of the process was quite low at room temperature (26% de) and enhanced at −10 °C (40% de). The formation of an additional five-membered ring, by bridging the oxygen at C3 and the nitrogen at C2, enhanced the rigidity of the auxiliary, resulting in a substantial improvement in the diastereoselectivity (56% to 94% de) [22]. Nevertheless, despite the rigidity of the chiral auxiliary **30a–d**, the diastereoselectivity of the oxidation is in general quite disappointing (Table 5.7, entries 4–6), which indicates the importance of a well-situated hydrogen donor heteroatom in the chiral auxiliary for directing the oxygen transfer. The α,β-epoxyamide can be separated from the sugar moiety with sodium borohydride in THF, as the corresponding epoxy alcohol, and the chiral auxiliary is recovered (Scheme 5.8).

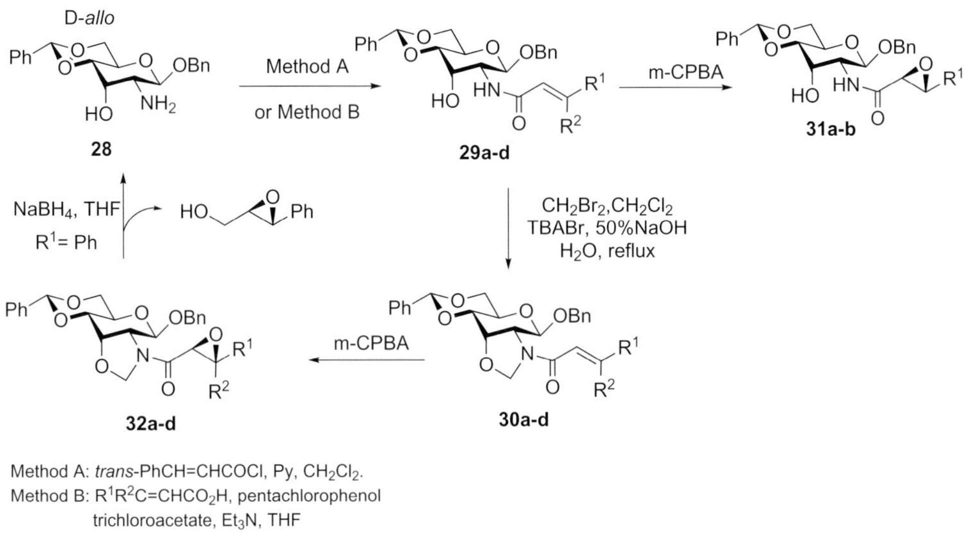
D-allo
Ph
O
O
O
OBn
HO
NH2
28
Method A
or Method B
29a-d
HN
R1
R2
m-CPBA
31a-b
NaBH4, THF
R1= Ph
HO
Ph
CH2Br2,CH2Cl2
TBABr, 50%NaOH
H2O, reflux
32a-d
30a-d
N
Method A: trans-PhCH=CHCOCl, Py, CH2Cl2.
Method B: R1R2C=CHCO2H, pentachlorophenol
trichloroacetate, Et3N, THF

Scheme 5.8 Diastereoselective epoxidation of α,β-unsaturated amides derived from carbohydrates.

Table 5.7 Diastereoselective epoxidation of α,β-unsaturated amides **29** and **30** with *m*-CPBA [22].

Entry	α,β-Unsaturated amide	R[1]	R[2]	Epoxide	Yield[a)] (%)	de[b)] (%)
1	**29a**	Ph	H	**31a**	80	26
2	**29b**	Me	H	**31b**	95	16
3	**30a**	Ph	H	**32a**	85	94
4	**30b**	Me	H	**32b**	80	44
5	**30c**	Et	H	**32c**	83	46
6	**30d**	Me	Me	**32d**	78	48

a) Reaction time: 2–3 days, at room temperature.
b) Determined by ^{1}H NMR.

5.3 Construction of Chiral Sulfur and Phosphorus Centers

The interest toward chiral tricoordinated sulfur compounds in general and sulfoxides in particular has shifted from purely academic, directed towards the study of their stereochemical behavior, to a well-established applied synthetic interest. This shift is due to the discovery at the beginning of the 1980s and the end of the 1990s that chiral sulfur derivatives, namely, chiral sulfoxides [23] and chiral sulfinamides [24], are efficient chiral auxiliaries able to bring about important asymmetric transformations. Additionally, numerous biologically significant molecules, including the best selling drug omeprazol, a proton pump inhibitor, comprise a chiral non-racemic sulfinyl group [23, 25]. Therefore, the last two decades have witnessed an exponential use of these chiral auxiliaries in asymmetric synthesis, establishing the chiral sulfinyl group as one of the most efficient and versatile chiral controllers in C–C and C–X bond formation. One of the most efficient and general methods developed so far for the synthesis of enantiopure sulfoxide is the so-called "DAG-methodology" introduced by the group of Fernández, Khiar, and Alcudia (Scheme 5.9) [26].

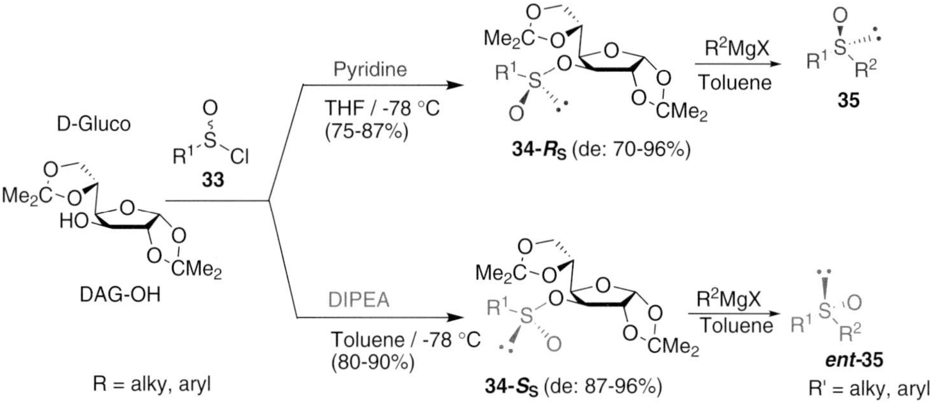

Scheme 5.9 Enantiodivergent synthesis of enantiopure sulfoxide using the DAG methodology.

The reaction of diacetone-D-glucose with various racemic alkyl and aryl sulfinyl chlorides in THF at −78 °C, using pyridine as base, gave the corresponding sulfinate esters **33** in high yield and selectivity in favor of the (R_S)-enantiomer. When Hunig's base (i-Pr$_2$NEt) (DIPEA) was used as catalyst the diastereoselectivity of the reaction was improved and a single isomer was obtained in most cases. Surprisingly, this isomer has the *opposite configuration* (S_S) at the sulfinyl stereocenter to the one obtained using pyridine as base (Table 5.8, entries 1 and 2). Another important feature of the DAG methodology is that the formation of sulfinate ester intermediates takes place with *dynamic kinetic resolution* of the racemic sulfinyl chlorides.

The reaction of a Grignard reagent with the intermediate sulfinate esters **33** leads smoothly to the corresponding sulfoxides **34** or *ent*-**34** in high yield and selectivity (Table 5.9). Hindered and unhindered dialkyl, alkyl aryl, and diaryl sulfoxides have been obtained in both enantiomeric forms [27]. These results, which show the high capacity of the DAG methodology for the synthesis of optically pure sulfoxides, confirm the stereo-directing effect of the achiral tertiary amine used in the first step. This stereo-directing effect allows the utilization of a single inducer of chirality, DAG, for the synthesis either (R_S)- or (S_S)-sulfinate ester in an *diastereodivergent* manner by simply changing the base from pyridine to -DIPEA-.

Table 5.8 Reaction of DAGOH with different sulfinyl chlorides.

$$\text{R}^1\text{S(O)Cl} + \text{DAG-OH} \xrightarrow[\text{-78 °C}]{\text{Base, THF or toluene}} \text{R}^1\text{S(O)ODAG} + \text{R}^1\text{S(O)ODAG}$$

rac-33 → 34-R_S + 34-S_S

Entry	R¹	Base	Yield[a] (%)	32-R_S : 32-S_S
1	Me	Pyridine	87	93 : 7
2	Me	i-Pr₂NEt	90	<2 : ≥98
3	Et	Pyridine	85	86 : 14
4	Et	i-Pr₂NEt	90	<2 : ≥98
5	n-Pr	Pyridine	75	85 : 15
6	n-Pr	i-Pr₂NEt	80	4 : 96
7	i-Pr	Pyridine	56	≥98 : <2
8	i-Pr	i-Pr₂NEt	50	<2 : ≥98
9	p-Tol	Pyridine	84	86 : 14
10	p-Tol	i-Pr₂NEt	87	6 : 94

a) Diastereomerically pure compounds after purification of the major isomers by chromatography or recrystallization.

Table 5.9 Synthesis of optically active sulfoxides, **35a–e** and **ent-35a–e**, from (R)- and (S)-DAG alkane- or arene-sulfinates, **34a–e**, and R²MgX.

Entry	Sulfinate	R¹	R² in R²MgX	Sulfoxide	Yield (%)	Configuration (ee %)
1	34a (R)	Me	pTol	35a	84	(R) 100
2	34a (S)	Me	p-Tol	ent-33a	90	(S) 100
3	34b (R)	Et	p-Tol	35b	96	(R) 99
4	34b (S)	Et	p-Tol	ent-33b	90	(S) 99
5	34c (R)	n-Pr	p-Tol	35c	88	(R) 100
6	34c (S)	n-Pr	p-Tol	ent-33c	89	(S) 100
7	34d (R)	i-Pr	p-Tol	35d	98	(R) 100
8	34d (S)	i-Pr	p-Tol	ent-33d	89	(S) 100
9	34e (R)	p-Tol	Et	35e	87	(S) 70
10	34e (S)	p-Tol	Et	ent-33e	80	(R) 100

The DAG methodology has been used in the synthesis of both enantiomers of oxisuran a synthetic immunosupressor drug [28], the antibiotic sparsomycin [29], and the antitumoral sulforaphane [30]. Application of the DAG methodology is not limited to chiral sulfoxides but can be used for the synthesis of other interesting chiral sulfinyl derivatives. Accordingly, the addition of LiHMDS to diastereomerically pure sulfinate esters leads to optically pure sulfinamides [31], which are efficient chiral auxiliaries for the synthesis of amine-containing chiral compounds [24].

Additionally, the same group has found (Scheme 5.10) that the reaction of diacetone-D-glucose or dicyclohexylidene-D-glucose with racemic 1,2-bis-sulfinyl chloride **36** in the presence of pyridine or Hunig's base affords C_2-symmetric **37**-(R_S,R_S) 1,2-bis-sulfinate esters and C_2-symmetric **37**-(S_S,S_S) sulfinate esters respectively, in high yield and high diastereoselectivity [30].

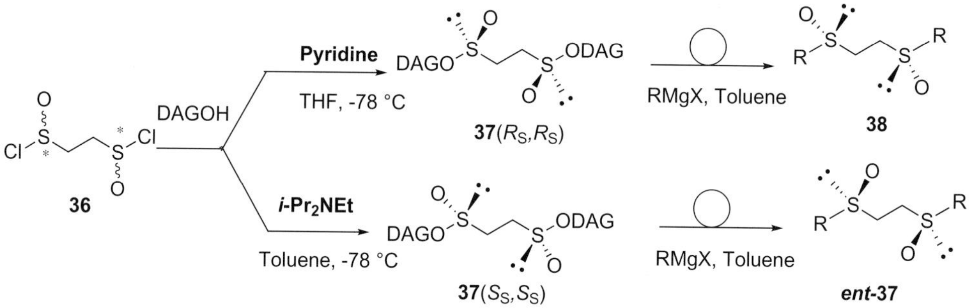

Scheme 5.10 Enantiodivergent synthesis of enantiopure C_2-symmetric bis-sulfoxides using the DAG methodology.

The reaction occurs with dynamic kinetic resolution of the racemic bis-sulfinyl chloride, with higher diastereoselectivity than the monosulfinate synthesis as a consequence of Horeau's principle. Reaction of Grignard reagents on the crude sulfinate ester mixture leads to the corresponding C_2-symmetric bis-sulfoxide **38** or **ent-38**, together with the *meso* compound, which are easily separable by column chromatography [31].

A study on the impact of the amine base on the diastereoselectivity has shown that neither the basicity nor the hybridization of the amine is important, and that the only meaningful parameter is its steric hindrance. A recent theoretical study at the ONIOM (Becke3LYP:UFF) level demonstrates that the base used plays a dual role, first by acting as catalyst in the racemization of the sulfinyl chloride and second by establishing a strong hydrogen bond between the OH group at C3 of the auxiliary and the sulfinyl oxygen in the pentavalent sulfurane intermediate. The analysis of the optimized geometries revealed that in the case of using pyridine as catalyst, the most sterically relevant group around sulfur is the substituent of the substrate, while in the case of using collidine as catalyst the most hindered group around sulfur is collidine itself. This leads to an inversion of distribution of the steric hindrance around sulfur that induces the reversal of the stereochemical outcome [32].

5.4
Chiral Phosphorus Compounds

P-chiral phosphine oxides, which are synthetically important in their own right [33], are also excellent precursors of P-chiral phosphine and diphosphines, which are important ligands in organic and organometallic catalysis [34]. Taking into account that the methods used for substrate driven asymmetric synthesis of sulfinate esters work equally well for phosphinate esters [35], it was anticipated that the DAG methodology would be useful for the synthesis of chiral phosphorus compounds. Indeed, the reaction of DAG-OH or dicyclohexylidene-D-glucose (DCG-OH) with racemic methyl(phenyl)phosphonic chloride **39** in toluene and using NEt_3 as base gave the phosphinate esters **40a** and **40b**, respectively, with the absolute configuration *(S)* at phosphorus, in nearly quantitative yield and 94% de. The change of achiral base from hindered triethylamine to non-hindered aromatic pyridine induces a change in diastereoselectivity of the process as the (R_p)-phosphinate **40** ester was predominantly obtained, although in significantly lower diastereomeric excess. Nevertheless, simple column chromatography allows the synthesis of diastereomerically pure (R_p)-sulfinate esters **40** in 65% yield (Scheme 5.11).

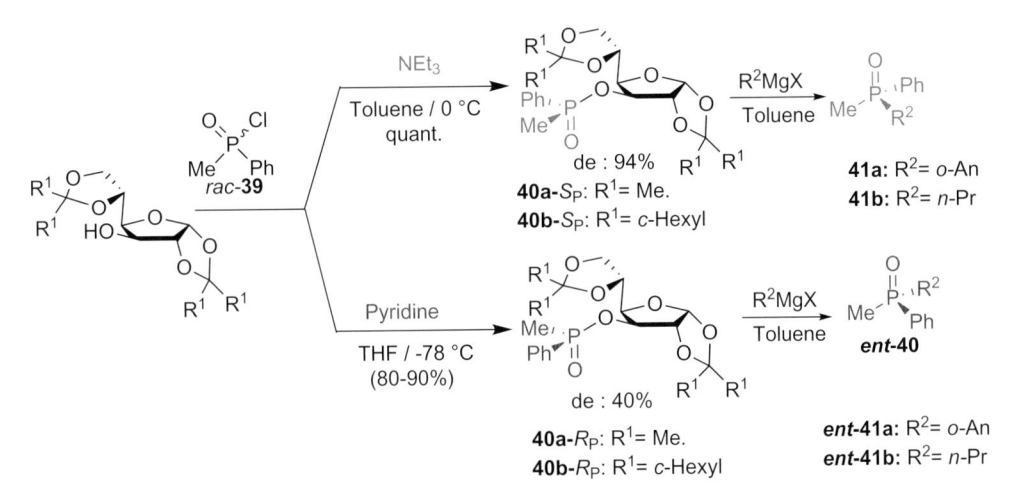

Scheme 5.11 Enantiodivergent synthesis of enantiopure phosphine oxides using the DAG methodology.

These phosphinate esters were shown to be excellent phosphinylating agents, as the reaction with Grignard reagents in toluene at room temperature afforded enantiomerically pure phosphine oxides **41** and ***ent*-41** in excellent yields with recovery of the chiral auxiliary (Table 5.10) [36].

Kolodiazhnyi has found that the base-induced reversal in the asymmetric induction also works for the synthesis of P-chiral phosphinites [37]. The reaction of racemic arylchlorophosphines with DAGOH in the presence of DABCO or pyridine leads to (S_p)-phosphinite **43** in quantitative yield, mainly as a single diastereoisomer. The change in base to the less hindered pyridine led to formation of the (R_p)-diastereoisomer **43** although in only 20% de. Upon treatment with organolithium reagents, phosphinites **43** were transformed directly into P-chiral phosphines **44** with complete inversion of configuration at the phosphorus atom (Scheme 5.12). Alternatively, phosphinites **43** were transformed into the

Table 5.10 Synthesis of phosphine oxides from diastereomerically pure phosphinate esters 40R$_P$ and 40S$_P$.

Entry	Phosphinate	R^1	R^2 in R^{2}MgX	Phosphine oxide	Yield[a] (%)	Configuration
1	40a-(S$_p$)	Me	o-An	41a	95	(S)
2	40a-(S$_p$)	Me	n-Pr	41b	75	(R)
3	40b-(S$_p$)	c-Hexyl	o-An	41a	85	(S)
4	40b-(S$_p$)	c-Hexyl	n-Pr	41b	80	(R)
5	40b-(R$_p$)	c-Hexyl	o-An	ent-41a	91	(R)
6	40b-(R$_p$)	c-Hexyl	n-Pr	ent-41b	70	(S)

a) All reactions were conducted in toluene at room temperature for 2–3 h.

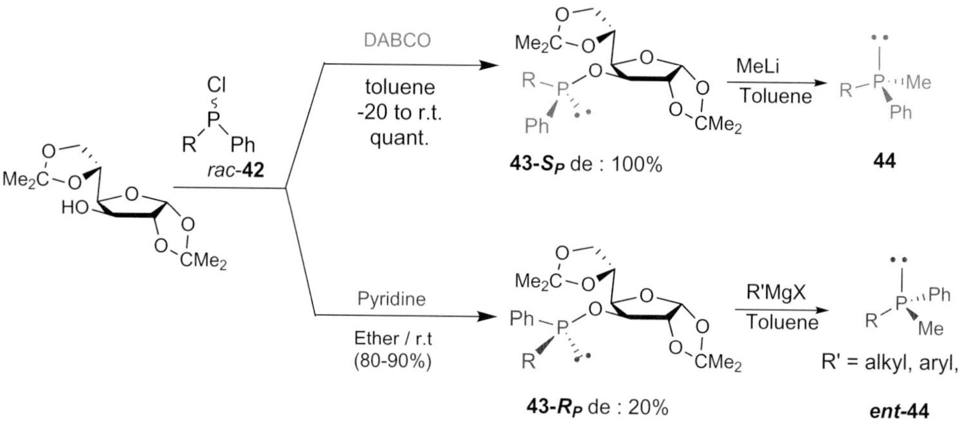

Scheme 5.12 Enantiodivergent synthesis of P-chiral phosphine using diacetone-D-glucose as chiral auxiliary.

corresponding phosphinate esters by oxidation with tBuOOH or into thiophosphinate upon treatment with S$_8$.

5.5
Concluding Remarks

The examples discussed above highlight the importance of carbohydrate-based auxiliaries in the synthesis of enantiopure cyclopropanes, epoxides, sulfoxides, P-chiral phosphine oxides, and phosphines. The Simmons–Smith cyclopropanation of alkene-tethered carbohydrates pioneered by Charette's group has demonstrated itself as a powerful approach to structurally diverse cyclopropane based compounds. Research carried out in this area has solved the two problems associated with the use of carbohydrates as chiral auxiliaries, namely, the synthesis of synthesis of both enantiomers of a desired cyclopropane subunit and recovery of the chiral auxiliary. With regard to epoxidation, even though the advances made in this area using carbohydrates as chiral auxiliaries are relevant, this method is far from competing with the efficiency of catalytic approaches developed so far. In the case of chiral sulfoxides, for which a general enantioselective approach is still missing, the DAG methodology is one of the best methods developed so far for the synthesis of this interesting class of molecules. Using diacetone-D-glucose as a single chiral auxiliary, both enantiomers of a large number of chiral sulfinyl derivatives have been obtained with good yields and enantioselectivities in an enantiodivergent manner, by a simple change of the tertiary amine used to catalyze the reaction. Another important feature of the DAG methodology is that the formation of sulfinate ester intermediates takes place with *dynamic kinetic resolution* of the starting sulfinyl chlorides. While not as general as for the synthesis of chiral sulfinyl derivatives, the DAG methodology works equally well for the enantiodivergent synthesis of P-chiral phosphine oxides and phosphines. The wide range of structures reported to date demonstrates that the basis for the synthesis of this interesting class of molecules is now established and further challenges are ready to be undertaken in the near future.

References

1. (a) Djerassi, C. and Doss, G.A. (1990) *New J. Chem.*, **14**, 713; (b) Salaün, J. (1995) *Curr. Med. Chem.*, **2**, 511; (c) Salaün, J. (2000) *Top. Curr. Chem.*, **207**, 1; (d) Faust, R. (2001) *Angew. Chem. Int. Ed.*, **40**, 2251.
2. (a) de Meijere, A. (1979) *Angew. Chem. Int. Ed. Engl.*, **18**, 809; (b) Arlt, D., Jautelat, M., and Lantzsch, R. (1981) *Angew. Chem. Int. Ed. Engl.*, **20**, 703; (c) de Meijere, A. and Wessjohann, L. (1990) *Synlett*, 20; (d) Wiberg, K.B. (1996) *Acc. Chem. Res.*, **29**, 229.
3. For recent reviews on the stereoselective synthesis of cyclopropanes see: (a) Lebe, H., Marcoux, J.-F., Molinaro, C., and Charette, A.B. (2003) *Chem. Rev.*, **103**, 977; (b) Pellissier, H. (2009) *Tetrahedron*, **65**, 2839.
4. (a) Simmons, H.E. and Smith, R.D. (1958) *J. Am. Chem. Soc.*, **80**, 5323; (b) Simmons, H.E. and Smith, R.D. (1959) *J. Am. Chem. Soc.*, **81**, 4256.
5. Charette, A.B., Côté, B., and Marcoux, J.-F. (1991) *J. Am. Chem. Soc.*, **113**, 8166–8167.
6. Charette, A.B., Turcotte, N., and Marcoux, J.-F. (1994) *Tetrahedron Lett.*, **35**, 513.
7. Charette, A.B. and Côté, B. (1993) *J. Org. Chem.*, **58**, 933.
8. Charette, A.B. and Côté, B. (1995) *J. Am. Chem. Soc.*, **117**, 12721.
9. Kang, J., Lim, G.J., Yoon, S.K., and Kim, M.Y. (1995) *J. Org. Chem.*, **60**, 564.
10. Vega-Pérez, J.M., Periñan, I., Vega, M., and Iglesias-Guerra, F. (2008) *Tetrahedron: Asymmetry*, **19**, 1720.
11. Vega-Pérez, J.M., Periñan, I., Vega, M., and Iglesias-Guerra, F. (2009) *Tetrahedron: Asymmetry*, **20**, 1065.
12. Ferreira, V.F., Leao, R.A.C., da Silva, F.C., Pinheiro, S., Lhoste, P., and Sinou, D. (2007) *Tetrahedron: Asymmetry*, **18**, 1217.
13. A review on enantioselective asymmetric epoxidation: Xia, Q.-H., Ge, H.-Q., Ye, C.-P., Liu, Z.-M., and Su, K.-X. (2005) *Chem. Rev.*, **105**, 1603; review on chiral-auxiliary-controlled asymmetric epoxidation: Adamas, W. and Zhang, A. (2005) *Synlett*, 1047.
14. (a) Hodgson, D.M., Gibbs, A.R., and Lee, G.P. (1996) *Tetrahedron*, **52**, 14361; (b) Jacobsen, E.N. (2000) *Acc. Chem. Res.*, **33**, 421; (c) Adam, W. and Bargon, R.M. (2004) *Chem. Rev.*, **104**, 251.

15 Johnson, R.A. and Sharpless, K.B. (1993) in *Catalytic Asymmetric Synthesis* (ed. I. Ojima), VCH, New York, p. 103.

16 Jacobsen, E.N. (1993) in *Catalytic Asymmetric Synthesis* (ed. I. Ojima), VCH, New York, p. 159.

17 Katsuki, T. (1995) *Coord. Chem. Rev.*, **140**, 189.

18 (a) Charette, A.B. and Côté, B. (1993) *Tetrahedron: Asymmetry*, **4**, 2283; for a related work see also: (b) Giuseppe, B., Catelani, G., Chiappe, C., D'andrea, F., and Grigó, G. (1997) *Tetrahedron: Asymmetry*, **8**, 765.

19 (a) Vega-Pérez, J.M., Candela, J.I., Blanco, E., and Iglesias-Guerra, F. (2002) *Tetrahedron: Asymmetry*, **13**, 2471; (b) Takasu, A., Bando, T., Morimoto, Y., Shibata, Y., and Hirabayashi, T. (2005) *Biomacromolecules*, **6**, 1707.

20 Vega-Pérez, J.M., Candela, J.I., Romero, I., Blanco, E., and Iglesias-Guerra, F. (2000) *Eur. J. Org. Chem.*, **13**, 3949.

21 Vega-Pérez, J.M., Vega, M., Blanco, E., and Iglesias-Guerra, F. (2007) *Tetrahedron: Asymmetry*, **18**, 1850.

22 Vega-Pérez, J.M., Vega, M., Blanco, E., and Iglesias-Guerra, F. (2001) *Tetrahedron: Asymmetry*, **12**, 3189.

23 For recent reviews on chiral sulfoxides see: (a) Fernández, I. and Khiar, N. (2003) *Chem. Rev.*, **103**, 3651; (b) Pellissier, H. (2006) *Tetrahedron*, **62**, 5559; (c) Carreño, C., Hernández-Torres, G., Ribagorda, M., and Urbano, A. (2009) *Chem. Commun.*, 6129.

24 For recent reviews on chiral sulfinamides see: (a) Davis, F.A. (2006) *J. Org. Chem.*, **71**, 8993; (b) Robak, M.T., Hrbage, M.A., and Ellman, J.A. (2010) *Chem. Rev.*, **75**, 3147.

25 (a) Bentley, R. (2005) *Chem. Soc. Rev.*, **34**, 609; (b) Legros, J., Dehli, J.R., and Bolm, C. (2005) *Adv. Synth. Catal.*, **347**, 19.

26 Fernández, I., Khiar, N., Llera, J.M., and Alcudia, F. (1992) *J. Org. Chem.*, **57**, 6789.

27 (a) Alcudia, F., Fernández, I., Khiar, N., and Llera, J.M. (1993) *Phosphorus Sulfur Silicon Relat. Elem.*, **74**, 309; (b) Khiar, N., Fenández, I., and Alcudia, F. (1993) *Phosphorus Sulfur Silicon Relat. Elem.*, **73**, 405; (c) Khiar, N., Fernández, I., and Alcudia, F. (1994) *Tetrahedron Lett.*, **35**, 5719.

28 El Ouazzani, H., Khiar, N., Fernández, I., and Alcudia, F. (1997) *J. Org. Chem.*, **62**, 287.

29 Khiar, N., Werner, S., Mallouk, S., Alcudia, A., and Fernández, I. (2009) *J. Org. Chem.*, **74**, 6002.

30 Khiar, N., Alcudia, F., Espartero, J.-L., Rodríguez, L., and Fernández, I. (2000) *J. Am. Chem. Soc.*, **122**, 7598.

31 Khiar, N., Araújo, C.S., Alcudia, F., and Fenández, I. (2002) *J. Org. Chem.*, **67**, 345.

32 (a) Balcells, D., Maseras, F., and Khiar, N. (2004) *Org. Lett.*, **6**, 2197–2200; (b) Balcells, D., Ujaque, G., Fernández, I., Khiar, N., and Maseras, F. (2006) *J. Org. Chem.*, **71**, 6388–6396; (c) Balcells, D., Ujaque, G., Fernández, I., Khiar, N., and Maseras, F. (2007) *Adv. Synth. Catal.*, **349**, 2103–2110.

33 Pietrusiewicz, K.M. and Zablocka, M. (1994) *Chem. Rev.*, **94**, 1375.

34 (a) Knowles, W. (2002) *Angew. Chem. Int. Ed.*, **41**, 1999; (b) Krepy, K.V.L. and Imamoto, T. (2003) *Adv. Synth. Catal.*, **345**, 79; (c) Dubrovina, N.V. and Borner, A. (2004) *Angew. Chem. Int. Ed.*, **43**, 5883.

35 Fernández, I., Khiar, N., Roca, A., Benabra, A., Alcudia, A., Espartero, J.-L., and Alcudia, F. (1999) *Tetrahedron Lett.*, **40**, 2029.

36 Benabra, A., Alcudia, A., Khiar, N., Fernández, I., and Alcudia, F. (1996) *Tetrahedron: Asymmetry*, **7**, 3353.

37 Kolodiazhnyi, O.I. and Grshkun, E.V. (1996) *Tetrahedron: Asymmetry*, **7**, 967.

**Part II
Carbohydrate Reagents**

6
Hydride Reductions and 1,2-Additions of Nucleophiles to Carbonyl Compounds Using Carbohydrate-Based Reagents and Additives

Omar Boutureira and Benjamin G. Davis

6.1
Introduction

The enantioselective addition of organometallic reagents to carbonyl compounds is one of the most important processes in organic synthesis. In particular, the use of carbohydrates as a source of chirality to control the outcome of such transformations has emerged as a versatile and powerful synthetic tool for the preparation of enantiomerically pure compounds. Carbohydrates are a naturally occurring, inexpensive, renewable, readily available source of chirality that contain a high density of stereogenic centers and may provide advantageously functional groups that can in principle be systematically modified for altered enantiomeric differentiation. The present chapter will deal with the application of carbohydrates, in particular as chiral reagents and additives that have already proven to be effective chiral inducers in asymmetric hydride reductions and 1,2-additions of nucleophiles to carbonyl compounds [1–8].

6.2
Hydride Reductions

The enantioselective reduction of carbonyl compounds continues to attract the attention of organic chemists. Since enantiomerically pure secondary alcohols are often important building blocks for the preparation of biologically active products, numerous carbohydrate-based asymmetric reducing agents have been reported [9–12]. These reagents are mainly chirally modified aluminum and boron hydrides.

6.2.1
Modified Aluminohydrides

Scheme 6.1 summarizes the carbohydrate additives, available in a few steps from D-glucose and D-xylose, respectively, employed for the asymmetric reduction of carbonyl compounds using the archetypal hydride source LiAlH$_4$. The first examples for this transformation were reported by Landor and coworkers [13–15]. Initially, monosaccharide derivatives **1–3** were chosen as chiral ligands for the stereoselective reduction of differently substituted ketones (e.g., aryl, alkyl, and acetylenic). Despite the breadth of an initial survey, 3-O-benzyl-1,2-O-cyclohexylidene-α-D-glucofuranose complex **2d**, which proved to be one of the best asymmetric inducers, created stereoselectivity of only 40% ee.

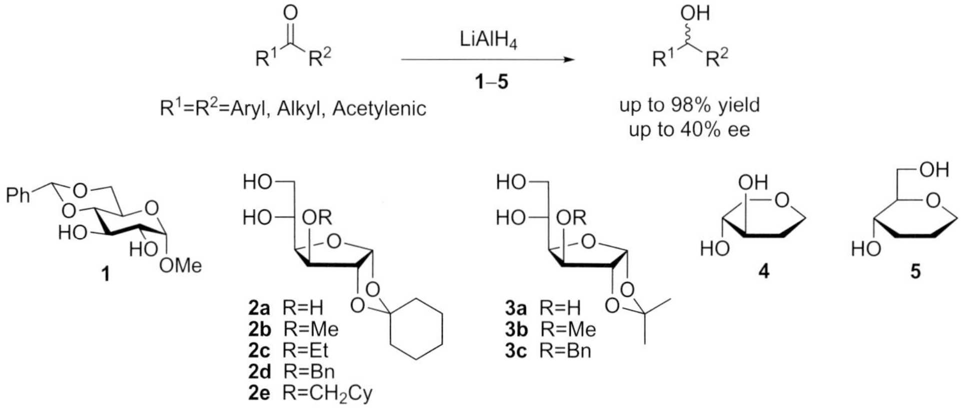

Scheme 6.1 Carbohydrate ligands employed in the stereoselective reduction of carbonyl compounds using LiAlH$_4$.

In a subsequent study, the selectivity was substantially improved (up to 70% for the opposite enantiomer) by the addition of one equivalent of EtOH. To rationalize the observed enantioselectivity the authors proposed the following model (Figure 6.1) [16, 17]. The hydrides available in complex **6b** (X = H^2) are diastereotopic and therefore non-equivalent. H^1 is shielded by the 3-O-benzyl group and it is therefore H^2 that is preferentially transferred to ketones to give secondary (S)-alcohols.

Accordingly, when ethanol is added to the complex, the most reactive hydride H^2 is removed and the reduction is accomplished by H^1, yielding the corresponding alcohol with opposite configuration [(R)-enantiomer]. Moreover, another important factor in this reaction is the presence of an aromatic group at C3 of the additive, which probably interacts with the aromatic moiety of some aromatic substrates and would explain the low asymmetric induction observed with certain aryl ketones [18].

In addition to more common ketone substrates, the same authors also investigated the reduction of isoelectronic ketoximes, ketoxime-O-alkyl ethers, and phenylazomethines (N-phenyl imines) with hydride-monosaccharide complex **6b**,

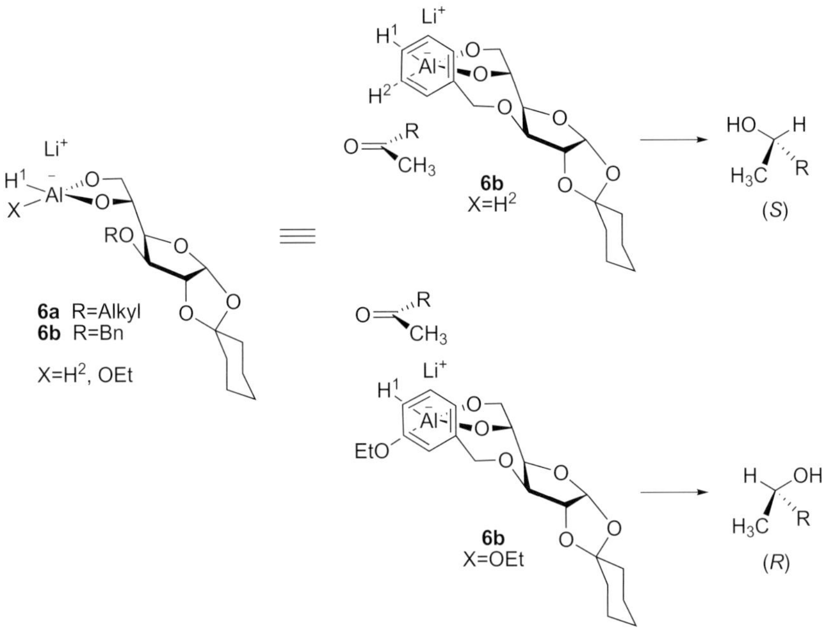

Figure 6.1 Cyclic diol complexes derived from ligands **2** and **3** formed during the stereoselective reduction of carbonyl compounds using LiAlH$_4$.

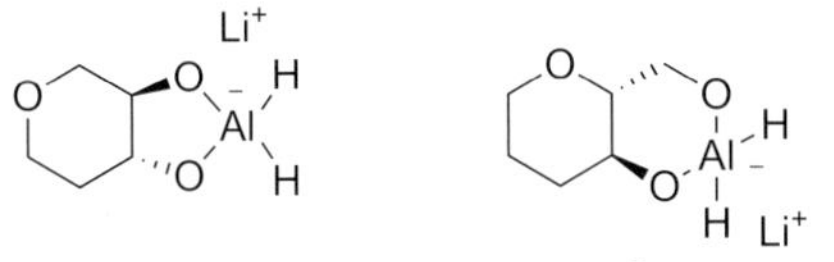

Figure 6.2 Cyclic diol complexes derived from ligands **4** and **5** formed during the stereoselective reduction of carbonyl compounds using LiAlH$_4$.

which yielded the corresponding optically active primary (up to 56% ee) and secondary amines (up to 24% ee) with moderate enantioselectivities [18–21]. Table 6.1 summarizes the best results obtained for the asymmetric reduction of carbonyl and other isoelectronic compounds with hydride-monosaccharide complex **6b**.

Low enantioselectivity (<16% ee) in acetophenone reduction (64% yield) was also induced by chiral diols **4** and **5**, obtained from D-xylose and D-glucose, even though the two residual hydrides of corresponding complexes **7** and **8** are diastereotopic (Figure 6.2) [22].

6.2.2
Modified Borohydrides

The first general strategy employed for the preparation of chirally modified borohydrides is the reaction of simple monosaccharides with metal borohydrides, in

Table 6.1 Summary of the best results obtained on the asymmetric reduction of carbonyl and other isoelectronic compounds with hydride-monosaccharide complex **6b**.

Entry	R$_1$	Y	R$^{2a)}$	6b (equiv.)	Additive	Yield (%)	ee (%)
1	Et	O	Ph	1	—	83	40 (S)
2	Me	O	Ph	2	EtOH	60–80	70 (R)
3	Cy	NOH	Me	2	—	>70	56 (S)
4	Me	NOH	Ph	2	EtOH	50–60	18 (R)
5	Cy	NOTHP	Me	2	—	N/a	50 (S)
6	Cy	NOMe	Me	2	EtOH	50–60	19 (R)
7	Me	NPh	Ph	1	—	N/a	24 (S)

a) Cy = cyclohexyl; THP = tetrahydropyranyl.

particular sodium borohydride but also lithium borohydride. Scheme 6.2 summarizes the carbohydrate ligands employed for the asymmetric reduction of carbonyl compounds using NaBH$_4$. Initially, the corresponding complexes between NaBH$_4$ and sugar ligands **3a, 9–13** reduced ketones in good yields (up to 100%) and moderate enantioselectivities (<40% ee) [23]. Later, the addition of one equivalent of carboxylic acid as additive improved the selectivity (up to 83% ee) under reaction conditions reported by Hirao [24–26] and Morrison [27].

According to these authors, the reaction between NaBH$_4$ and a carboxylic acid followed by the addition of a monosaccharide ligand (R*OH) initially forms monoalkoxy(acyloxy)borohydride **14b**. The subsequent addition of a second unit of ligand results in an overall evolution of two equivalents of hydrogen and the formation of chiral bis(alkoxy)(acyloxy)borohydride complex **14c** (Scheme 6.3). The selectivity for this system varied depending on the time of the substrate addition. Thus, complex **14b** is likely to be the most reactive and unselective species whereas complex **14c** gave rise to higher enantioselectivities and lower yields.

The asymmetric reduction of aromatic ketones with complexes formed between NaBH$_4$ and various Lewis acids (e.g., ZnCl$_2$, AlCl$_3$, MgCl$_2$, and BaCl$_2$) in the presence of sugar ligands **3a, 9a, 10, 12,** and **13** was also studied (Scheme 6.2). The best results in terms of stereoselectivity (up to 88%) were obtained with the system ZnCl$_2$–D-glucofuranose ligand **9a** [28, 29]. Interestingly, the reduction with NaBH$_4$ and ZnCl$_2$ afforded alcohols with (S)-configuration, while the reduction with

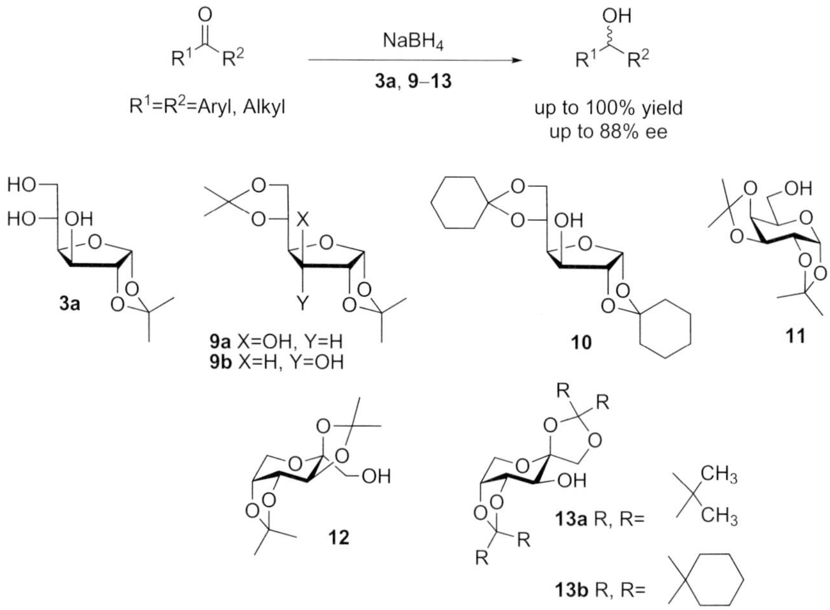

Scheme 6.2 Carbohydrate ligands employed in the stereoselective reduction of carbonyl compounds using NaBH$_4$.

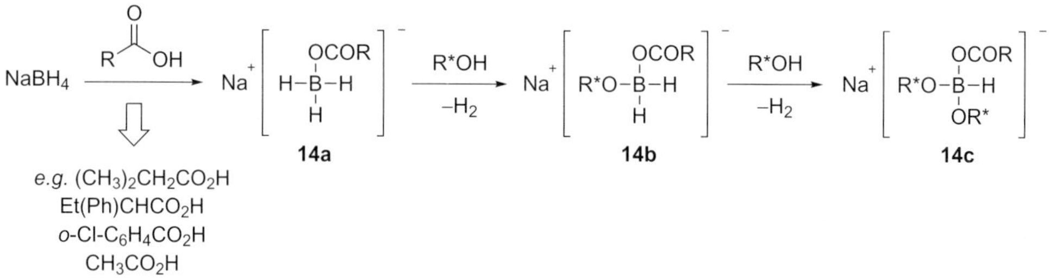

Scheme 6.3 Preparation of carbohydrate ternary complexes and reactive species present in the stereoselective reduction of carbonyl compounds with NaBH$_4$/carboxylic acid.

NaBH$_4$ alone or in combination with AlCl$_3$, BaCl$_2$, and MgCl$_2$ gave the same alcohol with (R)-configuration. Table 6.2 summarizes the best results obtained on the asymmetric reduction of carbonyl compounds with sodium borohydride and monosaccharide ligand **9a**.

Some heteroaromatic ketones can form stoichiometric inclusion complexes with cyclodextrins, which provide a weak chiral environment for their stereoselective (up to 27% ee) reduction with NaBH$_4$ [30]. Dendrimers having a chiral surface environment with internal cavities can also be used for the efficient enantioselective reduction of simple aryl and alkyl ketones with NaBH$_4$, as reported by Schmitzer and coworkers (Table 6.3) [31, 32]. In particular, the use of 3 mol.% of

Table 6.2 Summary of the best results obtained in the asymmetric reduction of carbonyl compounds with sodium borohydride and monosaccharide ligand **9a**.

Entry	R¹	R²	NaBH$_4$ (equiv.)	9a (equiv.)	Additive (equiv.)	Yield (%)	ee (%)
1	Et	Ph	1	2	—	78	39 (R)
2	Et	Ph	1	2	(Me)$_2$CHCO$_2$H (1.2)	55	83 (R)
3	Et	Ph	1	2	ZnCl$_2$ (0.5)	66	88 (S)
4	Et	Ph	1	2	AlCl$_3$ (0.3)	55	54 (R)

Table 6.3 Selected examples of the asymmetric borohydride reduction of ketones catalyzed by amphiphilic dendrimers.

Entry	R¹	R²[a]	T (°C)	Solvent	Yield (%)	ee (%)
1	Cy	Ph	25	THF	97	97 (S)
2	Cy	Ph	25	H$_2$O	95	50 (S)
3	Bu	Ph	25	THF	96	100 (S)
4	Me	Ph	0	THF	94	99 (S)
5	Me	Ph	25	H$_2$O	95	50 (S)
6[b]	Me	Ph	25	H$_2$O	92	98 (S)
7	Me	p-Br-Ph	25	THF	94	78 (S)
8	Me	C$_5$H$_{11}$	−20	THF	94	28 (S)
9	Me	C$_5$H$_{11}$	−80	THF	96	96 (S)
10	Me	o-Py	0	THF	90	90 (S)

a) Py = pyridine.
b) Using fourth-generation dendrimer (G4).

chiral amphiphilic compound **15**, prepared from D-glucolactone and a third-generation polyamidoamine (PAMAM) dendrimer as scaffold, permitted the reduction of butyl, propyl, and propyl phenyl ketones in good yields (90–94%) and excellent selectivities (up to >99% ee). Moreover, the $NaBH_4$–dendrimer complex was insoluble and could be recovered by filtration.

The second general approach for the preparation of chirally modified borohydrides is the formation of carbohydrate-modified boranes with subsequent reduction. This strategy provides a general method for the effective synthesis and characterization of the single species responsible for the reduction process. Scheme 6.4 summarizes the preparation of the most common carbohydrate-based hydride complexes employed for the asymmetric reduction of carbonyl compounds.

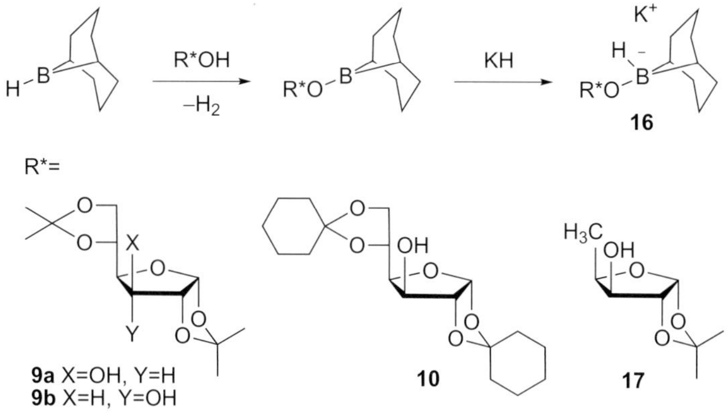

Scheme 6.4 Carbohydrate-based complexes and ligands employed in the stereoselective reduction of carbonyl compounds. Complex **16** derived from ligand **9a** was developed by Brown et al. while those complexes derived from ligands **9b, 10**, and **17** were developed by Cho et al.

In 1986, Brown et al. [33] reported the first synthesis of one of the most powerful boron-based asymmetric reducing agents with general structure **16**. Reagent **18**, commonly called K-Glucoride (Table 6.4) possessing just one reactive hydride, was prepared by reaction of the corresponding borane-containing glucose ligand **9a** with KH in excess. This reagent showed extraordinary consistency and predictable stereochemistry in the reduction of hindered cyclic and aromatic ketones (up to >99% ee) and α-keto esters (up to 98% ee) [34–37] although variable results in the reduction of imines and imides (0–84% ee) were observed (Table 6.4) [38–40].

Later, Cho et al. [41] prepared other chiral borohydrides derived from D-allose, D-glucose, D-xylose, and derivatives **9b, 10**, and **17** to improve the selectivity in the reduction of aliphatic ketones. The best results (80% ee with hindered cyclic aliphatic ketones and up to 99% ee with aromatic ketones) were obtained with derivative **19** (K-Xylide). The same system was successfully applied to the reduction

Table 6.4 Summary of the best results obtained in the asymmetric reduction of carbonyl and other isoelectronic compounds with K-Glucoride **18**.

Entry	R^1	Y	R^2	18 (equiv.)	Yield (%)	ee (%)
1	t-Bu	O	Ph	1.1	93	97 (R)
2	Me	O	o-Me-Ph	1.1	93	91 (R)
3	CO_2Me	O	i-Pr	1.1	83	98 (S)
4	2	tert	Butylcyclohexanone	1.1	96	>99
5	$CH_2N(-CH_2-)_5$	O	Ph	1.1	82	60 (S)
6	Me	NBn	Et	1.1	55	5
7	Me	$NP(O)Ph_2$	α-naphthyl	1.2	82	77 (R)
8	Me	$NP(O)Ph_2$	Cy	1.2	95	84 (R)

Table 6.5 Selected examples of the asymmetric reduction of carbonyl compounds with K-Xylide **19**.

Entry	R^1	R^2	19 (equiv.)	Yield (%)	ee (%)
1	2	2-Dimethylcyclopentanone	1.1	92	80 (R)
2	t-Bu	Ph	1.1	95	99 (R)
3	$ClCH_2$	Ph	1.1	99	92 (S)
4	Me	1,3-dioxane	1.1	72	>99 (S)

of α-keto acetals with very high asymmetric induction (up to 99%) depending on the substrate (Table 6.5) [42].

Brown also studied the synthesis of chiral dialkoxy monoalkylboranes and their corresponding hydrides (Scheme 6.5). Although yields were high (up to 99%), the stereoselectivity was variable (5–74% ee). The best results were obtained with borohydride derived from D-mannitol **20** in the reduction of acetophenone (74%

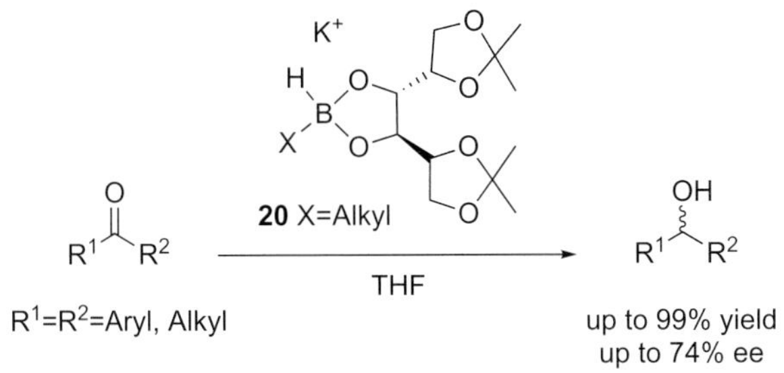

Scheme 6.5 D-Mannitol-based dialkoxy monoalkylborohydrides developed by Brown et al. employed in the stereoselective reduction of carbonyl compounds.

ee) and cyclohexenone (71% ee) – a notoriously difficult substrate – without affecting the double bond [43].

6.3
1,2-Additions of Nucleophiles to Carbonyl Compounds

The enantioselective 1,2-addition of nucleophiles to carbonyl compounds is one of the most important C–C bond forming reactions in asymmetric synthesis. Since enantiomerically pure secondary and tertiary alcohols/amines are important building blocks for the preparation of biologically active products, a great number of carbohydrate-based asymmetric methods for the generation of these stereocenters have been reported [44–52]. Secondary and tertiary carbon-atom stereocenters can be generated easily in most cases by using the appropriate chiral auxiliary, reagent, or catalyst. In this context, although it is widely recognized that there is strong potential application of carbohydrate-based catalysts in the stereoselective 1,2-addition of nucleophiles to carbonyl compounds, the scope of the present section has been restricted. Therefore, this part will mainly focus on the development of synthetic protocols involving the stoichiometric use of carbohydrate reagents and additives.

Among the different approaches reported for the synthesis of chiral secondary and tertiary alcohols/imines, the simplest method is the enantioselective addition of organometallic reagents to a carbon–heteroatom double bond (e.g., C=O, C=N) as depicted in Scheme 6.6. The early examples for this transformation were reported by Inch and coworkers [53]. Initially, monosaccharide derivatives **9a–11** and **21** were premixed with the corresponding Grignard reagents and added to aldehydes and ketones. Although yields were high (up to 95%), the stereoselectivity was variable (9–70% ee).

The best result was obtained from the reaction of CH$_3$MgBr (3.5 equiv.) and cyclohexyl phenyl ketone (1 equiv.) with glucofuranose ligand **9a** (2 equiv.) (Table 6.6). Sugar derivatives **10**, **11**, and **21** also proved effective in promoting the asymmetric addition of Grignard reagents to carbonyl compounds although with

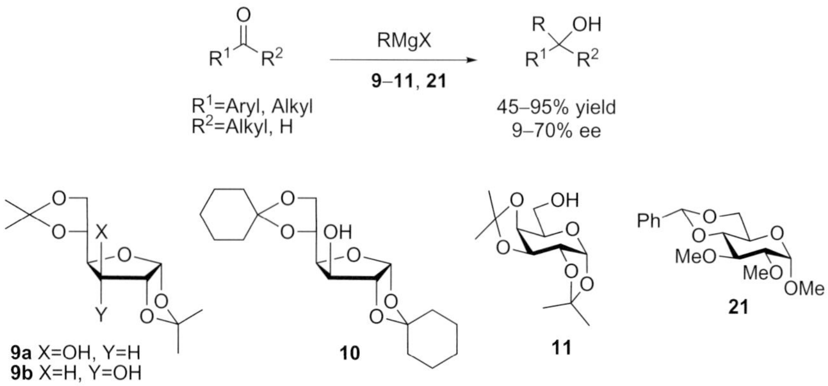

6 Hydride Reductions and 1,2-Additions of Nucleophiles to Carbonyl Compounds

$$R^1\text{-CO-}R^2 \xrightarrow[9-11,\,21]{RMgX} R^1R^2C(R)(OH)$$

R^1=Aryl, Alkyl
R^2=Alkyl, H

45–95% yield
9–70% ee

9a X=OH, Y=H
9b X=H, Y=OH

10

11

21

Scheme 6.6 Carbohydrate additives employed in the stereoselective 1,2-addition of Grignard reagents to carbonyl compounds.

moderate selectivities (up to 50%). Later, the enantioselective addition of complex magnesium alkoxides derived from sugar ligand **9b** to ketones was also studied. Although the role of the carbohydrate ligands was difficult to explain, the authors were able to suggest a predictive model for the additions, hypothesizing a bimolecular mechanism involving the ketone and the carbohydrate–magnesium complex [54]. Another individual example of the use of carbohydrate additives for the stereoselective addition of organometallic reagents to ketones involves the use of polystyrene resin anchored carbohydrates as asymmetric media [55].

In another example of the use of carbohydrates as chiral additives Ribeiro et al. [56] employed carbohydrate-derived ligands **9**, **11–13a**, **22**, and **23** for the enantioselective Reformatsky reaction, which is one of the most synthetically useful methods for the preparation of β-hydroxy esters (Scheme 6.7). The best results

Table 6.6 Summary of the best results obtained for the 1,2-addition of Grignard reagents to carbonyl compounds with **9a**.

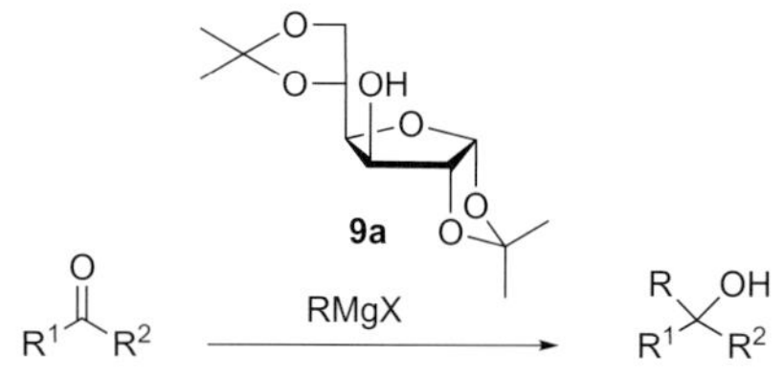

Entry	R^1	R^2	RMgX (equiv.)	9a (equiv.)	Yield (%)	ee (%)
1	Cy	Ph	MeMgBr (3.5)	2	95	70 (R)
2	Cy	Ph	MeMgI (3.5)	2	88	65 (R)
3	Cy	Me	PhMgBr (3.5)	2	50	28 (S)
4	H	Ph	EtMgBr (3.5)	2	81	24 (R)

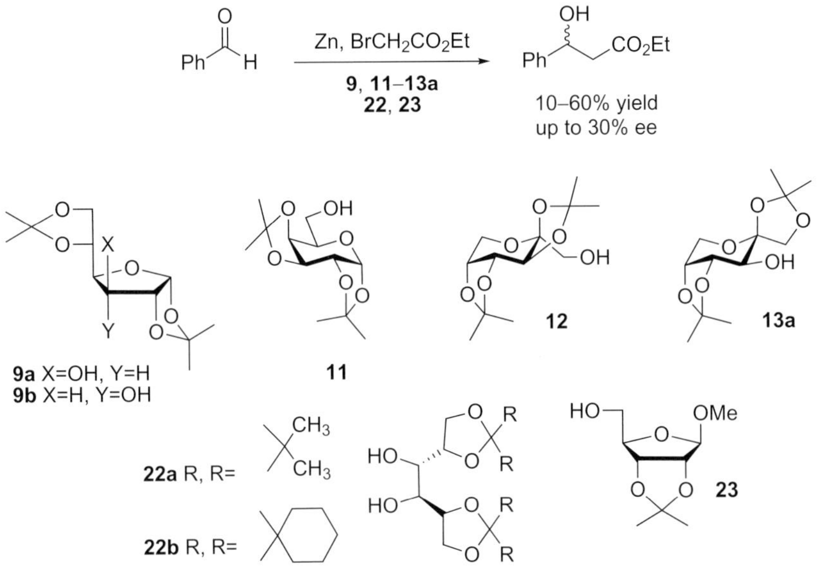

Scheme 6.7 Carbohydrate ligands employed in the stereoselective Reformatsky reactions.

were obtained with D-mannitol ligand **22a** (1–3 equiv.) although the enantioselectivity remained low (up to 30%).

Davis and coworkers have developed a family of glucosamine-derived aminoalcohol ligands **24–27** to promote asymmetric transformations (Scheme 6.8). In particular, ligands **24a–c** and **25–27** (1–3 equiv.) were used for asymmetric Reformatsky reactions, providing good to modest selectivities (42–74% ee) (Table 6.7) [57].

Moreover, before developing the corresponding catalytic version [58], the same authors also reported the use of stoichiometric amounts of ligands **24–27** (1–2 equiv.) for the asymmetric alkynylation of aldehydes, which provided excellent results in most cases (>99% ee) [58]. General trends could be discerned from these results. First, cyclic amine substituents at C2 gave enantioselectivities and yields higher than those of acyclic ones. Second, β-*gluco* ligands gave higher selectivity than their α-anomers, whereas α-*manno* ligands gave good yields but low enantioselectivity, and α-*allo* ligands gave only poor yields and low enantioselectivity (Table 6.8).

More recently, Appelt and Braga [59] reported an enantioselective metal-mediated (Zn and Sn) addition of allyl bromide to aldehydes with inexpensive and easily accessible saccharose **28** and β-cyclodextrin **29** as ligands (Table 6.9). The reaction proceeded in short reaction times, under simple operational conditions, and the desired secondary homoallylic alcohols were obtained in moderate to excellent yields (up to 88%) and useful levels of enantiomeric excess (up to 93%). The best results in the allylation of p-substituted benzaldehydes were obtained with either ligand **28** or **29** in combination with Zn as the metal. Furthermore,

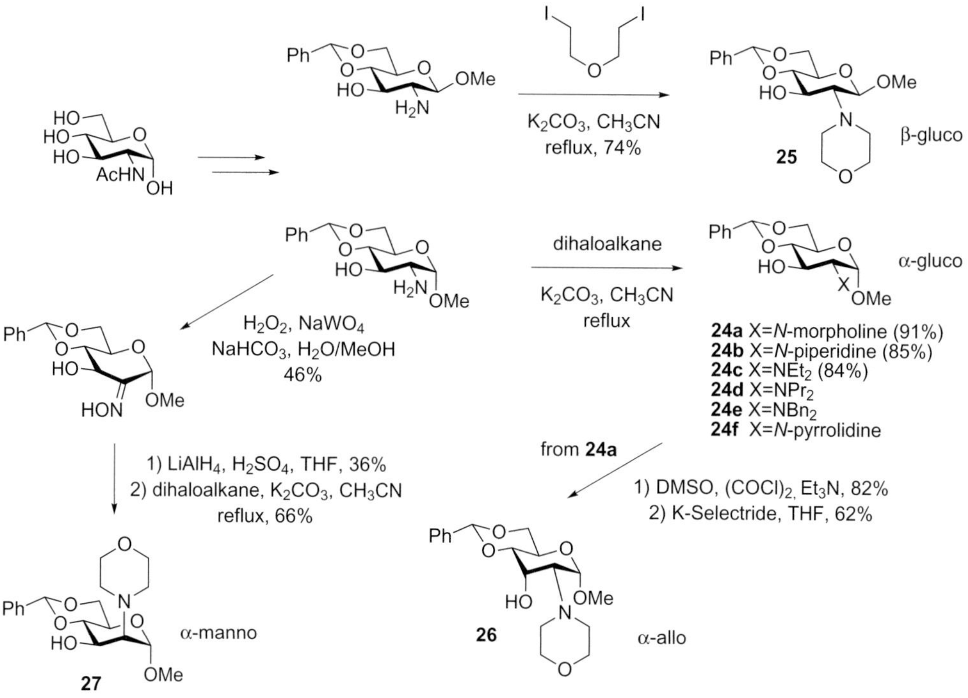

6 Hydride Reductions and 1,2-Additions of Nucleophiles to Carbonyl Compounds

Scheme 6.8 Synthesis of glucosamine-derived aminoalcohol ligands **24–27** reported by Davis et al.

Table 6.7 Reformatsky reaction of benzaldehyde with ligands **24a–c** and **25–27**.

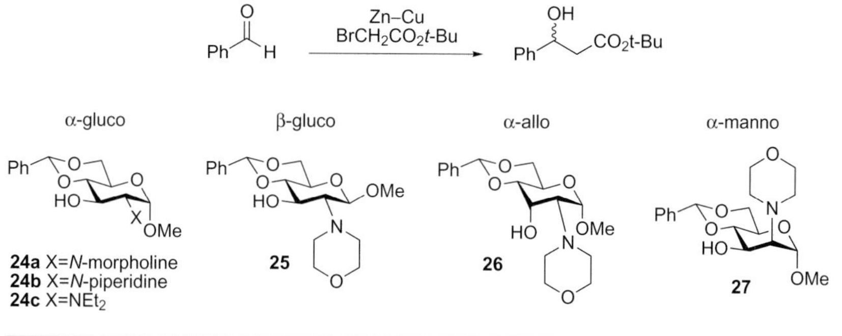

Entry	Ligand (equiv.)	Enolate (equiv.)	PhCHO (equiv.)	Yield (%)	ee (%)
1	24a (1)	3	1	20	42 (S)
2	24b (1)	3	1	N/a	N/a
3	24c (1)	3	1	25	15 (S)
4	24c (1)	3	1	20	35 (S)
5	24c (1)	6	1	57	16 (S)
6	24c (1)	6	1	30	19 (S)
7	24c (2)	6	1	44	42 (S)
8	25 (2)	6	1	21	33 (S)
9	26 (2)	6	1	10	74 (R)
10	27 (2)	6	1	N/a	N/a

6.3 1,2-Additions of Nucleophiles to Carbonyl Compounds

Table 6.8 Summary of the best results obtained in the asymmetric alkynylation of aldehydes with ligands **24–27**.[a]

$$R^1\text{CHO} + R^2\text{-}\!\!\equiv\!\!\text{-H} \xrightarrow{\text{Zn(OTf)}_2,\ \text{Et}_3\text{N}} R^1\text{-CH(OH)-C}\!\equiv\!\text{C-}R^2$$

α-gluco: **24a** X=N-morpholine; **24b** X=N-piperidine; **24c** X=NEt₂; **24d** X=NPr₂; **24e** X=NBn₂; **24f** X=N-pyrrolidine

β-gluco: **25**

α-allo: **26**

α-manno: **27**

Entry	Ligand	R₁CHO	R₂C≡CH	T (°C)	Yield (%)	ee (%)
1	24a	Cy	Ph	40	95	97 (R)
2	24b	Cy	Ph	40	73	81 (R)
3	24c	Cy	Ph	40	38	79 (R)
4	24d	Cy	Ph	40	37	54 (R)
5	24e	Cy	Ph	40	96	39 (R)
6	24f	Cy	Ph	40	58	99 (R)
7	25	Cy	Ph	40	94	99 (R)
8	26	Cy	Ph	40	8	35 (R)
9	27	Cy	Ph	40	92	22 (R)
10	25	i-Pr	Ph	23	70	97
11	25	c-Pr	Ph	23	71	94
12	25	t-Bu	Ph	40	62	97
13	24a	n-C₆H₁₃	Ph	40	55	92
14	25	Ph(CH₂)₂	Ph	23	15	98
15	24a	Ph	Ph	40	50	91
16	25	p-Me-Ph	Ph	23	38	97
17	25	p-OMe-Ph	Ph	50	12	N/a
18	25	p-Cl-Ph	Ph	23	36	98
19	24a	PhCH=CH	Ph	40	17	N/a
20	25	Cy	Ph(CH₂)₂	23	89	95
21	25	t-Bu	Ph(CH₂)₂	23	77	>99
22	25	Cy	Me₂COH	50	71	99
23	25	Cy	Me₂COTMS[b]	50	68	97
24	25	Cy	HO(CH₂)₂	50	90	98
25	25	Cy	Et₃Si	23	11	>99
26	25	Cy	Et₃Si	50	42	98

a) Zn(OTf)₂/ligand/Et₃N/R₂C≡CH/R₁CHO = 1.1 : 1.2 : 1.2 : 1.2 : 1.0 in toluene.
b) TMS = trimethylsilyl.

Table 6.9 Summary of the best results in the enantioselective allylation of aldehydes with ligands **28** and **29**.

saccharose **28**   β-cyclodextrin **29**

Entry	Ligand (equiv.)	Metal (equiv.)	R	T (°C)	Yield (%)	ee (%)
1	28 (2.0)	Zn (1)	H	Room temp.	18	28 (R)
2	28 (1.2)	Zn (1)	H	0	26	55 (R)
3	28 (1.2)	In (1)	H	Room temp.	14	3 (R)
4	28 (1.2)	Sn (1)	H	Room temp.	19	71 (R)
5	28 (1.2)	Sn (1)	H	0	25	81 (R)
6	29 (1.0)	Zn (1)	H	Room temp.	56	70 (R)
7	29 (1.0)	Zn (1)	H	0	36	23 (R)
8	28 (1.2)	Zn (1)	OMe	Room temp.	60	87 (R)
9	28 (1.2)	Zn (1)	Me	Room temp.	70	75 (R)
10	28 (1.2)	Zn (1)	Cl	Room temp.	88	85 (R)
11	28 (1.2)	Sn (1)	OMe	Room temp.	67	16 (R)
12	28 (1.2)	Sn (1)	Me	Room temp.	61	66 (R)
13	28 (1.2)	Sn (1)	Cl	Room temp.	81	68 (R)
14	29 (1.0)	Zn (1)	OMe	Room temp.	91	93 (R)
15	29 (1.0)	Zn (1)	Me	Room temp.	67	72 (R)
16	29 (1.0)	Zn (1)	Cl	Room temp.	75	63 (R)

notably, no special conditions such as inert atmosphere or dried solvents are required for the reaction, since all reactions were carried out in an open vessel, using ACS grade solvent.

References

1 Reissig, H.U. (1992) *Angew. Chem. Int. Ed. Engl.*, **31**, 288–290.

2 Kunz, H. and Ruck, K. (1993) *Angew. Chem. Int. Ed. Engl.*, **32**, 336–358.

3 Kunz, H. (1995) *Pure Appl. Chem.*, **67**, 1627–1635.
4 Hultin, P.G., Earle, M.A., and Sudharshan, M. (1997) *Tetrahedron*, **53**, 14823–14870.
5 Hollingsworth, R.I. and Wang, G. (2000) *Chem. Rev.*, **100**, 4267–4282.
6 Totani, K., Takao, K., and Tadano, K. (2004) *Synlett*, 2066–2080.
7 Boysen, M.M.K. (2007) *Chem. Eur. J.*, **13**, 8649–8659.
8 Fu, Y.Q., An, Y.J., and Tao, J.C. (2008) *Chin. J. Org. Chem.*, **28**, 44–51.
9 Vinogradov, M.G., Gorshkova, L.S., Pavlov, V.A., Mikhalev, O.V., Chel'tsova, G.V., Razmanov, I.V., Ferapontov, V.A., Malyshev, O.R., and Heise, G.L. (2000) *Russ. Chem. Bull.*, **49**, 460–465.
10 Sailes, H. and Whiting, A. (2000) *J. Chem. Soc., Perkin Trans. 1*, 1785–1805.
11 Cho, B.T. (2002) *Aldrichim. Acta*, **35**, 3–16.
12 Daverio, P. and Zanda, M. (2001) *Tetrahedron: Asymmetry*, **12**, 2225–2259.
13 Landor, S.R., Miller, B.J., and Tatchell, A.R. (1966) *J. Chem. Soc. (C)*, 1822–1825.
14 Landor, S.R., Miller, B.J., and Tatchell, A.R. (1966) *J. Chem. Soc. (C)*, 2280–2282.
15 Landor, S.R., Miller, B.J., and Tatchell, A.R. (1971) *J. Chem. Soc. (C)*, 2339–2341.
16 Landor, S.R., Miller, B.J., and Tatchell, A.R. (1967) *J. Chem. Soc. (C)*, 197–201.
17 Landor, S.R., Miller, B.J., and Tatchell, A.R. (1967) *J. Chem. Soc. (C)*, 1159–1163.
18 Landor, S.R., Chan, Y.M., Sonola, O.O., and Tatchell, A.R. (1984) *J. Chem. Soc., Perkin Trans. 1*, 493–496.
19 Landor, S.R., Sonola, O.O., and Tatchell, A.R. (1974) *J. Chem. Soc., Perkin Trans. 1*, 1902–1904.
20 Landor, S.R., Sonola, O.O., and Tatchell, A.R. (1978) *J. Chem. Soc., Perkin Trans. 1*, 605–608.
21 Landor, S.R., Sonola, O.O., and Tatchell, A.R. (1984) *Bull. Chem. Soc. Jpn.*, **57**, 1658–1661.
22 Snatzke, G., Raza, Z., Habus, I., and Sunjic, V. (1988) *Carbohydr. Res.*, **182**, 179–196.
23 Hirao, A., Mochizuki, H., Nakahama, S., and Yamazaki, N. (1979) *J. Org. Chem.*, **44**, 1720–1722.
24 Hirao, A., Nakahama, S., Mochizuki, H., Itsuno, S., and Yamazaki, N. (1980) *J. Org. Chem.*, **45**, 4231–4233.
25 Hirao, A., Mochizuki, H., Zoorob, H.H.A., Igarashi, I., Itsuno, S., Ohwa, M., Nakahama, S., and Yamazaki, N. (1981) *Agric. Biol. Chem.*, **45**, 693–697.
26 Hirao, A., Itsuno, S., Owa, M., Nagami, S., Mochizuki, H., Zoorov, H.H.A., Niakahama, S., and Yamazaki, N. (1981) *J. Chem. Soc., Perkin Trans. 1*, 900–905.
27 Morrison, J.D., Grandbois, E.R., and Howard, S.I. (1980) *J. Org. Chem.*, **45**, 4229–4231.
28 Hirao, A., Nakahama, S., Mochizuki, D., Itsuno, S., Ohowa, M., and Yamazaki, N. (1979) *J. Chem. Soc., Chem. Commun.*, 807–808.
29 Hirao, A., Ohwa, M., Itsuno, S., Mochizuki, H., Nakahara, S., and Yamazaki, N. (1981) *Bull. Chem. Soc. Jpn.*, **54**, 1424–1428.
30 Goldberg, Y., Abele, E., Rubina, K., Popelis, Y., and Shimanska, M. (1993) *Chem. Heterocycl. Compd.*, **29**, 1399–1404.
31 Schmitzer, A., Perez, E., Rico-Lattes, I., and Lattes, A. (1999) *Tetrahedron Lett.*, **40**, 2947–2950.
32 Schmitzer, A.R., Franceschi, S., Perez, E., Rico-Lattes, I., Lattes, A., Thion, L., Erard, M., and Vidal, C. (2001) *J. Am. Chem. Soc.*, **123**, 5956–5961.
33 Brown, H.C., Park, W.S., and Cho, B.T. (1986) *J. Org. Chem.*, **51**, 1934–1936.
34 Brown, H.C., Cho, B.T., and Park, W.S. (1986) *J. Org. Chem.*, **51**, 3396–3398.
35 Brown, H.C., Park, W.S., Cho, B.T., and Ramachandran, P.V. (1987) *J. Org. Chem.*, **52**, 5406–5412.
36 Brown, H.C., Cho, B.T., and Park, W.S. (1988) *J. Org. Chem.*, **53**, 1231–1238.
37 Cho, B.T. and Chun, Y.S. (1992) *Tetrahedron: Asymmetry*, **3**, 341–342.
38 Hutchins, R.O., Abdel-Magid, A., Stercho, Y.P., and Wambsgans, A. (1987) *J. Org. Chem.*, **52**, 702–704.
39 Miller, S.A. and Chamberlin, A.R. (1990) *J. Am. Chem. Soc.*, **112**, 8100–8112.
40 Cho, B.T. and Chun, Y.S. (1992) *Tetrahedron: Asymmetry*, **3**, 1583–1590.
41 Cho, B.T. and Chun, Y.S. (1992) *Tetrahedron: Asymmetry*, **3**, 73–84.

42 Cho, B.T. and Chun, Y.S. (1994) *Tetrahedron: Asymmetry*, **5**, 1147–1150.
43 Brown, H.C., Cho, B.T., and Park, W.S. (1987) *J. Org. Chem.*, **52**, 4020–4024.
44 Cintas, P. (1991) *Tetrahedron*, **47**, 6079–6111.
45 Enders, D. and Reinhold, U. (1997) *Tetrahedron: Asymmetry*, **8**, 1895–1946.
46 Enders, D. and Shilvock, J.P. (2000) *Chem. Soc. Rev.*, **29**, 359–373.
47 Knauer, S., Kranke, B., Krause, L., and Kunz, H. (2004) *Curr. Org. Chem.*, **8**, 1739–1761.
48 Chung, C.W.Y. and Toy, P.H. (2004) *Tetrahedron: Asymmetry*, **15**, 387–399.
49 Ding, H. and Friestad, G.K. (2005) *Synthesis*, 2815–2829.
50 Zani, L. and Bolm, C. (2006) *Chem. Commun.*, 4263–4275.
51 Ribeiro, C.M.R. and de Farias, F.M.C. (2006) *Mini-Rev. Org. Chem.*, **3**, 1–10.
52 Braun, M. (1987) *Angew. Chem. Int. Ed. Engl.*, **26**, 24–37.
53 Inch, T.D., Lewis, G.J., Sainsbury, G.L., and Sellers, D.J. (1969) *Tetrahedron Lett.*, **10**, 3657–3660.
54 Baggett, N. and Simmonds, R.J. (1982) *J. Chem. Soc., Perkin Trans. 1*, 197–200.
55 Zoorob, H.H. (1987) *Egypt. J. Chem.*, **29**, 199–202.
56 Ribeiro, C.M.R., Santos, E.D.S., Jardim, A.H.D.O., Maia, M.P., da Silva, F.C., Moreira, A.P.D., and Ferreira, V.F. (2002) *Tetrahedron: Asymmetry*, **13**, 1703–1706.
57 Emmerson, D.P.G., Hems, W.P., and Davis, B.G. (2005) *Tetrahedron: Asymmetry*, **16**, 213–221.
58 Emmerson, D.P.G., Hems, W.P., and Davis, B.G. (2006) *Org. Lett.*, **8**, 207–210.
59 Appelt, H.R., Limberger, J.B., Weber, M., Rodrigues, O.E.D., Oliveira, J.S., Lüdtke, D.S., and Braga, A.L. (2008) *Tetrahedron Lett.*, **49**, 4956–4957.

7
Aldol-Type Reactions

Inmaculada Fernández, Noureddine Khiar, Ana Alcudia, Maria Victoria García and Rocío Recio

7.1
Introduction

The aldol reaction can be considered as one of the most useful synthetic methods for a C–C bonding process [1], as the configuration of two new asymmetric centers can be controlled in most cases with high stereoselectivity.

In this sense, titanium complexes with chiral ligands have been shown to be good reagents for the stereoselective addition of C nucleophiles to carbonyl compounds. The use of these complexes for the chiral modification of enolates is an interesting approach to enantioselective aldol reactions as the chiral ligands can be easily recovered, and no additional manipulations on the products are needed.

7.2
Titanium Lewis Acids for Enolate Formation

Natural carbohydrates can be considered as privileged monodentate ligands for the preparation of chiral titanium reagents. In this sense, titanium complex **3** is a successful example of a carbohydrate reagent especially valuable in aldol reactions. It was first prepared by Duthaler and coworkers from CpTiCl$_3$ (**1**) and diacetone-D-glucose (DAGOH) **2** (Scheme 7.1) [2].

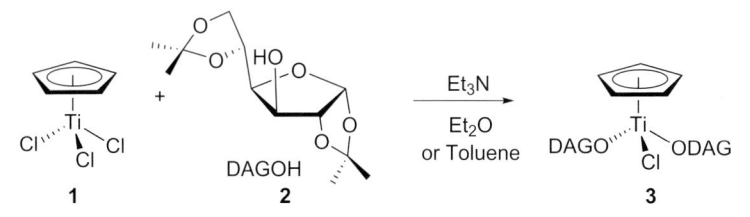

Scheme 7.1 Synthesis of titanium complex **3**.

Carbohydrates – Tools for Stereoselective Synthesis, First Edition. Edited by Mike Martin Kwabena Boysen.
© 2013 Wiley-VCH Verlag GmbH & Co. KGaA. Published 2013 by Wiley-VCH Verlag GmbH & Co. KGaA.

The best results were obtained in Et$_2$O or toluene as solvent; the HCl evolved is neutralized with a slight excess of Et$_3$N. Complex **3** can be isolated under anhydrous conditions and stored in solution.

Other stable analogs can be prepared in a similar way, by using different acetal protection of the D-glucose derived ligands [3]. However, it must be taken into account that small changes of the sugar skeleton lead to extremely labile compounds, such as those derived from α-L-idofuranose, D-xylose, or D-allose [4].

The titanium complex **3** has been applied by Duthaler and Hafner to solve one of the most important challenges with the aldol reaction, namely, the addition of acetyl groups with the stereocontrolled formation of one new stereogenic center. For this purpose, transmetalation of the Li enolate of *tert*-butyl acetate **4** by treatment with complex **3**, at –30 °C in toluene, yields the titanium enolate **5**, which adds with high enantioselectivity (90–96% ee) and good chemical yields to different aldehydes (Scheme 7.2) [5]. The favored attack at the *Re* face of the aldehyde is independent of temperature.

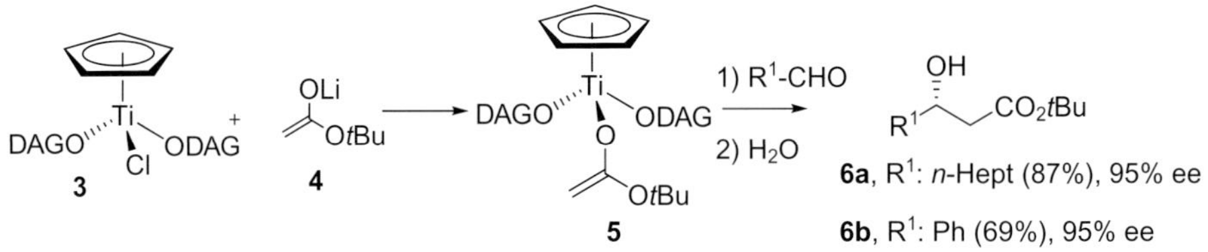

Scheme 7.2 Addition of an acetyl group in the aldol reaction with titanium enolate **5**.

The enantioselectivities obtained with the chiral acetate enolate **5** are higher than those obtained by others methods at room temperature and are matched only by those obtained with 2,4-dialkylborolane developed by Masamune [6] and Reetz [7].

The Duthaler–Hafner acetate aldol reaction has been applied as one of the key steps in the total asymmetric synthesis of different biologically active compounds, such as the aggregation pheromone of various species of bark beetles, (–)-(S)-ipsenol **7** (Figure 7.1) [8]. This has also been the strategy applied more recently by Kirschning *et al.* in the synthesis of the ansatrienol derivative **8**, a known precursor of ansamycin antibiotics, for introduction of the stereocenter at C3 (Figure 7.1) [9]. Cossy *et al.* have prepared the novel immunosuppressant FR252921, **9**, using the chiral titanium complex **3** to control the configuration of the stereogenic center at C18 [10].

In the case of substituted enolates, the situation is more complex as two new stereogenic centers are simultaneously generated. The double bond geometry of the enolate (*E* or *Z*) is directly related to the relative configuration (*syn* or *anti*) of the aldol products when the reaction takes place via a cyclic transition state [11]. Interestingly, with achiral titanium enolates, the *syn* diastereomers are mainly

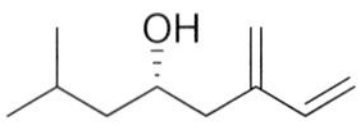

7
(*S*)-(-)-Ipsenol
(Pheromone)

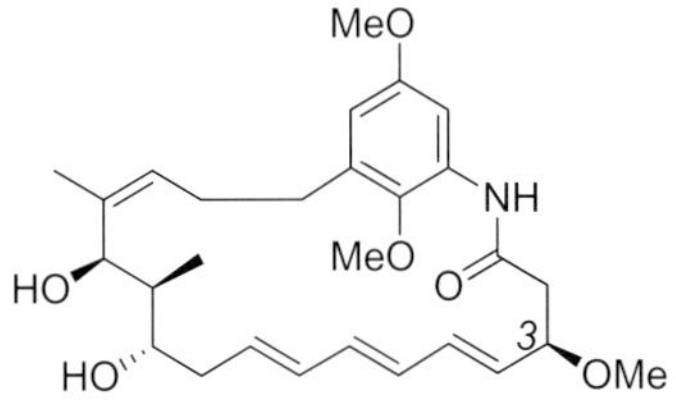

8
Ansamycin antibiotic

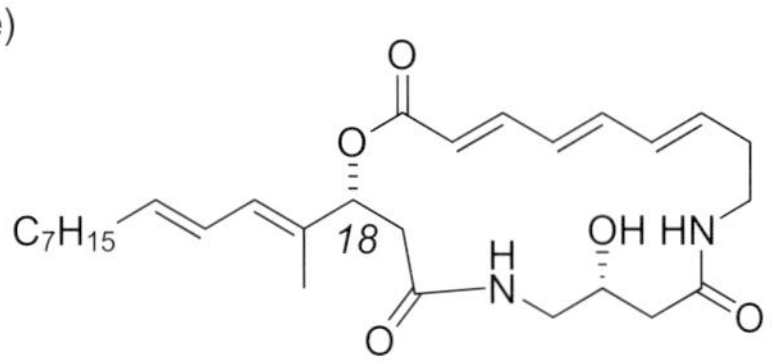

9
FR252921
(Immunosuppressant)

Figure 7.1 Biologically active compounds prepared using the Duthaler–Hafner acetate aldol reaction.

formed, if no cyclopentadienyl ligands are involved [12], in contrast to the *anti*-aldols obtained with the Li enolate.

Duthaler and coworkers have demonstrated that, in the case of propionyl-enolates with chiral Ti-complexes, both *(Z)* and *(E)* enolates are available and add to aldehydes in a highly diastereo- and enantioselective manner [5, 13].

Transmetalation of lithium enolate **10**, derived from 2,6-dimethylphenylpropionate, with the titanium complex **3** at –78 °C yields, after 24 h, the corresponding *(E)*-configurated enolate **11** (Scheme 7.3). This *(E)*-enolate added to the *Re* side of aldehydes, affording various *syn*-aldols **13**, with high dia- and enantioselectivity (84–94% ds, 91–96% ee, Scheme 7.3). In a similar way, racemic *anti*-aldols (±)-**14** can be obtained from the achiral Li enolate **10** (72–96% ds, Scheme 7.3). However, in contrast to this unstable Li enolate **10**, the Ti enolate **11** isomerizes at −30°C, probably via a titanium-bound ketene intermediate [14], to the thermodynamically more stable *(Z)*-enolate **12** (Scheme 7.3), which afforded *anti*-aldols **14** of high optical purity (94–98% ee) and, in most cases, with acceptable-to-good diastereoselectivity (62–78% ds). In this case, important exceptions were methacrolein and benzaldehyde, that is, branched unsaturated and aromatic aldehydes, which yielded larger amounts of *syn*-epimers of lower optical purity.

In this way, both epimers that result from the attack of **11** and **12** to the *Re* side of aldehydes can be obtained from the same precursor, according to the reaction conditions, using the cheap and easily accessible diacetone-D-glucose as the only chiral inductor.

The stereochemical outcome of the aldol additions of both enolates has been explained by assuming in both cases a boat conformation for the transition states, **ET-11** for the *(E)*-enolate and **ET-12** for the *(Z)*-enolate (Figure 7.2) [14].

Scheme 7.3

10: OLi enolate with 2,6-dimethylphenyl ester

3: Cp–Ti(Cl)(DAGO)(ODAG)

Conditions: −78 °C, 24 h.

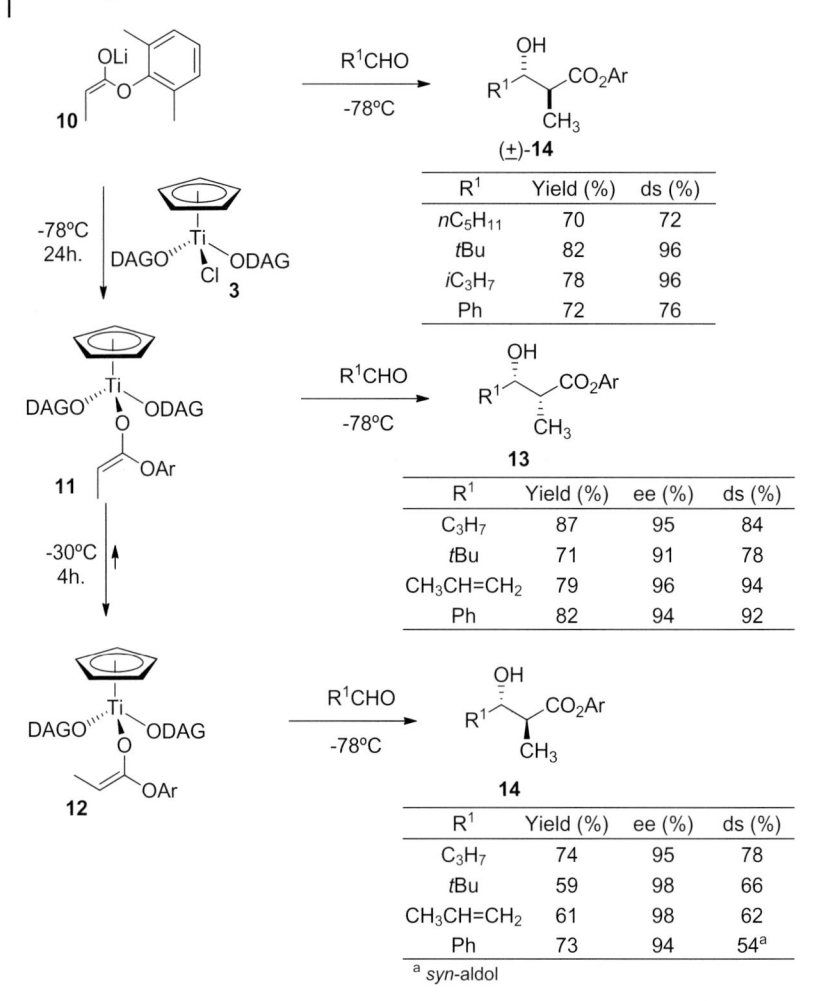

11: Cp–Ti(DAGO)(ODAG) with O–C=CH–CH₃ and OAr (E-enolate)

R¹CHO, −78 °C → (±)-**14**

R¹	Yield (%)	ds (%)
nC₅H₁₁	70	72
tBu	82	96
iC₃H₇	78	96
Ph	72	76

R¹CHO, −78 °C → **13**

R¹	Yield (%)	ee (%)	ds (%)
C₃H₇	87	95	84
tBu	71	91	78
CH₃CH=CH₂	79	96	94
Ph	82	94	92

−30 °C, 4 h.

12: Cp–Ti(DAGO)(ODAG) Z-enolate

R¹CHO, −78 °C → **14**

R¹	Yield (%)	ee (%)	ds (%)
C₃H₇	74	95	78
tBu	59	98	66
CH₃CH=CH₂	61	98	62
Ph	73	94	54[a]

[a] *syn*-aldol

Scheme 7.3 Synthesis and reactivity of chiral *(E)*- and *(Z)*-titanium enolates **11** and **12**.

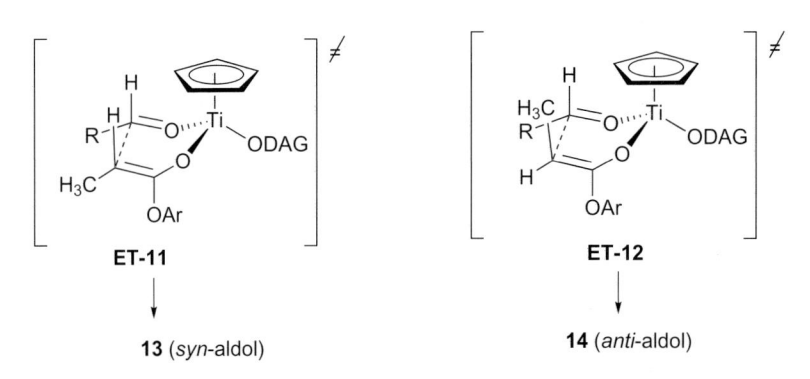

ET-11 → **13** (*syn*-aldol)

ET-12 → **14** (*anti*-aldol)

Figure 7.2 Transition states for the aldol additions of *(E)*-**11** and *(Z)*-**12** enolates.

7.2 Titanium Lewis Acids for Enolate Formation | 147

Consistent with these findings, *Re* facial and *anti*-selective aldol addition was also observed for the *(Z)*-configurated Ti enolate **16** of N-propionyl-oxazolidinone **15** (Scheme 7.4) [13b].

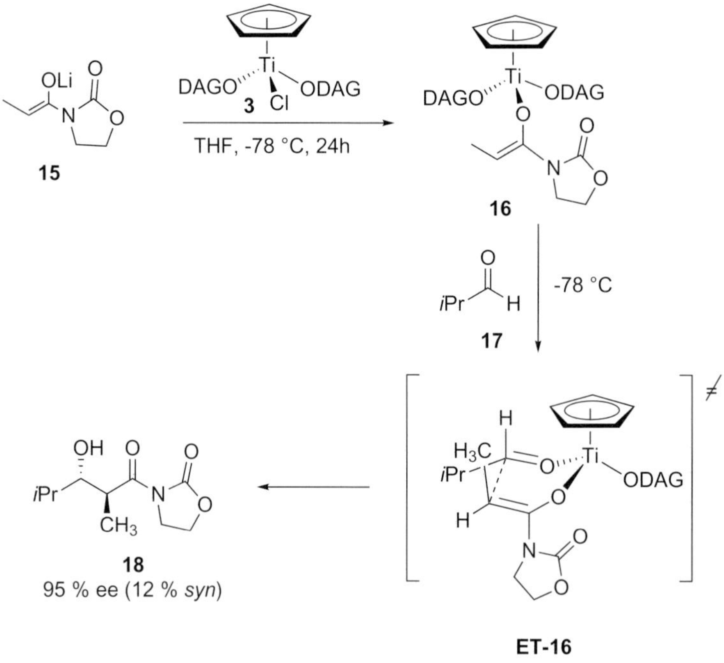

Scheme 7.4 Aldol addition of the *(Z)*-titanium enolate **16**.

A clear drawback of Ti-complex **3** is that the corresponding enantiomer is not readily available, as the non-natural L-glucose is very expensive, and different attempts to prepare pseudoenantiomeric Ti complexes with other carbohydrates as chiral ligands have been unsuccessful. In some cases, the titanium enolate derived from *(R,R)*-tartrate **19** (Scheme 7.5) constitutes an alternative answer to this problem, but the stereoselectivity in the aldol reaction is in general lower than that obtained with **3**. Consequently, the enantiomers of **13** and **14** are not easily available in optically pure form [2, 3].

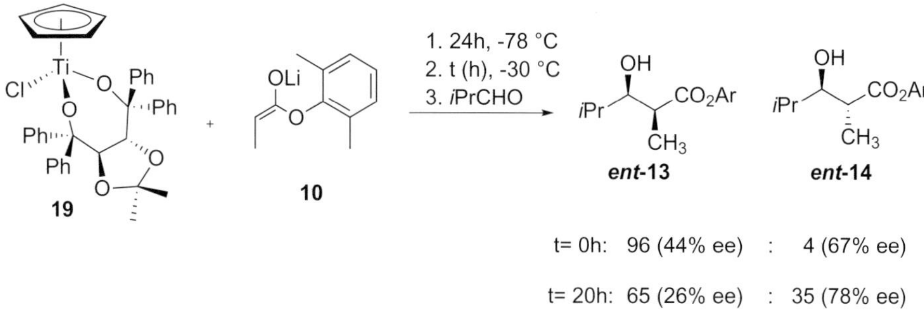

t= 0h: 96 (44% ee)   :   4 (67% ee)
t= 20h: 65 (26% ee)   : 35 (78% ee)

Scheme 7.5 Aldol reaction with the titanium enolate derived from *(R,R)*-tartrate **19**.

Table 7.1 Synthesis of D-amino acids with the chiral glycine titanium enolate **21**.

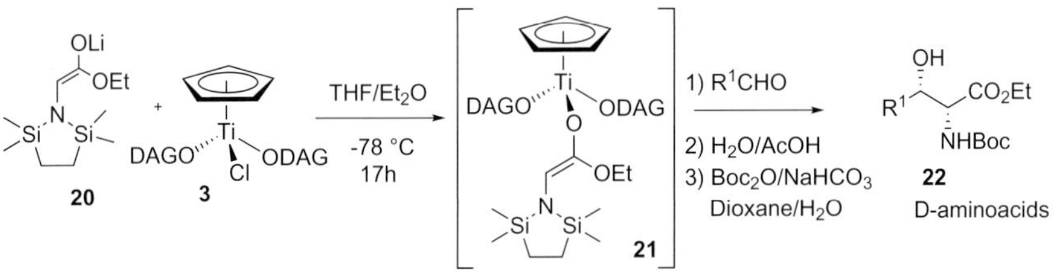

R¹	Product	Yield (%)	ee (%)	ds (%)
CH_3	22a	53	97	>98
nC_3H_7	22b	66	98	>98
tBu	22c	96	96	>96
$CH_2=CH$	22d	48	97	>97
Ph	22e	97	97	>96

Duthaler has also applied his chiral Ti-complex **3** in the asymmetric addition of glycine enolate and analogues to aldehydes to give β-hydroxy-α-amino acids, a very important synthetic process as β-hydroxylated amino acids are biologically interesting compounds. In this sense, the chiral titanium enolate **21**, obtained by transmetalation of the Li enolate of the silyl-protected glycine ester **20** with **3** (Table 7.1), has been demonstrated to be a highly stereoselective reagent in aldol reactions. It is a very mild reagent that adds to the *Re* side of different aldehydes, affording, after mild acidic hydrolysis, the *(R)*-configured *threo*-β-hydroxy-α-amino acid ethyl esters in a highly diastereo- and enantioselective manner [2, 3, 5].

Chiral enolate **21** is the only reagent that can compete in terms of diastereo- and enantioselectivity with the gold-catalyzed addition of isocyanoacetate to aldehydes [15].

The enantiomeric L-amino acids **ent-22** were obtained, as in the case of aldol reactions, by transmetalation of **20** with the threitol complex **19**, with a lower stereoselectivity than observed before. In this case, however, the stereoselectivity could be considerably increased (from 81 to 94% ee), by using the Li enolate of the analogous glycine *tert*-butyl ester derivative **23** (R = tBu in Scheme 7.6) [2].

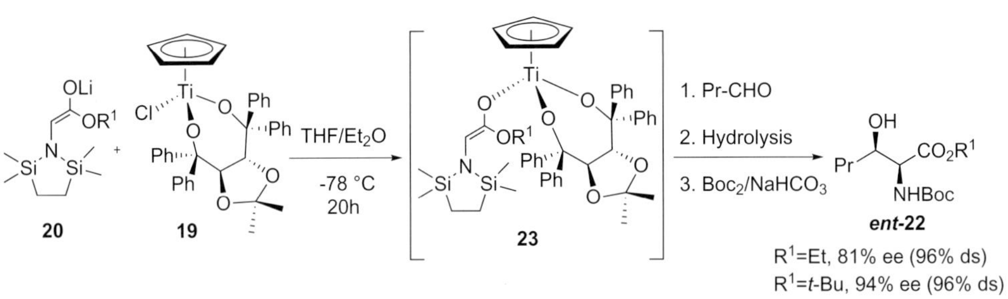

Scheme 7.6 Synthesis of L-amino acids with the chiral titanium complex **23**.

7.3
1,2-Additions of Nucleophiles to Carbonyl Compounds Using Stoichiometric Reagents

The stereocontrolled addition of allylic groups to aldehydes is without doubt a very important process in organic synthesis. The resulting homoallylic alcohols containing one or two new stereogenic centers are very valuable synthetic intermediates, as many possibilities for further transformations are provided by the hydroxyl group as well as the double bond [16].

7.3.1
Allyl-Titanium Reagents

With achiral allyl-titanium reagents high regio- and diastereocontrol has been observed. A six-center cyclic chair-like transition state has been proposed to explain the diastereoselectivity and the higher reactivity of allyl-titanium compounds compared to the alkyl derivatives [17]. Therefore, the high enantioselectivity observed in the nucleophilic additions to carbonyl compounds when chirality is incorporated in the ligands of the allyl-titanium reagent is not surprising. The chiral allyl-titanium reagents can be easily prepared by transmetalation of the corresponding titanium halide-complex with allylmagnesium chloride. Thus, the reaction of diacetone-D-glucose derivative 3 with allylmagnesium chloride gives allyl-titanate 25 [18]. The reaction of 25 with aldehydes at –78 °C affords homoallylic alcohols with good chemical yield (58–85%) and high enantioselectivity (86–94% ee), resulting from the addition to the *Re* side of aldehyde. Upon aqueous work-up, both the chiral inductor and the cyclopentadienyl titanium complex can be recovered and recycled (Table 7.2) [18].

Table 7.2 Chiral allyl-titanate **25**, synthesis and stereoselectivity in the addition to aldehydes.

R^1	Product	Yield (%)	Configuration	ee (%)
Ph	26a	85	(R)	90
9-Anthryl	26b	80	(R)	94
Vinyl	26c	61	(R)	86
Et	26d	67	(S)	93
n-Pr	26e	78	(S)	93
i-Pr	26f	67	(R)	90
Cyclohexyl	26g	78	(R)	92
t-Bu	26h	58	(R)	88

7 Aldol-Type Reactions

Several allyl reagents prepared from titanium chlorides with different acetal protection of their glucose ligands showed similar stereoselectivity as **25**, but compounds derived from others carbohydrates, such as L-idose, afforded in general unselective reagents [2, 3]. Other chiral titanium complexes related to **25** were prepared and tested in their reaction with benzaldehyde, including derivatives with substituents on the cyclopentadienyl moiety (**27a**, Scheme 7.7), yielding the corresponding homoallylic alcohol with lower enantioselectivity.

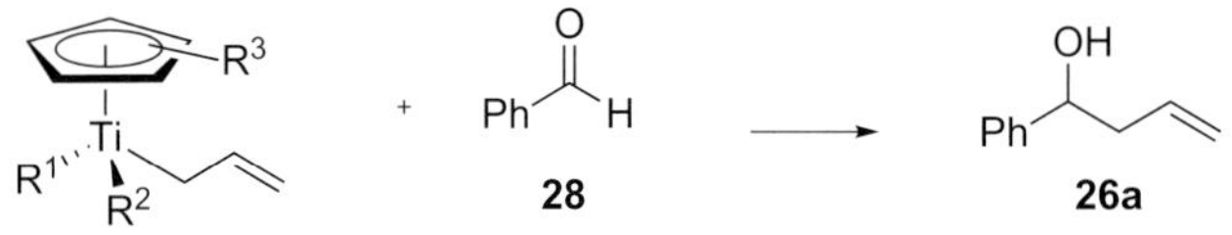

25: R^1, R^2 = ODAG, R^3 = H — — — — — — ▶ 90% ee (R)
27a: R^1, R^2 = ODAG, R^3 = SiMe$_3$, pentamethyl — — — — — — ▶ 85–90% ee (R)
27b: R^1 = ODAG, R^2 = Cl, R^3 = H — — — — — — ▶ 21% ee (R)
27c: R^1 = ODAG, R^2 = Allyl, R^3 = H — — — — — — ▶ 47% ee (R)

Scheme 7.7 Stereoselective addition to benzaldehyde of different chiral allyl titanium complexes, **25**, **27a–c**.

Little change was observed when the cyclopentadienyl ligand was substituted by the penta-TMS or the pentamethyl derivative, but the enantioselectivity dramatically decreased when one of the DAG ligands was replaced by chloride or another allyl group (**27b**, **27c**, Scheme 7.7) [2, 19].

Taking into account that the chiral inductor diacetone-glucose is readily available only in the D form, the enantiofacial discrimination in the allyltitanation of aldehydes is again restricted to one enantiomer, which is obtained from the *Re* side attack. With the aim of discovering another auxiliary, which might induce the opposite *Si*-face selectivity, several carbohydrates derivatives were tested. Therefore, various other chiral cyclopentadienyl(dialkoxy)titanium complexes were prepared and screened for their ability to reverse the stereoinduction in the enantioselective allyltitanation of benzaldehyde but only low stereoselectivities were obtained (Scheme 7.8) [19]. In this sense, the complex of diacetone-D-idose, which differs only in the configuration of C(5) from diacetone-D-glucose, turned out to be much more labile, the allylated derivative was much less stereoselective, and induced the opposite chirality (Scheme 7.8).

Other chiral ligands were also considered; the Ti complex with a ligand derived from tartaric acid **30** is the complementary reagent to **3** as it transfers allyl groups with higher *Si*-face selectivity to aldehydes (Scheme 7.9).

The possible mechanism of asymmetric induction in the allylation of aldehydes with complex **25** has been established based on crystal structure and NMR analyses. The two diacetone-D-glucose ligands do indeed form a highly asymmetric environment around the titanium chloride bond of **3**. However, the asymmetric distortion of the titanium coordination geometry seems to be essential for

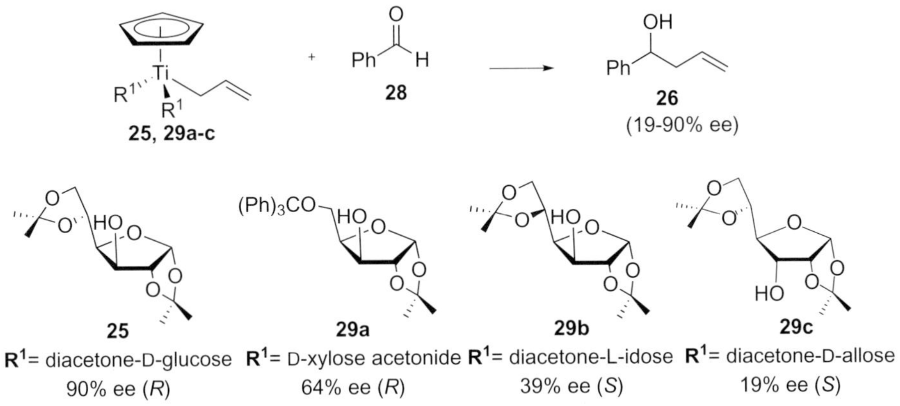

7.3 1,2-Additions of Nucleophiles to Carbonyl Compounds Using Stoichiometric Reagents | 151

Scheme 7.8 Enantioselective allyltitanation of benzaldehyde with chiral cyclopentadienyl(dialkoxy)titanium complexes **25, 29a–c**.

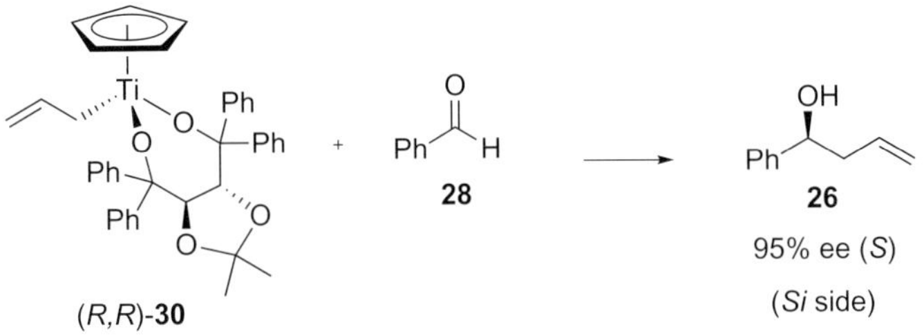

Scheme 7.9 Allyltitanation of benzaldehyde with chiral complex **30**, derived from tartaric acid.

enantioface discrimination, rather than direct interactions of reactants with the chiral ligand [19].

7.3.2
Allylsilicon Reagents

The asymmetric addition reaction of allylsilanes to carbonyl compounds promoted by a Lewis acid, known as the Hosomi–Sakurai reaction, can be considered as one of the most important methods developed for the stereoselective synthesis of chiral homoallylic alcohols [20]. For this reason research efforts have concentrated on the synthesis of chiral allylsilanes with the stereogenic center not residing in the allyl moiety. In this sense, several allylsilyl ethers of L-arabinose **34** and **35** were prepared by Shing, as indicated in Scheme 7.10, and used in the acid-mediated reaction with aldehydes to determine steric and coordinative effects on the asymmetric induction [21].

Table 7.3 summarizes the Lewis acid effect on the stereochemical outcome of the reaction of allylsilanes **34** and **35** with aldehydes. As can be seen, BF_3 leads to higher enantioselectivity than $TiCl_4$ or $SnCl_4$, although homoallylic alcohols were

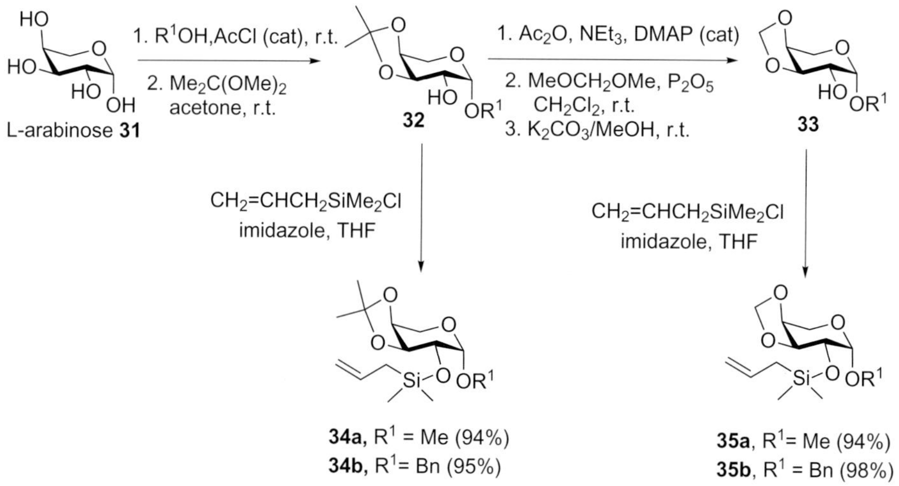

Scheme 7.10 Synthesis of allylsilyl ethers of L-arabinose, **34** and **35**.

obtained with only moderate enantioselectivity in all cases. Allylsilane reagent **34b**, which carries the bulkiest group, gave the highest enantioselectivity with *n*-heptanal in the presence of BF_3 [21].

The results show that a sterically demanding group in the chiral auxiliary could have an important influence on the asymmetric induction, despite being remote from the reaction site. All the reactions of allylsilane **34b** with different aliphatic aldehydes in the presence of BF_3 gave similar results [36–45% ee *(S)*, with moderate to high chemical yields (54–72%)]. According to the *(S)* configuration of the new stereogenic center of the homoallylic alcohols, it has been proposed that all reactions proceed via an intermediate where the *Re* face of the complex [22] of the

Table 7.3 Effect of Lewis acid on the stereochemical outcome of the reaction of allylsilanes **34** and **35** with aldehyde **36**.

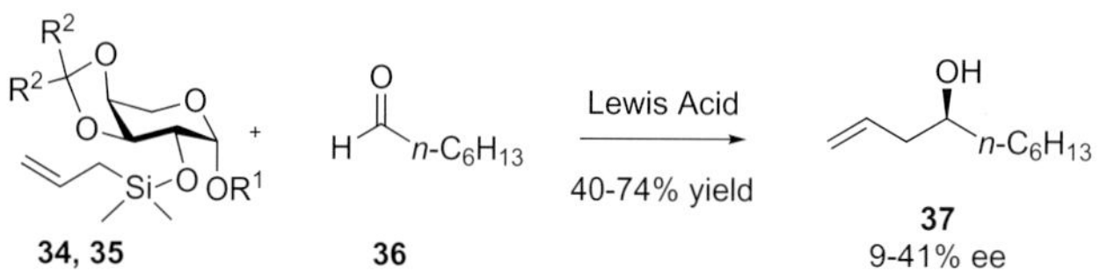

Silane	34a (R^1 = Me, R^2 = Me)			34b (R^1 = Bn, R^2 = Me)			35a (R^1 = Me, R^2 = H)			35b (R^1 = Bn, R^2 = H)		
Lewis acid	BF_3	$TiCl_4$	$SnCl_4$	BF_3	$TiCl_4$	$SnCl_4$	BF_3	$TiCl_4$	$SnCl_4$	BF_3	$TiCl_4$	$SnCl_4$
Yield (%)	48	40	71	63	65	74	57	47	55	77	62	71
ee (%)	25	15	9	41	11	9	13	6	0	33	15	9

aldehyde with BF$_3$ is attacked by the allyl moiety of the enantiopure allylsilane. This can be considered to be the first demonstration of a remote steric effect of the chiral substituent on the asymmetric Hosomi–Sakurai reaction.

References

1 Heathcock, C.H. (1984), in *Asymmetric Synthesis*, vol. 3, ch. 2 (ed. J.D. Morrison), Academic Press, Orlando, pp. 111–212.
2 Duthaler, R.O., Hafner, A., and Riediker, M. (1990) *Pure Appl. Chem.*, **62**, 631–642.
3 (a) Duthaler, R.O. and Hafner, A. (1992) *Chem. Rev.*, **92**, 807–832; (b) Kunz, H. and Rück, K. (1993) *Angew. Chem. Int. Ed. Engl.*, **32**, 336–358; (c) Boysen, M.M.K. (2007) *Chem. Eur. J.*, **13**, 8648–8659.
4 (a) Duthaler, R.O., Hafner, A., and Riediker, M. (1991), in *Organic Synthesis Via Organometallics* (eds K.H. Dötz and R.W. Hoffmann), Vieweg, Braunschweig, pp. 285–309; (b) Hafner, A., Duthaler, R.O., Marti, R., Rihs, G., Rothe-Streit, P., and Schwarzenbach, F. (1992) *J. Am. Chem. Soc.*, **114**, 285–309.
5 Duthaler, R.O., Herold, P., Lottenbach, W., Oertle, K., and Riediker, M. (1989) *Angew. Chem. Int. Ed. Engl.*, **28**, 495–497.
6 (a) Masamune, S., Sato, T., Kim, B.M., and Wollmann, T.A. (1986) *J. Am. Chem. Soc.*, **108**, 8279–8281; (b) Sort, R. and Masamune, S. (1987) *Tetrahedron Lett.*, **28**, 2841–2844.
7 (a) Reetz, M.T., Kunish, F., and Heitmann, P. (1986) *Tetrahedron Lett.*, **27**, 4721–4724; (b) Reetz, M.T. (1988) *Pure Appl. Chem.*, **60**, 1607–1614; (c) Reetz, M.T., Rivadeneira, E., and Niemeyer, C. (1990) *Tetrahedron Lett.*, **31**, 3863–3866.
8 Oertle, K., Beyeler, H., Duthaler, R.O., Lottenbach, W., Riediker, M., and Steiner, E. (1990) *Helv. Chim. Acta*, **73**, 353–357.
9 Kashin, D., Meyer, A., Wittenberg, R., Schoning, K.-U., Kamlage, S., and Kirschning, A. (2007) *Synthesis*, 304–319.
10 Amans, D., Bellosta, V., and Cossy, J. (2007) *Org. Lett.*, **9**, 4761–4764.
11 (a) Li, Y., Paddon-Row, M.N., and Houk, K.N. (1990) *J. Org. Chem.*, **55**, 481–493; (b) Yamago, S., Machii, D., and Nakamura, E. (1991) *J. Org. Chem.*, **56**, 2098–2106; (c) Denmark, S.E. and Henke, B.R. (1991) *J. Am. Chem. Soc.*, **113**, 2177–2194.
12 (a) Reetz, M.T. and Peter, R. (1981) *Tetrahedron Lett.*, **22**, 4691–4694; (b) Reetz, M.T., Steinbach, R., and Kessler, K. (1982) *Angew. Chem. Int. Ed. Engl.*, **21**, 1899–1905; (c) Kuwajima, I. and Nakamura, E. (1985) *Acc. Chem. Res.*, **18**, 181–187; (d) Harrison, C.R. (1987) *Tetrahedron Lett.*, **28**, 4135–4138; (e) Panek, J.S. and Bula, O.A. (1988) *Tetrahedron Lett.*, **29**, 1661–1664.
13 (a) Bold, G., Duthaler, R.O., and Riediker, M. (1989) *Angew. Chem. Int. Ed. Engl.*, **28**, 497–498; (b) Duthaler, R.O., Herold, P., Wyler-Helfer, S., and Riediker, M. (1990) *Helv. Chim. Acta*, **73**, 659–673.
14 Vuitel, L. and Jacot-Guillarmod, A. (1972) *Synthesis*, 608–610.
15 (a) Ito, Y., Sawamura, M., and Hayashi, T. (1986) *J. Am. Chem. Soc.*, **108**, 6405–6406; (b) Hayashi, T., Uozumi, Y., Yamazaki, A., Sawamura, M., Hamashino, H., and Ito, Y. (1991) *Tetrahedron Lett.*, **32**, 2799–2802; (c) Pastor, S.D. and Togni, A. (1989) *J. Am. Chem. Soc.*, **111**, 2333–2334.
16 (a) Hoffmann, R.W. (1982) *Angew. Chem. Int. Ed. Engl.*, **21**, 555–566; (b) Yamamoto, Y. and Maruyama, K. (1982) *Heterocycles*, **18**, 357–386; (c) Hoffman, R.W. (1987) *Angew. Chem. Int. Ed. Engl.*, **26**, 489–503; (d) Yamamoto, Y. (1987) *Acc. Chem. Res.*, **20**, 243–249; (e) Hoppe, D. (1984) *Angew. Chem. Int. Ed. Engl.*, **23**, 932–948; (f) Mulzer, J., Kattner, L., Strecker, A.R., Schröder, C., Buschmann, J., Lehmann, C., and Luger, P. (1991) *J. Am. Chem. Soc.*, **113**, 4218–4229.
17 (a) Sato, F., Iida, K., Iijima, S., Moriya, H., and Sato, M. (1981) *J. Chem. Soc.*,

Chem. Commun., 1140–1141; (b) Kobayashi, Y., Umeyama, K., and Sato, F. (1984) *J. Chem. Soc., Chem. Commun.*, 621–623; (c) Reetz, M.T. and Sauerwald, M. (1984) *J. Org. Chem.*, **49**, 2292–2293; (d) Seebach, D. and Widler, L. (1982) *Helv. Chim. Acta*, **65**, 1085–1089; (e) Weidmann, B. and Seebach, D. (1983) *Angew. Chem. Int. Ed. Engl.*, **22**, 31–45; (f) Reetz, M.T. (1985) *Pure Appl. Chem.*, **57**, 1781–1788; (g) Hoppe, D., Gonschorrek, C., Schmidt, D., and Egert, E. (1987) *Tetrahedron*, **43**, 2457–2466; (h) Ikeda, Y., Furuta, K., Meguriya, N., Ikeda, N., and Yamamoto, H. (1982) *J. Am. Chem. Soc.*, **104**, 7663–7665; (i) Martin, S.F. and Li, W. (1989) *J. Org. Chem.*, **54**, 6129–6133.

18 Riediker, M. and Duthaler, R.O. (1989) *Angew. Chem. Int. Ed. Engl.*, **28**, 494–495.

19 Hafner, A., Duthaler, R.O., Marti, R., Rihs, G., Rothe-Streit, P., and Schwarzenbach, F. (1992) *J. Am. Chem. Soc.*, **114**, 2321–2336.

20 Chan, T.H. and Wang, D. (1992) *Chem. Rev.*, **92**, 995–1006.

21 (a) Shing, T.K.M. and Li, L.-H. (1997) *J. Org. Chem.*, **62**, 1230–1233; (b) Hultin, P.G., Earle, M.A., and Sudharshan, M. (1997) *Tetrahedron*, **53**, 14823–14870.

22 Corey, E.J., Loh, T.-P., Sarshar, S., and Aziioara, M. (1992) *Tetrahedron Lett.*, **33**, 6945–6948.

Part III
Carbohydrate Ligands

8
Hydrogenation Reactions

Carmen Claver, Sergio Castillón, Montserrat Diéguez and Oscar Pàmies

8.1
Introduction

Asymmetric hydrogenation is one of the most important reactions catalyzed by metal transition complexes modified by chiral ligands. Both academic and industrial research groups have studied and developed this reaction during recent decades. Many intermediates and building blocks that are key intermediates in organic synthesis are obtained through this reaction [1]. Several metals, mainly Ru, Rh, and Ir, as well as different strategies are involved in the asymmetric hydrogenation of different substrates through asymmetric catalysis [2]. A wide number of ligands belonging to different families of chiral compounds – such as binap and related atropoisomeric derivatives, phospholanes, ferrocenylphosphines, oxazoline ligands, and so on – have been explored and applied in this reaction. Among the wide variety of families of ligands, carbohydrate derivatives present particular properties because of their unique structural features. Carbohydrates have been used for decades as starting materials to synthesize enantiomerically pure compounds. More recently, carbohydrate derivatives have also been used to obtain chiral ligands for catalysis [3]. Carbohydrate derivatives have some distinct advantages as ligands: the starting materials are cheap and available in bulk quantities and the ligands can be easily constructed in modules. The presence of hydroxyl functions in carbohydrate backbones facilitates the introduction of phosphorus donor sites such as phosphinites, phosphonites, and phosphines. On the other hand, they present other particularities. For instance, although for other type of chiral ligands C_2-symmetry is highly advantageous, carbohydrates, in contrast, posses C_1-symmetry, and thus most ligands prepared from them also have C_1-symmetry. Several reviews deal with the application of carbohydrate-derived ligands in asymmetric catalysis [3]. In particular, some are specific for the application of carbohydrate derivative ligands in asymmetric hydrogenation [3d,e].

8.2
Hydrogenation of C=C and C=N Bonds

8.2.1
Hydrogenation of C=C Bonds

The hydrogenation of functionalized carbon–carbon double bonds is widely used to prepare high-value compounds that can be employed as building blocks in asymmetric synthesis (Scheme 8.1). The hydrogenation of dehydroamino acid derivatives such as *(Z)*-amidocinnamic acids and esters provides access to unnatural amino acids and amines that are useful intermediates for the pharmaceutical and agrochemical industries. These products are also extensively used as benchmark substrates for asymmetric hydrogenation. Rhodium and ruthenium complexes containing chiral ligands with phosphorus and nitrogen donor centers have proven to be the best catalysts for asymmetric hydrogenation of substrates bearing a polar coordinating group next to the C=C bond. Excellent activities and enantioselectivities have been achieved over recent decades for the asymmetric hydrogenation of dehydroamino acids and other functionalized substrates [1, 2].

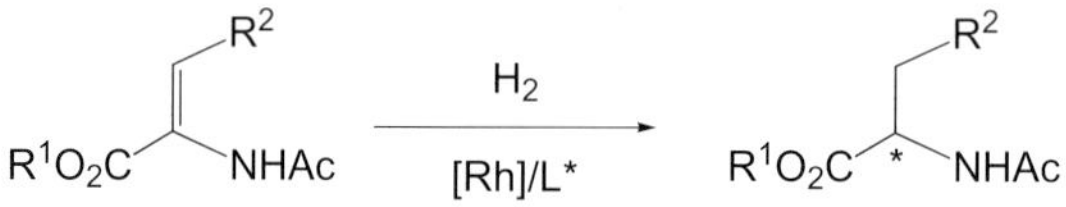

a $R^1 = H; R^2 = Ph$
b $R^1 = Me; R^2 = Ph$
c $R^1 = H; R^2 = H$
d $R^1 = Me; R^2 = H$

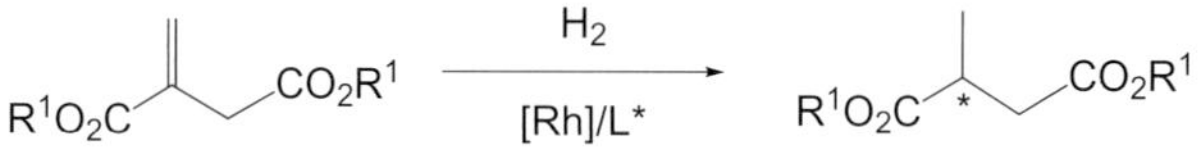

a $R^1 = H$
b $R^1 = Me$

Scheme 8.1 Hydrogenation of dehydroamino acids and itaconites.

8.2.2
Hydrogenation of C=N Bonds

Asymmetric hydrogenation of unfunctionalized olefins and C=N double bonds using chiral transition metal complexes as catalysts constitutes one of the most promising methods for the preparation of optically active hydrocarbons and amine building blocks (Scheme 8.2). Although significant progress has been made in the catalytic asymmetric reduction of unfunctionalized olefins and imines over recent years, these asymmetric hydrogenations remain challenging [4, 5]. Imine hydrogenation, for instance, has some serious drawbacks: coordination can take place through both the nitrogen atom and the double bond, and both the substrate and catalyst intermediates are unstable under catalytic conditions. Homogeneous catalysts can complex both the imine substrate and the amine product. Consequently, catalytic activity is often low. Contrary to the case of the asymmetric hydrogenation of functionalized substrates, iridium complexes are the best catalysts for unfunctionalized olefins and imines.

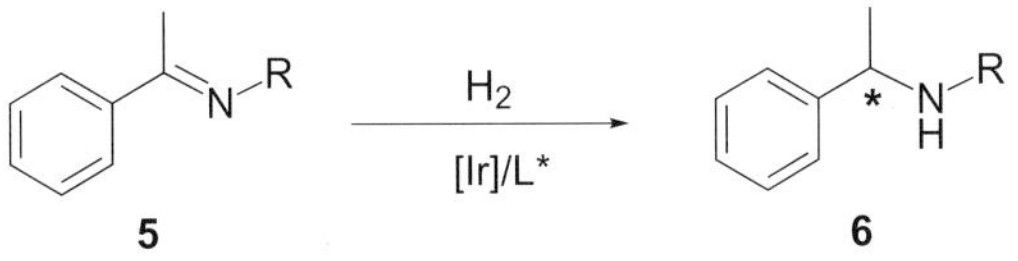

Scheme 8.2 Imine hydrogenation.

a R= Ph
b R= CH$_2$Ph

This chapter discusses the most important results and applications of metal transition catalysts containing carbohydrate-derived ligands in asymmetric hydrogenation of both C=C and C=N substrates. The following sections are organized according the nature of the phosphorous ligands.

8.3
P-Donor Ligands

In the early days, diphosphines were the most widely employed phosphorous ligands and have provided the best catalysts for asymmetric hydrogenation. However, they are generally difficult to synthesize and prone to oxidation. In recent years, chiral diphosphites, diphosphinites, and diphosphonites have emerged as new types of ligands. The most important advantage of these new ligands over diphosphines is their preparation, which has fewer steps and is easier. Another advantage is that they are less sensitive to air and other oxidizing agents than phosphines, although they are more prone to hydrolysis. The availability of chiral alcohols enables the preparation of a whole series of chiral ligands that can be screened in the search for high activity and selectivity. Diphosphite, diphosphinite, and diphosphonite ligands are not only readily available, they also have a modular nature that enables fine tuning of their steric and electronic properties. For these reasons, chiral diphosphite and diphosphinite ligands in particular are extremely attractive for catalysis [6].

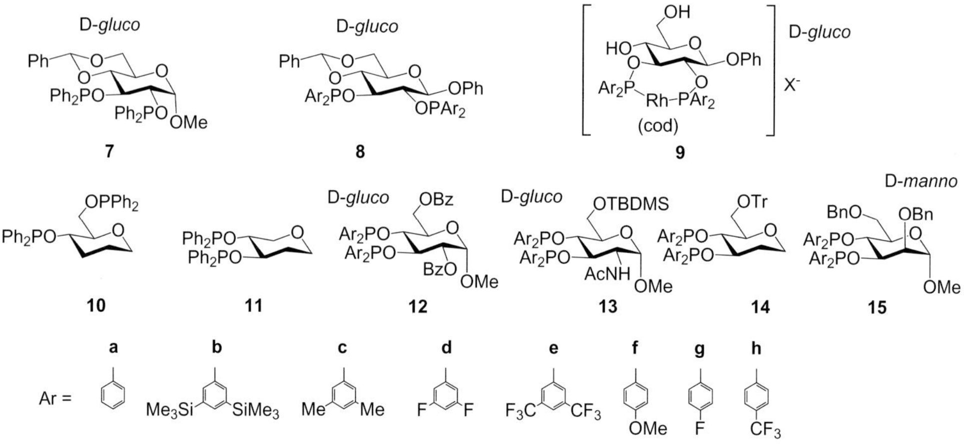

Figure 8.1 Diphosphinite ligands **7–15**.

8.3.1
Phosphinite Ligands

Together with the phosphines used by Sinou and Descotes [7] in the late 1970s, (Section 8.3.3) diphosphinites derived from carbohydrates were the first phosphorous ligands used in asymmetric hydrogenation. The first of this type, ligand diphosphinite **7**, was described by Cullen [8] and Thompson [9] (Figure 8.1). This ligand provided enantioselectivities up to 80% in the hydrogenation of α-acetamidoacrylic acids and their esters (entry 12, Table 8.1). Selke *et al.* systematically studied the effect of substituents at the anomeric position, and found that β configured ligands **8** increased the ee in the hydrogenation of acetamidocinnamic derivatives **1a** and **1b** in Scheme 8.1 to 96% and 91%, respectively (entries 1 and 5, Table 8.1) [10]. They also demonstrated the importance of the all-equatorial arrangement in vicinal diphosphinite hexapyranoside ligands for high enantioselectivities in rhodium-catalyzed asymmetric hydrogenation. Ligands **8**, derived from D-glucose, were easily obtained by reacting a dichloromethane solution of phenyl 4,6-O-benzylidene-β-D-glucopyranose with two equivalents of the corresponding chlorophosphine in the presence of pyridine as base (Scheme 8.3).

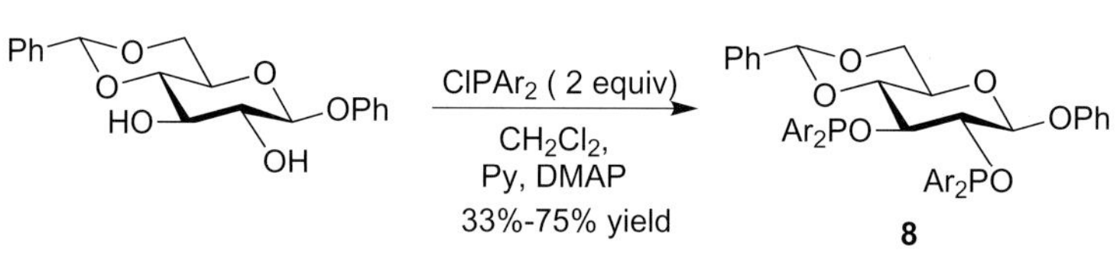

Scheme 8.3 Synthesis of pyranoside diphosphinite ligands **8**.

8.3 P-Donor Ligands

Table 8.1 Hydrogenation of unsaturated acids and esters with diphosphinite ligands.

$$R^1O_2C-C(NHAc)=CHR^2 \xrightarrow[\text{[Rh]/L*}]{H_2} R^1O_2C-C^*H(NHAc)-CH_2R^2$$

1 → 2

Entry	Product	Ligand	Conversion (%)	ee (%)
1	2a (R²=Ph, R¹=H; HO₂C–*C(NHAc)–CH₂Ph)	8a	100	96 (S)
2		8c	100	99 (S)
3		10	53	63 (R)
4		11	79	90 (S)
5		8a	100	91 (S)
6		8b	100	98 (S)
7	2b (R²=Ph, R¹=Me; MeO₂C–*C(NHAc)–CH₂Ph)	8c	100	94 (S)
8		9[a]	100	95 (S)
9		12a	100	93 (R)
10		13a	100	98 (R)
11		16b	100	88[b] (S)
12	2c (HO₂C–*C(NHAc)–CH(CH₃)₂)	7a	100	80 (S)
13	2d (MeO₂C–*C(NHAc)–CH(CH₃)₂)	17a	100	93 (S)
14		18a[c]	100	61 (S)
15		19a	100	83 (S)
16		20a[c]	100	49 (S)

a) Catalyst supported on silica.
b) 99.9% in the presence of sodium dodecyl sulfate.
c) Reaction in water.

Subsequently, RajanBabu improved the performance of diphosphinite **8** by modifying its electronic properties. Introducing different substituents in the P-aryl groups (**8b–h**) made these ligands excellent inductors of enantioselectivity. For instance, in the hydrogenation of several *(Z)*-acetamidocinnamic acids and esters the enantioselectivities were in the range 97–99% when 3,5-dimethylphenyl groups were present (entry 2, Table 8.1) [11]. The structure of the precatalyst and intermediates has been studied to rationalize the effect of electronic density from the ligand in the Rh-catalyzed asymmetric hydrogenation [11].

Complex **9** [12] (Ar = Ph) contains a ligand with a diphosphinite function at positions 2,3 and two free hydroxyl groups in positions 4 and 6 of a β-pyranoside. When this complex was supported on silica 95% ee was achieved in the hydrogenation of dehydroamino acid **1b** using methanol as solvent (entry 8, Table 8.1). In this reaction, the enantioselectivity depended on the nature of the solvent.

One limitation of ligands prepared from carbohydrates is that only one enantiomer is easily accessible while the other is expensive or even unavailable. Sunjic et al. [13], however, observed that diphosphinite **10**, derived from D-glucose, behaved as a pseudo-enantiomer of **11**, derived from D-xylose. Indeed, it provides the opposite enantiomer in the rhodium-catalyzed hydrogenation of (Z)-acetamidocinnamic acid (**1a**), although the ee was higher (90%, S) for the Rh/**11** than for the Rh/**10** catalytic system (63% ee, R) (entries 4 and 3, respectively, Table 8.1). In the same way, RajanBabu observed that pyranoside derivatives prepared from D-glucose and D-glucosamine containing phosphorus bonded to the 3,4-hydroxyl groups (**12–15**) could be considered as pseudo-enantiomers of the D-glucose derived ligands with phosphorus bonded to the 2,3-hydroxyl groups (**8**). In fact, in the hydrogenation of (Z)-α-acetamidocinnamic acids the Rh/**12–15** catalytic system provided the (R)-enantiomer, while the Rh/**8** catalytic system gave the (S)-enantiomer. Enantioselectivities were higher than 95% when these catalytic systems were used (entries 9 and 10, Table 8.1) [14]. Scheme 8.4 shows the steps involved in the synthesis of diphosphinite ligands **12** and **13**, derived from D-glucose and D-glucosamine, respectively.

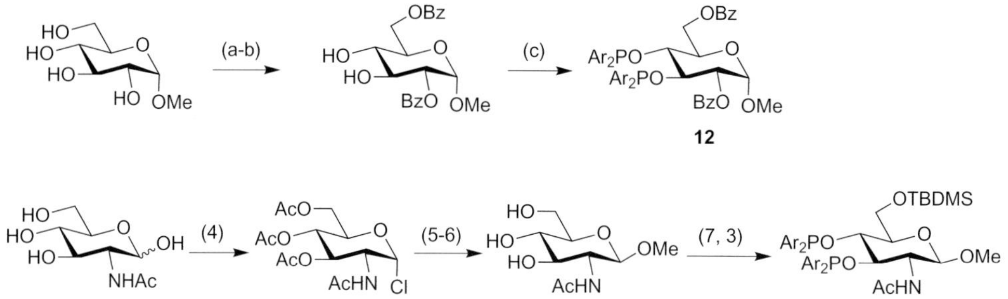

Scheme 8.4 Synthesis of diphosphinite ligands **12** and **13**: (a) (Bu$_3$Sn)$_2$O/toluene; (b) PhCOCl/toluene (65% combined yield of a and b); (c) Ar$_2$PCl/CH$_2$Cl$_2$/Py/DMAP (~91% yield); (4) MeCOCl/CH$_2$Cl$_2$ (58% yield); (5) Ag$_2$CO$_3$/CH$_2$Cl$_2$/MeOH/mol. sieves 4Å (78% yield); (6) MeOH/Biorad AGMP(OH$^-$) resin (95% yield); (7) TBDMSCl/DMF/imidazole (yield not reported).

Uemura prepared the rhodium complexes **16a,b**, which incorporated diphosphinite ligands derived from α,α-trehalose (**16a**) and β,β-trehalose (**16b**) (Figure 8.2) [15]. Complex **16b**, with a ligand derived from β,β-trehalose, is closely related to **9** (Figure 8.1). In **16b** the anomeric phenyl group is replaced by a glucose moiety, which increases the solubility in water. Complex **16b** provided an ee of 88% when methyl (Z)-α-acetamidocinnamate (**1b**) was hydrogenated in water, but the stereoselectivity increased to 99.9% in the presence of sodium dodecyl sulfate (entry 11, Table 8.1). When the reaction was performed in the biphasic system AcOEt/water, the ee was 87% and the catalyst could be recovered simply by phase separation at the end of the reaction. The ligands derived from β,β-trehalose gave better results than those derived from α,α-trehalose, in agreement with what was observed for α-configured ligand **7** and β-configured **8**.

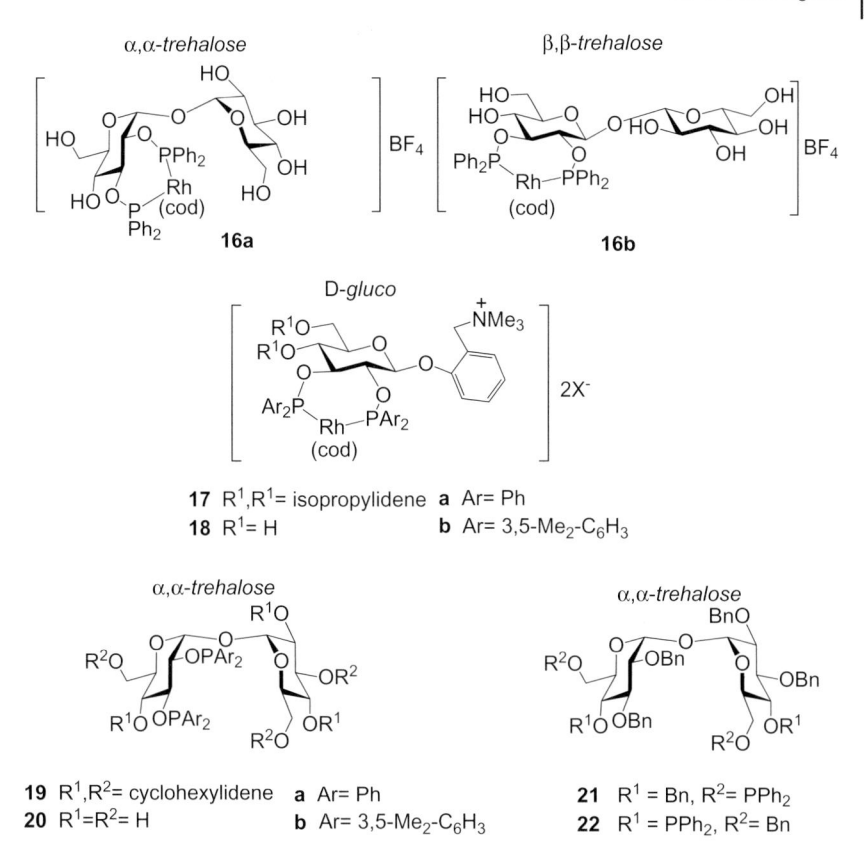

Figure 8.2 Rh-complexes **16–18** and trehalose diphosphinite ligands **19–22**.

Ligands in which the hydroxyl groups were unprotected were also used to perform catalytic hydrogenation in water [16]. RajanBabu has prepared a series of diphosphinites **19–22** derived from α,α-trehalose in which the phosphinite functions were at positions 2,3 of a sugar unit (**19, 20**) or at positions 6,6′ (**21**) or 4,4′ (**22**) of both monosaccharide units. The catalytic system Rh/**19** (R^1, R^2 = cyclohexylidene) provided an ee of 83% in the hydrogenation of methyl α-acetamidoacrylate (entry 15, Table 8.1), but it dropped to 49% when the reaction was performed in water using the catalytic system Rh/**20** ($R^1 = R^2 = H$). One possible explanation for this decrease in enantioselectivity is the intervention of protonolysis of the putative Rh–C bond before the final reductive elimination [17]. Later, complex **16b** was modified by replacing the second glucose unit with an ammonium substituent in the anomeric phenyl group of the ligand, leading to the preparation of rhodium complexes **17** and **18** [16]. Complex **17** was tested in the hydrogenation of dehydroamino acids and ees in organic solvents were slightly lower than those obtained for the corresponding complexes without the ammonium group (entry 13, Table 8.1). Although the solubility of this complex in water was sufficient to complete the hydrogenation reaction, it required long reaction times and the ees were very

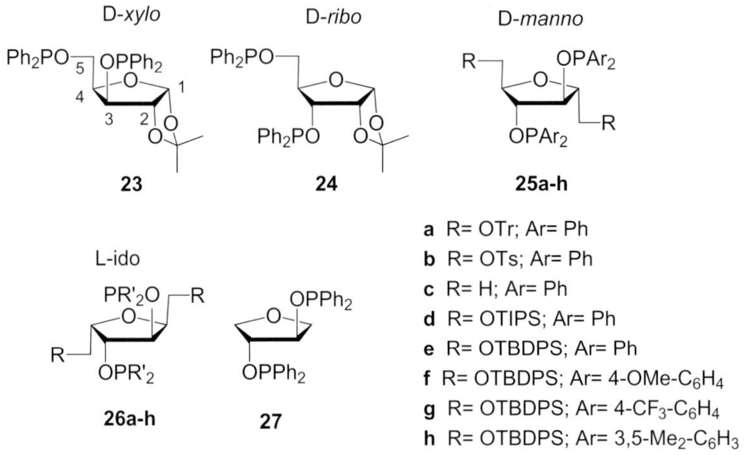

Figure 8.3 Diphosphinite ligands **23–27**.

low. Complex **18** was used in the hydrogenation of methyl (Z)-α-acetamidoacrylate (**1d**) in water as the solvent to give 61% ee. When the product was left in contact with water for long periods of time under the reaction conditions the enantioselectivity decreased rapidly. These complexes also showed similar trends to that observed for ligand **20**.

Diphosphinite derivatives with a furanoside backbone **23** and **24** (Figure 8.3) have been used in rhodium- and iridium-catalyzed asymmetric hydrogenation of prochiral substrates. Curiously, the enantiomeric excess in the hydrogenation of methyl α-acetamidoacrylate (**1d**) was 76% (R) with the Rh/**24** catalytic system and 78% (R) with the Ir/**23** catalytic system. The enantiomeric excess was dependent on both the absolute configuration of the C3 stereocenter of the carbohydrate backbone and on the nature of the metal precursor [18].

C_2-symmetric diphosphinites **25a–d** and **26a–d**, developed by Castillón et al., have been prepared from D-glucosamine and D-glucitol [19]. These ligands were used in the Rh-catalyzed hydrogenation of methyl acetamidocinnamate (**1b**), methyl acetamidoacrylate (**1d**), and dimethyl itaconate (**3b**) (Table 8.2). Catalytic systems containing ligand **25c** afforded the best results with an ee of 93% in the hydrogenation of methyl acetamidoacrylate (**1d**) (entry 5, Table 8.2). Ligand **27**, which does not contain substituents at positions 2 and 5 of the tetrahydrofuran ring, only gave an ee of 18% (entry 7, Table 8.2). This indicates that stereogenic centers that are not directly bonded to the coordinating atoms also have a strong influence on selectivity. Substituents in **25** and **26** also affect the stereoselectivity. The ees were lower for dimethyl itaconate (**3b**) than for the other substrates but unexpectedly **25a** and **26a** gave the product with opposite stereochemistry. The configuration of the major enantiomer seems to be determined by the configuration of substituents at carbon atoms 2 and 5 (entries 8 and 9, Table 8.2). Scheme 8.5 shows the synthesis of diphosphinite ligands **25** and **26** from D-glucosamine and D-glucitol, respectively.

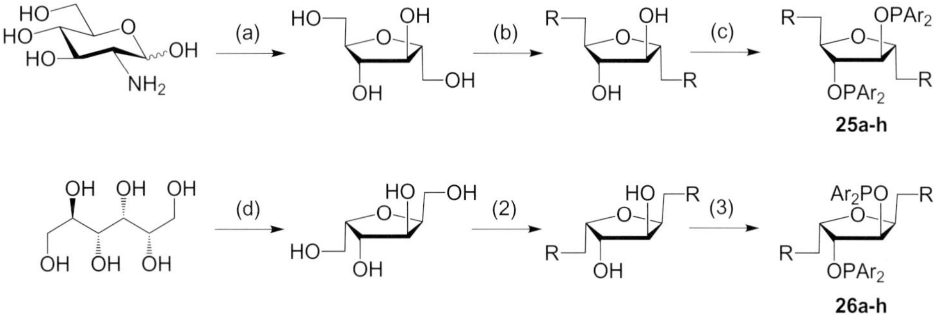

Scheme 8.5 Synthesis of ligands **25** and **26**: (a) (1) H$_2$O; (2) NaNO$_2$/AcOH, $T < 2\,°C$; (3) NaBH$_4$/H$_2$O (80% overall yield); (b) RCl/DMF/imidazole or RCl/CH$_2$Cl$_2$/Py (70–72% yield); (c) Ar$_2$PCl/CH$_2$Cl$_2$/NEt$_3$ (60–91% yield); (d) (1) xylene/methanesulfonic acid; (2) acetone/H$_2$SO$_4$; (3) THF/H$_2$O, Amberlite IR-120 (H$^+$) (3% overall yield).

Table 8.2 Hydrogenation of unsaturated esters and imines with diphosphinite ligands **25–27**.

Entry	Product	Ligand	Conversion (%)	ee (%)
1	**2b** (MeO$_2$C-*-NHAc, Ph)	25b	96	81 (R)
2	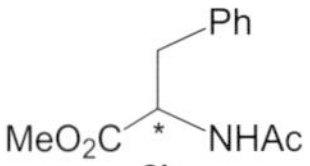	26b	100	59 (R)
3		27	100	27 (R)
4	**2d** (MeO$_2$C-*-NHAc, iPr)	25b	100	85[a] (R)
5	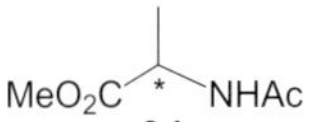	25c	100	87[b] (R)
6		26b	100	59 (R)
7		27	100	18 (R)
8	**4b** (MeO$_2$C-*-CO$_2$Me, iPr)	25a	100	48 (S)
9	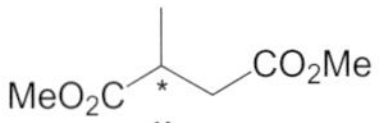	26a	100	53 (R)
10		26b	100	9 (R)
11		27	100	4 (R)
12	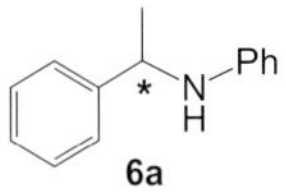 **6a** (PhCH(*)N(H)Ph)	25f	100	70 (+)
13	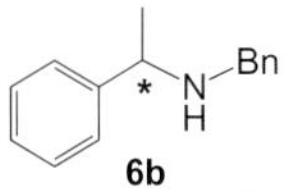 **6b** (PhCH(*)N(H)Bn)	25h	97	76 (−)

a) 91% ee at −25 °C.
b) 93% ee at −25 °C.

Phosphinite xylose derivatives (**23, 24**) (Figure 8.3), have been used as ligands in iridium-catalyzed hydrogenation of imines although they provide only moderate ee; N-(phenylethylidene)aniline **5a** and N-(phenylethylidene)benzylamine **5b** (Scheme 8.2) were used as model substrates. The important fact is that the enantioselectivity depends on the fine tuning of the structural parameters of the ligand. These catalytic systems were active at 50 bar of hydrogen and 25 °C. In the asymmetric hydrogenation of N-arylimines, results were poorly reproducible. However, in the hydrogenation of **5a** (Scheme 8.2) using the complex [Ir(COD)(**23**)]BF$_4$ as catalyst precursor, enantioselectivities up to 57% were achieved. The best enantioselectivity was obtained at 10 bar [20].

One of the advantages of the diphosphinite ligands is their modular nature, which allows different backbones as well as different substituents groups. Diphosphinites **25e–h** (Figure 8.3), modified with different electron-donating or electron-withdrawing groups on the aryl residue, have been used in combination with [Ir(COD)Cl]$_2$ or as cationic iridium complexes formed from [Ir(COD)$_2$]BF$_4$. Cationic iridium complexes gave rise to catalytic systems that were more active than the neutral iridium complexes, promoting the hydrogenation of N-(phenylethylidene)benzylamine **5b** with conversions of between 70% and 100%. The use of additives was, in general, detrimental to both the conversion and the enantioselectivity. The system based on **25f** (Ar = 4-OMe–C$_6$H$_4$), which is a stronger electron donor than **25g** (Ar = 4-CF$_3$–C$_6$H$_4$), gave lower conversion than the systems based on ligand **25g**. Concerning the enantioselectivity, however, the best results were with ligand **25h** (Ar = 3,5-Me$_2$–C$_6$H$_3$) (76% ee) in the hydrogenation of N-(phenylethylidene)aniline (**6a**), and in the hydrogenation of N-(phenylethylidene)benzylamine (**6b**) (entries 12 and 13, Table 8.2) the results were best with ligand **25f** (Ar = 4-OMe–C$_6$H$_4$), (70% ee) [21].

8.3.2
Phosphite Ligands

While diphosphinites derived from carbohydrates were successfully applied in metal-catalyzed asymmetric hydrogenation, the early application of carbohydrate phosphite ligands only provided low-to-moderate enantioselectivities. The use of diphosphite ligands in asymmetric catalysis was first reported by Brunner [22] and Wink [23]. These authors used diphosphite ligands derived from D-mannitol and tartaric acid in the Rh-catalyzed hydrogenation of enamides and obtained low enantioselectivities (1–34% ee).

In 1999, Selke *et al.* reported the use of a series of diphosphite ligands **28** with a glucopyranoside backbone (Figure 8.4) in the Rh-catalyzed hydrogenation of methyl *(Z)-N*-acetamidocinnamate (**1b**), showing rather low enantioselectivities (ees up to 13%) [24]. This contrasts with the excellent results obtained with the related diphosphinite ligands **8** (Figure 8.1).

Diphosphite ligands **29–34** derived from D-(+)-xylose and D-(+)-glucose with a furanoside backbone (Figure 8.4), developed by Claver *et al.*, have proved to be highly efficient for the Rh-catalyzed hydrogenation [25]. The new ligands were

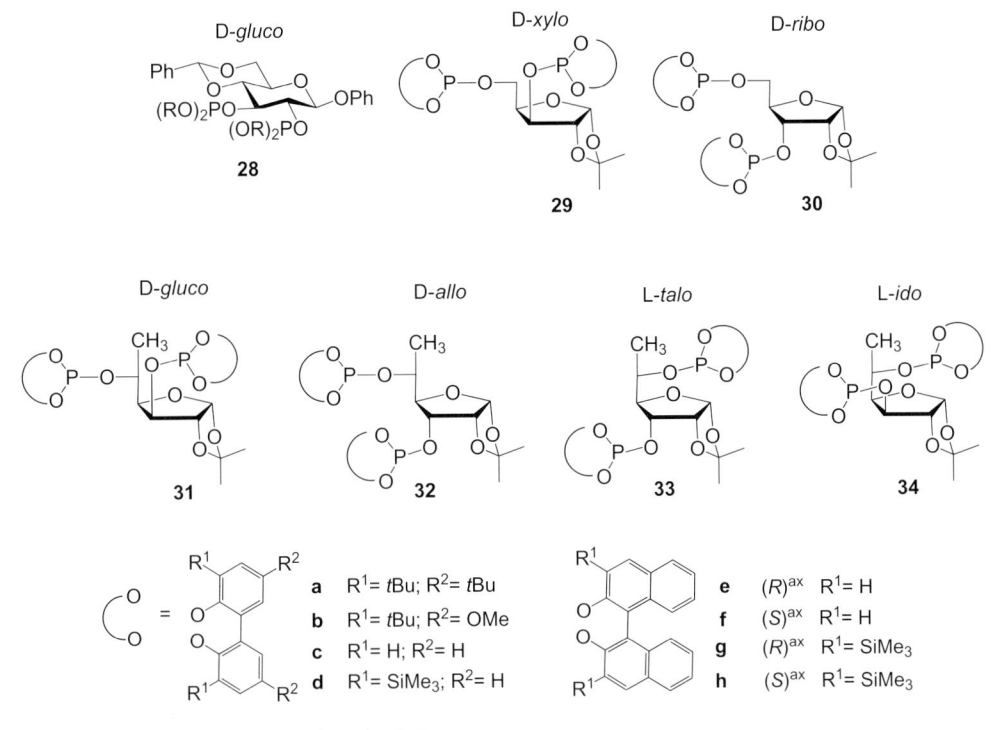

Figure 8.4 Diphosphite modular ligands **28–34**.

synthesized very efficiently in one step from the corresponding diols, which were easily prepared on a large scale from D-(+)-xylose and D-(+)-glucose using standard procedures (Scheme 8.6). Reaction of the corresponding diol with two equivalents of the desired *in-situ* formed phosphorochloridite in the presence of base afforded the required ligands as white air-stable solids in moderate-to-good overall yield (50–67%). The interesting feature of these ligands is their modular character, which allows sufficient flexibility to fine-tune (i) the different configurations of the carbohydrate backbone (C3 and C5) and (ii) the steric and electronic properties of the diphosphite substituents (**33a–h**). Excellent enantioselectivities (ee up to 99%) and activities were achieved in the Rh-catalyzed hydrogenation of dimethyl itaconate (Table 8.3). Systematic variation of stereocenters C3 and C5 at the ligand backbone showed that enantiomeric excesses depended strongly on the absolute configuration of C3 and slightly on that of the stereocenter carbon C5. Enantioselectivities were best with ligands **32** with a 6-deoxy-allo scaffold.

Varying the chirality at the axial chiral binaphthyl substituents in ligands **32** showed that the sense of the enantiodiscrimination is predominantly controlled by the configuration of the biaryls at the phosphite moieties (entries 6 and 7, Table 8.3). Bulky substituents at the *ortho*-positions of the biaryl diphosphite moieties have a positive effect on enantioselectivity. Enantiomeric excess was highest for

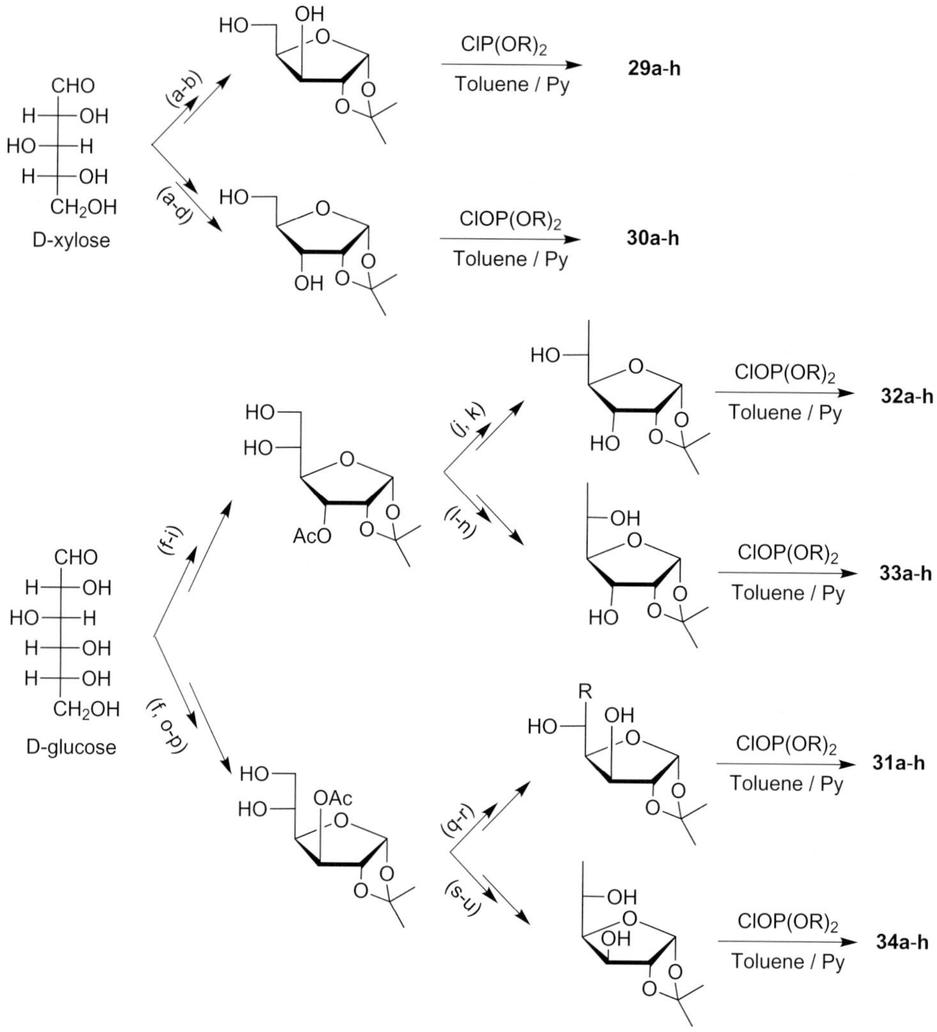

Scheme 8.6 Synthesis of furanoside diphosphite ligands **29–34**: (a) I₂/acetone (95% yield); (b) H₂SO₄/CH₃OH (90% yield); (c) BzCl/CH₂Cl₂/Py, −20 °C to room temp. (85% yield); (d) PCC/AcONa/CH₂Cl₂ then NaBH₄/EtOH/H₂O, −15 °C (70% yield); (e) NH₄OH/MeOH (70% yield); (f) I₂/acetone (98% yield); (g) PCC/AcONa/CH₂Cl₂ then NaBH₄/EtOH/H₂O, −15 °C (82% yield); (h) Ac₂O/Py/CH₂Cl₂, 0 °C to room temp. (91% yield); (i) AcOH/H₂O (88% yield); (j) TsCl/Py/CH₂Cl₂, −20 °C to room temp. (88%); (k) NaOMe/CH₂Cl₂ then LiAlH₄/THF, 60 °C (92% yield); (l) Ac₂O/Py/CH₂Cl₂, 0 °C to room temp. (90% yield); (m) Tf₂O/Py/CH₂Cl₂ (57% yield); (n) NaOMe/CH₂Cl₂ then LiAlH₄/THF, 60 °C (79% yield); (o) Ac₂O/Py/CH₂Cl₂, 0 °C to room temp. (87% yield); (p) AcOH/H₂O (84% yield); (q) TsCl/Py/CH₂Cl₂, −20 °C to room temp. (81%); (r) NaOMe/CH₂Cl₂ then LiAlH₄/THF, 60 °C (92% yield); (s) Ac₂O/Py/CH₂Cl₂, 0 °C to room temp. (92% yield); (t) Tf₂O/Py/CH₂Cl₂ (54% yield); (u) NaOMe/CH₂Cl₂ then LiAlH₄/THF, 60 °C (78% yield).

8.3 P-Donor Ligands

Table 8.3 Hydrogenation of dimethyl itaconate **3b** using diphosphite ligands **29–34**.[a]

$$\text{MeO}_2\text{C}\diagdown\!\!\!\diagup\text{CO}_2\text{Me} \quad \xrightarrow[\text{[Rh]/L*}]{\text{H}_2} \quad \text{MeO}_2\text{C}\diagdown\overset{*}{\diagup}\text{CO}_2\text{Me}$$
3b → **4b**

Entry	Ligand	Conversion (%)	ee (%)
1	29b	12	22 (R)
2	30b	28	64 (R)
3	31b	90	90 (R)
4	31c	82	85 (R)
5	31d	100	97 (R)
6	31e	50	50 (S)
7	31f	46	52 (R)
8	31g	100	90 (S)
9	31h	100	92 (R)
10	32b	100	2 (R)
11	33b	87	67 (R)
12	34b	73	29 (R)
13[b]	31d	100	99 (R)

a) Conditions: catalyst: 0.01 mmol, L/Rh = 1:1, substrate/Rh = 100, P_{H2} = 5 bar, T = 5 °C, t = 8 h.
b) T = 5 °C, P_{H2} = 30 bar, t = 4 h.

allofuranoside ligand **32d**, which has *o*-trimethylsilyl substituents on the biphenyl moieties (entries 5 and 13). It was also found that deoxygenation of position 6 significantly increased the activity (entries 3–12 versus 1 and 2). This set of ligands was also applied in the Rh-catalyzed hydrogenation of methyl N-acetylaminoacrylate (**1d**) and methyl (*Z*)-(*N*)-acetylaminocinnamate (**1b**). The results followed the same trend as those for dimethyl itaconate (**3b**), but the activities were somewhat higher [25c].

Diphosphite ligands **35–38** with C_2-symmetry and a tetrahydrofuran backbone, related to the diphosphinites **25** and **26**, have been synthesized starting from D-glucosamine and D-glucitol (Figure 8.5). Cationic rhodium complexes of **35–38** were prepared by reacting [Rh(COD)$_2$]BF$_4$ with the respective ligands. The structure of complexes with **35** and **37** were determined by single-crystal X-ray diffraction. These rhodium complexes were tested in the asymmetric hydrogenation of methyl acetamidoacrylate (**1d**). The conversions and enantioselectivities were much lower than the obtained with the corresponding diphosphinites **25** and **26**, and were mainly influenced by substitution in the biphenyl moiety and by the configuration of the remote centers in positions 2 and 5 of the tetrahydrofuran ring [26].

Although it has been generally accepted that bidentate ligands are the most appropriate for metal-catalyzed enantioselective hydrogenation, in recent years it

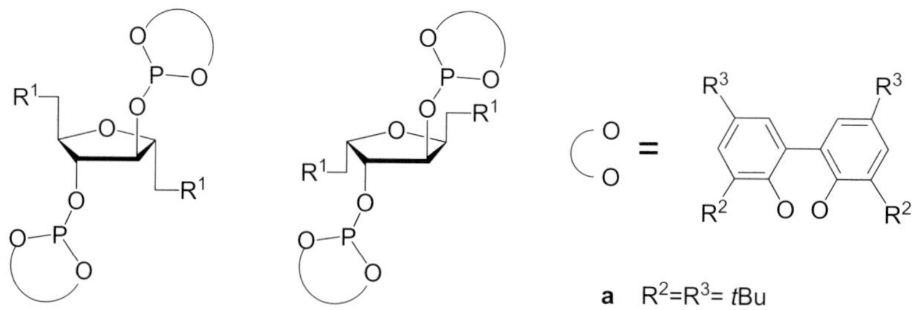

Figure 8.5 C_2-symmetric diphosphite ligands **35–38**.

has been shown that some monophosphorus ligands are very efficient for Rh-catalyzed asymmetric hydrogenation [27, 28].

Monophosphite ligands, often containing a binaphthol moiety, have been used for the Rh-catalyzed asymmetric hydrogenation of vinyl carboxylates and enamides [29, 30]. The results reported by Reetz and coworkers for the hydrogenation of vinyl carboxylates using ligands **39a,b** and **41–43a,b** show that there is a cooperative effect between the configuration of the binaphthyl moieties and the configuration of the sugar backbone (entries 1–4, Table 8.4). The results were best with phosphite **39b**, prepared from *(R)*-BINOL and a D-(+)-glucose derivative (up to 94% ee) [29].

Chen et al. [30, 31] prepared monophosphite ligands **39c,d** and **44–47** (Figure 8.6) derived from carbohydrates and BINOL *(R)*-octahydro-1,1'-binaphthalene-2,2'-dioxy moiety, *(R)*-H8-BINOL, by monoesterification of the corresponding carbohydrate-derived hydroxy-compound with PCl_3 with subsequent esterification with *(R)*-H8-BINOL. These ligands were tested in the rhodium-catalyzed asymmetric hydrogenation of dimethyl itaconate (**3b**) (entries 5–14, Table 8.4) and enamides (entries 15–17, Table 8.4), providing enantioselectivities of over 99% ee [31].

The hydrogenation results reported for ligands **39**, **41**, **44**, and **46** (Figure 8.6) indicate that the enantiomeric excess depends strongly on the configuration of carbon atom C3. Ligands **41** and **46** with an *(R)*-configuration produce much higher enantioselectivity than ligands **39** and **44** with opposite configuration. In this case, their results also suggest that there is a cooperative effect between the configuration of the binaphthyl moieties and the configuration of the carbohydrate backbone. The best enantioselectivities (>99% ee) were therefore obtained with ligands **41b** and **46b**.

Concerning the hydrogenation of C=N bonds, carbohydrate diphosphite derivatives **29a** (Figure 8.4) and **48a** (Figure 8.7) related to the previously described diphosphinite **23** (Figure 8.3) have been used as ligands for the hydrogenation of the N-(phenylethylidene)aniline (**5a**) and N-(phenylethylidene)benzylamine (**5b**) (Scheme 8.2) [20]. The complex [Ir(cod)$_2$]BF$_4$/**49a** selectively provided the

Table 8.4 Hydrogenation of hex-1-en-2-yl benzoate, dimethyl itaconate, and N-acetylphenylethenamine with monophosphite ligands **39–47**.

Entry	Product	Ligand	Conversion (%)	ee (%)
1	n-C₄H₉–*–OBz	39b	100	86[a] (S)
2		41b	100	39 (S)
3	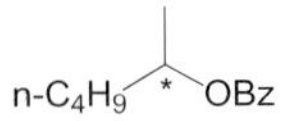	42b	100	68 (S)
4		43b	100	51 (S)
5	MeO₂C–*–CO₂Me **4b**	39a	100	99 (S)
6		39b	100	93 (R)
7		40a	100	97 (S)
8		41a	100	78 (S)
9		44a	100	83 (S)
10	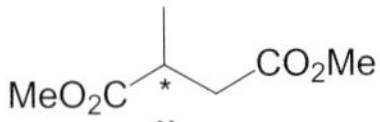	45a	100	92 (S)
11		46a	100	99 (S)
12		46b	100	>99 (R)
13		46d	100	>99 (R)
14		47b	100	99 (S)
15	Ph–*–NHAc	39b	100	94 (S)
16	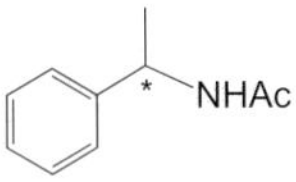	47b	100	95 (S)
17		47d	100	99 (S)

a) 94% (S) using hex-1-en-2-yl furan-2-carboxylate.

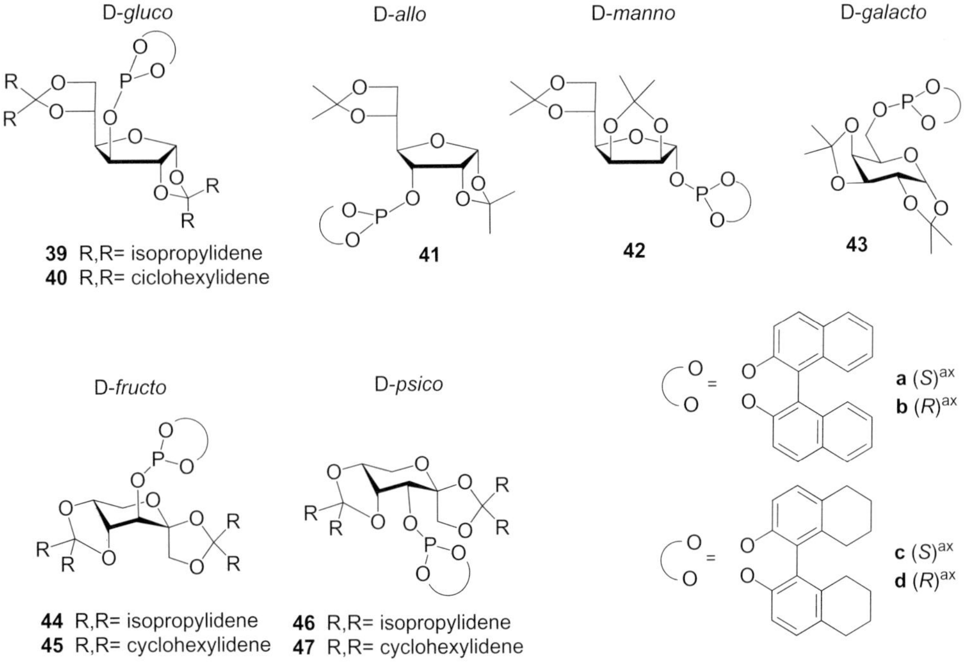

D-gluco

39 R,R= isopropylidene
40 R,R= ciclohexylidene

D-allo

41

D-manno

42

D-galacto

43

D-fructo

44 R,R= isopropylidene
45 R,R= cyclohexylidene

D-psico

46 R,R= isopropylidene
47 R,R= cyclohexylidene

a (S)^{ax}
b (R)^{ax}

c (S)^{ax}
d (R)^{ax}

Figure 8.6 Monophosphite ligands **39–47**.

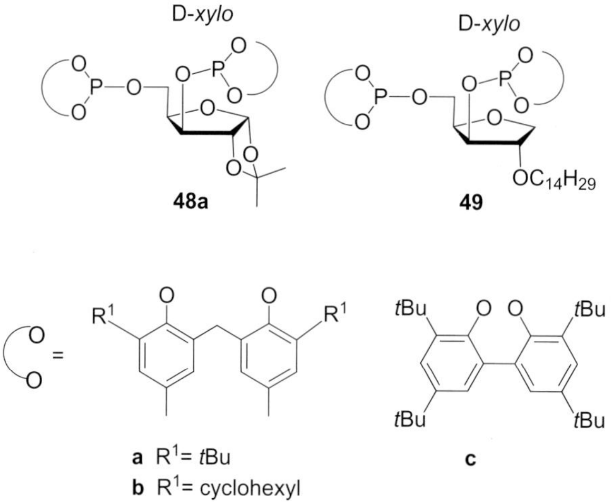

Figure 8.7 Diphosphite ligands **48** and **49**.

corresponding amine but did not achieve any enantioselectivity. However, when Bu$_4$NI was used as additive the ee increased to 46%. Iodine also improved the enantiomeric excess, but not as much as Bu$_4$NI. Other additives such as phthalimide and benzylamine resulted in deactivation of the catalytic system. The presence of bulky *tert*-butyl groups in the *ortho*-positions of the biphenyl moiety has an extremely positive effect on enantioselectivity, as had been previously found in the hydrogenation of dehydroamino acids derivatives [20, 25]. These moderate results of enantioselectivity and the sensibility of the diphosphite ligands to hydrolysis could explain the limited use of diphosphites in the hydrogenation of imines.

A more recent application of diphosphites derived from a carbohydrate is the stabilization of metal nanoparticles [32]. Palladium, rhodium, and ruthenium, nanoparticles have been prepared in the presence of substoichiometric amounts of carbohydrate-derived diphosphites through reduction of organometallic complexes under H$_2$ pressure. Again, the modular character of the carbohydrate diphosphite derivatives offers the possibility of tuning the environment at the surface of the nanoparticles through electronic or steric modifications. These variations can influence the catalytic properties when the metal nanoparticles are active in catalysis. Ruthenium nanoparticles synthesized from [Ru(COD)(COT)] under 3 bar of H$_2$, in the presence of xylose-derived diphosphite **29a** (Figure 8.4), **48a**, and **49** (Figure 8.7) (Ru/L = 1:0.1), are active in arene hydrogenation. Isolated as black powders, TEM analysis reveals the presence of small particles 1–4 nm in size, which have been tested in the hydrogenation of 3-methylanisole (**50**) and 2-methylanisole (**51**) (Scheme 8.7).

Interestingly, modification of the diphosphites allows control of the size and dispersity of the nanoparticles. Thus, modification of the diol moiety and, even more significantly, the introduction of a long lipophilic chain stabilize smaller and

(a)

Scheme 8.7 Hydrogenation of anisole derivatives.

better dispersed nanoparticles that are more soluble in organic media and more actives in hydrogenation. Moreover, the Ru/**49a** nanocatalyst provides high activity and diastereoselectivity (cis/trans ratio), although no significant enantioselectivity has been observed [33].

8.3.3
Phosphine Ligands Including Mixed Donor Ligands

The first chiral ligands derived from carbohydrates used in asymmetric hydrogenation were diphosphines. In the late 1970s Sinou and Descotes [7] synthesized the monophosphines **52** and **53** (Figure 8.8) from D-xylose and D-glucose. They were all tested in the asymmetric rhodium-catalyzed hydrogenation of α-acetamidocinnamic and α-acetamidoacrylic acids. Starting from D-xylose, Brunner [22] prepared hydroxyl/phosphine **54** and the phosphine/phosphinite ligand **55**.

A new series of furanoside backbone diphosphines, **56–58**, have been developed and used for the Rh asymmetric hydrogenation of dehydroamino acid and itaconic acid derivatives (Figure 8.9) [34].

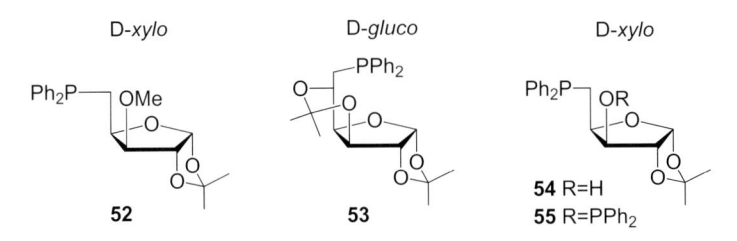

Figure 8.8 Phosphines **52–55**.

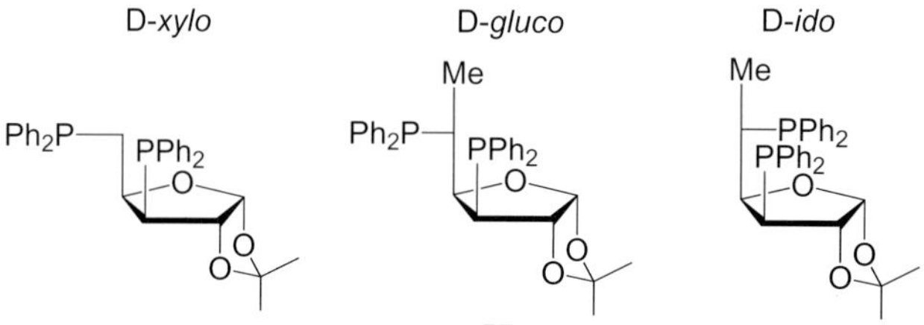

Figure 8.9 Diphosphines **56–58**.

Table 8.5 Hydrogenation of unsaturated acids and esters with diphosphine ligands **56–58**.

Entry	Product	Ligand	Conversion (%)	ee (%)
1	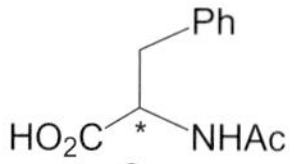 **2a** HO$_2$C–*–NHAc (Ph)	56	100	89 (S)
2	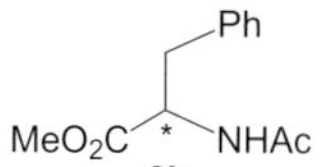 **2b** MeO$_2$C–*–NHAc (Ph)	57	100	98 (S)
3		58	100	53 (S)
4	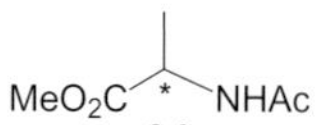 **2d** MeO$_2$C–*–NHAc	56	100	92 (S)
5		57	100	98 (S)
6		58	100	53 (S)
7	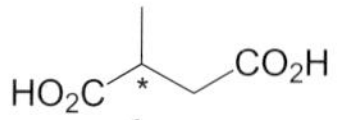 **4a** HO$_2$C–*–CO$_2$H	56	100	62 (R)

The difference between ligands **57** and **58** on one hand and ligand **56** on the other hand is the presence of a new stereogenic center introduced at C5 in **57** and **58**. The results indicate that the methyl substituent at C5 significantly increased activity (TOFs were approximately double for ligands **57** and **58**). The configuration at C5 also influences strongly the enantioselectivity. The catalytic system Rh/**57**, with a (R) configuration at C5, proved to be the most efficient, providing an ee of 98% in the hydrogenation of methyl α-acetamidocinnamate (**1b**), and acetamidoacrylate (**1d**) (entries 2 and 5, Table 8.5), while the system Rh/**58** only gave 53% ee (entries 3 and 6, Table 8.5).

More recently, phosphorus functionalities have been incorporated into cyclodextrins (ligand **59**, Figure 8.10) to take advantage of the properties of cyclodextrins as water-soluble chiral supports. Ligand **59** [35] contains phosphine in two of the

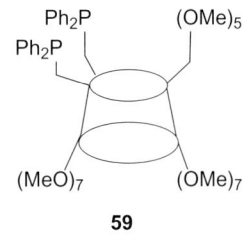

Figure 8.10 Cyclodextrin containing diphosphines (**59**).

positions 6 of a β-cyclodextrin. The Rh/**59** catalytic system has been tested in the hydrogenation of substrates **1a–d** and **3a,b** (Scheme 8.1), affording up to 92% ee, but only organic solvents were used for these reactions [35].

Several types of mixed carbohydrate ligands have been developed for application in asymmetric hydrogenation catalysis. In particular, phosphine–phosphite and phosphite–phosphoroamidite ligands have produced excellent results [3a].

Phosphine–phosphites **60a–d** derived from xylose were used as ligands in the Rh-catalyzed asymmetric hydrogenation of several α,β-unsaturated carboxylic acid derivatives (up to >99% ee) under very mild reaction conditions (Figure 8.11) [36]. Scheme 8.8 shows the synthesis of ligands **60**. The variation of the biphenyl substituents in the phosphite moiety greatly affected the enantioselectivity. The best enantioselectivity was obtained using ligand **60b**, which contains bulky *tert*-butyl groups in the *ortho* and *para* positions of the biphenyl moiety. The results also indicate that the sense of stereoinduction is mainly controlled by the configuration of the axially chiral phosphite moiety. Both enantiomers can therefore be obtained with high enantioselectivities. Notably, these phosphite–phosphine ligands showed higher degrees of enantioselectivity and higher reaction rates than their corresponding diphosphine (**56**) and diphosphite (**29**) analogues under the same reaction conditions. ^{31}P-{^{1}H} NMR and kinetic studies on intermediates of the catalytic cycle show that the [Rh(P-P′)(enamide)]BF$_4$ (P-P′ = phosphite–phosphine) species is the resting state and that the rate dependence is first order in rhodium and hydrogen pressure and zero order in enamide concentration [37].

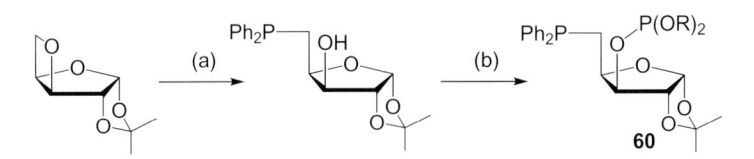

Scheme 8.8 Synthesis ligands **60**: (a) KPPh$_2$/DMF (80% yield); (b) (OR)$_2$PCl/Py/toluene (45–71% yield).

Phosphinite–phosphite ligands **61e,f** modified with different substituents (Figure 8.11) have shown considerable activity and selectivity together with higher ee than the related diphosphite ligands. When the catalytic precursor [Ir(COD)$_2$]

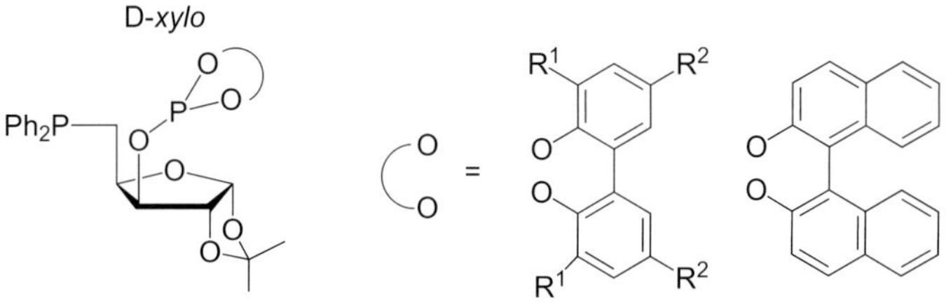

60		Substrate 1b % Conv	Substrate 1b % ee	Substrate 1d % Conv	Substrate 1d % ee
a	$R^1 = R^2 = H$	100	88.2 (*S*)	100	84.1 (*S*)
b	$R^1 = R^2 = t$Bu	100	>99 (*R*)	100	98.8 (*R*)
c	(*S*)^{ax}	100	98.3 (*S*)	100	91 (*S*)
d	(*R*)^{ax}	100	97.6 (*R*)	100	94.3 (*R*)

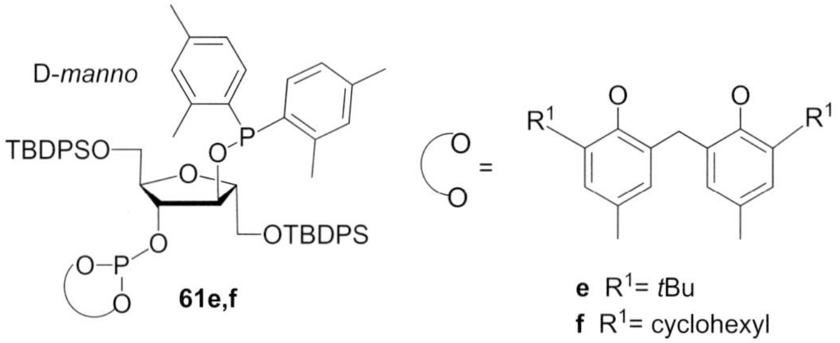

61e,f

e R^1= *t*Bu
f R^1= cyclohexyl

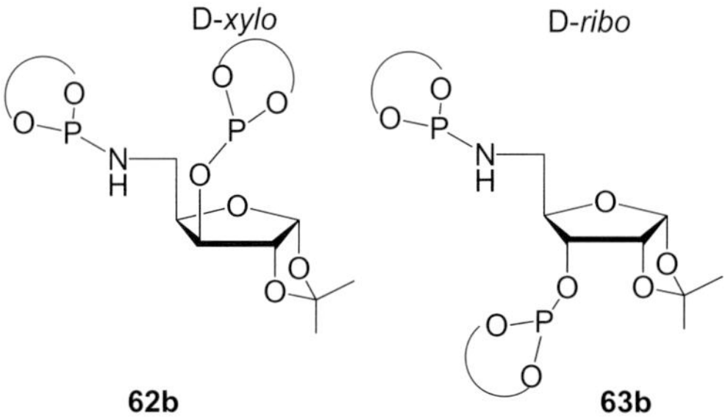

g R^1 = *t*Bu, R^2 = OMe

62b
62g

63b
63g

Figure 8.11 Phosphine–phosphite **60**, phosphinite–phosphite **61**, and phosphite–phosphoramidites **62** and **63**.

BF_4/**61e** was used in the hydrogenation of *N*-(phenylethylidene)benzylamine (**5b**) (Scheme 8.2) the ee was 76% [21].

Among the different mixed ligands, phosphite–phosphoroamidite ligands **62–63b,g** (Figure 8.11) – easily prepared in a few steps from commercial D-(+)-xylose – have been successfully applied in the asymmetric hydrogenation reaction. High activity and enantioselectivity up to >99% ee were obtained in asymmetric hydrogenation of a series of α,β unsaturated carboxylic acid derivatives. It is remarkable that these phosphite–phosphoroamidite ligands showed a much higher degree of enantioselectivity and higher reaction rates than their corresponding diphosphite analogues under similar reaction conditions [38].

8.4
P–N Donors

Although O–S, N,S and P,N ligands have been synthesized and used in other asymmetric reactions [3a], heterodonor ligands derived from carbohydrates have been scarcely used in asymmetric hydrogenation, with the exception of P,N ligands. The most important application of P,N ligands in asymmetric hydrogenation is in the asymmetric hydrogenation of C=N and C=C unfunctionalized substrates. In this context, Pfaltz and others have used phosphine–oxazoline ligands as chiral mimics of Crabtree's catalyst. They have been successfully used for the asymmetric hydrogenation of imines [39] and a limited range of alkenes [40]. Later, iridium complexes with chiral P,N ligands were found to be efficient catalysts for the hydrogenation of unfunctionalized olefins [41].

Carbohydrate phosphite–oxazolines developed by Diéguez et al. have been shown to be a new class of highly versatile ligands appropriate for asymmetric catalysis [42]. A phosphite–oxazoline ligand library 64–67 (Figure 8.12) can easily be prepared in a few steps from commercial D-glucosamine (Scheme 8.9) and applied in the iridium-catalyzed hydrogenation of unfunctionalized alkenes [43].

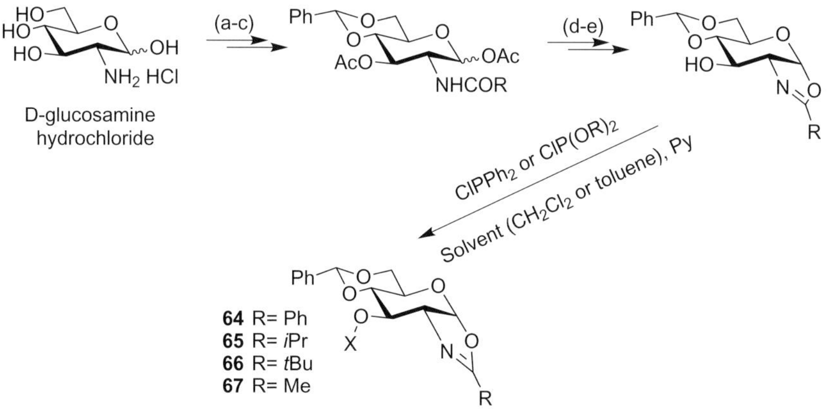

Scheme 8.9 Synthesis of pyranoside phosphite/phosphinite–oxazoline ligands 64–67: (a) (RCO)$_2$O/NaOMe/MeOH or RCOCl/NaHCO$_3$(aq.); (b) PhCHO/ZnCl$_2$; (c) Ac$_2$O/Py; (d) SnCl$_4$/CH$_2$Cl$_2$; (e) K$_2$CO$_3$/MeOH.

These ligands can easily be tuned in the oxazoline and biaryl phosphite moieties so that their effect on catalytic performance can be explored. By carefully selecting the ligand components, excellent activities and enantioselectivities (92–>99%) are

178 | 8 Hydrogenation Reactions

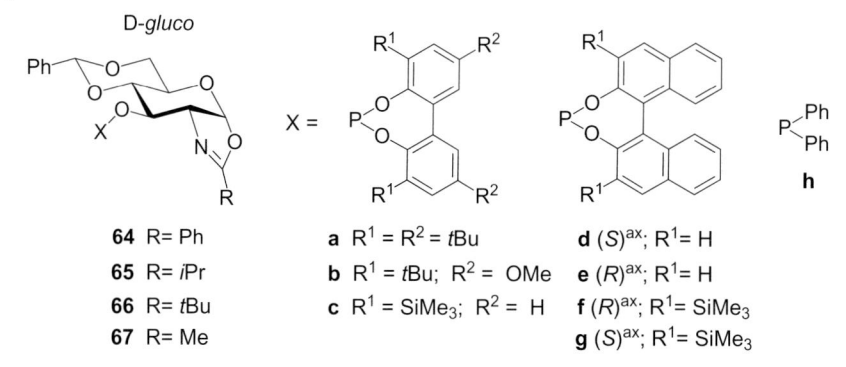

Figure 8.12 Phosphite/phosphinite–oxazoline ligands **64–67**.

64 R= Ph
65 R= iPr
66 R= tBu
67 R= Me

a $R^1 = R^2 = tBu$
b $R^1 = tBu$; $R^2 = OMe$
c $R^1 = SiMe_3$; $R^2 = H$

d $(S)^{ax}$; $R^1 = H$
e $(R)^{ax}$; $R^1 = H$
f $(R)^{ax}$; $R^1 = SiMe_3$
g $(S)^{ax}$; $R^1 = SiMe_3$

64c; 100% Conv, >99% (R)
64c; 100% Conv, >99% (R)
64c; 85% Conv, 99% (R)
64a; 100% Conv, 95% (S)

64a; 100% Conv, 92% (R)
64a; 100% Conv, 94% (R)
64c; 100% Conv, 99% (S)
64c; 100% Conv, 97% (S)

Figure 8.13 Summary of the best results for the hydrogenation of several minimally functionalized olefins using ligands **64–67**.

obtained at low catalyst loadings with simple disubstituted olefins [43] (Figure 8.13).

8.5
P–S Donors

Concerning the use of P–S carbohydrate derivatives in asymmetric hydrogenation, thioether–phosphinite ligands (P–SR; R = Ph, iPr, and Me), containing a furanoside backbone, **68–70** (Figure 8.14), provide rhodium and iridium catalysts for the

68 R = Ph
69 R = iPr
70 R = Me

Figure 8.14 Thioether–phosphinite ligands **68–70**.

asymmetric hydrogenation of α-acylaminoacrylates and itaconic acid derivatives (entries 1–3, 9–14, Table 8.6). High enantiomeric excesses (up to 96%) were obtained [44].

The enantiomeric excesses depend strongly on the steric properties of the substituent in the thioether moiety, the metal source, and the substrate structure. A bulky group in the thioether moiety along with the metal Rh had a positive effect on enantioselectivity. Results for α-acylaminoacrylates derivatives were satisfactory, while enantiomeric excesses for itaconic acid derivatives were low [44].

Phosphinite thioglycosides **71–74** (Figure 8.15) have been used in the rhodium-catalyzed asymmetric hydrogenation of enamides for the synthesis of proteinogenic and nonproteinogenic *(R)*-amino acids in high enantiomeric excesses (entries 4–8, Table 8.6) [45, 46]. The rhodium complexes employed in these processes are cationic [Rh(COD)(P,S)]SbF$_6$ complexes containing phosphinite thioglycosides ligands. As mentioned in Section 8.3.1, a drawback when using carbohydrates as chiral ligands in asymmetric catalysis is the difficulties in accessing an enantiomeric ligand and pseudo, therefore, the enantiomeric products. To solve this problem α-D-arabinose **74**, a cheap commercially available D-pentopyranose, has been used as a pseudo-mirror image of β-galactose.

Table 8.6 Hydrogenation of unsaturated esters with phosphinite–thioether ligands **68–74**.

Entry	Product	Ligand	Conversion (%)	ee (%)
1		68	100	68 (R)
2		69	100	93 (R)
3		70	100	45 (R)
4	MeO$_2$C–*–NHAc (Ph) **2b**	72	20	8 (S)
5		73a	100	94 (S)
6		73b	100	92 (S)
7		73c	100	92 (S)
8		74	100	94 (R)
9		68	100	68 (R)
10	MeO$_2$C–*–NHAc **2d**	69	100	93[a] (R)
11		70	100	54 (R)
12		68	100	57 (S)
13	MeO$_2$C–*–CO$_2$Me **4b**	69	100	64 (S)
14		70	100	35 (S)

a) 96% ee at 0 °C.

D-galacto   D-galacto   D-arabino

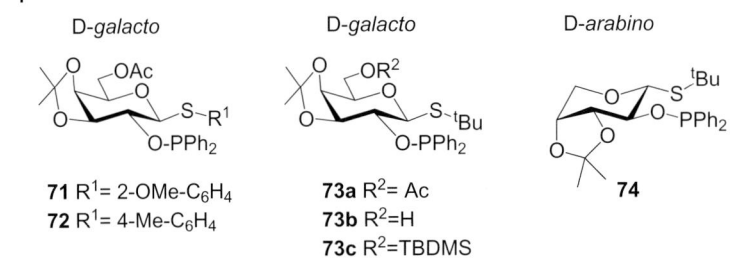

71 R¹= 2-OMe-C$_6$H$_4$
72 R¹= 4-Me-C$_6$H$_4$

73a R²= Ac
73b R²=H
73c R²=TBDMS

74

Figure 8.15 Thioether–phosphinite ligands **71–74**.

The 2-phosphinite *tert*-butyl-thioarabinoside **74** has been synthesized in four high-yielding steps from arabinose tetraacetate. The cationic Rh(I) complex in the hydrogenation of methyl cinnamate afforded the *(R)*-isomer in 94% ee. Thus, even though belonging to the D-series, both ligands behave as pseudo enantiomers [45]. Structural modification through preparation of compounds derived from 2-diphenylphosphinite-3,4-*O*-isopropyliden thiogalactosides **71–74** allows optimization of the corresponding rhodium catalysts, and both enantiomers of important proteogenic and non-proteogenic *(R)*-amino acid derivatives such as D- and L-DOPA can be prepared in quantitative yields with enantioselectivities of 97% and 98%, respectively [46].

Acknowledgments

We would like to thank the Spanish Government for providing grants Consolider Ingenio Intecat-CSD2006-0003, CTQ2010-14938/BQU, CTQ2008-1569/BQU, CTQ2010-15835/BQU, and 2008PGIR/07 to O. Pàmies and 2008PGIR/08 to M. Diéguez, the Catalan Government for grant 2009SGR116, and the ICREA Foundation for providing M. Diéguez and O. Pàmies with financial support through the ICREA Academia awards.

References

1 (a) Blaser, H.U. and Schmidt, E. (2004) *Asymmetric Catalysis on Industrial Scale: Challenges, Approaches and Solutions*, Wiley-VCH Verlag GmbH, Weinheim;
(b) Jacobsen, E.N., Pfaltz, A., and Yamamoto, H. (eds) (1999) *Comprehensive Asymmetric Catalysis*, vol. 1, Springer, Berlin.
2 de Vries, J.G. and Elsevier, C.J. (eds) (2007) *Handbook of Homogeneous Hydrogenation*, Wiley-VCH Verlag GmbH, Weinheim.

3 (a) Diéguez, M., Pàmies, O., and Claver, C. (2004) *Chem. Rev.*, **104**, 3189;
(b) Castillón, S., Claver, C., and Diez, Y. (2005) *Chem. Soc, Rev.*, **34**, 702;
(c) Diéguez, M., Pàmies, O., Ruiz, A., Díaz, Y., Castillón, S., and Claver, C. (2004) *Coord. Chem. Rev.*, **248**, 2165;
(d) Diéguez, M., Pàmies, O., Ruiz, A., and Claver, C. (2004) *Methodologies in Asymmetric Catalysis*, ch. 11, American Chemical Society, Washington, DC;
(e) Pàmies, O., Diéguez, M., Ruiz, A.,

and Claver, C. (2004) *Chim. Oggi*, **22**, 12; (f) Boysen, M.M.K. (2007) *Chem. Eur. J.*, **13**, 8648; (g) Benessere, V., Del Litto, R., De Roma, A., and Ruffo, F. (2010) *Coord. Chem. Rev.*, **254**, 390; (h) Woodward, S., Diéguez, M., and Pàmies, O. (2010) *Coord. Chem. Rev.*, **254**, 2007.

4 (a) Blaser, H.-U. (2002) *Adv. Synth. Catal.*, **344**, 17; (b) Roseblade, S.J. and Pfaltz, A. (2007) *Acc. Chem. Res.*, **40**, 1402; (c) Zhou, Y.-G. (2007) *Acc. Chem. Res.*, **40**, 1357.

5 (a) Blaser, H.-U. and Spinder, F. (2007) *Handbook of Homogeneous Hydrogenation* (eds J.G. de Vries and C.J. Elsevier), Wiley-VCH Verlag GmbH, Weinheim, p. 1193; (b) Claver, C. and Fernandez, E. (2008) *Modern Reduction Methods* (eds P. Anderson and I. Munslow), Wiley-VCH Verlag GmbH, Weinheim, p. 237.

6 Börner, A. (ed.) (2008) *Phosphorous Ligands in Asymmetric Catalysis*, ch. 3, Wiley-VCH Verlag GmbH, Weinheim.

7 Lafont, D., Sinou, D., and Descotes, G. (1979) *J. Organomet. Chem.*, **169**, 87.

8 Cullen, W.R. and Sugi, Y. (1978) *Tetrahedron Lett.*, **19**, 1635.

9 Jackson, R. and Thompson, D.J. (1978) *J. Organomet. Chem.*, **159**, C29.

10 Selke, R., Schwarze, M., Baudisch, H., Grassert, I., Michalik, M., Oehme, G., Stoll, N., and Costisella, B. (1993) *J. Mol. Catal.*, **84**, 223.

11 RajanBabu, T.V., Radetich, B., You, K.K., Ayers, T.A., Casalnuovo, A.L., and Calabrese, J.C. (1999) *J. Org. Chem.*, **64**, 3429.

12 Selke, R. and Capka, M. (1990) *J. Mol. Catal.*, **63**, 319.

13 Habus, I., Raza, Z., and Sunjic, V. (1987) *J. Mol. Catal.*, **42**, 173.

14 RajanBabu, T.V., Ayers, T.-A., Halliday, G.A., You, K.K., and Calabrese, J.C. (1997) *J. Org. Chem.*, **62**, 6012.

15 Yonehara, K., Ohe, K., and Uemura, S. (1999) *J. Org. Chem.*, **64**, 9381.

16 Yan, Y.Y. and RajanBabu, T.V. (2001) *J. Org. Chem.*, **66**, 3277.

17 Shin, S. and RajanBabu, T.V. (1999) *Org. Lett.*, **1**, 1229.

18 Guimet, E., Dieguez, M., Ruiz, A., and Claver, C. (2004) *Tetrahedron: Asymmetry*, **15**, 2247.

19 Aghmiz, M., Aghmiz, A., Díaz, Y., Masdeu-Bultó, A., Claver, C., and Castillón, S. (2004) *J. Org. Chem.*, **69**, 7502.

20 Guiu, E., Muñoz, B., Castillón, S., and Claver, C. (2003) *Adv. Synth. Catal.*, **345**, 169.

21 Guiu, E., Aghmiz, M., Díaz, Y., Claver, C., Meseguer, B., Militzer, C., and Castillón, S. (2006) *Eur. J. Org. Chem.*, 627.

22 Brunner, H. and Pieroncyk, W. (1980), *J. Chem. Soc. Res. (S)*, 74.

23 Wink, D.J., Kwok, T.J., and Yee, A. (1990) *Inorg. Chem.*, **29**, 5006.

24 Kadyrov, R., Heller, D., and Selke, R. (1999) *Tetrahedron: Asymmetry*, **9**, 329.

25 (a) Pàmies, O., Net, G., Ruiz, A., and Claver, C. (2000) *Eur. J. Inorg. Chem.*, 1287; (b) Pàmies, O., Net, G., Ruiz, A., and Claver, C. (2000) *Tetrahedron: Asymmetry*, **11**, 1097; (c) Diéguez, M., Ruiz, A., and Claver, C. (2002) *J. Org. Chem.*, **67**, 3796; (d) Diéguez, M., Ruiz, A., and Claver, C. (2003) *Dalton Trans.*, 2957.

26 Axet, M.R., Benet-Buchholz, J., Claver, C., and Castillón, S. (2007) *Adv. Synth. Catal.*, **349**, 1983.

27 (a) Peña, D., Minnaard, A., de Vries, J.G., and Feringa, B.L.J. (2002) *J. Am. Chem. Soc.*, **124**, 14552; (b) Reetz, M.T., Sell, T., Meiswinkel, A., and Mehler, G. (2003) *Angew Chem. Int. Ed.*, **42**, 790.

28 (a) Claver, C., Fernandez, E., Gillon, A., Heslop, K., Hyett, D.J., Martorell, A., Orpen, A.G., and Pringle, P.G. (2000) *Chem. Commun.*, 961; (b) Reetz, M.T., Mehler, G., Meiswinkel, A., and Sell, T. (2002) *Tetrahedron Lett.*, **43**, 7941.

29 Reetz, M.T., Goosen, L.J., Meiswinkel, A., Paetzol, J., and Jensen, J.F. (2003) *Org. Lett.*, **5**, 3099.

30 Huang, H., Zheng, Z., Luo, H., Bai, C., Hu, X., and Chen, H. (2003) *Org. Lett.*, **5**, 4137.

31 Huang, H., Liu, X., Chen, H., and Zheng, Z. (2005) *Tetrahedron: Asymmetry*, **16**, 693.

32 Favier, I., Gómez, M., Muller, G., Axet, M.R., Castillón, S., Claver, C., Jansat, S., Chaudret, B., and Philippot, K. (2007) *Adv. Synth. Catal.*, **349**, 2459 (and references therein).

33 Gual, A., Axet, M.R., Philippot, K., Chaudret, B., Denicourt-Nowicki, B.A., Roucoux, A., Castillon, S., and Claver, C. (2008) *Chem. Commun*, 2759.

34 (a) Pàmies, O., Net, G., Ruiz, A., and Claver, C. (2000) *Eur. J. Inorg. Chem.*, 2011; (b) Diéguez, M., Pàmies, O., Ruiz, A., Castillón, S., and Claver, C. (2000) *Tetrahedron: Asymmetry*, **11**, 4701.

35 Wong, Y.T., Yang, C., Ying, K.-C., and Jia, G. (2002) *Organometallics*, **21**, 1782.

36 (a) Pàmies, O., Diéguez, M., Net, M.G., Ruiz, A., and Claver, C. (2000) *Chem. Commun.*, 2383; (b) Pàmies, O., Diéguez, M., Net, M.G., Ruiz, A., and Claver, C. (2001) *J. Org. Chem.*, **66**, 8364.

37 Similar kinetics have been observed in the Rh-catalyzed hydrogenation using other phosphine–phosphite ligands. See, for example: Deerenberg, S., Pàmies, O., Diéguez, M., Claver, C., Kamer, P.C.J. and van Leeuwen, P.W.N.M. (2001) *J. Org. Chem.*, **66**, 7626.

38 Diéguez, M., Ruiz, A., and Claver, C. (2001) *Chem. Commun.*, 2702.

39 Helmchen, G. and Pfaltz, A. (2000) *Acc. Chem. Res.*, **33**, 336.

40 (a) Liu, D., Tang, W., and Zhang, X. (2004) *Org. Lett.*, **6**, 513; (b) Drury, W.J., III, Zimmermann, N., Keenan, M., Hayashi, M., Kaiser, S., Goddard, R., and Pfaltz, A. (2004) *Angew. Chem. Int. Ed.*, **43**, 70; (c) Cheruku, P., Gohil, S., and Andersson, P.G. (2007) *Org. Lett.*, **9**, 1659.

41 (a) Cui, X. and Burgess, K. (2005) *Chem. Rev.*, **105**, 3272; (b) Roseblade, S.J. and Pfaltz, A. (2007) *Acc. Chem. Res.*, **40**, 1402; (c) Church, T.L. and Andersson, P.G. (2008) *Coord. Chem. Rev.*, **252**, 513.

42 (a) Mata, Y., Dieguez, M., Pamies, O., and Claver, C. (2005) *Adv. Synth. Catal.*, **347**, 1943; (b) Mata, Y., Dieguez, M., and Pamies, O. (2007) *Chem. Eur. J.*, **13**, 3296; (c) Mata, Y., Diéguez, M., Pamies, O., and Claver, C. (2005) *Org. Lett.*, **7**, 5597.

43 Diéguez, M., Mazuela, J., Pàmies, O., Verendel, J.J., and Andersson, P.G. (2008) *J. Am. Chem. Soc.*, **130**, 7208.

44 Guimet, E., Diéguez, M., Ruiz, A., and Claver, C. (2005) *Dalton Trans.*, 2557.

45 Kira, N., Suarez, B., Stiller, M., Valdivia, V., and Fernández, I. (2005) *Phosphorus Sulfur Silicon*, **180**, 1253.

46 Khiar, N., Navas, R., Suárez, B., Alvarez, E., and Fernández, I. (2008) *Org. Lett.*, **10**, 3697.

9
Hydroformylations, Hydrovinylations, and Hydrocyanations

Mike M.K. Boysen

9.1
Hydroformylation Reactions

Asymmetric hydroformylation is a highly useful process as it creates chiral aldehydes from simple alkene precursors, gaseous hydrogen, and carbon monoxide as inexpensive feedstock [1]. The reaction is catalyzed by rhodium(I) complexes and leads to the desired, branched chiral product **2** but may also give rise to achiral unbranched aldehyde **3** as an undesired side product (Scheme 9.1). Therefore, a chiral catalyst has not only to provide a high level of stereoinduction but also to exert efficient control over the regioselectivity of this transformation [1, 2].

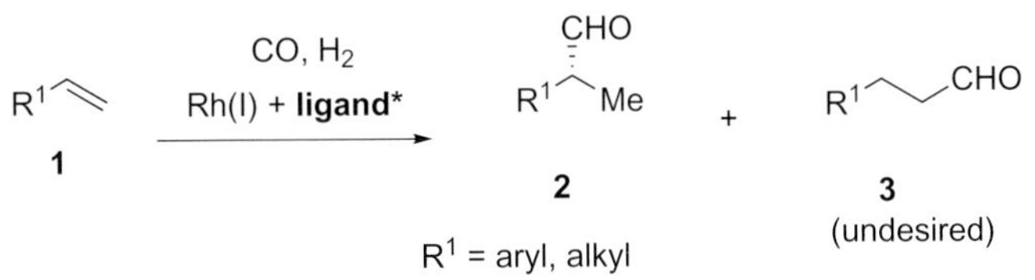

Scheme 9.1 Chiral branched aldehydes and achiral unbranched aldehydes from rhodium-catalyzed asymmetric hydroformylation reaction of alkenes.

The most common ligands for rhodium(I)-catalyzed hydroformylation reactions are phosphorus based: the ligand BINAPHOS, which was introduced by Takaya and Nozaki [3], contains one phosphite and one phosphine as donor sites and was the first ligand to offer high levels of stereoinduction. With regard to carbohydrate-based ligands for hydroformylation, the vast majority of these feature either phosphite or phosphinite donor sites, as they can be easily installed in a carbohydrate framework by reacting carbohydrate hydroxy groups with an appropriate chlorophosphite or phosphine chloride, respectively. Apart from these ligands, a few carbohydrate phosphines and carbohydrate-derived P-S-donor ligands have been reported.

Carbohydrates – Tools for Stereoselective Synthesis, First Edition. Edited by Mike Martin Kwabena Boysen.

9.1.1
Diphosphite Ligands

Bidentate phosphites, which were the first carbohydrate-derived ligands applied in asymmetric hydroformylation, are also the most important and most successful group of sugar-based tools for this reaction. In 1995 the group of van Leeuwen introduced bidentate ligands from simple monosaccharides such as D-xylose, D-mannose, D-glucose, and D-galactose with phosphite donor sites containing a biphenyl motif (Scheme 9.2, 4–9) [4]. Further, ligands with mixed phosphite donor sites, one with a biphenol group and another derived from substituted phenols were prepared.

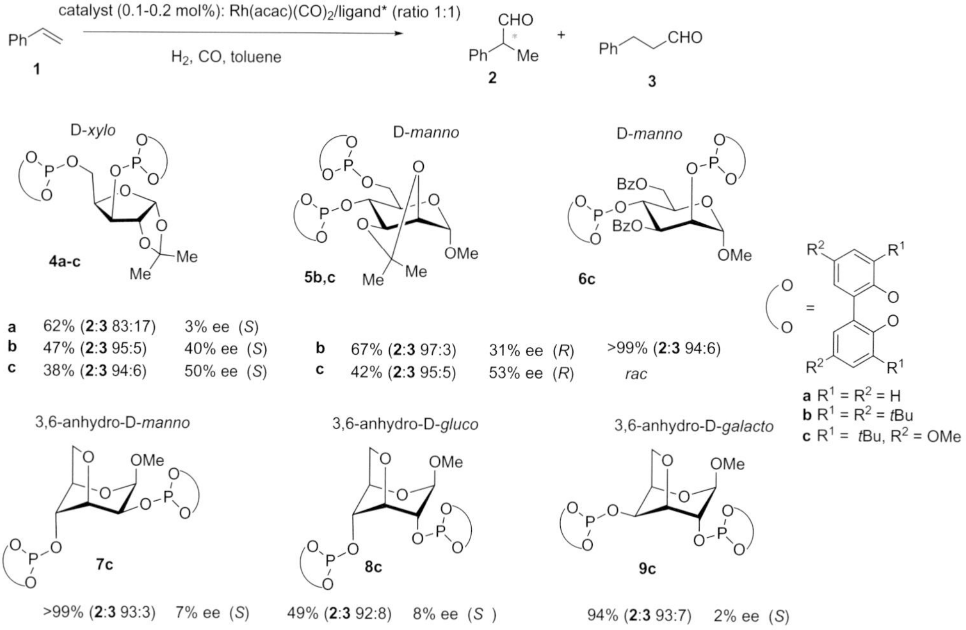

Scheme 9.2 First examples of asymmetric hydroformylation of styrene using carbohydrate diphosphite ligands.

As it turned out, all tested ligands gave good to excellent regioselectivity in favor of the desired chiral regioisomer. However, both, conversion of the starting material and the enantioselectivity, proved to be strongly dependent on the architecture of the carbohydrate scaffold (Scheme 9.2). The highest stereoselectivities were observed for *xylo*-configured furanose ligands with 3,5-diphosphite groups (4) and *manno*-configured pyranose ligands with phosphites in the 4- and 6-position (5), which led to 50% ee but notably to products with the opposite stereochemistry. Substitution of the biphenyl moieties with bulky *tert*-butyl in the *ortho* position and methoxy residues in the *para* position to the phenolic oxygens were a prerequisite for stereoselectivities of around 50% ee, while ligands with unsubstituted

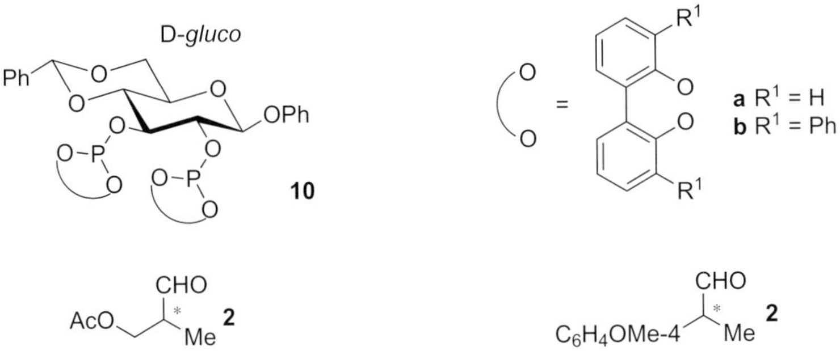

10a 63% (**2**:**3** 59:41) 1% ee
10b 99% (**2**:**3** 82:18) 32% ee

10a 65% (**2**:**3** 93:7) *rac*
10b 65% (**2**:**3** 93:7) 4% ee

Figure 9.1 Asymmetric hydroformylation of allyl acetate and 4-methoxystyrene using D-glucose-derived diphosphinites **10.**

biphenyl groups or four *tert*-butyl substituents gave significantly lower ees. In addition, the positioning of the phosphinites within the carbohydrate framework proved to be crucial. The shift of a phosphinite group from position 6 to position 2 in *manno*-configured ligand **6** resulted in a total loss of stereoinduction, and the conformationally rigid ligands **7–9**, in which the relative spatial orientation of the phosphinite groups is altered, also did not offer high levels of ee. The conversions for all examples ranged from 40 to 99%; however, unfortunately, higher ees were always accompanied with lower conversion.

Later, Selke *et al.* reported a family of glucose-based diphosphites containing different phenol residues. Figure 9.1 summarizes the results obtained with ligands **10** in the asymmetric hydroformylation of allyl acetate, and methoxystyrene [5]. While diphosphinite ligands closely related to **10** have been extremely successful in asymmetric hydrogenation reactions (cf. Chapter 8 on hydrogenation reactions), diphosphite **10b** led to only 32% ee in the hydroformylation of allyl acetate (Figure 9.1), highlighting again the importance of the location of the phosphites within the carbohydrate scaffold and the overall structure of it.

Based on the work of van Leeuwen [4], in 2000 Ruiz prepared ribofuranose ligands **11**, which are closely related to *xylo*-configured ligands **4** and gave the hydroformylation product of styrene in 53% ee, in the opposite configuration to *xylo*-ligands **4** (Figure 9.2). They also performed experiments with ligand **4c** and substituted styrenes as substrates and obtained the respective products in ees of around 50%. However, in all cases, the conversion remained below 30% [6]. Shortly afterwards Diéguez reported *gluco*-configured furanose ligand **12c** [7], which is yet again very similar to *xylo*-ligands **4**; however, it contains a stereocenter in the furanose side chain rather than a simple hydroxymethyl group. This ligand did not only lead to a high selectivity for the branched product (above 97%) but also to high conversions of above 80% and very good stereoselectivities around 90% ee for styrene as well as some derivatives.

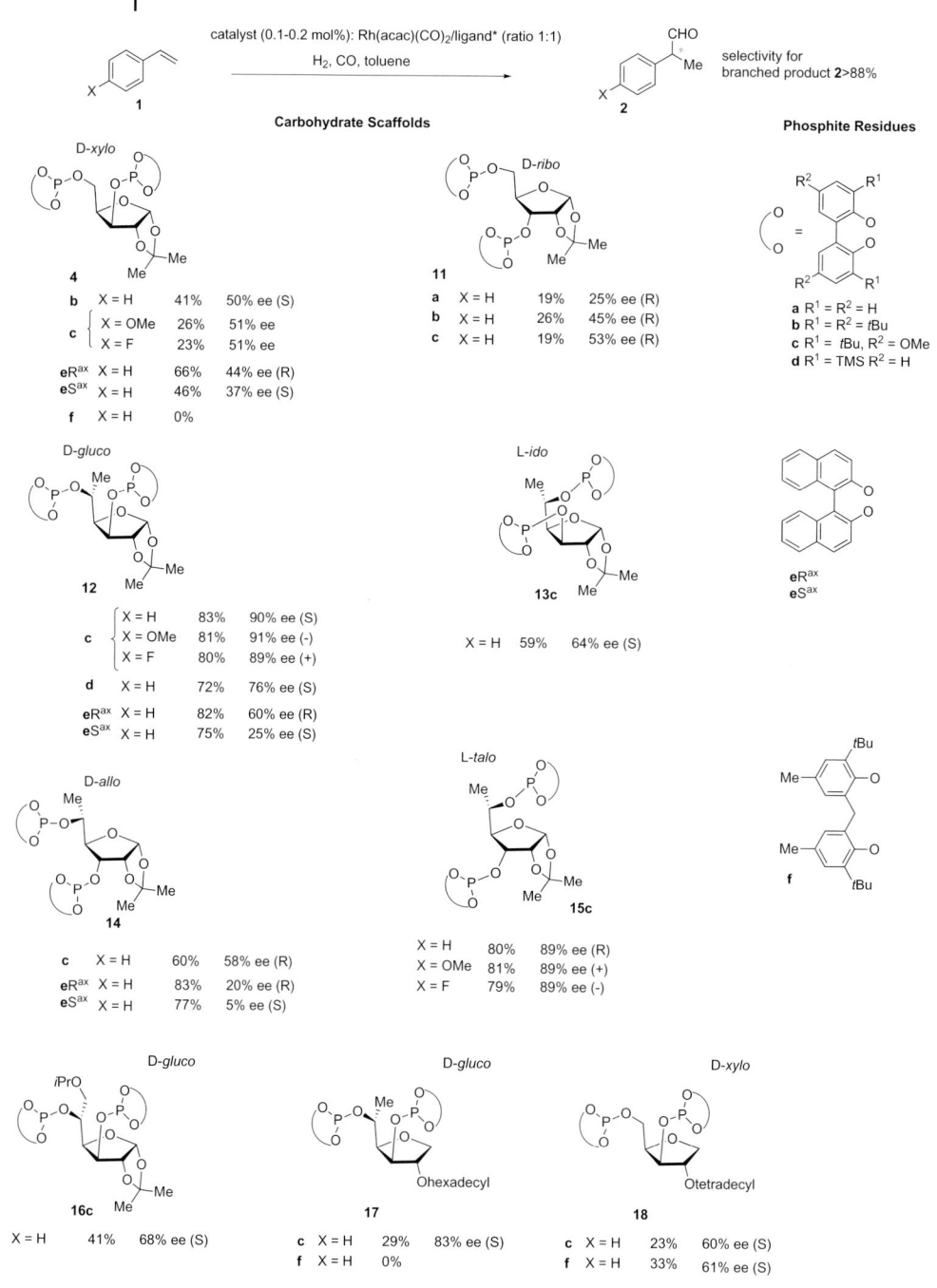

9 Hydroformylations, Hydrovinylations, and Hydrocyanations

Figure 9.2 Asymmetric hydroformylation of styrene and styrene derivatives with furanoside ligands.

9.1 Hydroformylation Reactions

In a very detailed study Diéguez and Claver systematically varied the configurations of endocyclic furanose position 3 and exocyclic pyranose position 5, thus creating three new ligand scaffolds that are diastereomers of successful ligand **12c** [8]. Scheme 9.3 summarizes the synthesis of all four diastereomeric ligand scaffolds with D-*gluco,* L-*ido,* D-*allo,* and L-*talo* configuration (**12c–15c**). For this route, D-glucose was first transformed into its 1:2,5:6-diisopropylidene derivative **20** as a key intermediate. This was subsequently elaborated into D-*gluco* configured precursor **21** and D-*allo* configured precursor **26** by known methods [9]. The *gluco* configured diol **23** and the *allo* configured diol **28** were obtained in three straightforward steps from precursors **21** and **26**, respectively. The synthesis of L-*ido* diol **25b** from D-*gluco* precursor **21** and L-*talo* diol **30** from D-*allo* compound **26** involved four steps, consisting of inversion of the stereocenter at C5 via an S_N2 reaction. All four diols where then transformed into the corresponding diphosphite ligands by treatment with chlorophosphites **24a** and **24b**, yielding a total of eight ligands (**12b,c–15b,c**).

Hydroformylation experiments with ligands **13–15** (Figure 9.2) confirmed the previous finding that biphenyl substitution with *tert*-butyl groups in the *ortho* position and methoxy groups in the *para* position improved stereoselectivity. Furthermore, it was again observed that the configuration of the carbohydrate scaffold has a substantial influence on the stereoselectivity, and this time strong matched/mismatched effects could be observed for the respective configurations of the stereocenters at positions 3 and 5. The absolute configuration of the product was predominantly determined by the configuration of the endocyclic stereocenter at position 3. For D-*gluco*-configured ligand **12c** and L-*talo*-configured ligand **15c** the configuration of the exocyclic stereocenter at position 5 enhanced the enantioselectivity and, therefore, can be regarded as matched to the one at position 3. Both ligands gave the hydroformylation products of styrene as well as two styrene derivatives in high conversion and ee and, importantly, in opposite absolute configuration. Thus these ligands, which are both accessible from D-glucose, act as an efficient pair of *pseudo*-enantiomers. In contrast, the use of D-*allo* ligand **14c** and L-*ido* ligand **13c** resulted in a severe drop of stereoselectivity. Therefore, these ligands can be viewed as a mismatched combination of the stereocenters at positions 3 and 5.

Diéguez [10] explored structural changes around the phosphorus donors and incorporated both enantiomers of axially chiral BINOL into the phosphinite residues of selected ligands. As the results obtained with the diastereomeric *xylo*-configured ligands **4eR^{ax}** and **4eS^{ax}** clearly demonstrated, the sense of asymmetric induction was governed by the configuration of the respective binaphthyl residue, which completely overrode the influence of the carbohydrate configuration (Figure 9.2). The same phenomenon was observed for *gluco-* and *allo*-configured ligands **12eR^{ax},S^{ax}** and **14eR^{ax},S^{ax}** with axially chiral residues. The ee remained in all cases well below the high levels obtained with *gluco-* and *talo*-ligands **12c** and **15c**. Ligand **12d**, with an achiral biphenyl unit containing two *o*-TMS groups, was also inferior to the ligands **12c** and **15c**. A more recent study by Claver and

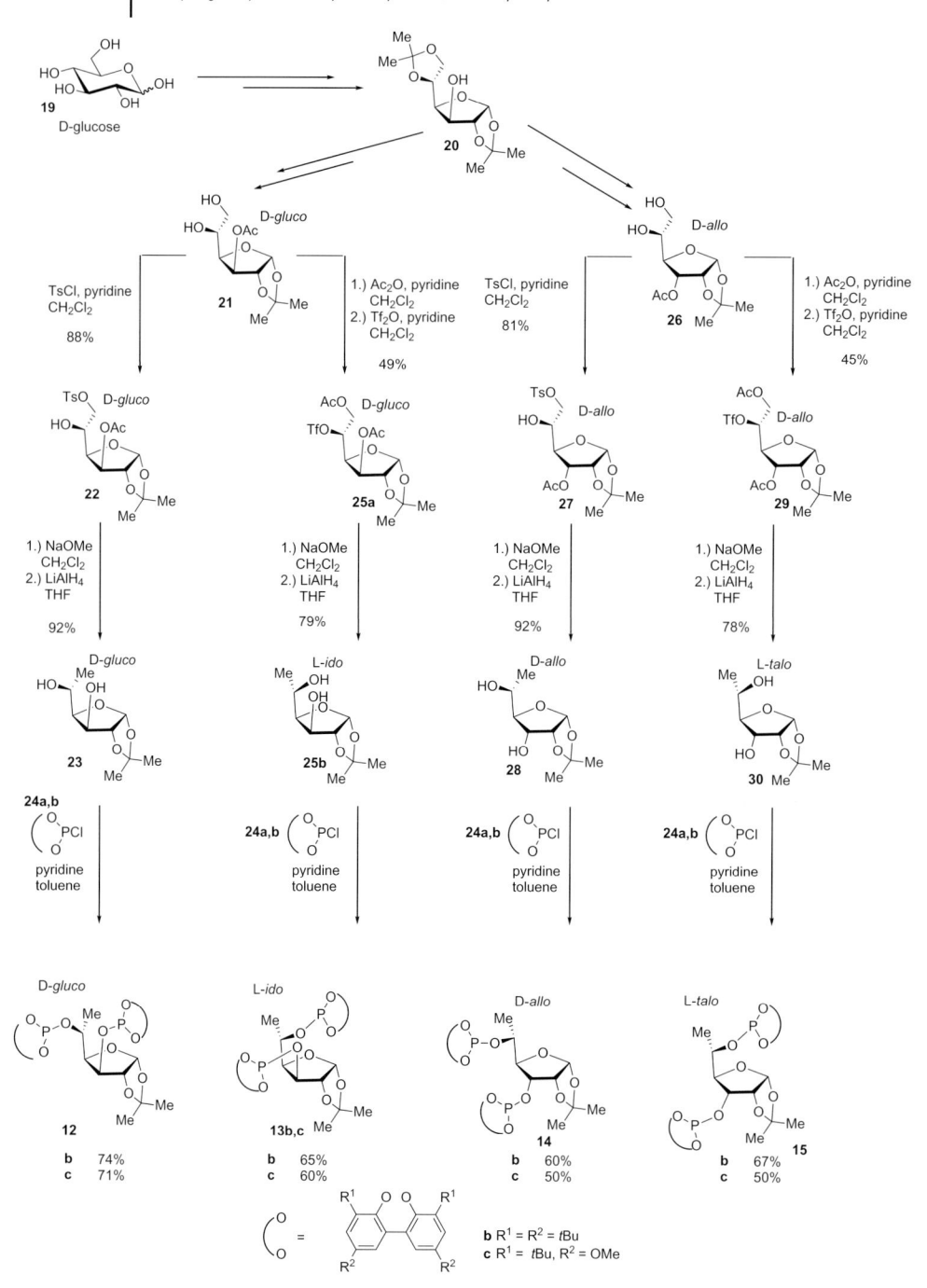

Scheme 9.3 Synthesis of four diastereomeric hexofuranoside scaffolds from D-glucose for diphosphite ligand design.

Castillón [11] elucidated the influence of the 1,2-acetonide and the nature of the substituent at C5 in ligands **12** (Figure 9.2). Replacement of the 1,2-acetonide in **12c** with a 2-*O*-alkyl ether resulted in *gluco*-ligand **17c**, which gave 83% ee under optimized conditions, and thus remained below that of **12c**, while introduction of an isopropoxy group at C5, as in ligand **16c**, resulted in an obvious loss of stereoselectivity. The *xylo*-configured ligand **18c** gave slightly higher ee than its acetonide-containing counterpart **4c**. Replacement of the optimized biphenyl residue of the phosphite with a methylene-bridged bis-phenol led to a complete loss of catalytic activity in almost all cases.

An especially interesting application of carbohydrate diphosphites was reported by Diéguez and Claver in 2005, when they employed ligands **4b** and **12b** in the asymmetric hydroformylation of heterocyclic olefins [12]. 2,5-Dihydrofuran (**31**) is a challenging substrate as hydroformylation does not exclusively yield the expected 3-tetrahydrofuran carbaldehyde **32** but may also gives rise to the undesired 2-regioisomer **33** (Scheme 9.4). This side product is formed from 2,3-dihydrofuran **34**, which in turn results from isomerization of 2,5-dhydrofuran (**31**) mediated by the rhodium-based hydroformylation catalyst [13]. When 2,3-dihydrofuran (**34**) is employed as substrate, again 3- and the 2-carbaldehydes **32** and **33** may be obtained as products (Scheme 9.4). The mechanistic rationale behind the hydroformylation of substrates **31** and **34** is shown in Scheme 9.5. Thus, to obtain good results for **31** in this transformation and to avoid product mixtures, the isomerization of **31** to **34** via β-hydride elimination from intermediate **35** has to be suppressed. For the reaction of **34**, a good control of the regiochemistry of the hydrorhodation step, leading to intermediates **35** and **36**, respectively, is essential.

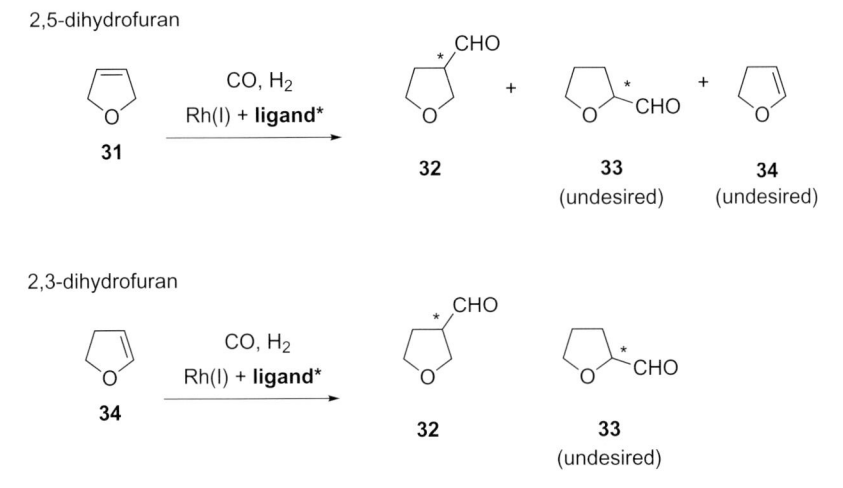

Scheme 9.4 Products and side products of the asymmetric hydroformylation of 2,5- and 2,3-dihydrofuran.

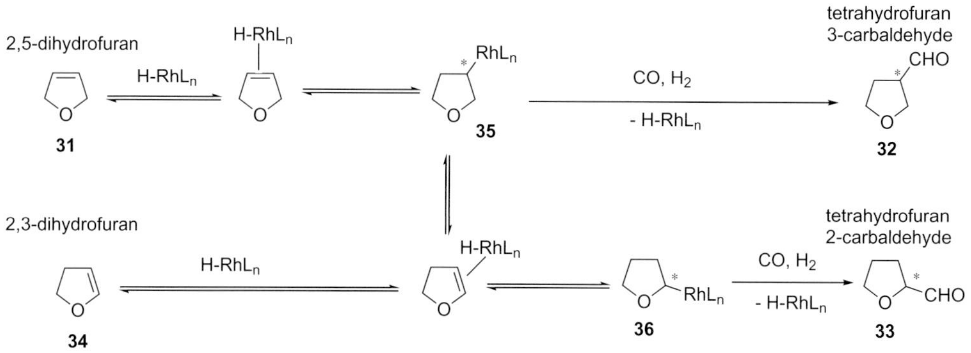

Scheme 9.5 Putative mechanism for rhodium-catalyzed hydroformylation and isomerization of dihydrofurans **31** and **34**.

Both rhodium complexes of **4b** and **12b** promoted the reaction with 2,5-dihydrofuran (**31**) with quantitative conversion of the starting material (Table 9.1). *gluco*-Ligand **12b** gave an excellent ratio of hydroformylation product **32** to isomerization products (ratio **32**:**33** of 99:1; yield of side product **34**: 1%) as well as an unprecedentedly high stereoselectivity of 74% ee in favor of the *(S)*-enantiomer (Table 9.1, entry 8), while *xylo*-ligand **4b** yielded exclusively the desired regioisomer **32** but gave rise to substantial substrate isomerization (12% yield of side product **34**) and only 53% ee (Table 9.1, entry 2). When 2,3-dihydrofuran (**34**) was employed as substrate (Table 9.2), no isomerization was observed for ligands **4b** and **12b** and the regioselectivity was almost identical for both (**32**:**33** approx. 75:25). In terms of stereoselectivity, again *gluco*-ligand **12b** with 75% ee (Table 9.2, entry 3) gave a better result than **4b** with 48% ee (Table 9.2, entry 1), but, more importantly, the change of substrate from **31** to **34** led to the *(R)*-configured hydroformylation product **32**. Thus, both enantiomers of **32** can be accessed by simply switching the substrate and without the need to resort to pseudo-enantiomeric ligands. In a recent study, Pàmies and Diéguez examined ligand scaffolds **4**, **11**, and **12–15** with different phosphite substituents for the hydroformylation of **31** and **34** [14]. Interestingly, for the heterocyclic substrate **31** (Table 9.1), ligands with phosphinites containing four bulky *tert*-butyl groups on the biphenyl unit gave the best stereoselectivities. None of the other ligand scaffolds employed could outperform *gluco*-ligand **12b** (Table 9.1, entries 1–7 and entries 9–13 vs. entry 8). For comparison, **37b**, a congener of **12b** with a very bulky OTBS group at C6, was tried but this also gave lower ee (cf. Table 9.1, entries 8 and 14).

Very similar trends were observed for substrate **34**, which are summarized in Table 9.2. Again, *gluco*-ligand **12b** gave the best stereoselectivity of all tested ligands (75% ee), in favor of the *(R)*-enantiomer of the product (Table 9.2, entry 3). In all cases the regioselectivity of the hydroformylation reaction, however, remained only moderate.

Table 9.1 Asymmetric hydroformylation of 2,5-dihydrofuran **31** with carbohydrate ligands.

catalyst (0.25 mol%): Rh(acac)(CO)$_2$/ligand* (ratio 1:2), H$_2$, CO, toluene

31 → **32** + **33** (undesired) + **34** (undesired)

Entry	Ligand		Yield 32 + 33 (%)	Yield 34 (%)	Ratio 32:33	ee (%)
1	D-xylo	4a, R^1 = R^2 = H	55	6	99:1	<5
2		4b, R^1 = R^2 = tBu	88	12	100:0	53 (S)
3		4c, R^1 = tBu, R^2 = OMe	66	7	98:2	15 (S)
4		4d, R^1 = TMS, R^2 = H	89	11	99:1	34 (S)
5	D-ribo	11b, R^1 = R^2 = tBu	91	9	95:5	24 (R)
6		11c, R^1 = tBu; R^2 = OMe	100	—	77:23	5 (S)
7	D-gluco	12a, R^1 = R^2 = H	64	8	99:1	<5
8		12b, R^1 = R^2 = tBu	99	1	99:1	74 (S)
9		12c, R^1 = tBu, R^2 = OMe	86	12	98:2	43 (S)
10		12d, R^1 = TMS, R^2 = H	98	2	99:1	63 (S)
11	L-ido	13b, R^1 = R^2 = tBu	94	6	98:2	25 (R)

(Continued)

Table 9.1 (Continued)

Entry	Ligand		Yield 32 + 33 (%)	Yield 34 (%)	Ratio 32:33	ee (%)
12	D-allo	14b, $R^1 = R^2 = t$Bu	92	8	98:2	47 (R)
13	L-talo	15b, $R^1 = R^2 = t$Bu	75	15	96:4	27 (R)
14	D-gluco	37b, $R^1 = R^2 = t$Bu	92	8	97:3	61 (S)

Table 9.2 Asymmetric hydroformylation of 2,3-dihydrofuran **34** with carbohydrate ligands.

catalyst (0.25 mol%): Rh(acac)(CO)$_2$/ligand* (ratio 1:2), H$_2$, CO, toluene

34 → **32** + **33** (undesired)

Entry	Ligand		Yield 32+33 (%)	Ratio 32:33	ee (%)
1	D-xylo	**4b** R^1 = R^2 = tBu	88	78:22	48 (R)
2	D-gluco	**12a** R^1 = R^2 = H	80	75:25	<5
3		**12b** R^1 = R^2 = tBu	100	76:24	**75 (R)**
4		**12c** R^1 = tBu, R^2 = OMe	100	74:26	49 (R)
5		**12d** R^1 = TMS, R^2 = H	100	73:27	61 (R)
6	L-ido	**13b** R^1 = R^2 = tBu	92	69:31	21 (S)

(Continued)

Table 9.2 (Continued)

catalyst (0.25 mol%): Rh(acac)(CO)₂/ligand* (ratio 1:2), H₂, CO, toluene

34 → **32** + **33** (undesired)

Entry	Ligand		Yield 32+33 (%)	Ratio 32:33	ee (%)
7	D-allo **14b**	R¹ = R² = tBu	100	73:27	48 (S)
8	L-talo **15b**	R¹ = R² = tBu	97	70:30	29 (S)
9	D-gluco **37b**	R¹ = R² = tBu	100	72:28	48 (R)

Ligands **4b** and **12b** were also tried on acetylated 2,5-dihydropyrrole **38**, giving the desired 3-pyrrolidine carbaldehyde **39** in excellent regioselectivity and in 49% ee and 71% ee respectively (Scheme 9.6). Hydroformylation of seven-membered heterocyclic alkenes, which can be obtained by acetalization of *cis*-but-2-ene-1,4-diol, was also explored by the same authors [13]. Table 9.3 presents the results obtained with *cis*-4,7-dihydro-1,3-dioxepines **41a,b**. While in almost all cases no isomerization of substrates **41a,b** was observed, the stereoselectivity remained moderate to low for most ligands tested in this series. The highest ee was achieved with the 6-*O*-TBDPS-derivative **37b** with 47% ee for formaldehyde acetal **41a** and 55% ee for acetonide **41b** (Table 9.3, entries 11 and 12). These results are, however, still among the best obtained for these dihydrodioxepines.

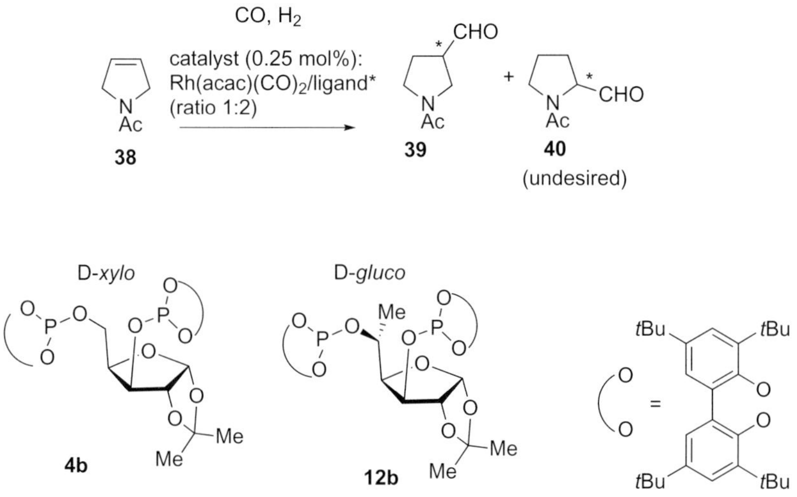

	39	39:40	
4b	quant.	99:1	49%ee (+)
12b	quant.	100:0	71%ee (+)

Scheme 9.6 Products and side products of the asymmetric hydroformylation of acetylated 2,5-dihydropyrrole **38**.

Claver and Castillón prepared C_2-symmetric phosphite ligands **44–48** with a furanosidic scaffold from glucosamine and glucitol [15]. Though none of the structures achieved high ees, the results obtained with them in asymmetric hydroformylations (Scheme 9.7) are interesting. Firstly, the study revealed that the direction of the asymmetric induction is not exclusively governed by the configurations of the furanoside sites C3 and C4 carrying the phosphite donors but that the more remote stereocenters at C1 and C5 have a significant impact as well (ligands **44** and **46** as well as ligands **45** and **47** give products with opposite configuration). Secondly, the size of the substituents at C1 and C5 also strongly influences the sense of the asymmetric induction (**44** and **45** as well as ligands **46** and **47** give

Table 9.3 Asymmetric hydroformylation of cis-4,7-dihydro-1,3-dioxepines **41a,b** using carbohydrate ligands.

Entry	Ligand		Substrate	Yield 42 (%)	Yield 41 (%)	ee (%)
1	D-xylo	**4b** $R^1 = R^2 = tBu$	$R^3 = H$	59	—	40 (+)
2			$R^3 = Me$	86	—	51 (S)
3	D-ribo	**11b** $R^1 = R^2 = tBu$	$R^3 = H$	48	26	30 (+)
4	D-gluco	**12a** $R^1 = R^2 = H$	$R^3 = H$	54	—	<5
5		**12b** $R^1 = R^2 = tBu$	$R^3 = H$	85	—	23 (+)
6		**12c** $R^1 = tBu, R^2 = OMe$	$R^3 = H$	75	—	18 (+)
7		**12d** $R^1 = TMS, R^2 = H$	$R^3 = H$	83	—	22 (+)
8	L-ido	**13b** $R^1 = R^2 = tBu$	$R^3 = H$	90	—	35 (+)

Table 9.3 (Continued)

catalyst (0.25 mol%): Rh(acac)(CO)$_2$/ligand* (ratio 1:2), H$_2$, CO, toluene

41 → **42** + **43** (undesired)

a R^3 = H
b R^3 = H

42: a R^3 = H, b R^3 = H

Entry	Ligand		Substrate	Yield 42 (%)	Yield 41 (%)	ee (%)
9	D-allo	14b, R^1 = R^2 = tBu	R^3 = H	79	—	37 (−)
10	L-talo	15b, R^1 = R^2 = tBu	R^3 = H	51	27	30 (+)
11	D-gluco		R^3 = H	93	—	47 (+)
12	(TBDPSO ligand)	37b, R^1 = R^2 = tBu	R^3 = Me	94	—	55 (S)

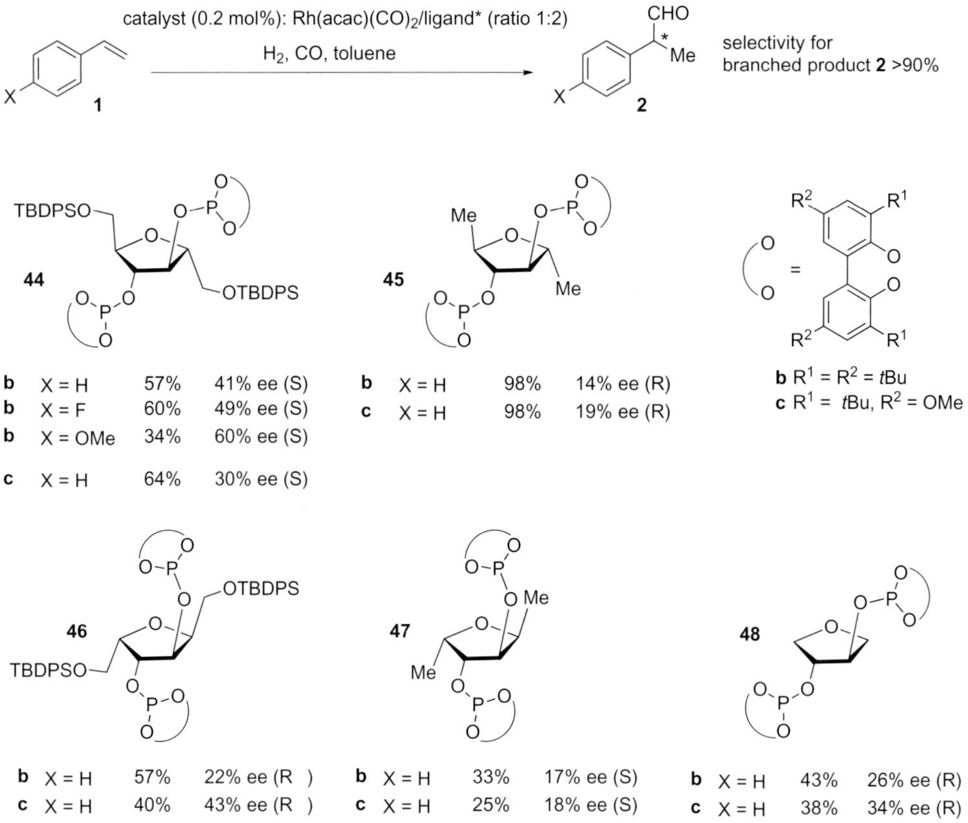

Scheme 9.7 Asymmetric hydroformylation of styrene and derivatives with C_2-symmetric carbohydrate phosphite ligands.

enantiomeric products). With substituted styrenes, the ee increased slightly for ligand **44b**.

9.1.2
Diphosphinite Ligands

While bidentate carbohydrate phosphinite ligands are highly successful in asymmetric hydrogenation reactions, they have only met with rather limited success in asymmetric hydroformylation. The group of RajanBabu [16] reported the use of D-*gluco* configured ligands **49** carrying substituted phenyl residues on the phosphinite donor sites in the reaction with vinyl-naphthalenes as substrates (Scheme 9.8). The efficiency of the asymmetric discrimination with ligands **49** turned out to be highly sensitive to the solvent and the nature of the substituents on the phosphinite donors. The highest ee values were achieved in apolar solvents such as hexane and with ligand **49b** containing electron-withdrawing CF_3 groups, while polar solvents such as THF and electron-donating substituents on the phosphinite such as methyl groups led to poor enantioselectivity. Overall, the stereoselectivities obtained with ligand **49b** under optimized conditions did not exceed 51% ee.

Scheme 9.8 Asymmetric hydroformylation of vinyl-naphthalenes using carbohydrate diphosphinite ligands **49**.

Diéguez and coworkers explored furanosidic *xylo-* and *ribo* ligands **50** and **51**, which are the diphosphinite counterparts to diphosphites **4** and **11** [17] (Figure 9.3). For ligands **50** and **51**, stereoselectivities above 50% ee were only observed with 4-methoxystyrene, while styrene itself and its electron-deficient 4-fluoro derivative gave almost racemic products. Surprisingly, the absolute configuration of the major enantiomer was the same for both ligands, irrespective of their furanoside scaffold, which is in striking contrast to the findings with the corresponding diphosphites **4** and **11**, which act as pseudo-enantiomers. With vinyl-naphthalene substrates both ligands gave stereoselectivities of 53% ee.

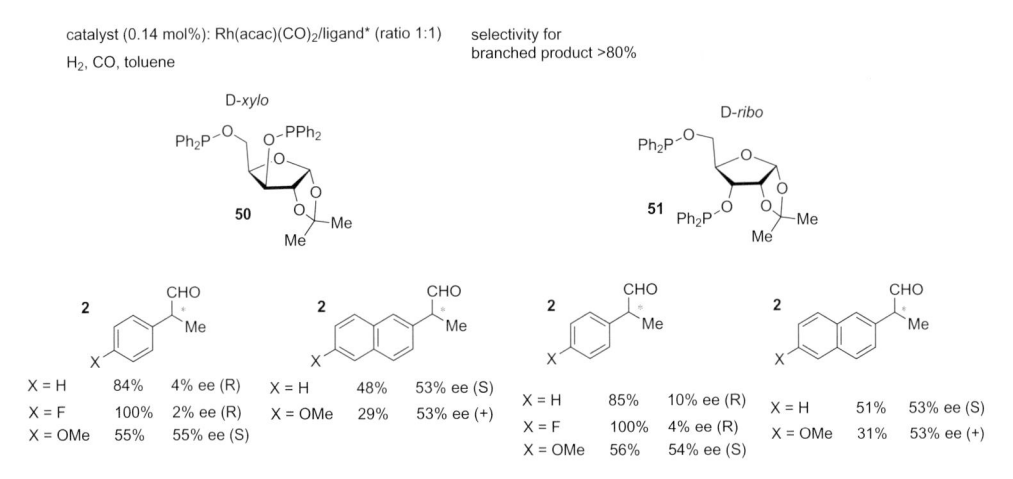

Figure 9.3 Asymmetric hydroformylation of vinyl arenes with carbohydrate diphosphinite ligands **50** and **51**.

9.1.3
Diphosphine Ligands

Unlike carbohydrate phosphite and phosphinite ligands, which are obtained by the reaction of chlorophosphites or phosphine chlorides with carbohydrate alcohols, the synthesis of carbohydrate phosphine ligands is much more complicated, as new carbon–phosphorus bonds have to be installed. This may be one of the reasons why examples of these ligands and their application are comparatively rare. Lu and coworkers reported diphosphine **52** derived from D-glucose [18], which gave the hydroformylation product of allyl acetate in 93% ee and high yield. The stereoselectivities obtained for styrene and norbornene were significantly lower. The authors explain the high stereoselectivity achieved for allyl acetate by invoking a hydrogen bond between the acetyl group of the substrate and the free 3-hydroxy group on the carbohydrate ligand. The results are shown in Figure 9.4.

Diphosphines with a furanosidic scaffold were introduced by Diéguez and Ruiz [19]. Ligands **53** and **54** (Figure 9.5) are structurally closely related to the corre-

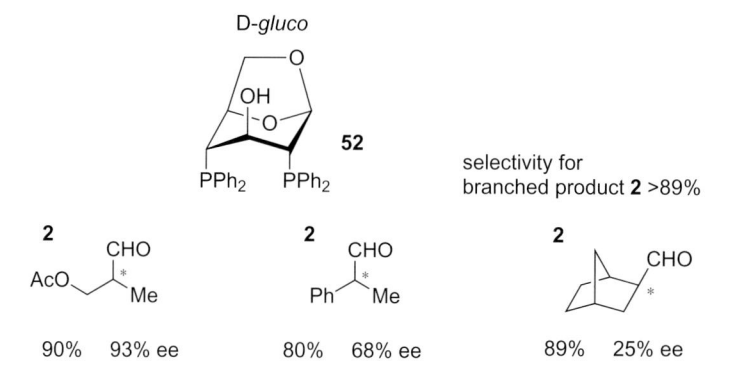

Figure 9.4 Asymmetric hydroformylation of various olefinic substrates with rigid carbohydrate diphosphine **52**.

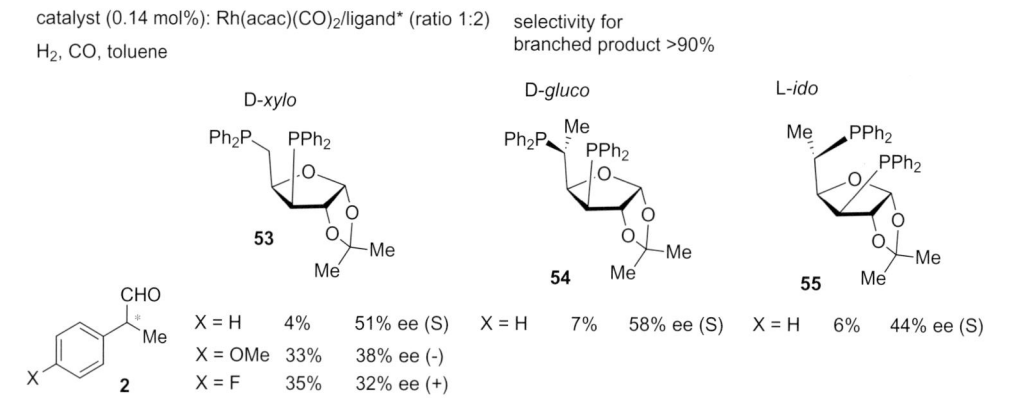

Figure 9.5 Asymmetric hydroformylation of styrenes with carbohydrate diphosphines **53–55**.

sponding diphosphites **4** and **12**. However, in contrast to ligand **12c**, which offered high stereoselectivities of around 90% ee, ligands **53–55** gave rise to only moderate selectivities.

9.1.4
Mixed Phosphorus Donor and P-S Donor Ligands

Ligands containing two different phosphorus donor sites have only been sparingly applied in hydroformylation reactions, even though they were among the first ligands of the type to be tested for this transformation. Börner and coworkers reported a family of structurally simple phosphine–phosphite furanoside ligands with one diphenylphosphino group and phosphites derived from both enantiomers of axially chiral BINOL and from simple catechol [20]. The authors tested ligands **56** with a 2,3-*cis,* and ligands **57** with a 2,3-*trans,* relationship between the two donor sites. While all ligands gave good conversion, the regioselectivity was in all cases moderate to poor. For the ligands with 2,3-*cis* architecture, **56eR^{ax}** and **56eS^{ax}**, the stereodirecting influence of the axially chiral BINOL unit overrode that exerted by the carbohydrate framework, but matched/mismatched effects between BINOL and carbohydrate configuration were apparent. With the 2,3-*trans* architecture, both ligands **57eR^{ax}** and **57eS^{ax}** produced predominantly the same enantiomer irrespective of the BINOL configuration, but again strong matched/mismatched effects were observed. Even in the matched cases, the stereoselectivities did not exceed 45% ee, and tests with catechol phosphite ligand **56g** proved again that the phosphite donor has to have bulky residues to achieve any stereoinduction (Figure 9.6).

The group of Ruiz explored D-*xylo* configured furanosides **58** with a phosphine donor at position 6 and various phosphite donors at position 3 [21]. These ligands, which are again related to xylose diphosphites **4**, only offered moderate ees, when the phosphite part contained bulky groups; the introduction of BINOL as an additional chiral motif had hardly no effect on the ee (Figure 9.6). One report on carbohydrate-based phosphite–phosphoramidite ligands was published by Diéguez and coworkers, who prepared ligands **59** and **60** with a D-xylo- and a D-ribofuranosidic scaffold [22]. Moderate stereoselectivities were obtained for *xylo*-ligands **59** and again the ee was significantly influenced by the substituents of the phosphoramidite and phosphite donors. Ligands **60** with *ribo*-configuration gave modest ee but the same enantiomer of the product (Figure 9.6).

So far only one report on mixed P-S donor ligands based on carbohydrates has appeared. The structures contained various alkylthio moieties at position 5 and biphenyl-derived phosphites in position 3 of a xylose. While these ligands promoted the reaction in good conversion they, unfortunately, offered no stereoselectivity [23].

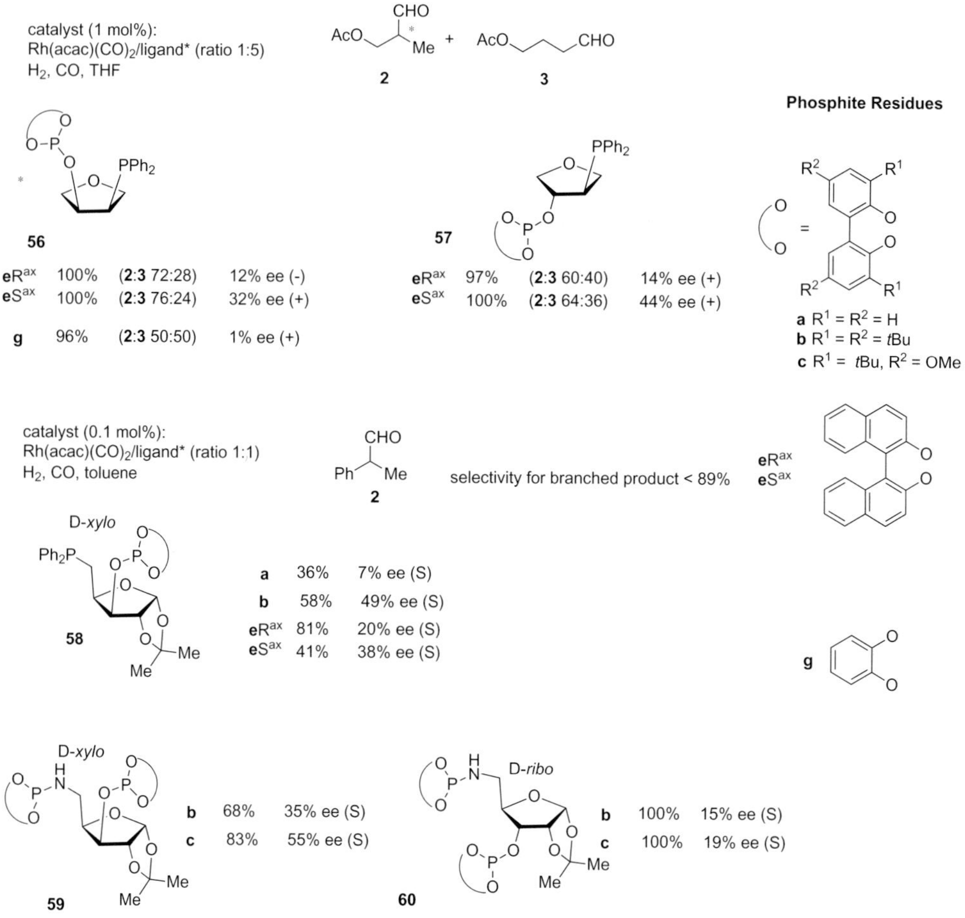

Figure 9.6 Asymmetric hydroformylation of allyl acetate and styrene using phosphine–phosphite and phosphoramidite–phosphite ligands with a furanoside architecture.

9.2
Hydrovinylation Reactions

Hydrovinylation is the formal addition of one hydrogen atom and a vinyl group across an olefinic double bond. As with hydroformylation, asymmetric hydrovinylation [24, 25] uses simple organic precursors to create synthetically valuable chiral target molecules and is therefore a highly attractive C–C-bond forming process. After pioneering work by Wilke in this field [24], nickel-based catalysts derived from allylnickel chloride dimer were identified as advantageous for the reaction, which typically couples vinylarenes **61** with ethene, resulting in chiral 3-aryl-1-butenes **62** (Scheme 9.9). However, the process may also yield achiral

9.2 Hydrovinylation Reactions

Scheme 9.9 Nickel-catalyzed asymmetric hydrovinylation of vinylarenes.

internal alkenes **63** as undesired side products, which are formed via isomerization of the desired chiral vinyl compound **62** by the catalyst. Further, alkene oligomerization can be a serious issue. For the process to be of any preparative value, ligands for chiral hydrovinylation catalysts do not only have to provide high asymmetric induction but must also suppress the these side reactions, making ligand design for this reaction a formidable challenge.

9.2.1
Monophosphinite and Monophosphite Ligands

RajanBabu and coworkers prepared carbohydrate-based monophosphinites, which were among the first ligands offering both high product yield and high stereoselectivity in asymmetric hydrovinylation [26–28]. Scheme 9.10 shows the synthesis of *gluco*- and *allo*-configured ligands **67** and **68** as representative examples:

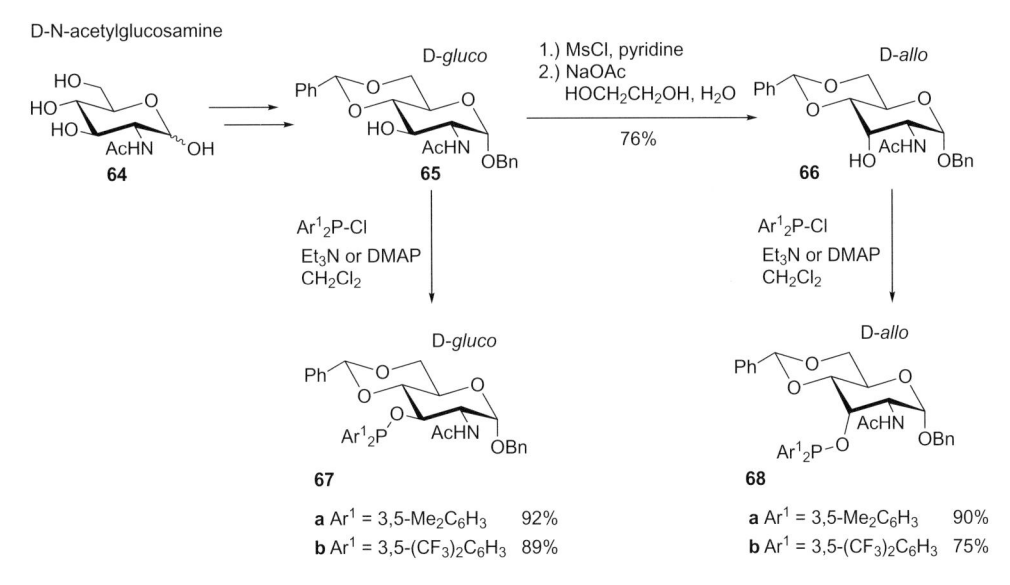

67
a Ar^1 = 3,5-Me$_2$C$_6$H$_3$ 92%
b Ar^1 = 3,5-(CF$_3$)$_2$C$_6$H$_3$ 89%

68
a Ar^1 = 3,5-Me$_2$C$_6$H$_3$ 90%
b Ar^1 = 3,5-(CF$_3$)$_2$C$_6$H$_3$ 75%

Scheme 9.10 Preparation of monophosphinite ligands **67** and **68** from D-*N*-acetylglucosamine.

The *gluco*-configured alcohol **65** was prepared as key intermediate from *N*-acetylglucosamine [29]; epimerization to *allo*-configured compound **65** was effected via known methods [30]. Both alcohols were subsequently transformed into the corresponding ligands by reaction with diarylphosphine chlorides under basic conditions, giving ligands **67** and **68** in good overall yields [26].

In asymmetric hydrovinylation with styrene derivatives, a total of five different carbohydrate scaffolds was evaluated. Further, the impact of electronic effects was explored by introduction of electron-donating and electron-withdrawing substituents of the phosphinite aryl residues. The results of this catalytic study are summarized in Scheme 9.11. Most of the ligands **69–71** gave the desired product in moderate to good isolated yield and *(S)*-configuration, while the stereoselectivity varied dramatically and clear trends regarding optimal carbohydrate architecture and substitution of the phosphinite residues emerged. For α-D-*gluco* ligands **69**, the stereoselectivity ranged from poor to moderate and generally increased, when phosphinites with electron-withdrawing CF_3-groups were employed, while in case of the β-D-*gluco* ligands **67** the ee remained low irrespective of the phosphinite substituents. In stark contrast, the D-*allo* ligands **68** in all but one case gave good to excellent ee. The best results were obtained with ligand **68a**, containing electron-donating methyl residues on the phosphinite aryl groups, which achieved stereoinduction levels of 74–89% ee for styrene and two derivatives. When the *N*-acetyl group of ligand **68a** was exchanged for a benzoate (ligand **70a**) or a trifluoroacetate (ligand **71a**), the stereoselectivity in the reaction with styrene even increased, but both ligands suffered from poor product yields due to increased formation of isomerized and polymeric by-products. Apart from the monophosphinites, RajanBabu also tried *gluco* and *allo* monophosphite ligands **72c** and **73c,d** containing a chiral BINOL or a chiral diol motif as an additional stereodirecting element; however, none of these could compete in terms of stereoselectivity with *allo*-configured phosphinite **68** (Scheme 9.11) [26].

Starting from the hydrovinylation product of 4-bromostyrene (**74**) RajanBabu prepared the important drug *(R)*-ibuprofen **77** [26]. This stereoselective synthesis is shown in Scheme 9.12. Despite the early success with ligand **68a**, carbohydrate ligands have until today not found any further application in hydrovinylation reactions. Tests with more complex substrates are few and far between and have up to now only yielded disappointing results [31].

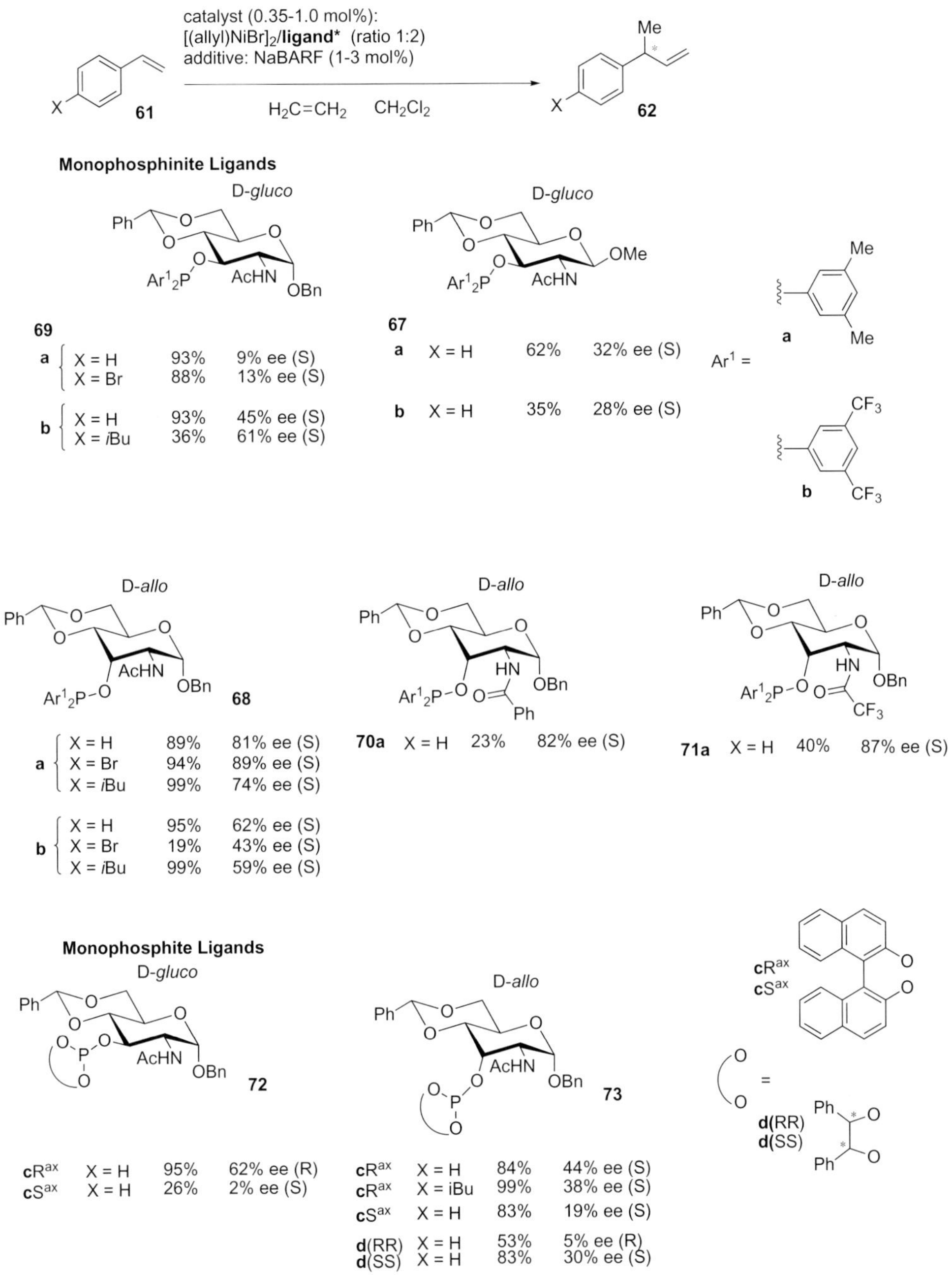

9.2 Hydrovinylation Reactions | 205

catalyst (0.35-1.0 mol%):
[(allyl)NiBr]$_2$/**ligand*** (ratio 1:2)
additive: NaBARF (1-3 mol%)

$H_2C=CH_2$ CH_2Cl_2

61 → **62**

Monophosphinite Ligands

69 D-gluco (Ph, Ar$^1_2$P, AcHN, OBn)

a: X = H, 93%, 9% ee (S); X = Br, 88%, 13% ee (S)
b: X = H, 93%, 45% ee (S); X = iBu, 36%, 61% ee (S)

67 D-gluco (Ph, Ar$^1_2$P, AcHN, OMe)

a: X = H, 62%, 32% ee (S)
b: X = H, 35%, 28% ee (S)

Ar^1 =
a: 3,5-dimethylphenyl (Me, Me)
b: 3,5-bis(trifluoromethyl)phenyl (CF$_3$, CF$_3$)

68 D-allo (Ph, Ar$^1_2$P-O, AcHN, OBn)

a: X = H, 89%, 81% ee (S); X = Br, 94%, 89% ee (S); X = iBu, 99%, 74% ee (S)
b: X = H, 95%, 62% ee (S); X = Br, 19%, 43% ee (S); X = iBu, 99%, 59% ee (S)

70a D-allo (Ph, Ar$^1_2$P-O, HN-Ph, OBn)
X = H, 23%, 82% ee (S)

71a D-allo (Ph, Ar$^1_2$P-O, HN-CF$_3$, OBn)
X = H, 40%, 87% ee (S)

Monophosphite Ligands

72 D-gluco (Ph, phosphite, AcHN, OBn)

cR^{ax} X = H 95% 62% ee (R)
cS^{ax} X = H 26% 2% ee (S)

73 D-allo (Ph, phosphite, AcHN, OBn)

cR^{ax} X = H   84% 44% ee (S)
cR^{ax} X = iBu 99% 38% ee (S)
cS^{ax} X = H   83% 19% ee (S)
d(RR)   X = H   53% 5% ee (R)
d(SS)   X = H   83% 30% ee (S)

cR^{ax}
cS^{ax}
= binaphthyl phosphite

d(RR)
d(SS)
= Ph/Ph TADDOL-like

Scheme 9.11 Asymmetric hydrovinylation of styrene derivatives with monodentate carbohydrate phosphinite and carbohydrate phosphite ligands.

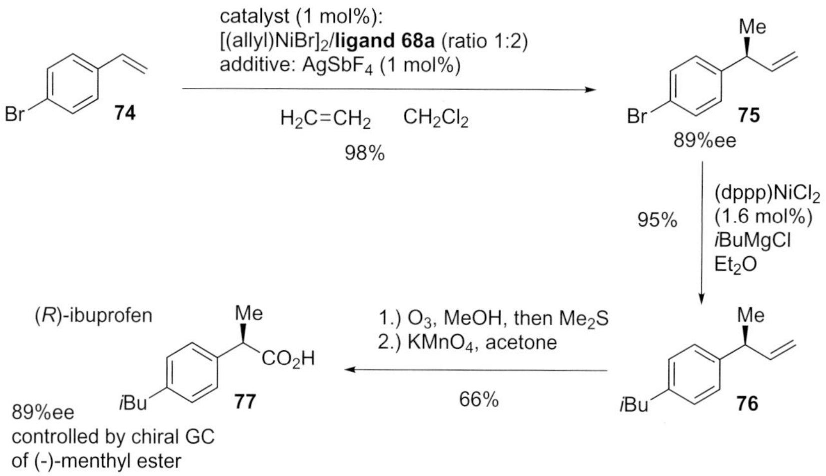

9 Hydroformylations, Hydrovinylations, and Hydrocyanations

Scheme 9.12 Asymmetric synthesis of (R)-ibuprofen via asymmetric hydrovinylation of 4-bromostyrene.

9.3
Hydrocyanation Reactions

The catalytic addition of hydrogen cyanide to an olefin, also called hydrocyanation [32], offers potentially easy access to nitriles, which can be elaborated into a host synthetically valuable compounds such as aldehydes, carboxylic acids, and amines. Therefore, an efficient asymmetric version of this transformation is highly desirable. The reaction proceeds mainly with activated alkenes such as vinylarenes and dienes as substrates and nickel(0) compounds make the most important catalysts (Scheme 9.13). As with the previously described addition processes across olefinic double bonds, an undesired achiral side product (**80**) may be formed along with the chiral addition product (**79**) and, thus, apart from high stereoselectivity, the catalyst has to provide a control of the regioselectivity.

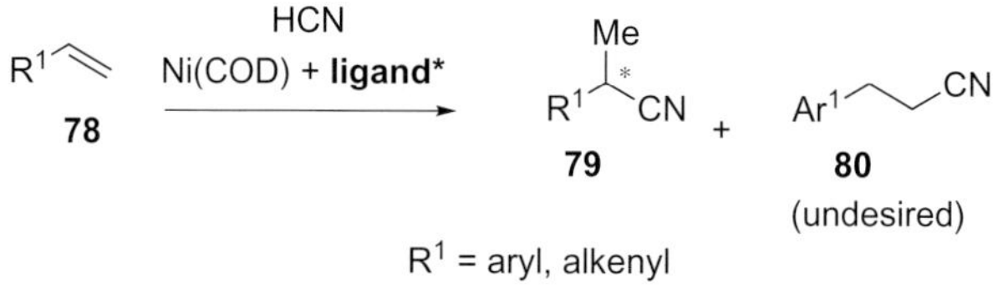

Scheme 9.13 Nickel-catalyzed asymmetric hydrocyanation of activated alkene substrates.

9.3.1
Diphosphinite Ligands

After the initial work by Elms and Jackson on asymmetric hydrocyanation of norbornene as a substrate [33], it was again the group of RajanBabu who employed a highly useful series of ligands based on carbohydrates, giving unprecedented levels of asymmetric induction [27, 34]. Inspired by the early work on carbohydrate ligands, they tried known phenyl-substituted diphosphinite **49a** [35] and prepared diphosphinite ligands **49a–d** based on D-glucose by reacting the corresponding 2,-diol with diaryl(chloro)phosphines carrying different substituents on the aromatic residues (Scheme 9.14) [36]. These ligands were subsequently tested in the reaction of 2-vinylnaphthalenes as substrates [33, 35] – the results of these studies are summarized in Scheme 9.15. The level of stereoinduction proved to be strongly dependent on the electronic nature of the phosphinite substituents on ligand scaffold **49**. While ligand **49a** carrying simple phenyl groups gave only low to moderate ee, electron-poor phosphinites dramatically increased the stereoselectivity, to up to 91% ee in the case of the strongly electron-withdrawing CF_3 groups on **49b**, while the fluoro-modified ligand **49c**, gave slightly less ee. In contrast, the electron-donating methyl groups in **49d** led to a severe drop for the stereoselectivity. With these findings, the group of RajanBabu was one of the first to illustrate that stereoinduction processes are not solely governed by steric shielding but can also be significantly influenced by electronic factors, which adds an additional dimension to ligand design [37].

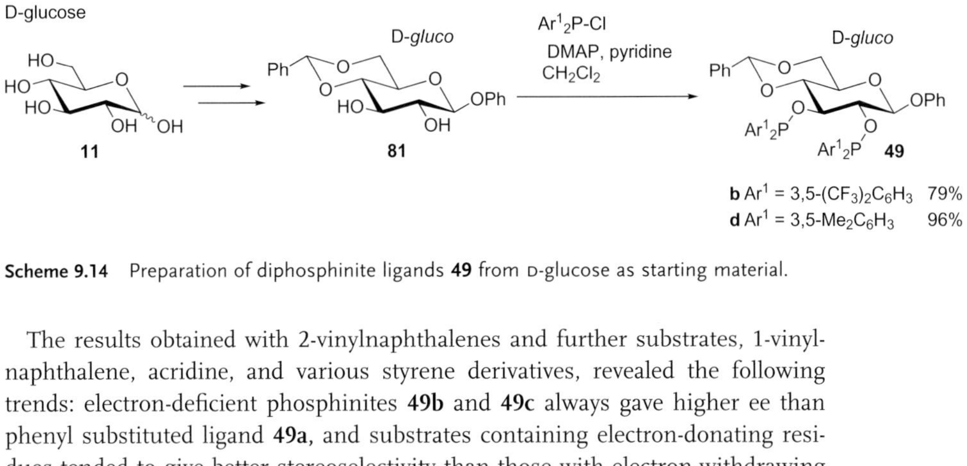

Scheme 9.14 Preparation of diphosphinite ligands **49** from D-glucose as starting material.

The results obtained with 2-vinylnaphthalenes and further substrates, 1-vinylnaphthalene, acridine, and various styrene derivatives, revealed the following trends: electron-deficient phosphinites **49b** and **49c** always gave higher ee than phenyl substituted ligand **49a**, and substrates containing electron-donating residues tended to give better stereoselectivity than those with electron-withdrawing groups (Scheme 9.15) [35]. In a more recent study, the same authors used ligands **49a–d** in the asymmetric hydrocyanation of 1,3-dienes (Scheme 9.15) [38]. In all cases the simple diphenyl phosphinite ligand **49a** led to the lowest observed ees, but, surprisingly, electron-deficient trifluoromethyl- or fluoro-modified ligands (**49b** and **49c**) as well as electron-rich methyl-derivative **49d** led to comparable stereoselectivities (65–80% ee) as best results for these substrates. The asymmetric hydrocyanation of a geraniol-derived diene led with high regioselectivity to a product containing a quaternary stereocenter; however, the asymmetric induction remained rather low at 19% ee.

To access both enantiomers of their hydrocyanation products, RajanBabu and coworkers endeavored to find ligands that might act as pseudo-enantiomers to *gluco*-ligands **49a–d**. Using a scaffold derived from the inexpensive

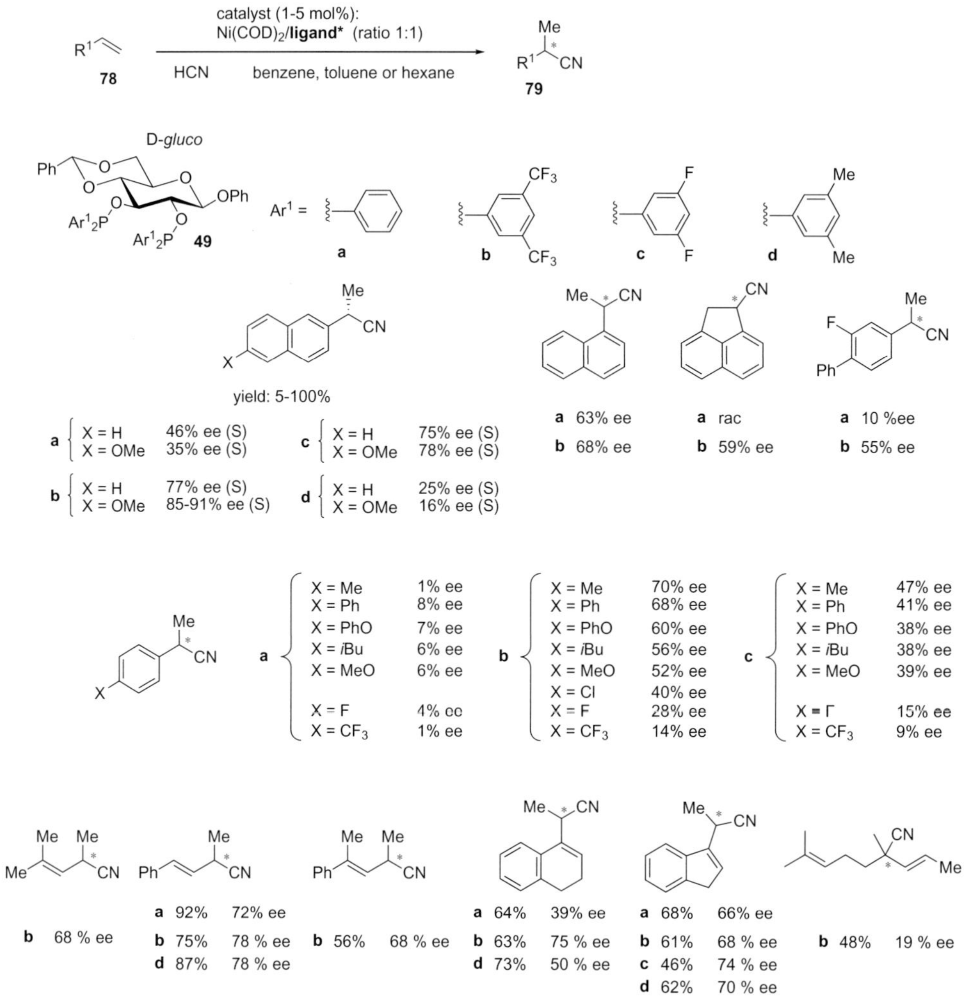

Scheme 9.15 Asymmetric hydrocyanation of vinylarenes using bidentate carbohydrate phosphinite ligands **49a–d** derived from D-glucose.

monosaccharide D-fructose [36, 39], ligands **82** were prepared, which were able to switch the direction of the asymmetric induction for the hydrocyanation of a vinylnaphthalene derivative (Table 9.4). Again, a strong electronic influence of the substituents on the phosphinite moieties was observed and, once again, the electron-withdrawing residues on the phosphinites led to higher asymmetric induction (entries 1–5). As the fructose scaffold of ligands **82** allowed a sequential attachment of two different phosphinite units at positions 3 and 4, a series of electronically unsymmetrical ligands was prepared as well. These novel ligands revealed a surprising result: when the (trifluoromethyl)phenyl phosphinite at position 2 of ligand **82bb** was replaced with a simple phenyl phosphinite (ligand **82ab**)

Table 9.4 Asymmetric hydrocyanation of vinylarenes using D-fructose-derived ligands **82**, which act as pseudo-enantiomers to D-glucose derived ligands **49a–d**.

Entry	Ligand	Ar¹	Ar²	ee (%)
1	82aa	phenyl	phenyl	43
2	82bb	3,5-(CF$_3$)$_2$-C$_6$H$_3$	3,5-(CF$_3$)$_2$-C$_6$H$_3$	56
3	82cc	3,5-F$_2$-C$_6$H$_3$	3,5-F$_2$-C$_6$H$_3$	45
4	82dd	3,5-Me$_2$-C$_6$H$_3$	3,5-Me$_2$-C$_6$H$_3$	40
5	82ee	4-MeO-C$_6$H$_4$	4-MeO-C$_6$H$_4$	25
6	82ab	phenyl	3,5-(CF$_3$)$_2$-C$_6$H$_3$	58
7	82ba	3,5-(CF$_3$)$_2$-C$_6$H$_3$	phenyl	89 (95 at 0 °C)
8	82ac	phenyl	3,5-F$_2$-C$_6$H$_3$	40

(Continued)

Table 9.4 (Continued)

Entry	Ligand	Ar¹	Ar²	ee (%)
9	82ca	3,5-F₂-C₆H₃	C₆H₅	63
10	82bc	3,5-(CF₃)₂-C₆H₃	3,5-F₂-C₆H₃	78
11	82cb	3,5-F₂-C₆H₃	3,5-(CF₃)₂-C₆H₃	42
12	82bd	3,5-(CF₃)₂-C₆H₃	3,5-Me₂-C₆H₃	78
13	82be	3,5-(CF₃)₂-C₆H₃	4-MeO-C₆H₄	84
14	82bf	3,5-(CF₃)₂-C₆H₃	4-F-C₆H₄	88

the stereoselectivity increased slightly from 56% ee to 58% ee (entry 2 vs. entry 6). The attachment of a (trifluoromethyl) phenyl phosphinite at position 4 and a phenyl phosphinite at position 3 (ligand **82ba**) led to a dramatic increase of the stereoselectivity to 89% ee, which could even be improved to 95% ee by lowering the reaction temperature (entry 7). The stereoselectiVITIES OBtained for other combinations of phosphinites on scaffold **82** are summarized Table 9.4 (entries 8–14). As can be seen, a (trifluoromethyl) phenyl phosphinite at position 4 is a prerequisite for high stereoinduction, while position 3 may carry a broader range of phosphinites. The optimal residue for this position is, however, the simple diphenyl phosphinite. Compounds **82** are an interesting example of electronically unsymmetrical bidentate ligands that also demonstrate that the relative location of the electronically differentiated donor sites is of crucial importance for the stereoinduction process [38].

RajanBabu et al. highlighted the large preparative utility of their asymmetric hydrocyanation process by demonstrating a brief and efficient stereoselective synthesis of the broadly applied analgesic naproxen (Scheme 9.16) [40].

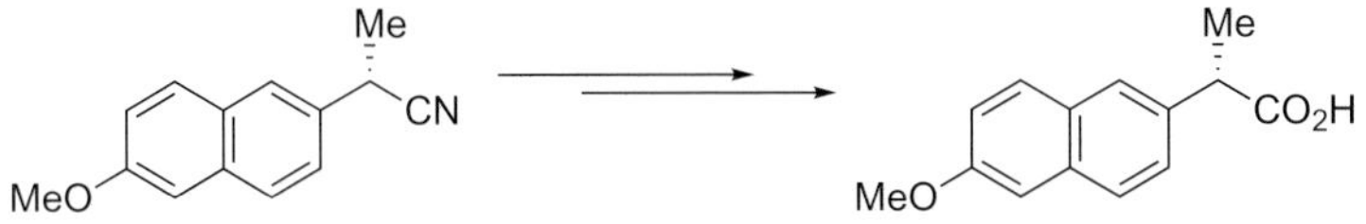

Scheme 9.16 Stereoselective synthesis of the analgesic drug naproxen via asymmetric hydrocyanation.

9.4
Hydrosilylation Reactions

Up to now, there have only been a few reports on the rhodium-catalyzed asymmetric hydrosilylation of aryl methyl ketoses using carbohydrate-derived ligands, which are summarized below (Scheme 9.17).

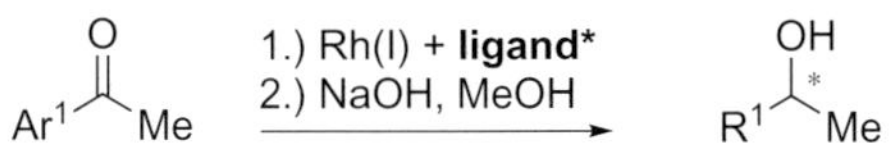

Scheme 9.17 General reaction conditions for the asymmetric hydrosilylation of aryl methyl ketones **83**.

9.4.1
Monophosphite Ligands

Pizzano [41] and coworkers explored structurally simple, monodentate phosphite ligands **85a,b** containing three glucofuranose residues as well as some ligands with only one carbohydrate residue (Figure 9.7). Their studies showed that significant ees were only obtained when the ligands contained three carbohydrate units and

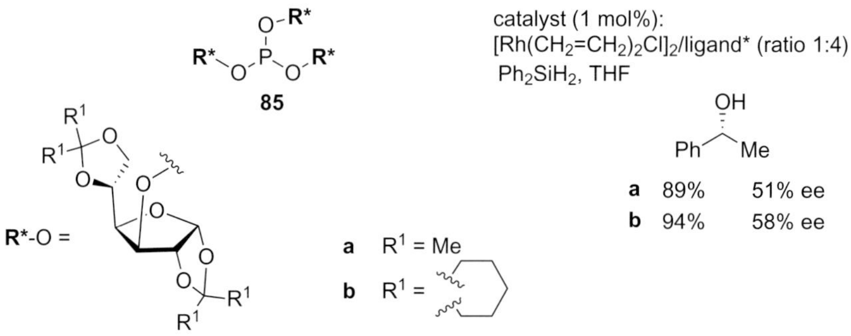

Figure 9.7 Asymmetric hydrosilylation of acetophenone with monodentate carbohydrate–phosphite ligands with a furanoside architecture.

the ligand-to-Rh(I) ratio was 3 : 1. However, even with the best ligand **85b**, the stereoselectivity did not exceed 60% ee.

9.4.2
Diphosphite and Phosphite–Phosphoroamidite Ligands

The group of Diéguez [42] has explored the hydrosilylation of various methyl ketones using their series of furanose diphosphites **4, 11,** and **12–15** as well as mixed phosphite–phosphoramidite ligand **59**, which have also found application in asymmetric hydroformylation reactions. The authors found that, as in hydroformylation, the direction of the asymmetric induction depended on the furanose configuration at position 3, with D-*xylo*, D-*gluco*, and L-*ido* ligands **4, 12, 13,** and **59** giving the (*R*)-enantiomer of the hydrosilylation product of acetophenone (Ar^1 = Ph), while D-*ribo*, D-*allo*, and L-*talo* ligands **11, 14,** and **15** produced an excess of the (*S*)-enantiomer. As before, the efficiency of the stereoinduction process is strongly influenced by the substituents on the biphenyl residues incorporated into the phosphites. The only ligand giving an asymmetric induction above 50% ee was D-*xylo* configured diphosphite **4c**. The closely related, D-*xylo* configured phosphite–phosphoramidite **59**, on the other hand, gave nearly racemic products. For reactions on further substrates, the authors surprisingly chose ligand **4c** instead of the more successful congener **4a**. With this ligand, low to moderate selectivities were observed for various substituted acetophenones. The results are summarized in Scheme 9.18.

9.4.3
P-S Donor Ligands

The only example of carbohydrate-derived ligands of this type was again reported by Diéguez and Pàmies, who prepared a family of mixed phosphinite–thioether ligands [43]. While related structures did not give rise to any significant asymmetric induction in hydroformylation reactions, they have met with more success in asymmetric hydrosilylation (Scheme 9.19). The seven ligands **86a–g** were pre-

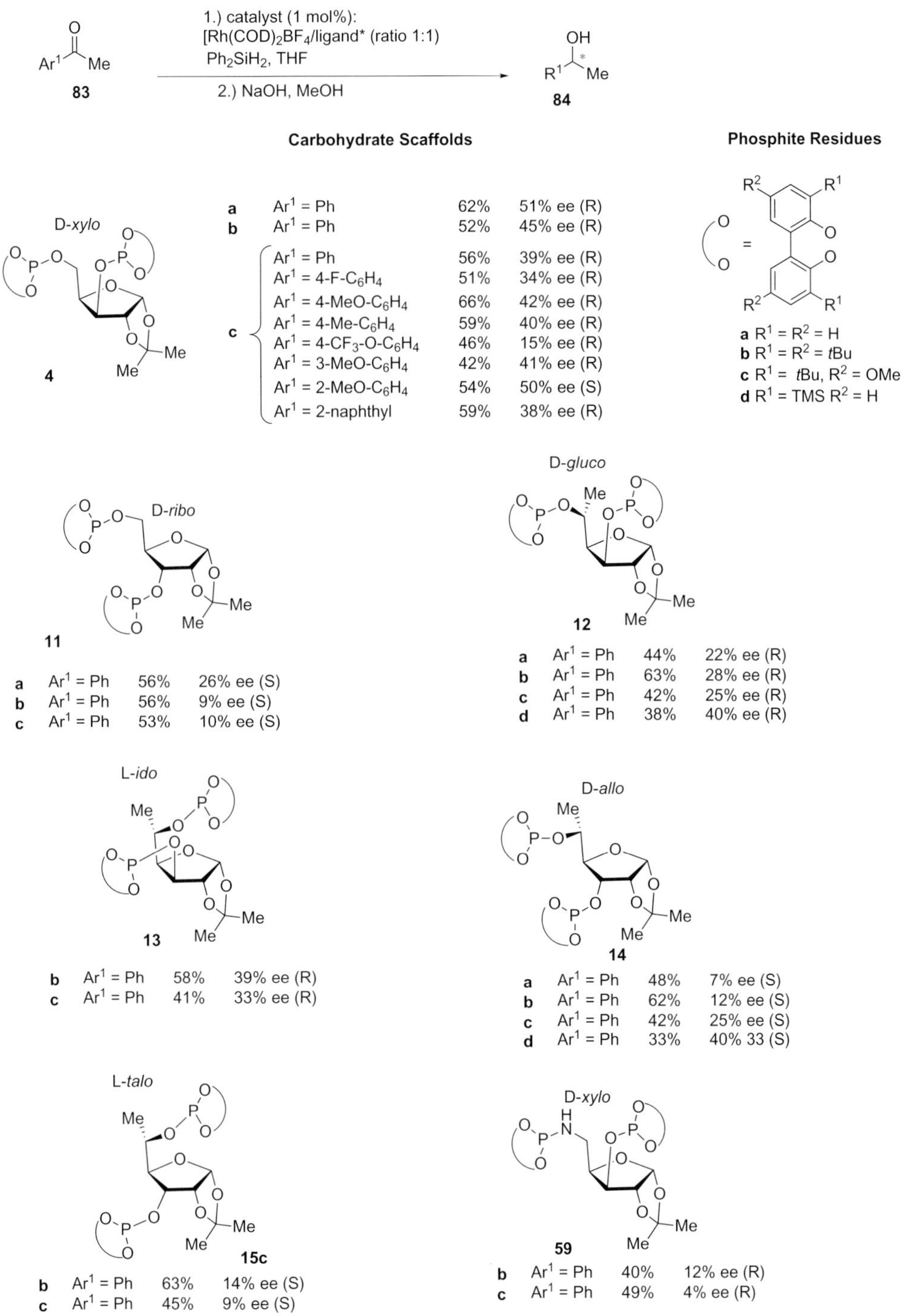

9.4 Hydrosilylation Reactions

Scheme 9.18 Asymmetric hydrosilylation of aryl methyl ketones with furanoside diphosphites and phosphite–phosphoramidites.

Ketone **83** (Ar^{1}COMe) → alcohol **84** (R^{1}CH(OH)Me)

1.) catalyst (1 mol%): [Rh(COD)$_2$BF$_4$/ligand* (ratio 1:1)], Ph$_2$SiH$_2$, THF
2.) NaOH, MeOH

Carbohydrate Scaffolds

4 (D-xylo)

a	Ar^1 = Ph	62%	51% ee (R)
b	Ar^1 = Ph	52%	45% ee (R)
c	Ar^1 = Ph	56%	39% ee (R)
	Ar^1 = 4-F-C$_6$H$_4$	51%	34% ee (R)
	Ar^1 = 4-MeO-C$_6$H$_4$	66%	42% ee (R)
	Ar^1 = 4-Me-C$_6$H$_4$	59%	40% ee (R)
	Ar^1 = 4-CF$_3$-O-C$_6$H$_4$	46%	15% ee (R)
	Ar^1 = 3-MeO-C$_6$H$_4$	42%	41% ee (R)
	Ar^1 = 2-MeO-C$_6$H$_4$	54%	50% ee (S)
	Ar^1 = 2-naphthyl	59%	38% ee (R)

Phosphite Residues

a R^1 = R^2 = H
b R^1 = R^2 = tBu
c R^1 = tBu, R^2 = OMe
d R^1 = TMS R^2 = H

11 (D-ribo)

a	Ar^1 = Ph	56%	26% ee (S)
b	Ar^1 = Ph	56%	9% ee (S)
c	Ar^1 = Ph	53%	10% ee (S)

12 (D-gluco)

a	Ar^1 = Ph	44%	22% ee (R)
b	Ar^1 = Ph	63%	28% ee (R)
c	Ar^1 = Ph	42%	25% ee (R)
d	Ar^1 = Ph	38%	40% ee (R)

13 (L-ido)

| b | Ar^1 = Ph | 58% | 39% ee (R) |
| c | Ar^1 = Ph | 41% | 33% ee (R) |

14 (D-allo)

a	Ar^1 = Ph	48%	7% ee (S)
b	Ar^1 = Ph	62%	12% ee (S)
c	Ar^1 = Ph	42%	25% ee (S)
d	Ar^1 = Ph	33%	40% 33 (S)

15c (L-talo)

| b | Ar^1 = Ph | 63% | 14% ee (S) |
| c | Ar^1 = Ph | 45% | 9% ee (S) |

59 (D-xylo)

| b | Ar^1 = Ph | 40% | 12% ee (R) |
| c | Ar^1 = Ph | 49% | 4% ee (R) |

9 Hydroformylations, Hydrovinylations, and Hydrocyanations

Ar¹–C(O)–Me **83**

1.) catalyst (1 mol%): [Rh(COD)$_2$BF$_4$/ligand* (ratio 1:1)] Ph$_2$SiH$_2$, THF
2.) NaOH, MeOH

→ Ar¹–C*H(OH)–Me **84**

Ligand **86** (D-xylo, R¹-S, O-PPh$_2$ scaffold with acetonide):

- **a** R¹ = Ph
- **b** R¹ = Me
- **c** R¹ = iPr
- **d** R¹ = tBu
- **e** R¹ = 4-MeC$_6$H$_4$
- **f** R¹ = 4-CF$_3$-C$_6$H$_4$
- **g** R¹ = 2,6-Me$_2$C$_6$H$_3$

	Ar¹	Yield	ee
a	Ph	68%	38% ee (R)
b	Ph	93%	25% ee (R)
c	Ph	97%	47% ee (R)
d	Ph	90%	86% ee (R)
e	Ph	82%	41% ee (R)
f	Ph	94%	37% ee (R)
g	Ph	92%	44% ee (R)

With ligand **d**:

Ar¹	Yield	ee
4-F-C$_6$H$_4$	72%	84% ee (R)
4-MeO-C$_6$H$_4$	98%	78% ee (R)
4-CF$_3$-O-C$_6$H$_4$	98%	72% ee (R)
3-MeO-C$_6$H$_4$	95%	87% ee (R)
2-MeO-C$_6$H$_4$	82%	90% ee (R)
2-naphthyl	85%	88% ee (R)

Scheme 9.19 Asymmetric hydrosilylation of aryl methyl ketones using phosphinite–thioether ligands **86**.

pared from xylose and consist of a 3-diphenyl-phosphinite and a thioether donor with varying substitution on position 5. Hydrosilylation experiments on acetophenone as the benchmark substrate showed a strong influence of the thioether substituent on the efficiency of the asymmetric induction. While aryl substitution and small alkyl groups led only to moderate selectivity, the bulky *tert*-butyl residue in ligand **86d** gave 86% ee under optimized conditions. Fair to very good selectivities were also obtained with this ligand and a selection of other aryl methyl ketones.

9.5 Conclusion

The examples discussed in this chapter demonstrate once again the versatility of carbohydrate ligands in catalytic asymmetric reactions. By variation of the carbohydrate scaffold and/or variation of the nature of the donor sites, small libraries of ligands are rapidly available and can be screened against a specific reaction problem. The findings of electronic ligand tuning, by modulating donor sites by aryl substituents attached to them, have added a further dimension to ligand design. As carbohydrates as multifunctional molecules are especially amenable for quick attachment of such modified donor groups, further interesting new ligand structures will no doubt emerge in the future.

References

1. For reviews see: (a) Gual, A., Godard, C., Castillón, S., and Claver, C. (2010) *Tetrahedron: Asymmetry*, **21**, 1135; (b) Claver, C., Diéguez, M., Pàmies, O., and Castillón, S. (2006) *Top. Organomet. Chem.*, **18**, 35; (c) Diéguez, M., Pàmies, O., and Claver, C. (2004) *Tetrahedron: Asymmetry*, **15**, 2113; (d) Breit, B. (2003) *Acc. Chem. Res.*, **36**, 264; (e) Agbossou, F., Carpentier, J.-F., and Mortreux, A. (1995) *Chem. Rev.*, **95**, 2485.
2. Breit, B. and Seiche, W. (2001) *Synthesis*, 1.
3. Sakai, N., Mano, S., Nozaki, K., and Takaya, H. (1993) *J. Am. Chem. Soc.*, **115**, 7033.
4. Buisman, G.J.H., Martin, M.E., Vos, E.J., Klootwijk, A., Kamer, P.C.J., and van Leeuwen, P.W.N.M. (1995) *Tetrahedron: Asymmetry*, **6**, 719.
5. Kadyrov, R., Heller, D., and Selke, R. (1998) *Tetrahedron: Asymmetry*, **9**, 329.
6. Pàmies, O., Net, G., Ruiz, A., and Claver, C. (2000) *Tetrahedron: Asymmetry*, **11**, 1097.
7. Diéguez, M., Pàmies, O., Ruiz, A., Castillón, S., and Claver, C. (2000) *Chem. Commun.*, 1607.
8. Diéguez, M., Pàmies, O., Ruiz, A., Castillón, S., and Claver, C. (2001) *Chem. Eur. J.*, **7**, 3086.
9. Schmidt, O.T. (1963) *Methods in Carbohydrate Chemistry* (eds R.L. Whistler and M.L. Wolfrom), Academic Press, New York, vol. 2, pp. 318–325.
10. Diéguez, M., Pàmies, O., Ruiz, A., Castillón, S., and Claver, C. (2002) *New J. Chem.*, **26**, 827.
11. Gual, A., Godard, C., Claver, C., and Castillón, S. (2009) *Eur. J. Org. Chem.*, 1191.
12. Diéguez, M., Pàmies, O., and Claver, C. (2005) *Chem. Commun.*, 1221.
13. (a) Polo, A., Real, J., Claver, C., Castillón, S., and Bayón, J.C. (1990) *J. Chem. Soc., Chem. Commun.*, 600; (b) Polo, A., Claver, C., Castillón, S., Ruiz, A., Bayón, J.C., Real, J., Mealli, C., and Masi, D. (1992) *Organometallics*, **11**, 3525.
14. Mazuela, J., Coll, M., Pàmies, O., and Diéguez, M. (2009) *J. Org. Chem.*, **74**, 5440.
15. Axet, M.R., Benet-Buchholz, J., Claver, C., and Castillón, S. (2007) *Adv. Synth. Catal.*, **349**, 1983.
16. RajanBabu, T.V. and Ayers, T.A. (1994) *Tetrahedron Lett.*, **35**, 4295.
17. Guimet, E., Parada, J., Diéguez, M., Ruiz, A., and Claver, C. (2005) *Appl. Catal. A*, **282**, 215.
18. Lu, S., Li, X., and Wang, A. (2000) *Catal. Today*, **63**, 531.
19. Diéguez, M., Pàmies, O., Net, G., Ruiz, A., and Claver, C. (2001) *Tetrahedron: Asymmetry*, **12**, 651.
20. Kless, A., Holz, J., Heller, D., Kadyrov, R., Selke, R., Fischer, C., and Börner, A. (1996) *Tetrahedron: Asymmetry*, **7**, 33.
21. Pàmies, O., Net, G., Ruiz, A., and Claver, C. (2001) *Tetrahedron: Asymmetry*, **12**, 3445.
22. Diéguez, M., Ruiz, A., and Claver, C. (2001) *Tetrahedron: Asymmetry*, **12**, 2827.
23. Pàmies, O., Diéguez, M., Net, G., Ruiz, A., and Claver, C. (2000) *Organometallics*, **19**, 1488.
24. (a) Bogdanović, B., Henc, B., Meister, B., Pauling, H., and Wilke, G. (1972) *Angew. Chem. Int. Ed. Engl.*, **11**, 1023; (b) Bogdanović, B., Henc, B., Löser, A., Meister, B., Pauling, H., and Wilke, G. (1973) *Angew. Chem. Int. Ed. Engl.*, **12**, 954.
25. For reviews see: (a) RajanBabu, T.V. (2009) *Synlett*, 853; (b) RajanBabu, T.V. (2003) *Chem. Rev.*, **103**, 2845; (c) Goößen, L.J. (2002) *Angew. Chem. Int. Ed.*, **41**, 3775.
26. Park, H. and RajanBabu, T.V. (2002) *J. Am. Chem. Soc.*, **124**, 734.
27. Park, H., Kumareswaran, R., and RajanBabu, T.V. (2005) *Tetrahedron*, **61**, 6352.
28. For a short review on carbohydrate ligands in asymmetric hydrovinylation and hydrocyanation reactions see: RajanBabu, T.V., Casalnuovo, A.L., Ayers, T.A., Noumura, N., Jin, J., Park, H., and Nandi, M. (2003) *Curr. Org. Chem.*, **7**, 301.
29. Kuhn, R., Baer, H.H., and Seelinger, A. (1958) *Liebigs Ann. Chem.*, **611**, 236.
30. Meyer zu Reckendorf, W. (1969) *Chem. Ber.*, **102**, 4207.
31. Smith, C.R., Lim, H.J., Zhang, A., and RajanBabu, T.V. (2009) *Synthesis*, 2089.

32 For a review see: Bini, L., Müller, C., and Vogt, D. (2010) *Chem. Commun.*, 8325.
33 Elms, P.S. and Jackson, W.R. (1982) *Aust. J. Chem.*, **50**, 5370.
34 RajanBabu, T.V. and Casalnuovo, A.L. (1992) *J. Am. Chem. Soc.*, **114**, 6265.
35 (a) Cullen, W.R. and Sugi, Y. (1978) *Tetrahedron Lett.*, **19**, 1635; (b) Jackson, R. and Thompson, D.J. (1978) *J. Organomet. Chem.*, **159**, C29; (c) Selke, R. (1979) *React. Kinet. Catal. Lett.*, **10**, 135; (d) Sinou, D. and Descotes, G. (1980) *React. Kinet. Catal. Lett.*, **14**, 463.
36 Casalnuovo, A.L., RajanBabu, T.V., Ayers, T.A., and Warren, T.H. (1994) *J. Am. Chem. Soc.*, **116**, 9869.
37 RajanBabu, T.V. and Casalnuovo, A.L. (1994) *Pure Appl. Chem.*, **66**, 1535.
38 Saha, B. and Rajan Babu, T.V. (2006) *Org. Lett.*, **8**, 4657.
39 RajanBabu, T.V. and Casalnuovo, A.L. (1996) *J. Am. Chem. Soc.*, **118**, 6325.
40 Nugent, W.A., RajanBabu, T.V., and Burk, M.J. (1993) *Science*, **259**, 479.
41 Suárez, A., Pizzano, A., Fernández, I., and Khiar, N. (2001) *Tetrahedron: Asymmetry*, **12**, 633.
42 Diéguez, M., Pàmies, O., Ruiz, A., and Claver, C. (2002) *Tetrahedron: Asymmetry*, **13**, 83.
43 Diéguez, M., Pàmies, O., and Claver, C. (2005) *Tetrahedron: Asymmetry*, **16**, 3877.

10
Carbohydrate-Derived Ligands in Asymmetric Tsuji–Trost Reactions

Montserrat Diéguez and Oscar Pàmies

10.1
Introduction

Metal-catalyzed asymmetric allylic substitution (metal = Pd, Mo, Ir, Cu, Ru, Rh, and Pt), which involves the attack of diverse nucleophiles at an allylic metal intermediate or S_N2'-type allylic substitution, is one of the most versatile routes for preparing optically active compounds [1]. Besides a high level of asymmetric induction, the advantages of this method are its tolerance of a wide range of functional groups and a great flexibility in the type of bonds that can be formed. For example, H-, C-, N-, O-, and S-centered nucleophiles can be employed. With regard to the metal source, various transition metal complexes catalyze allylic substitutions. However, the most widely used catalysts are palladium complexes. The catalytic cycle for Pd-catalyzed asymmetric allylic substitution is well established and involves olefin coordination, subsequent oxidative addition with cleavage of the leaving group, followed by nucleophilic addition, and, finally, dissociation of the substituted olefin (Scheme 10.1).

Except for the last step (dissociation of the olefin), the cycle of reactions offers various possibilities for enantiodiscrimination [1]. Scheme 10.2 shows two important classes of allylic substitutions that can be carried out enantioselectively with chiral catalysts. Type A reactions start from a racemic substrate (linear or cyclic) and proceed via symmetrical allyl systems. In this case, the enantioselectivity is determined by the regioselectivity of nucleophilic attack and therefore depends on the ability of the chiral ligand to differentiate between the two allylic termini [1]. In type B reactions, racemic or prochiral substrates possessing two identical geminal substituents at one of the allylic termini react via a π–allyl intermediate that can isomerize via the well-established π–σ–π mechanism [1]. In this case, enantioselection can occur either in the ionization step, leading to the allyl intermediate, or in the nucleophilic addition step [1]. For these latter substrates, not only does the enantioselectivity of the process need to be controlled but the regioselectivity is also a problem because a mixture of regioisomers may be obtained.

Carbohydrates – Tools for Stereoselective Synthesis, First Edition. Edited by Mike Martin Kwabena Boysen.
© 2013 Wiley-VCH Verlag GmbH & Co. KGaA. Published 2013 by Wiley-VCH Verlag GmbH & Co. KGaA.

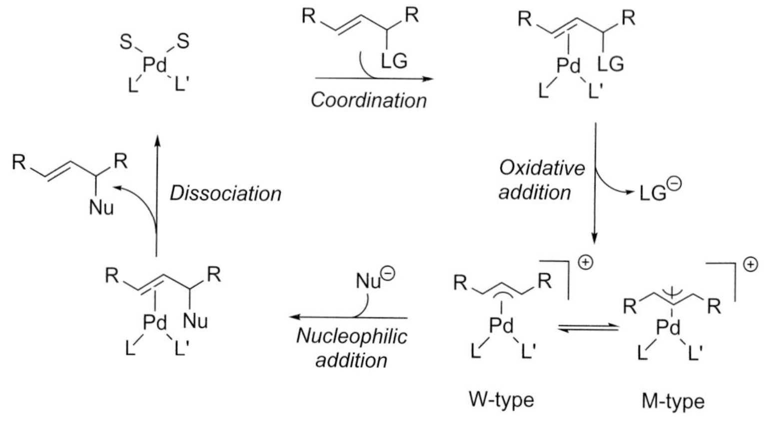
Coordination
Oxidative addition
$LG^{\ominus}$
Nucleophilic addition
$Nu^{\ominus}$
Dissociation
W-type
M-type

L,L'=mono- or bidentate ligand; S=solvent or vacant; LG=leaving group; Nu=Nucleophile

Scheme 10.1 Accepted mechanism for Pd-catalyzed allylic substitutions.

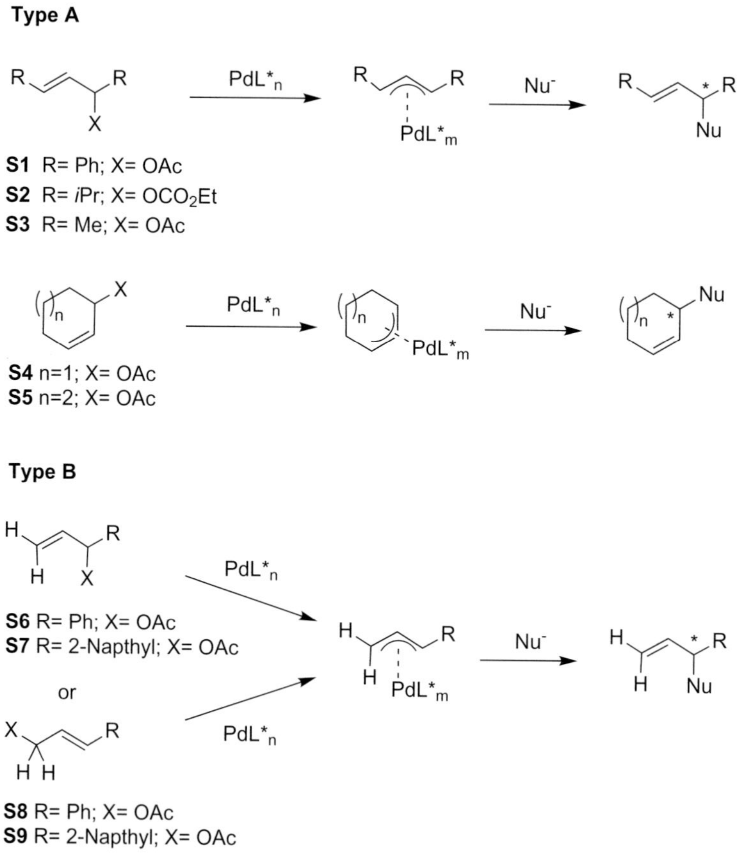
Type A
S1 R= Ph; X= OAc
S2 R= *i*Pr; X= OCO_2Et
S3 R= Me; X= OAc
S4 n=1; X= OAc
S5 n=2; X= OAc
Type B
S6 R= Ph; X= OAc
S7 R= 2-Napthyl; X= OAc
or
S8 R= Ph; X= OAc
S9 R= 2-Napthyl; X= OAc

Scheme 10.2 Two classes of asymmetric allylic substitution reactions.

10.1 Introduction

Most Pd catalysts developed to date favor the formation of an achiral linear product rather than the desired branched isomer [1]. Therefore, the development of highly regio- and enantioselective Pd catalysts is still a challenge. In contrast to Pd catalytic systems, Ir, Ru, W, and Mo catalysts provide very high selectivity for attack at the non-terminal carbon to give the chiral product [1, 2].

Since the first enantioselective catalytic process described by Trost in 1977, with moderate enantioselectivity [3], many catalytic systems have been tested. These have provided excellent enantiomeric excesses [1]. Unlike asymmetric hydrogenation processes, few diphosphines have provided good enantioselectivities in allylic substitutions. Though high ees could be obtained in certain cases, for instance with BINAP and CHIRAPHOS (Figure 10.1), the scope of standard diphosphines in this process seems limited [1].

One of the most versatile ligands for this process is a diphosphine **1** developed by Trost (Figure 10.2) [1c, 4]. The remarkable properties of this ligand are related

Figure 10.1 BINAP and CHIRAPHOS ligands.

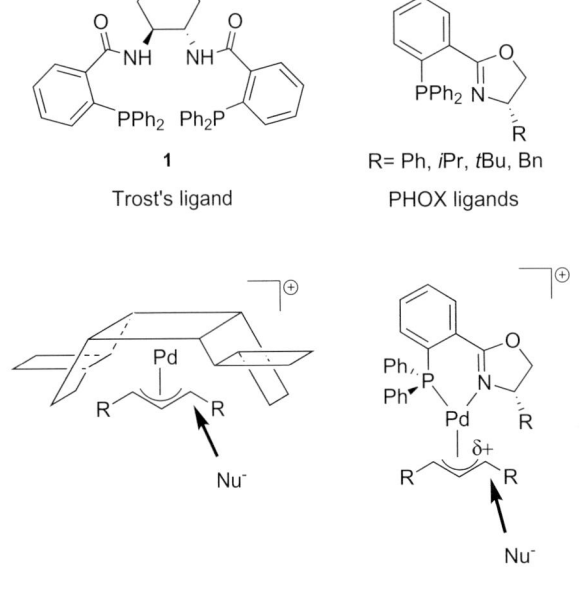

Figure 10.2 Two representative ligands developed for Pd-catalyzed allylic substitution reactions.

to the bite angle, which is larger than in unstrained Pd-diphosphine complexes. Consequently, the P-aryl groups generate a chiral cavity, in which the allyl system is embedded, that provides high ees for several sterically undemanding substrates. For diphosphines and other homodonor systems, the chiral discrimination is induced by the C_2 or C_1 symmetric backbone of the ligand.

The selection of chiral ligands for highly enantioselective allylic substitution has focused mainly on the use of mixed bidentate donors such as phosphorus–nitrogen, phosphorus–sulfur, and sulfur–nitrogen ligands [1, 5]. In this context, the phosphinooxazoline (PHOX) ligands represent, together with Trost's ligand, one of the most successful ligands developed for this process (Figure 10.2) [6]. The efficiency of such hard–soft heterodonor ligands has been mainly attributed to the different electronic effects of the donor atoms that predominantly produced the nucleophilic attack at one of the allyl carbon atoms (the one located *trans* to the best π-acceptor).

Other ligands, such as bidentate nitrogen and sulfur donors, have also exhibited very good catalytic behavior [1, 5].

Carbohydrate ligands have only recently shown their huge potential as a source of highly effective chiral ligands in Pd-catalyzed asymmetric allylic substitution reactions. Several types of ligands, mainly heterodonors, have been developed for this process and some of the results are among the best ever reported [7]. In the next section we summarize the most relevant catalytic data published for Pd-catalyzed allylic substitution with carbohydrate ligands.

10.2
Ligands

10.2.1
P-Donor Ligands

10.2.1.1 Phosphine Ligands

In 2000, C_1-symmetric diphosphine ligands **2–4**, with a furanoside backbone, were applied in Pd-catalyzed asymmetric allylic substitution reactions with moderate success (Figure 10.3) [8]. These ligands were prepared from D-(+)-xylose and

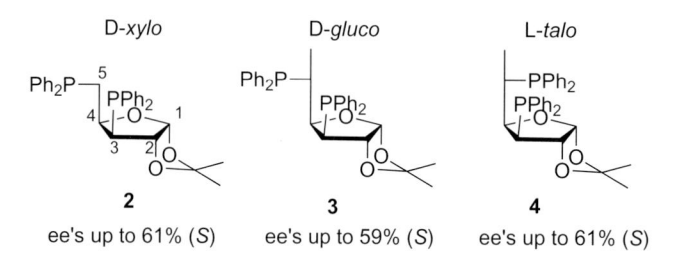

Figure 10.3 Furanoside diphosphine ligands **2–4** prepared from D-xylose and D-glucose. The maximum ee values reached are also shown.

D-(+)-glucose. Ligands **3** and **4** differ from ligand **2** at C5, where a new stereogenic center was introduced. The results in the allylic alkylation of dimethyl malonate to *(E)*-1,3-diphenylprop-2-enyl acetate (**S1**) showed that the configuration of C5 has no relevant influence on enantiodiscrimination (ees up to 61%). Moreover, for ligand **2** there was a strong solvent effect and enantioselectivities increased to 78% when THF was used.

In 2006, Ruffo and coworkers developed a modification of the Trost bis(phosphinoamides) ligands (**1**) using diamines based on D-glucose and D-mannose as chiral ligands (Scheme 10.3, ligands **5** and **6**). These sugar diamines were prepared using known procedures from *N*-acetyl-α-D-glucosamine and methyl-α-D-glucopyranoside for glucose and mannose derivatives, respectively [9]. Ligands **5** and **6** were obtained in high yield by reacting the 2,3-sugar diamine precursors with 2-(diphenylphosphino)benzoic acid in the presence of 1,3-dicyclohexylcarbodiimide (DCC) and 4-dimethylaminopyridine (DMAP) in dichloromethane (Scheme 10.3). These ligands provided high enantioselectivities in the Pd-catalyzed desymmetrization of *meso*-cyclopenten-2-ene-1,4-diol bis-carbamate (**S10**) (Scheme 10.4, up to 97% ee) [10]. Interestingly, both enantiomers of the product can be obtained in high enantioselectivities by switching from D-glucose (**5**) to D-mannose (**6**) derived ligands.

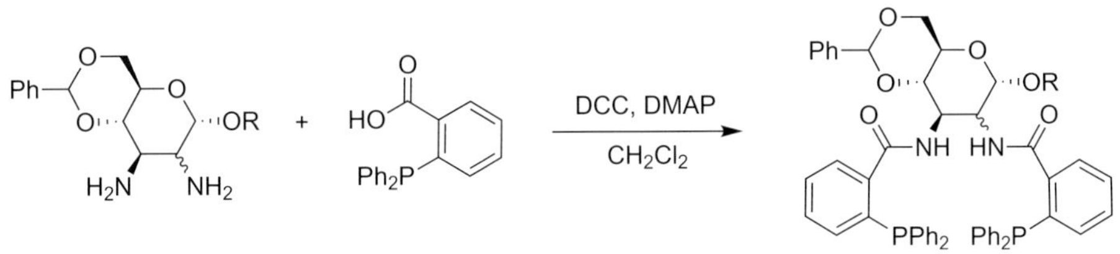

5 (R= Bn, 65% yield)
6 (R= Me, 75% yield)

Scheme 10.3 Synthesis of pyranoside–bis(phosphinoamides) **5** and **6**.

10.2.1.2 Phosphinite Ligands

RajanBabu and coworkers reported the use of the first carbohydrate-based diphosphinite ligands **7** for the allylic substitution reaction (Scheme 10.5) [11]. These ligands, derived from D-glucose, were easily obtained by reacting a dichloromethane solution of phenyl 4,6-*O*-benzylidene-β-D-glucopyranose with two equivalents of the corresponding chlorophosphine in the presence of pyridine as base (Scheme 10.5). Pyranoside diphosphinite ligands **7** were applied in the Pd-catalyzed asymmetric allylic alkylation of diethyl malonate to 1,3-diphenylprop-2-enyl acetate (**S1**) with low-to-moderate enantioselectivities (ees up to 59%,

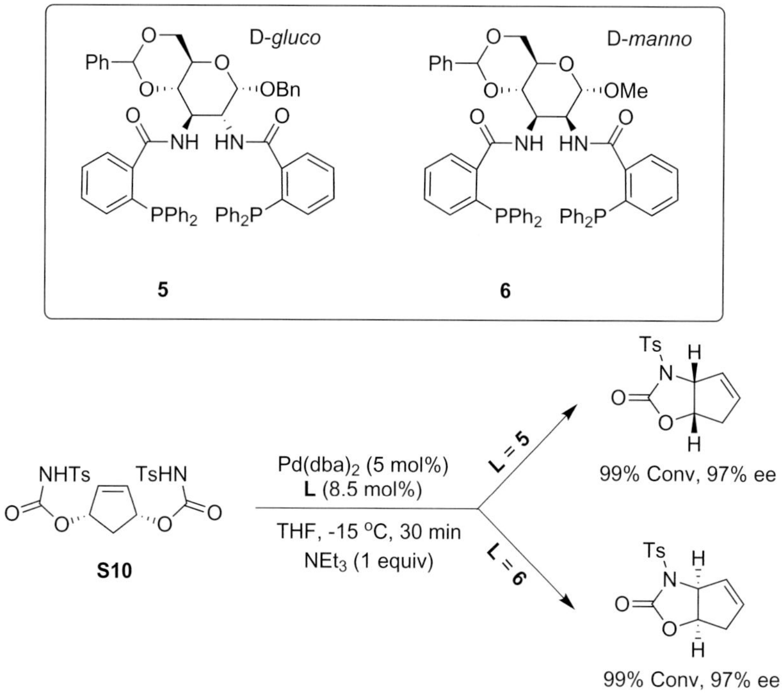

Scheme 10.4 Asymmetric desymmetrization of *meso*-cyclopenten-2-ene-1,4-diol bis-carbamate using pyranoside–bis(phosphinoamides) **5** and **6**.

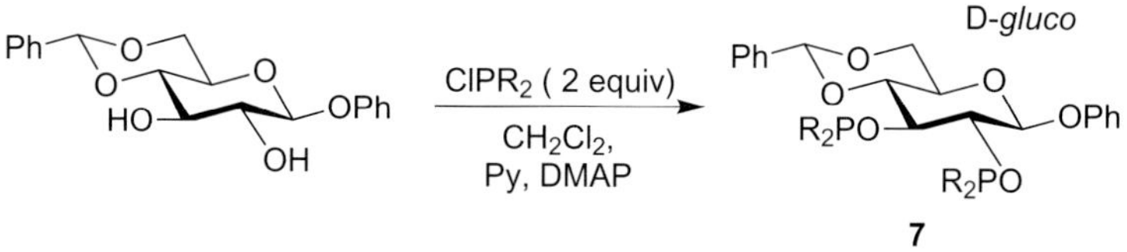

Scheme 10.5 Synthesis of pyranoside diphosphinite ligands **7a–i**.

a R= Ph; b R= 3,5-bis-TMS-C$_6$H$_3$; c R= 3,5-bis-*t*-Bu-4-OMe-C$_6$H$_2$; d R= 3,5-bis-CF$_3$-C$_6$H$_3$; e R= 4-F-C$_6$H$_4$; f R= 3,5-bis-F-C$_6$H$_3$; g R= 4-CF$_3$-C$_6$H$_4$; h R= Et; i R= Cy

Scheme 10.6) [11a]. Interestingly, the results indicated an unprecedented electronic effect. Electron-withdrawing and electron-rich diphosphinite ligands lead to products with opposite stereochemistry. Moreover, sterically bulky substituents have the same effect as electron-donating ones.

The second family developed were the furanoside diphosphinite ligands **8** and **9** with C_1 symmetry, modified at the C3 position of the carbohydrate backbones, prepared from D-xylose (Figure 10.4). Application of these ligands in the

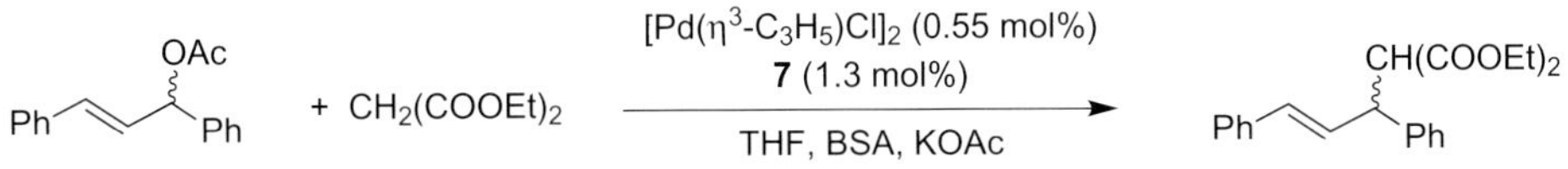

Scheme 10.6 Summary of the enantioselectivities obtained using ligands **7**. In all cases yields were nearly quantitative. Chemicals yields >95% in all cases.

Ligand	%ee
7a	0
7b	16 (R)
7c	25 (R)
7d	39 (R)
7e	17 (S)
7f	41 (S)
7g	55 (S)
7h	18 (R)
7i	59 (R)

Pd-catalyzed asymmetric allylic alkylation of linear (**S1**) and cyclic (**S4**) substrates provided low enantioselectivities (up to 31% ee) [12]. The results also showed that the absolute configuration of the carbon C3 of the ligand controlled the configuration of the allylic alkylation product.

10.2.1.3 Phosphite Ligands

In 2001, the first diphosphite ligand family (**10–16**) applied to Pd-catalyzed asymmetric allylic substitution reactions was reported (Figure 10.5) [8a, 13]. The new ligands were synthesized very efficiently in one step from the corresponding diols, which were easily prepared on a large scale from D-(+)-xylose and D-(+)-glucose using standard procedures (Scheme 10.7) [14]. Therefore, reacting the corresponding diol with two equivalents of the desired *in-situ* formed phosphorochloridite in the presence of base afforded the desired ligands as white air-stable solids in moderate to good overall yield (50–67%) [14]. The highly modular construction of these ligands allows sufficient flexibility to fine-tune (i) the various configurations of the carbohydrate backbone, (ii) the substituents on C5 (R = H, Me, OTBDPS), and (iii) the steric and electronic properties of the diphosphite substituents (**a–h**).

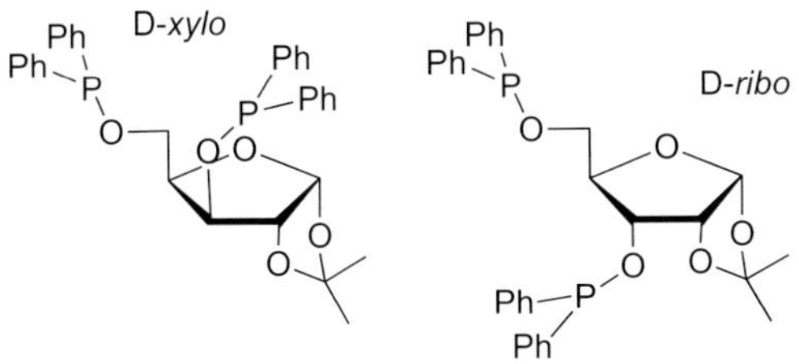

Figure 10.4 Furanoside diphosphinite ligands **8** and **9**.

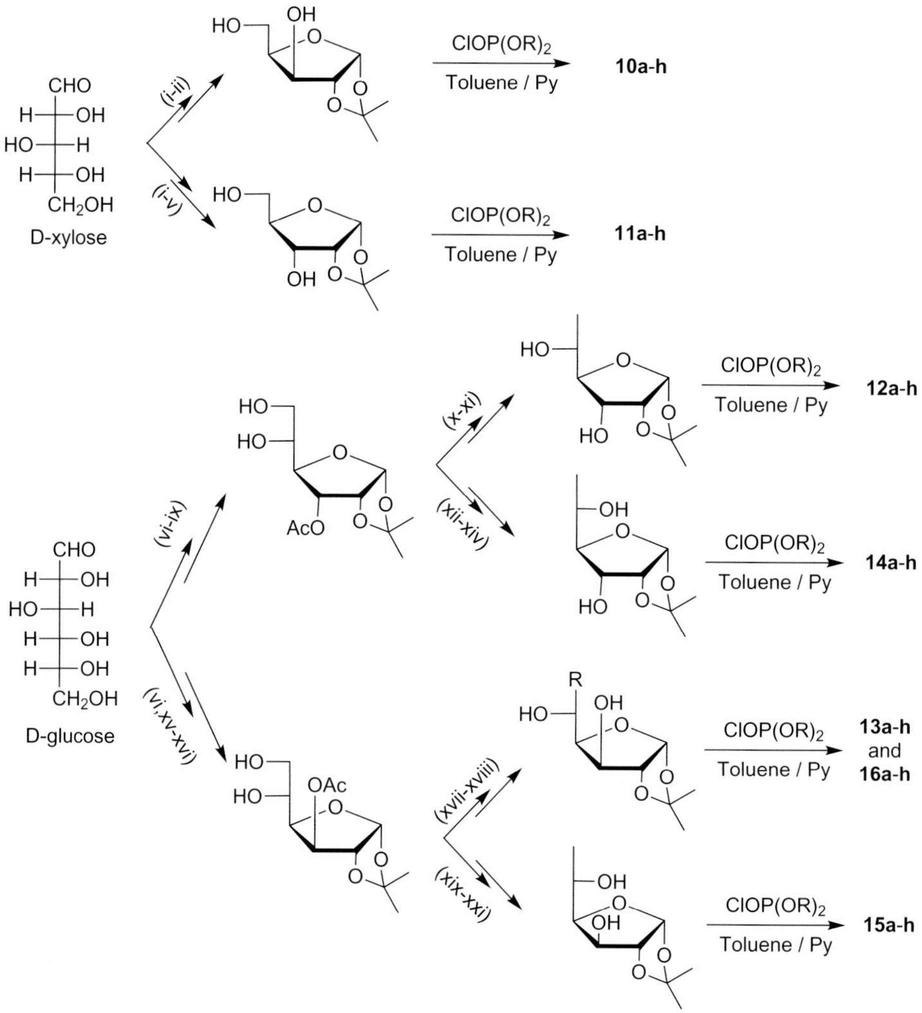

Scheme 10.7 Synthesis of furanoside diphosphite ligands **10–16**: (i) I$_2$/acetone (95% yield); (ii) H$_2$SO$_4$/CH$_3$OH (90% yield); (iii) BzCl/CH$_2$Cl$_2$/Py, −20 °C to room temp. (85% yield); (iv) PCC/AcONa/CH$_2$Cl$_2$ then NaBH$_4$/EtOH/H$_2$O, −15 °C (70% yield); (v) NH$_4$OH/MeOH (70% yield); (vi) I$_2$/acetone (98% yield); (vii) PCC/AcONa/CH$_2$Cl$_2$ then NaBH$_4$/EtOH/H$_2$O, −15 °C (82% yield); (viii) Ac$_2$O/Py/CH$_2$Cl$_2$, 0 °C to room temp. (91% yield); (ix) AcOH/H$_2$O (88% yield); (x) TsCl/Py/CH$_2$Cl$_2$, −20 °C to room temp. (88%); (xi) NaOMe/CH$_2$Cl$_2$ then LiAlH$_4$/THF, 60 °C (92% yield); (xii) Ac$_2$O/Py/CH$_2$Cl$_2$, 0 °C to room temp. (90% yield); (xiii) Tf$_2$O/Py/CH$_2$Cl$_2$ (57% yield); (xiv) NaOMe/CH$_2$Cl$_2$ then LiAlH$_4$/THF, 60 °C (79% yield); (xv) Ac$_2$O/Py/CH$_2$Cl$_2$, 0 °C to room temp. (87% yield); (xvi) AcOH/H$_2$O (84% yield); (xvii) TsCl/Py/CH$_2$Cl$_2$, −20 °C to room temp. (81%); (xviii) NaOMe/CH$_2$Cl$_2$ then LiAlH$_4$/THF, 60 °C (92% yield); (xix) Ac$_2$O/Py/CH$_2$Cl$_2$, 0 °C to room temp. (92% yield); (xx) Tf$_2$O/Py/CH$_2$Cl$_2$ (54% yield); (xxi) NaOMe/CH$_2$Cl$_2$ then LiAlH$_4$/THF, 60 °C (78% yield).

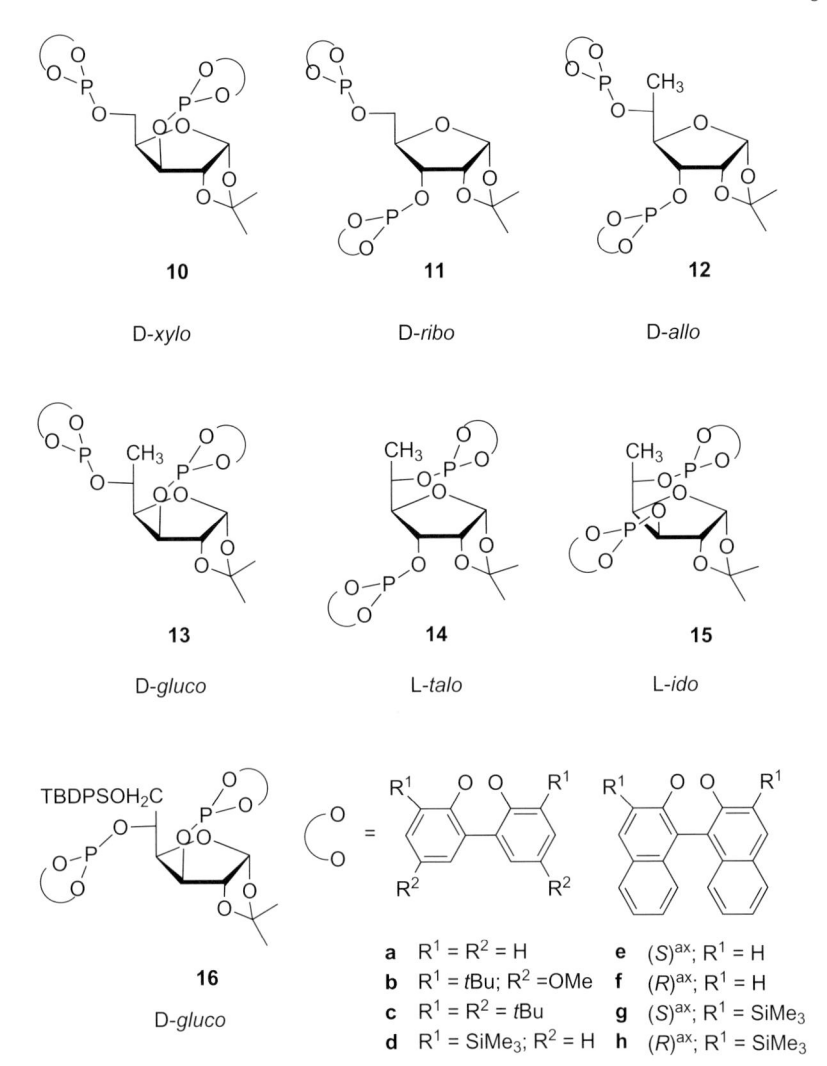

Figure 10.5 Furanoside diphosphite ligands **10–16**.

The many combinations they provide are the key to finding the most suitable ligands for each particular substrate. Therefore, this set of ligands has been successfully applied in the Pd-catalyzed allylic substitution of dimethyl malonate and benzylamine to several acyclic (**S1**, **S3**, **S6**, and **S7**) and cyclic (**S4**) allylic esters (Figure 10.6).

The results indicated that activities were best when the substituent at C5 was methyl and when the ligand contained biphenyl-derived phosphites with bulky substituents at the *ortho* positions located on the phosphites and electron-donating substituents at the *para* positions of the biphenyl moieties (i.e., **b** ~ **c** > **d** > **a**).

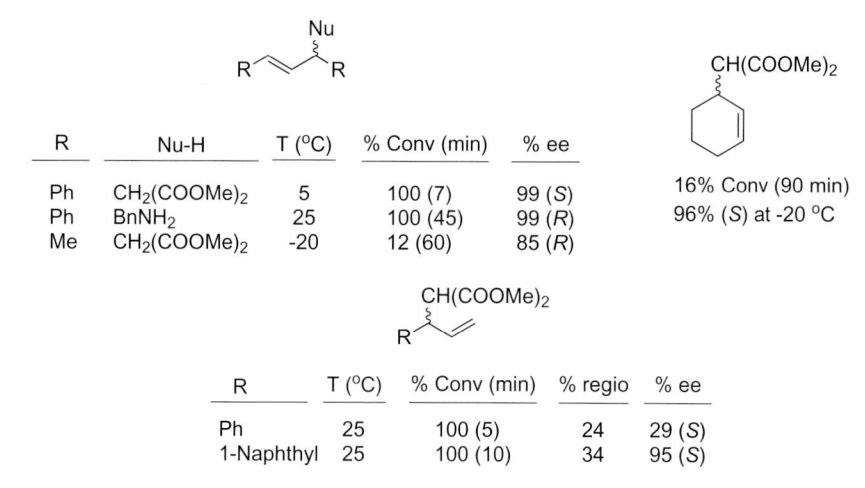

Figure 10.6 Summary of the best results obtained using ligand **13c**.

Ligand **13c** structure with tBu biphenyl phosphite groups.

R	Nu-H	T (°C)	% Conv (min)	% ee
Ph	CH$_2$(COOMe)$_2$	5	100 (7)	99 (S)
Ph	BnNH$_2$	25	100 (45)	99 (R)
Me	CH$_2$(COOMe)$_2$	-20	12 (60)	85 (R)

Cyclohexenyl CH(COOMe)$_2$: 16% Conv (90 min), 96% (S) at −20 °C

R	T (°C)	% Conv (min)	% regio	% ee
Ph	25	100 (5)	24	29 (S)
1-Naphthyl	25	100 (10)	34	95 (S)

Enantioselectivities were affected by the substituent at C5 and the phosphite moieties and by the configuration of C3 and C5 and the configurations of the biaryl moieties. Enantioselectivities were best with ligand **13c**, which has a glucofuranoside backbone and bulky *tert*-butyl substituents at both *ortho* and *para* positions of the biphenyl moieties. The results also indicated that nucleophilic attack takes place *trans* to the phosphite located on carbon atom C5 (Figure 10.7). Ligand **10c** was also used to stabilize Pd-nanoparticles. These particles catalyzed the allylic alkylation of *rac*-1,3-diphenylprop-2-enyl acetate with dimethyl malonate, leading to an almost total conversion of the *(R)*-enantiomer and almost no reaction with the *(S)*-enantiomer. This gives rise to 97% ee for the alkylation product and a kinetic resolution of the substrate recovered with ca. 90% ee [15].

Recently, furanoside ligands of C_2 symmetry **17** and **18** – systematically modified at positions 2 and 5 and in the biaryl phosphite ligands, and prepared from D-glucosamine and D-glucitol (Figure 10.8) – were successfully applied in the Pd-catalyzed allylic substitution reaction of 1,3-diphenylprop-2-enyl acetate (**S1**) using benzylamine and dimethyl malonate under standard conditions [16]. Ligand **17** provided high activity (TOFs > 22 000 h^{-1}) and excellent enantioselectivities [up to 99% ee, (S)].

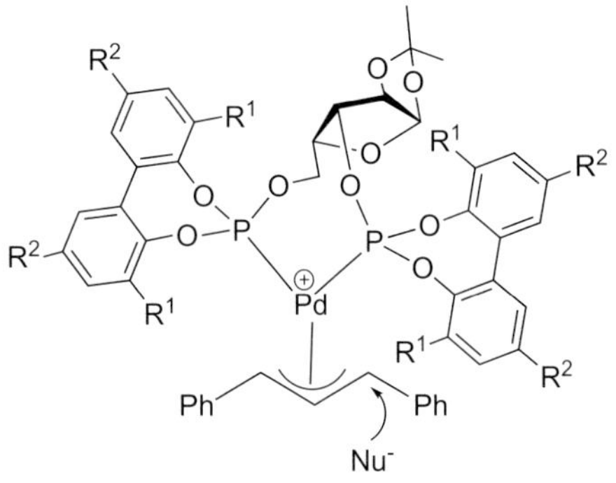

Figure 10.7 Key Pd-allyl intermediate containing the furanoside diphosphite ligand.

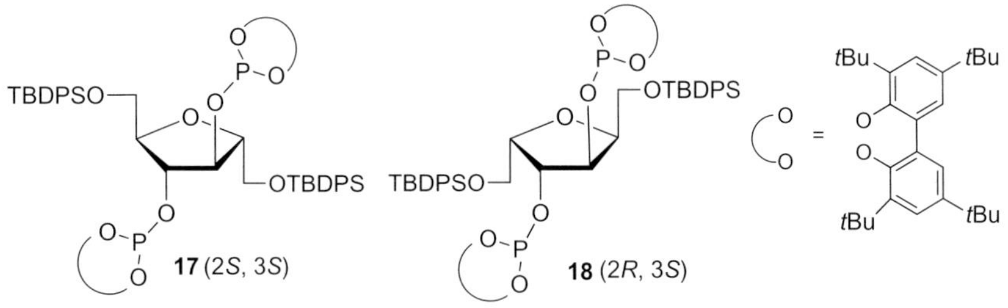

Figure 10.8 Diphosphite ligands **17** and **18**.

10.2.1.4 Phosphoroamidite Ligands

During recent decades, there has been a huge advance in the use of phosphoroamidite ligands for several asymmetric processes [17]. However, to the best our knowledge, only one family of diphosphoramidite ligands (**19**) based on carbohydrates has been successfully applied in asymmetric catalysis (Scheme 10.8) [18]. The new diphosphoramidite ligands **19** were synthesized very efficiently from 3,5-dideoxy-3,5-diamino-1,2-O-isopropylidene-ribofuranose by reacting two equivalents of the desired *in-situ* formed phosphorochloridite in the presence of pyridine. 3,5-Dideoxy-3,5-diamino-1,2-O-isopropylidene-ribofuranose was, in turn, easily prepared on a large scale from inexpensive D-(+)-xylose (Scheme 10.8). Good-to-excellent activities [TOFs up to 850 mol substrate x (mol Pd x h)$^{-1}$] and enantioselectivities (up to 95% ee) have been obtained in the Pd-catalyzed allylic alkylation for several di- (**S1**) and monosubstituted (**S6**) linear and cyclic (**S4**) substrates (Figure 10.9). The results indicate that catalytic performance is greatly affected by the substituents and the axial chirality of the biaryl moieties of the ligand. The study of 1,3-diphenyl and cyclohexenyl Pd-π-allyl intermediates indicates that the nucleophilic attack takes place predominantly at the allylic terminal carbon atom located *trans* to the phosphoroamidite moiety attached to C5.

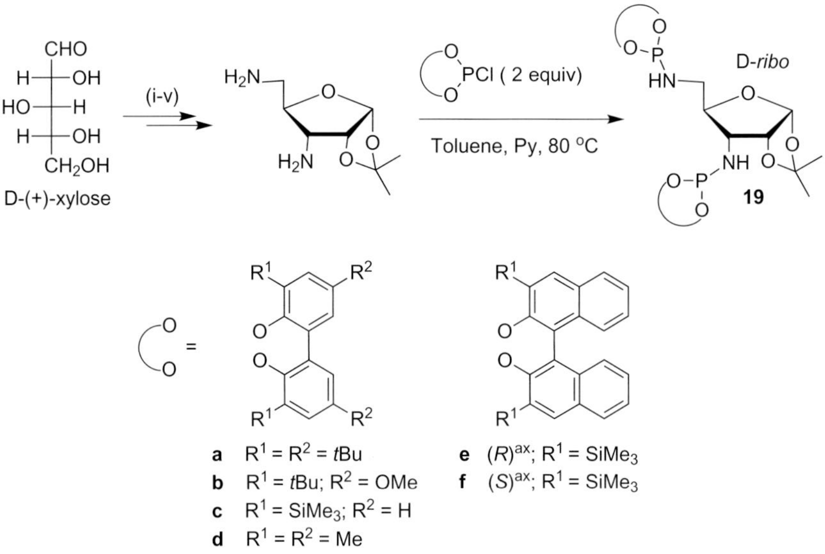
CHO
H OH
HO H
H OH
CH2OH
D-(+)-xylose
(i-v)
H2N
H2N
PCl (2 equiv)
Toluene, Py, 80 °C
D-ribo
HN
NH
19
R1
R2
a R1 = R2 = tBu
b R1 = tBu; R2 = OMe
c R1 = SiMe3; R2 = H
d R1 = R2 = Me
e (R)ax; R1 = SiMe3
f (S)ax; R1 = SiMe3

Scheme 10.8 Synthesis of furanoside diphosphoramidite ligands **19:** (i) I_2/acetone (95% yield); (ii) H_2SO_4/CH_3OH (90% yield); (iii) Tf_2O/Py/CH_2Cl_2, −20 °C (55% yield); (iv) NaN_3/DMF, 80 °C (76% yield); (v) PPh_3/THF/H_2O (77% yield).

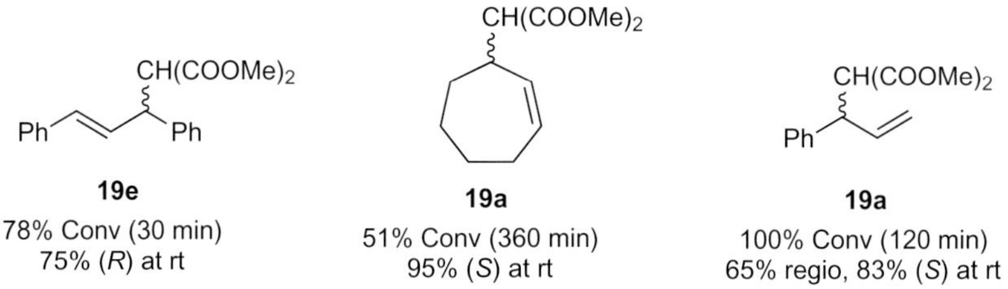
CH(COOMe)2
Ph
Ph
19e
78% Conv (30 min)
75% (R) at rt
CH(COOMe)2
19a
51% Conv (360 min)
95% (S) at rt
CH(COOMe)2
Ph
19a
100% Conv (120 min)
65% regio, 83% (S) at rt

Figure 10.9 Summary of the best results obtained using ligands **19.**

10.2.1.5 **P-P′ Ligands**

The first successful family of P-P′ carbohydrate ligands were the phosphite–phosphoroamidite ligands **20–23** (Scheme 10.9). These ligands were synthesized very efficiently from the corresponding easily accessible aminoalcohol sugar derivatives, which are easily made in a few steps from the corresponding D-xylose or D-glucose, by reaction with two equivalents of the appropriate *in-situ* formed phosphorochloridite [ClP(OR)$_2$; (OR)$_2$ = **a–f**] in the presence of pyridine (Scheme 10.9). Ligands **20–23** were successfully applied in Pd-asymmetric allylic substitution (up to 98% ee) [19]. Interestingly, this ligand family also provides high activity (because of the high π-acceptor capacity of the phosphoroamidite moiety) and high enantioselectivities for different substrate types [mono- (**S7** and **S9**) and disubstituted (**S1** and **S3**) linear and cyclic (**S4** and **S5**) substrates] (Figure 10.10). Related phosphine–phosphite ligands **24** (Figure 10.11) with a furanoside backbone have also

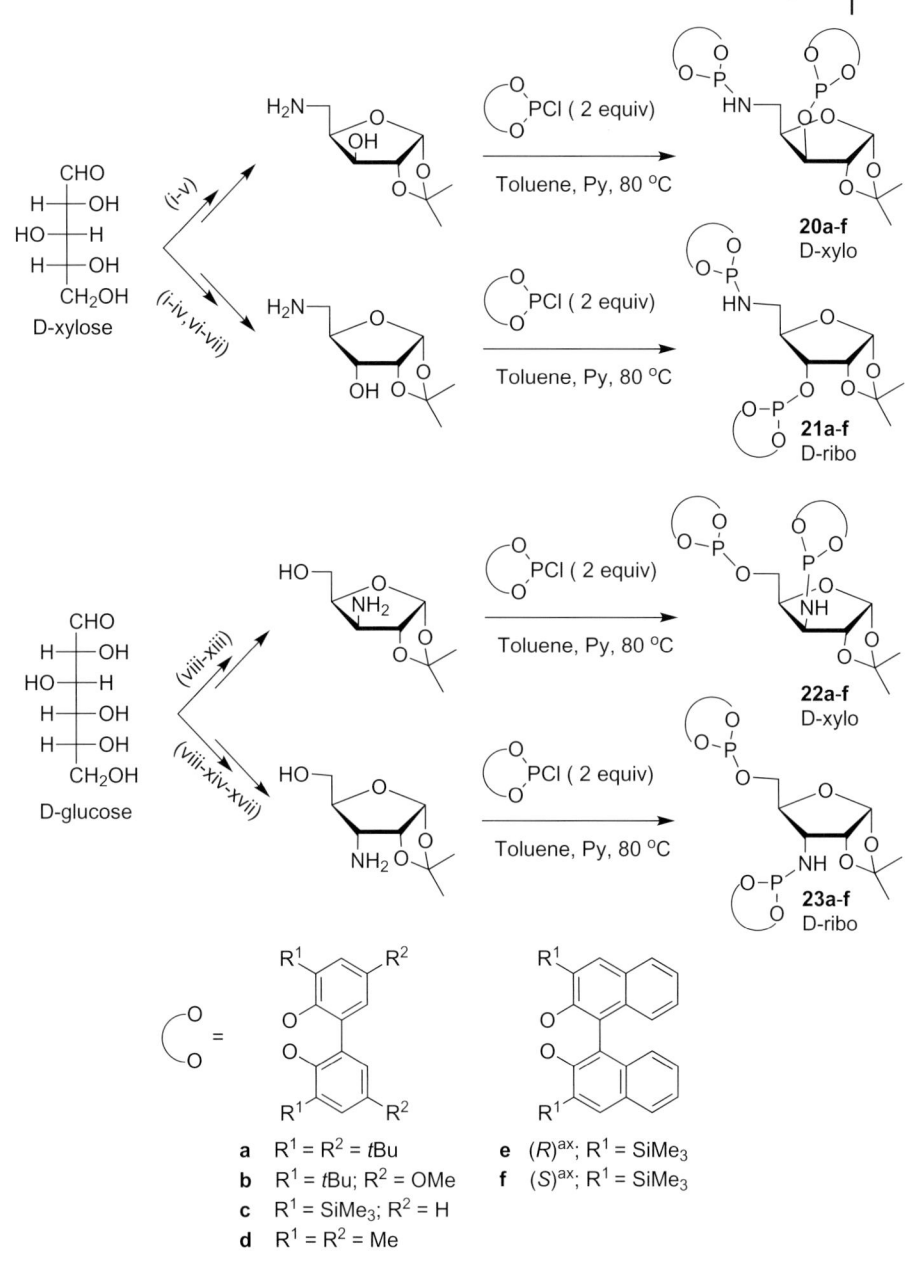

Scheme 10.9 Synthesis of furanoside phosphite–phosphoramidite ligands **20–23**: (i) I$_2$/acetone (95% yield); (ii) H$_2$SO$_4$/CH$_3$OH (90% yield); (iii) TsCl/CH$_2$Cl$_2$/Py, −20 °C to room temp. (89% yield); (iv) NaN$_3$/DMF (96% yield); (v) PPh$_3$/THF/H$_2$O (92% yield); (vi) PCC/AcONa/CH$_2$Cl$_2$ then NaBH$_4$/EtOH/H$_2$O, −15 °C (74% yield); (vii) PPh$_3$/THF/H$_2$O (90% yield); (viii) I$_2$/acetone (98% yield); (ix) PCC/AcONa/CH$_2$Cl$_2$ then NaBH$_4$/EtOH/H$_2$O, −15 °C (82% yield); (x) Tf$_2$O/Py/CH$_2$Cl$_2$, −10 °C (76% yield); (xi) NaN$_3$/BuNCl/DMF, 50 °C (86% yield); (xii) AcOH/H$_2$O then NaIO$_4$/NaHCO$_3$/H$_2$O (85% yield); (xiii) NaBH$_4$/EtOH then PPh$_3$/THF/H$_2$O (82% yield); (xiv) Tf$_2$O/Py/CH$_2$Cl$_2$, −10 °C (98% yield); (xv) NaN$_3$/BuNCl/DMF, 50 °C (70% yield); (xvi) AcOH/H$_2$O then NaIO$_4$/NaHCO$_3$/H$_2$O (91% yield); (xvii) NaBH$_4$/EtOH then PPh$_3$/THF/H$_2$O (82% yield).

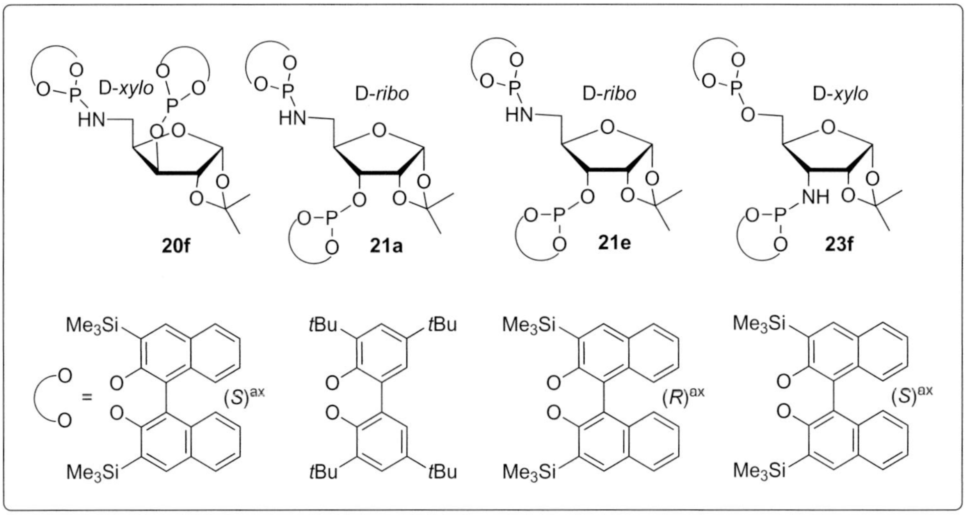

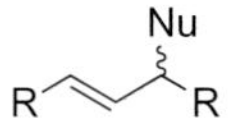

10 Carbohydrate-Derived Ligands in Asymmetric Tsuji–Trost Reactions

Figure 10.10 Summary of the best results obtained in the Pd-catalyzed allylic substitution of several substrates using ligands **20–23**.

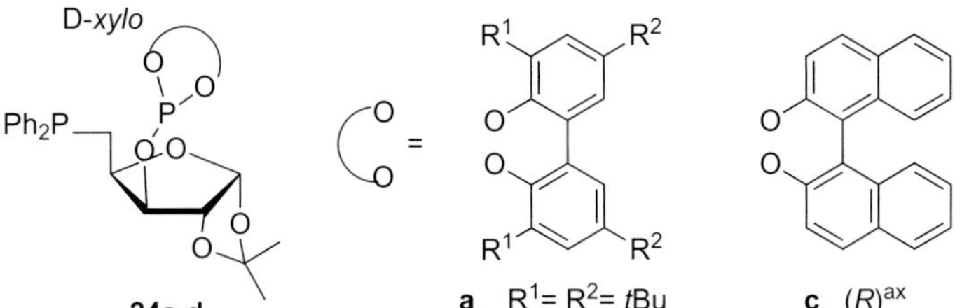

R	Nu-H	L	% Conv (time)	% ee
Ph	CH$_2$(COOMe)$_2$	20f	99 (15 min)	98 (S)
Ph	BnNH$_2$	20f	98 (6 h)	97 (R)
Me	CH$_2$(COOMe)$_2$	23f	51 (30 min)	84 (S)
Me	BnNH$_2$	23f	68 (4 h)	84 (R)

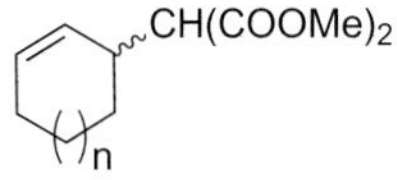

n	L	% Conv (time)	% ee
1	21a	100 (2 h)	85 (S)
2	21a	52 (6 h)	91 (S)

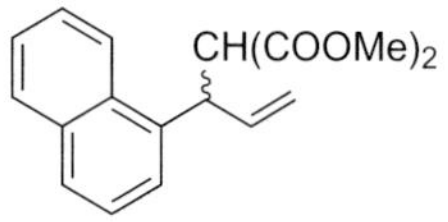

L	% Conv (time)	% regio	% ee
21e	100 (3 h)	70	90 (S)

Figure 10.11 Phosphine–phosphite ligands **24**.

a R^1 = R^2 = tBu
b R^1 = tBu; R^2 = OMe
c (R)^{ax}
d (S)^{ax}

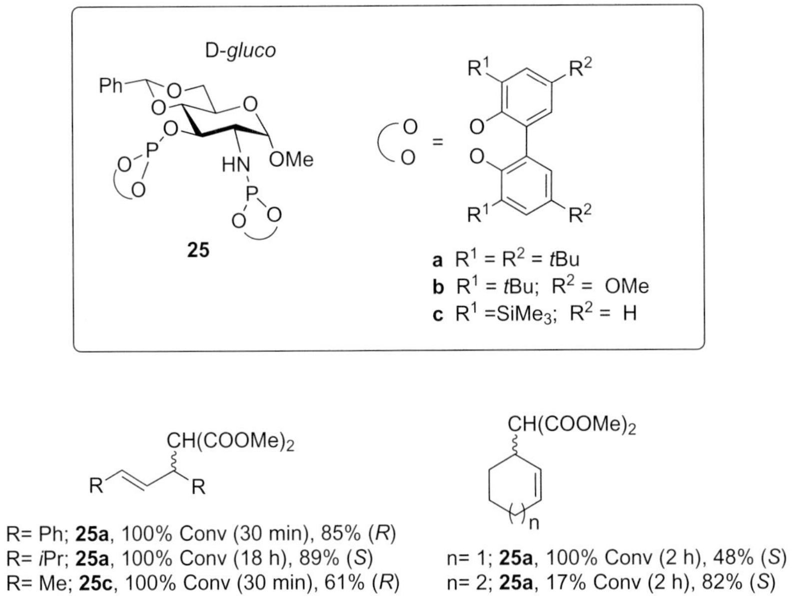
D-gluco
Ph
O
O
O
O
P
O
O
HN
OMe
P
O
O
25
R1
R2
O
O
=
O
O
R1
R2
a R1 = R2 = tBu
b R1 = tBu; R2 = OMe
c R1 =SiMe3; R2 = H
CH(COOMe)2
R
R
R= Ph; 25a, 100% Conv (30 min), 85% (R)
R= iPr; 25a, 100% Conv (18 h), 89% (S)
R= Me; 25c, 100% Conv (30 min), 61% (R)
CH(COOMe)2
n
n= 1; 25a, 100% Conv (2 h), 48% (S)
n= 2; 25a, 17% Conv (2 h), 82% (S)

Figure 10.12 Pyranoside phosphine–phosphite ligands **25:** summary of the best results obtained with several substrates.

been used in enantioselective Pd-catalyzed allylic alkylation and amination substitutions of **S1** as model reactions, providing ees of up to 42% *(S)* and 66% *(R)*, respectively [13b].

More recently, the phosphite–phosphoroamidite ligands **25** with a pyranoside backbone (Figure 10.12) have been developed for the Pd-catalyzed allylic substitution reaction of several substrates. Enantioselectivities up to 89% have been obtained for disubstituted linear and cyclic substrates (Figure 10.12) [20].

10.2.2
S-Donor Ligands

Sulfur donor ligands have been used much less frequently than phosphorus ligands in Tsuji–Trost reactions because a mixture of diastereomeric complexes may be formed upon coordination of the thioether ligand to the metal, which can lead to a decrease in stereoselectivity if the relative rates of the catalytically active intermediates are similar. Despite this problem, high enantiomeric excesses have been achieved [5]. In this context, Khiar and coworkers used a combinatorial approach to find the best dithioether ligand **26** from a library of 64 potential ligands made by combining four linkers, four sugar residues, and four protective groups (Figure 10.13) for the Pd-catalyzed allylic alkylation of dimethyl malonate to 1,3-diphenylprop-2-enyl acetate (**S1**) (up to 90% ee, Scheme 10.10) under standard conditions [21a]. To access both enantiomers of the alkylation product, the authors

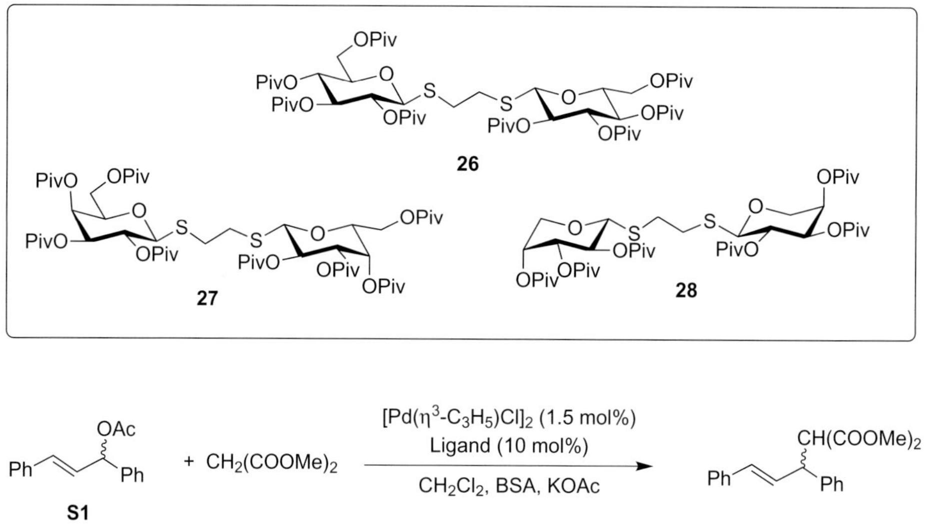

Scheme 10.10 Summary of the best results obtained in the Pd-catalyzed allylic alkylation of **S1** using dithioether ligands **26–28**.

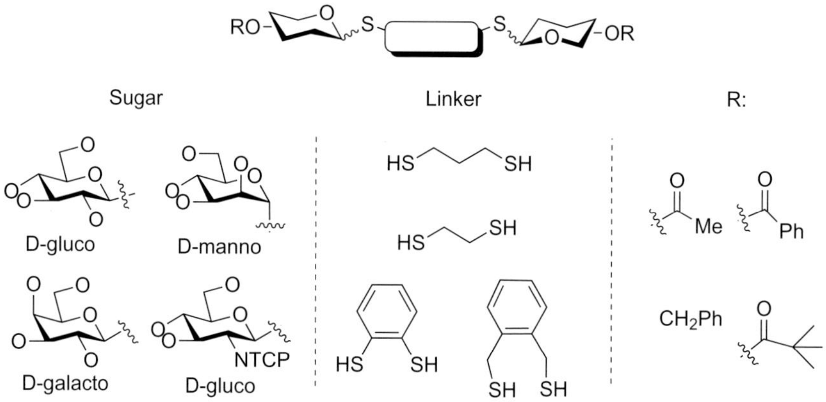

L	% ee
26	90 (S)
27	88 (S)
28	86 (R)

NTCP = Tetrachlorophthalimide

Figure 10.13 Dithioether ligand library studied by Khiar and coworkers.

successfully prepared *pseudo*-enantiomers **27** and **28**, derived from D-galactose and D-arabinose, respectively (Scheme 10.10) [21b].

The authors found that although three diastereomeric Pd(II) complexes may be formed from these ligands (one in which *syn* and two *anti*), only one was obtained in all cases, regardless of the sugar moiety and the protecting groups. This high diastereoselectivity for the complexation was attributed to the *exo*-anomeric effect [22]. Accordingly, in the Newman projection of the starting bis-thioglycoside in the conformation stabilized by the *exo* anomeric effect, it can be seen that only the coordination of pro-*(S)* lone pair of the sulfur to the Pd leads to a complex that retains the stabilization of the *exo*-anomeric effect while also being sterically favorable (Scheme 10.11). However, the observed enantiomeric discrimination in the catalytic reactions seems to be related to the steric shielding exerted by the protecting groups of the carbohydrate scaffolds and not to the high stereocontrol of the complexation of the sulfur atom directed by the *exo*-anomeric effect.

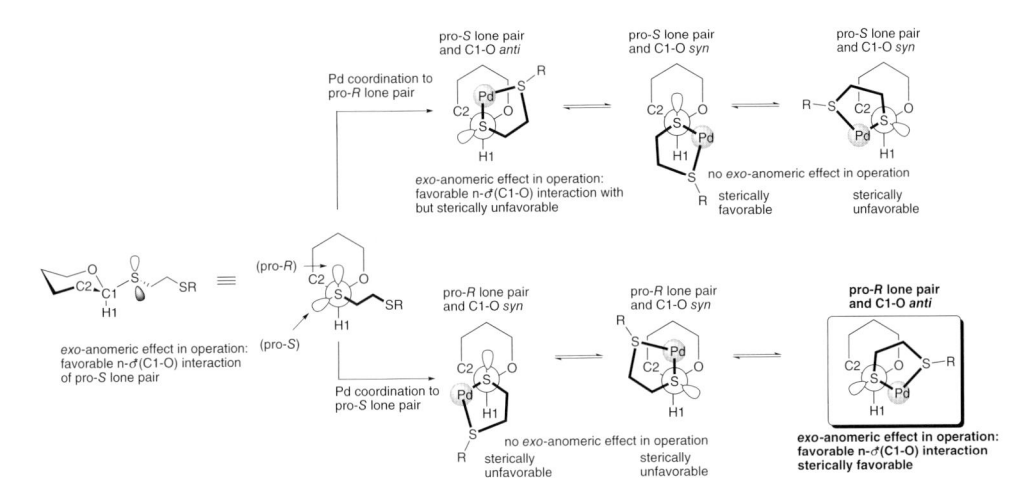

Scheme 10.11 Newman projections illustrating the *exo*-anomeric effect when the Pd is coordinated to the pro-*(S)* lone electron pair.

10.2.3
Heterodonor Ligands

10.2.3.1 P-S Ligands

Several combinations of *P,S*-donor ligands such as phosphine–thioethers, phosphinite–thioethers, phosphine–oxathianes, and phosphite–thioethers have been studied. In particular, the phosphine–thioethers, phosphinite–thioethers, and phosphine–oxathianes have proven to be effective in enantioselective Pd-catalyzed allylic substitutions.

The ferrocenylphosphine–thioglucoside ligand **29** (Figure 10.14) with multiple stereogenic units afforded an ee of 88% in the palladium allylic substitution of diethyl malonate with 1,3-diphenylprop-2-enyl acetate (**S1**) [23a]. However, when

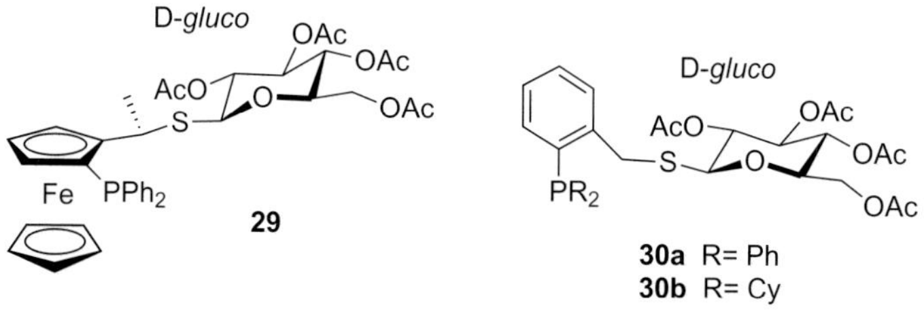

Figure 10.14 Thioether–phosphine ligands **29** and **30**.

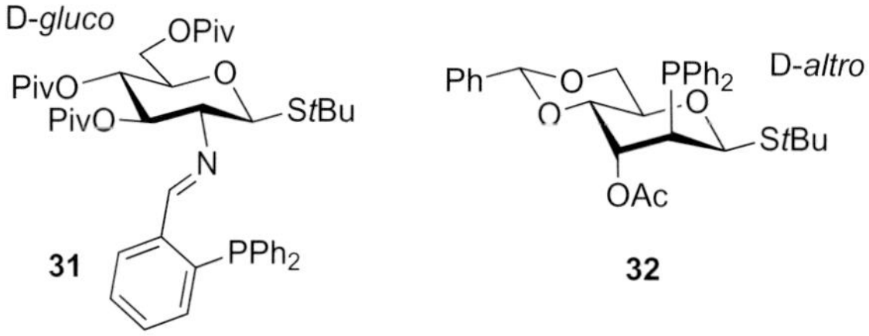

Figure 10.15 Phosphine–thioether ligands **31** and **32**.

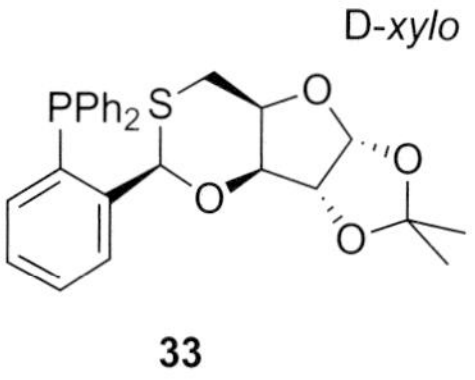

Figure 10.16 Phosphine–oxathiane ligand **33**.

the thiosugar moiety was the sole stereogenic unit on ligand **30** (Figure 10.14), enantioselectivities were only moderate (up to 64% ee) [23b].

Khiar and coworkers reported the successful use of phosphine–thioether ligand **31** (Figure 10.15) in the Pd-catalyzed asymmetric allylic alkylation of 1,3-diphenylprop-2-enyl acetate (**S1**) [yields up to 82% and up to 90% ee, (*S*)] [24a]. The same group also reported the application of ligand **32** (Figure 10.15), but with little success [yields up to 55% and up to 30% ee, (*R*)] [24b].

In 2003, a phosphine–oxathiane ligand **33**, derived from D-(+)-xylose, was developed for Pd-catalyzed allylic substitution reactions (Figure 10.16). Good enantioselectivities were obtained in the addition of dimethyl malonate and benzylamine to 1,3-diphenylprop-2-enyl acetate (**S1**) [ees up to 91% (*S*) and 94% (*R*), respectively] [25].

More recently, a series of phosphinite–thioether ligands with a furanoside backbone, **34a–g**, were easily prepared from D-xylose (Scheme 10.12). These ligands were applied in the Pd-catalyzed allylic substitution of mono- (**S6** and **S7**) and disubstituted (**S1–S3**) linear and cyclic (**S4** and **S5**) substrates (up to 95% ee)

Scheme 10.12 Synthesis of phosphinite–thioether ligands **34**: (i) I$_2$/acetone (95% yield); (ii) H$_2$SO$_4$/CH$_3$OH (90% yield); (iii) Tf$_2$O/CH$_2$Cl$_2$/Py, −20 °C (65% yield).

a R = Ph; **b** R = Me; **c** R = iPr; **d** R = tBu; **e** R = 4-Me-C$_6$H$_4$; **f** R = 4-CF$_3$-C$_6$H$_4$; **g** R = 2,6-di-Me-C$_6$H$_3$

(Figure 10.17) [26]. These ligands contained several thioether substituents with different electronic and steric properties. The authors found that this group had an important effect on catalytic performance. Thus, enantioselectivities were best when the bulkiest ligands **34c,d** were used. Replacement of the phosphinite group by a bulky biaryl phosphite leads to much lower enantioselectivity [13b].

At the same time, simple phosphinite–thioether ligands **35** and **36** with pyranoside backbones (Figure 10.18) were successfully applied in Pd-catalyzed allylic substitution of 1,3-diphenylprop-2-enyl acetate (**S1**) (up to 96% ee). Enantioselectivities were best when bulky *tert*-butyl substituents were present in the thioether moiety. Both enantiomers of the products were obtained by using pseudo-enantiomeric ligands **35a** and **36** [27].

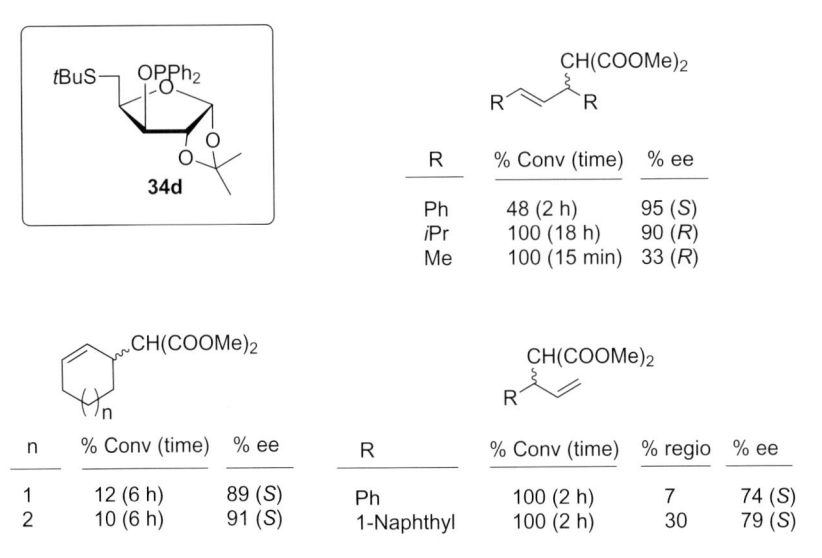

R	% Conv (time)	% ee
Ph	48 (2 h)	95 (S)
iPr	100 (18 h)	90 (R)
Me	100 (15 min)	33 (R)

n	% Conv (time)	% ee
1	12 (6 h)	89 (S)
2	10 (6 h)	91 (S)

R	% Conv (time)	% regio	% ee
Ph	100 (2 h)	7	74 (S)
1-Naphthyl	100 (2 h)	30	79 (S)

Figure 10.17 Summary of the best results obtained in the Pd-catalyzed allylic alkylation of several substrates using ligand **34d**.

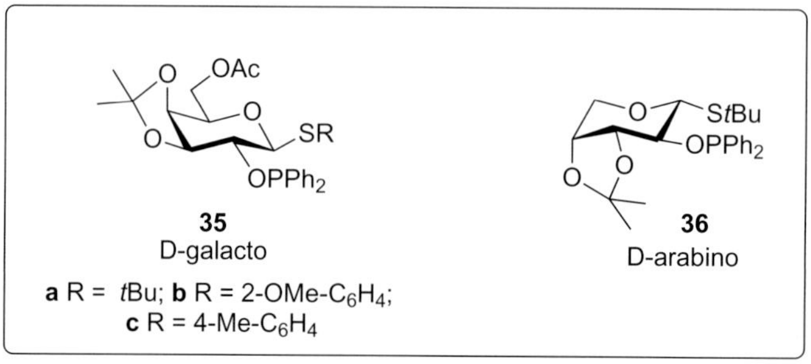

35
D-galacto

a R = tBu; b R = 2-OMe-C₆H₄;
c R = 4-Me-C₆H₄

36
D-arabino

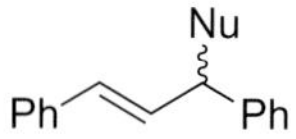

L	Nu-H	% ee
35a	CH₂(COOMe)₂	96 (S)
35a	BnNH₂	96 (R)
36	CH₂(COOMe)₂	84 (R)
36	BnNH₂	88 (S)

Figure 10.18 Phosphinite–thioether ligands **35** and **36**: summary of the best results obtained in the Pd-catalyzed allylic substitution of **S1**.

10.2.3.2 P-N Ligands

Several types of P,N-donor carbohydrate ligands have been developed for use in Pd-asymmetric allylic substitutions. In particular, many phosphorus–oxazoline ligands have produced excellent results.

Kunz and coworkers developed a phosphine–oxazoline ligand **37** derived from D-glucosamine for the Pd-catalyzed allylic alkylation of dimethyl malonate to symmetrically and non-symmetrically substituted allyl acetates with high enantioselectivities (up to 98% ee) (Figure 10.19) [28]. These results are in agreement with a nucleophilic attack *trans* to the phosphorus atom.

Uemura and coworkers have developed a series of phosphinite–oxazoline ligands **38** for the Pd-catalyzed allylic substitution reactions (Scheme 10.13) [29]. These ligands are prepared from the commercially available D-glucosamine hydrochloride. They showed high enantioselectivity with 1,3-diphenylprop-2-enyl acetate

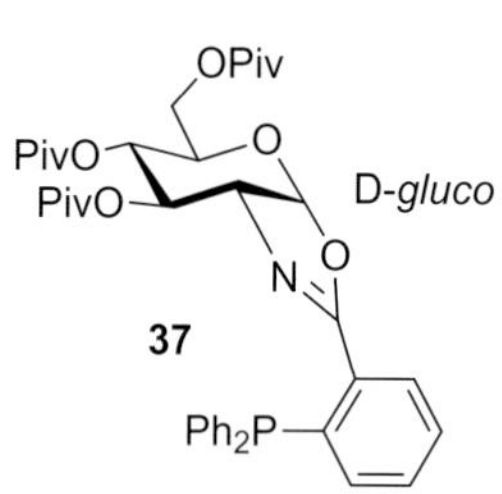

Figure 10.19 Phosphine–oxazoline ligand **37** developed by Kunz and coworkers.

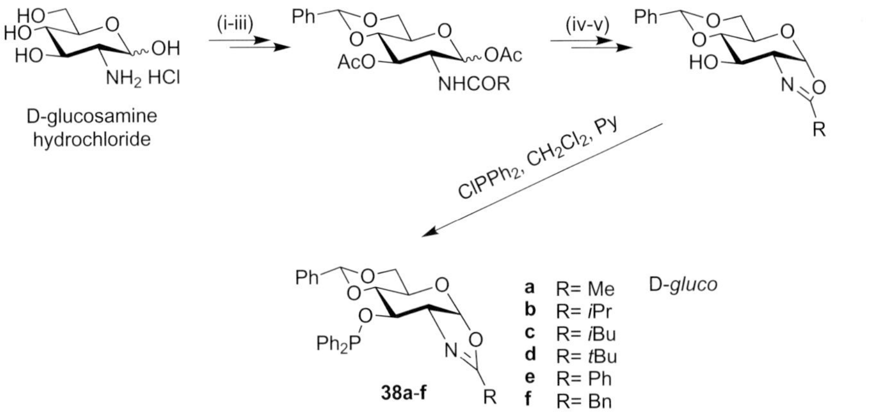

Scheme 10.13 Synthesis of pyranoside phosphinite–oxazoline ligands **38**: (i) (RCO)$_2$O/NaOMe/MeOH or RCOCl/NaHCO$_3$(aq.); (ii) PhCHO/ZnCl$_2$; (iii) Ac$_2$O/Py; (iv) SnCl$_4$/CH$_2$Cl$_2$; (v) K$_2$CO$_3$/MeOH.

(**S1**) as a substrate, but enantioselectivities were low to moderate for unhindered linear (**S3**) and cyclic (**S4**) substrates. The results of the allylic alkylation of dimethyl malonate with **S1** indicated that the best enantioselectivity was obtained with the smallest substituent on the oxazoline (R = Me, ligand **38a**, Scheme 10.14).

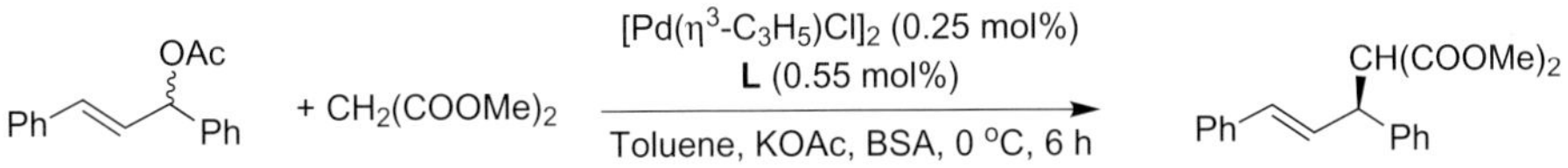

L	%Yield	%ee
38a	81	96 (S)
38b	99	90 (S)
38c	82	95 (S)
38d	91	83 (S)
38e	74	94 (S)
38f	88)	78 (S)

Scheme 10.14 Palladium-catalyzed asymmetric allylic alkylation of dimethyl malonate to 1,3-diphenylprop-2-enyl acetate using phosphinite–oxazoline ligands **38**.

Water-soluble ligand **39** (Figure 10.20), related to **38a**, was effective for the Pd-catalyzed allylic alkylation of different nucleophiles with 1,3-diphenylprop-2-enyl acetate (**S1**) in aqueous or biphasic media (up to 85% ee).

Recently, replacement of the phosphinite group in ligands **38** with a phosphite moiety led to the formation of phosphite–oxazoline ligands **40–43a–i** (Scheme 10.15) [30]. These ligands were prepared by reacting the corresponding oxazoline alcohol, described in Scheme 10.13, with one equivalent of the *in-situ* formed

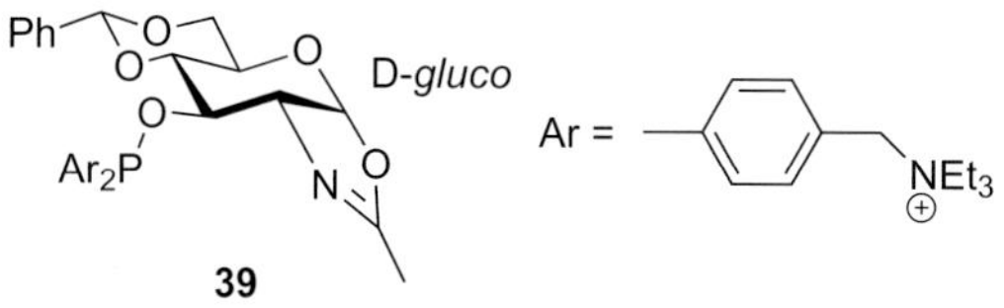
Ph
D-gluco
Ar =
Ar2P
NEt3
39

Figure 10.20 Water-soluble ligand **39.**

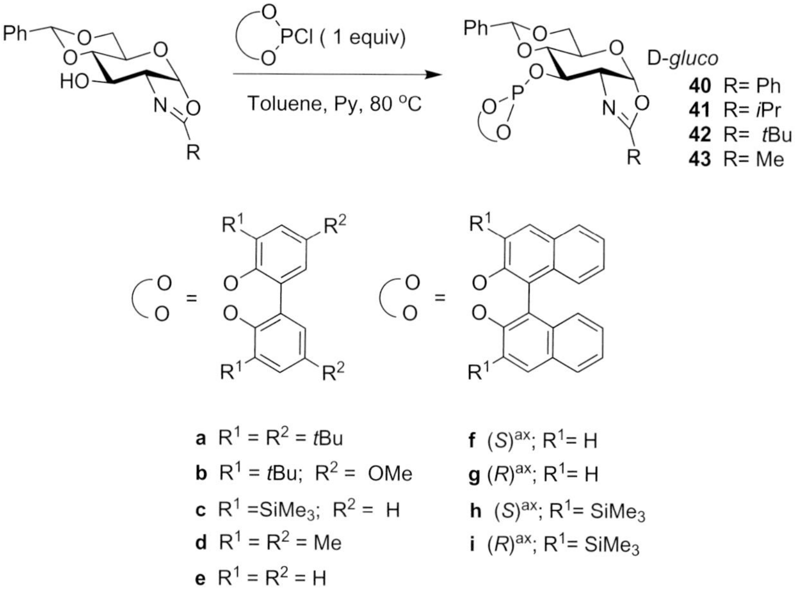
PCl (1 equiv)
Toluene, Py, 80 °C
D-gluco
40 R= Ph
41 R= iPr
42 R= tBu
43 R= Me
a R1 = R2 = tBu
b R1 = tBu; R2 = OMe
c R1 =SiMe3; R2 = H
d R1 = R2 = Me
e R1 = R2 = H
f (S)ax; R1= H
g (R)ax; R1= H
h (S)ax; R1= SiMe3
i (R)ax; R1= SiMe3

Scheme 10.15 Synthesis of phosphite–oxazoline ligands **40–43.**

phosphorochloridite under basic conditions (Scheme 10.15). The introduction of a biaryl phosphite moiety in the ligand design proved to be highly advantageous [31]: the new ligands **40–43** provided higher enantioselectivities and reaction rates than the related phosphinite–oxazoline ligands in the allylic substitution [up to 99% ee, TOFs up to 400 mol substrate x (mol Pd x h)$^{-1}$]. Moreover, the presence of a flexible phosphite moiety opens up the possibility of using the Pd–phosphite–oxazoline catalytic systems with a wide range of different substrate types (Figure 10.21). These ligands were also used to stabilize Pd nanoparticles [32].

In 2003, phosphine–oxazine ligands **44**, related to ligand **33**, was developed for the Pd-catalyzed allylic substitution of 1,3-diphenylprop-2-enyl acetate (**S1**) (Figure 10.22) using dimethyl malonate as nucleophile under standard conditions. Enantioselectivities up to 75% were obtained [25].

Several phosphine–imine ligands with a pyranoside backbone, **45–50**, have been developed for Pd-catalyzed allylic substitution reactions (Figure 10.23) [33]. The results indicated that having the imine–phosphine residue at C2 (ligands **49**)

10.2 Ligands

R	Nu-H	L	% Conv (time)	% ee
Ph	CH₂(COOMe)₂	40a	100 (6 h)	99 (S)
Ph	BnNH₂	40a	60 (24 h)	94 (R)
Me	CH₂(COOMe)₂	43h	88 (10 h)	89 (R)

n	L	% Conv (time)	% ee
1	40a	37 (36 h)	85 (R)
2	40a	43 (36 h)	92 (R)

L	% Conv (time)	% regio	% ee
40c	100 (1 h)	80	90 (S)

L	% Conv (time)	% ee
40a	80 (24 h)	91 (S)

Figure 10.21 Summary of the best results obtained in the Pd-catalyzed allylic substitution of several substrates using ligands **40–43**.

	R	% Yield	% ee
a	iPr	67	46(S)
b	Bn	64	75(S)
c	CH₂C₁₀H₇	75	49(S)
d	(R)-CH(CH₃)Ph	17	73(S)

Figure 10.22 Phosphine–oxazine ligands **44**; and yields and enantioselectivities obtained with them.

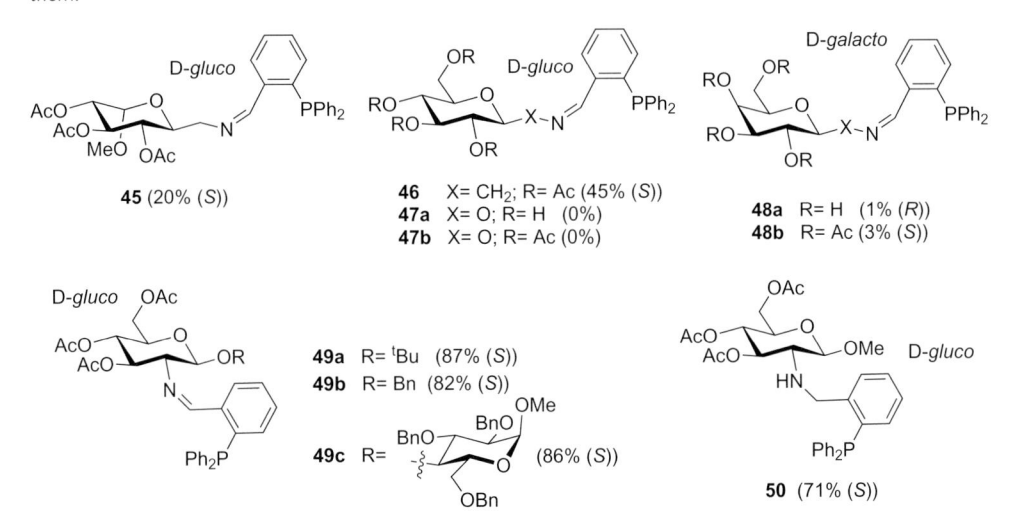

45 (20% (S))

46 X= CH₂; R= Ac (45% (S))
47a X= O; R= H (0%)
47b X= O; R= Ac (0%)

48a R= H (1% (R))
48b R= Ac (3% (S))

49a R= ᵗBu (87% (S))
49b R= Bn (82% (S))
49c R= (86% (S))

50 (71% (S))

Figure 10.23 Phosphine–imine **45–49** and phosphine–amine **50** ligands; and enantioselectivities obtained with them in the Pd-allylic alkylation of 1,3-diphenylprop-2-enyl acetate.

provided better enantioselectivities than when the residue is at C1 of the pyranoside backbone (ligands **45–48**). Notably, ligands with general structure **49** provided enantioselectivities, up to 99%, in the amination of 1,3-diphenylprop-2-enyl acetate (**S1**) using morpholine as nucleophile (Scheme 10.16) [33c]. Recently, the imine group in ligands with general structure **49** has been replaced by an amine group (ligand **50**, Figure 10.23), and also provided good results [33d].

10.2.3.3 P-O Ligands

Phosphine–amide ligands **51–56** (Figure 10.24) with a pyranoside backbone have been extensively studied for the Pd-catalyzed allylic alkylation of 1,3-diphenylprop-2-enyl acetate (**S1**) with dimethyl malonate under basic conditions [33c, 34]. The results clearly showed that the enantioselectivity is highly affected by the configuration of the anomeric carbon, the chelate ring size formed upon coordination to Pd, and the rigidity of the ligand. Ligands **51**, **55**, and **56**, which form a six-membered chelate ring and with a β anomeric carbon, afforded higher enantioselectivities than ligands **52** with an α anomeric carbon and **54**, which form a seven-membered chelate ring. Moreover, the results achieved with ligands **55** and **56** indicate a cooperative effect between the additional stereocenters in **55** and the carbohydrate backbone, which resulted in a matched combination for ligand (*S*)-**55**.

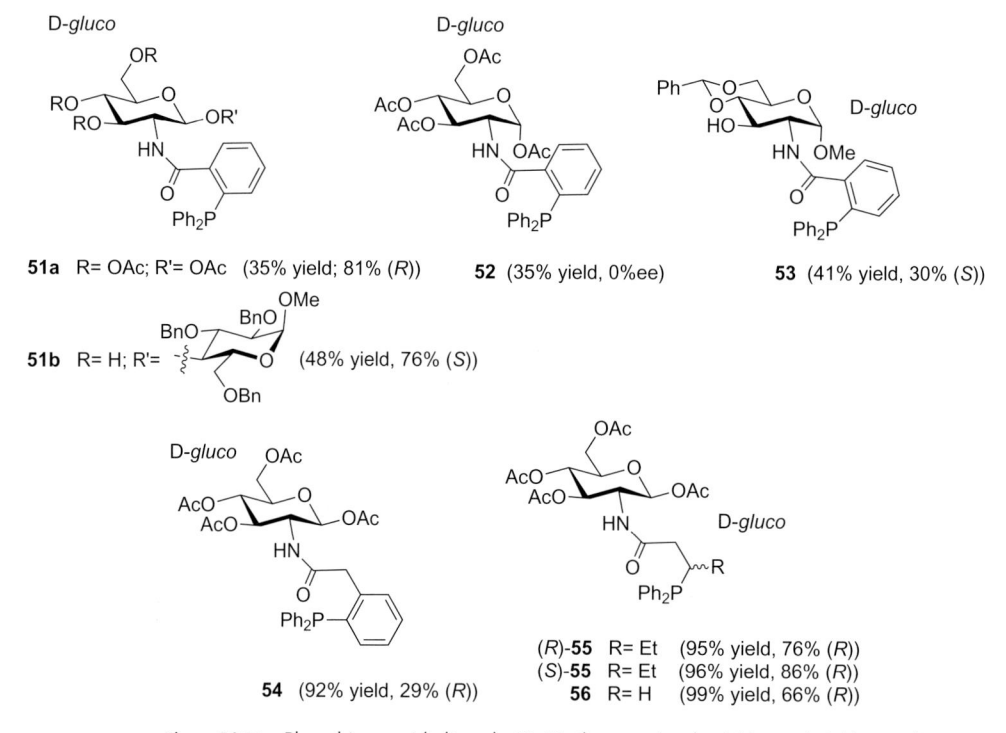

Figure 10.24 Phosphine–amide ligands **51–56**; the enantioselectivities and yields are shown in brackets.

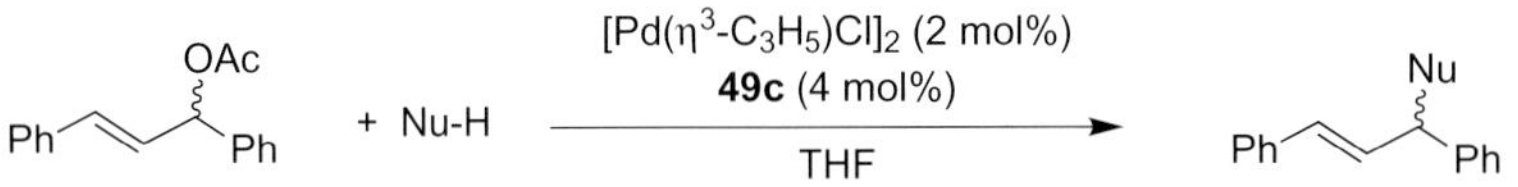

Scheme 10.16 Palladium-catalyzed allylic substitution of **S1** using several nucleophiles with ligand **49c**.

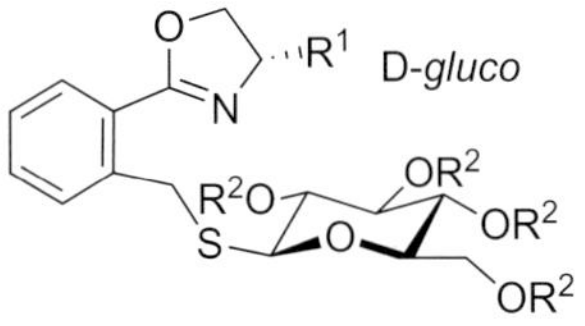

Nu-H	% Yield	% ee
$CH_2(COOMe)_2$	93	88 (S)
$MeCH(COOMe)_2$	-	46 (R)
$BnNH_2$	-	85 (R)
Morpholine	96	99 (R)

	R^1	R^2	% Yield	% ee
a	iPr	Ac	99	93 (S)
b	iPr	Piv	40	97 (S)
c	Ph	Ac	99	90 (S)
d	Ph	Piv	62	96 (S)

Figure 10.25 Thioether–oxazoline ligands **57**; the yields and enantioselectivities obtained in the Pd-catalyzed asymmetric allylic alkylation of dimethyl malonate to **S1** are also shown.

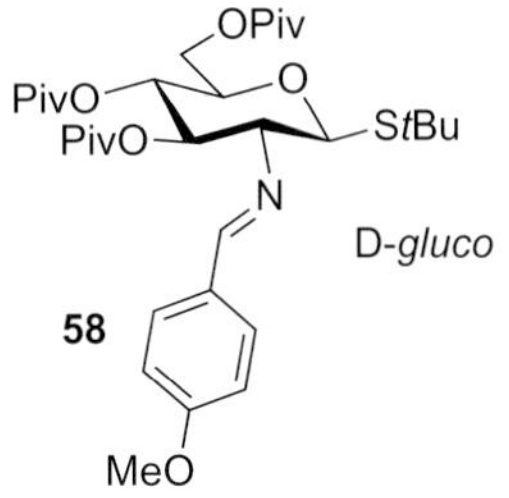

Figure 10.26 Thioether–imine ligand **58**.

10.2.3.4 N-S Ligands

Thioglucoside-derived ligands **57**, containing a chiral oxazoline moiety (Figure 10.25), have been used as ligands in the palladium-catalyzed allylic alkylation of diphenylprop-2-enyl acetate (**S1**), providing some of the best results achieved in this reaction with mixed N,S-donor ligands [35]. The effects of the thiosugar substituents on enantioselectivity were mild. The success of this kind of system seems to lie in the combination of thiosugar function and the proximity of all stereogenic units to the palladium allylic fragment, because the Pd–N distance is shorter than the Pd–P distance in related phosphino-thiosugar palladium complexes.

More recently, the pyranoside thioether–imine ligand **58** (Figure 10.26), related to P-S ligand **31**, was applied in the allylic alkylation of 1,3-diphenylprop-2-enyl acetate (**S1**), affording low enantioselectivity (up to 34% ee) [24a].

10.3
Conclusions

Carbohydrate ligands have become some of the most versatile ligands for the Tsuji–Trost reaction. The excellent results obtained can be expected to lead to new designs of carbohydrate ligands and their application to other asymmetric allylic substitution reactions catalyzed by metals other than Pd. Among these ligands, heterodonor donors have been widely used. They are mainly P,N-ligands but others containing donor atoms such as sulfur or oxygen have also emerged as suitable carbohydrate ligands. All these ligands were synthesized from simple monosaccharides such as xylose, fructose, glucose, and mannose among others. This has led to a wide range of sugar backbones with different electronic and steric properties. The properties of the ligand have led to excellent control of selectivity. This means that by appropriate ligand tuning (i) a ligand can be selected for each particular substrate and (ii) both enantiomers of the product are accessible.

Acknowledgments

We thank the Spanish Government (CTQ2010-15835, 2008PGIR/07 to O.P. and 2008PGIR/08 and ICREA Academia award to M.D. and to O.P.) and the Catalan Government (2009SGR116) for financial support.

References

1 (a) Jacobsen, E.N., Pfaltz, A., and Yamamoto, H. (1999) *Comprehensive Asymmetric Catalysis*, Springer, Berlin; (b) Tsuji, J. (1995) *Palladium Reagents and Catalysis, Innovations in Organic Synthesis*, John Wiley & Sons, Inc., New York; (c) Trost, B.M. and van Vranken, D.L. (1996) *Chem. Rev.*, **96**, 395; (d) Johannsen, M. and Jorgensen, K.A. (1998) *Chem. Rev.*, **98**, 1869; (e) Trost, B.M. and Crawley, M.L. (2003) *Chem. Rev.*, **103**, 292; (f) Lu, Z. and Ma, S. (2008) *Angew. Chem. Int. Ed.*, **47**, 258.

2 See, for instance: (a) Bruneau, C., Renaud, J.L., and Demersemen, B. (2006) *Chem. Eur. J.*, **12**, 5178; (b) Malkov, A.V., Gouriou, L., Lloyd-Jones, G.C., Starý, I., Langer, V., Spoor, P., Vinader, V., and Kočovský, P. (2006) *Chem. Eur. J.*, **12**, 6910; (c) Trost, B.M., Hildbrand, S., and Dogra, K. (1999) *J. Am. Chem. Soc.*, **121**, 10416; (d) Alexakis, A. and Polet, D. (2004) *Org. Lett.*, **6**, 3529.

3 Trost, B.M. and Strege, P.E. (1977) *J. Am. Chem. Soc.*, **99**, 1649.

4 Trost, B.M. (1996) *Acc. Chem. Res.*, **29**, 355.

5 (a) Masdeu-Bultó, A.M., Diéguez, M., Martin, E., and Gómez, M. (2003) *Coord. Chem. Rev.*, **242**, 159; (b) Martin, E. and Diéguez, M. (2007) *C. R. Chem.*, **10**, 188.

6 Helmchen, G. and Pfaltz, A. (2000) *Acc. Chem. Res.*, **33**, 336.

7 See for example: (a) Diéguez, M., Pàmies, O., and Claver, C. (2004) *Chem. Rev.*, **104**, 3189; (b) Diéguez, M., Pàmies, O., Ruiz, A., Díaz, Y., Castillón, S., and Claver, C. (2004) *Coord. Chem. Rev.*, **248**, 2165; (c) Castillón, S., Díaz, Y., and Claver, C. (2005) *Chem. Soc. Rev.*, **34**, 702; (d) Diéguez, M., Claver, C., and Pàmies, O. (2007) *Eur. J. Org. Chem.*, 4621; (e) Boysen, M.M.K. (2007) *Chem. Eur. J.*, **13**, 8648; (f) Benessere, V., Del Litto, R., De Roma, A., and Ruffo, F. (2010) *Coord. Chem. Rev.*, **254**, 390.

8 (a) Diéguez, M., Jansat, S., Gómez, M., Ruiz, A., Muller, G., and Claver, C. (2001) *Chem. Commun.*, 1132; (b) Pàmies, O., Ruiz, A., Net, G., Claver, C., Kalchhauser, H., and Widhalm, M. (2000) *Monatsh. Chem.*, **131**, 1173.

9 (a) Meyer zu Reckendorf, W., Weber, R., and Hehenberger, H. (1981) *Chem. Ber.*, **14**, 1306; (b) Guthrie, R.D. and Murphy, D. (1965) *J. Chem. Soc.*, 6956.

10 Ruffo, F., Del Lito, R., De Roma, A., D'Errico, A., and Magnolia, S. (2006) *Tetrahedron: Asymmetry*, **17**, 2265.

11 (a) Nomura, N., Mermet-Bouvier, Y.C., and RajanBabu, T.V. (1996) *Synlett*, 745; (b) Clyne, D.S., Mermet-Bouvier, Y.C., Nomura, N., and RajanBabu, T.V. (1999) *J. Org. Chem.*, **64**, 7601.

12 Guimet, E., Diéguez, M., Ruiz, A., and Claver, C. (2005) *Inorg. Chim. Acta*, **358**, 3824.

13 (a) Diéguez, M., Ruiz, A., and Claver, C. (2003) *Dalton Trans.*, 2957; (b) Pàmies, O., van Strijdonck, G.P.F., Diéguez, M., Deerenberg, S., Net, G., Ruiz, A., Claver, C., Kamer, P.C.J., and van Leeuwen, P.W.N.M. (2001) *J. Org. Chem.*, **66**, 8867; (c) Diéguez, M., Pàmies, O., and Claver, C. (2005) *Adv. Synth. Catal.*, **347**, 1257.

14 (a) Diéguez, M., Pàmies, O., Ruiz, A., Castillón, S., and Claver, C. (2001) *Chem. Eur. J.*, **7**, 3086; (b) Diéguez, M., Pàmies, O., Ruiz, A., and Claver, C. (2002) *New J. Chem.*, **26**, 829.

15 (a) Jansat, S., Gómez, M., Philippot, K., Muller, G., Guiu, E., Claver, C., Castillón, S., and Chaudret, B. (2004) *J. Am. Chem. Soc.*, **126**, 1592; (b) Favier, I., Gómez, M., Muller, G., Axet, M.R., Castillón, S., Claver, C., Jansat, S., Chaudret, B., and Philippot, K. (2007) *Adv. Synth. Catal.*, **349**, 2459.

16 Balanta, A., Favier, I., Teuma, E., Castillón, S., Godard, C., Aghmiz, A., Claver, C., and Gómez, M. (2008) *Chem. Commun.*, 6197.

17 See for example: (a) Feringa, B.L. (2000) *Acc. Chem. Res.*, **33**, 346; (b) Jagt, R.B.C., Toullec, P.Y., Geerdink, D., de Vries, J.G., Feringa, B. L., and Minnaard, A.J. (2006) *Angew. Chem. Int. Ed.*, **45**, 2789; (c) Tissot-Croset, K., Polet, D., and Alexakis, A. (2004) *Angew. Chem. Int. Ed.*, **43**, 2426; (d) Polet, D., Alexakis, A., Tissot-Croset, K., Corminboeuf, C., and Ditrich, K. (2006) *Chem. Eur. J.*, **12**, 3596; (e) Boele, M.D.K., Kamer, P.C.J., Lutz, M., Spek, A.L., de Vries, J.G., van Leeuwen, P.W.N.M., and van Strijdonck, G.P.F. (2004) *Chem. Eur. J.*, **10**, 6232; (f) Pàmies, O. and Diéguez, M. (2008) *Chem. Eur. J.*, **14**, 944; (g) Diéguez, M., Ruiz, A., and Claver, C. (2001) *Chem. Commun.*, 2702; (h) Biswas, K., Prieto, O., Goldsmith, P.J., and Woodward, S. (2005) *Angew. Chem. Int. Ed.*, **44**, 2232; (i) Alejakis, A., Benham, C., Rosset, S., and Humam, M. (2002) *J. Am. Chem. Soc.*, **124**, 5262.

18 Raluy, E., Diéguez, M., and Pàmies, O. (2007) *J. Org. Chem.*, **72**, 2842.

19 (a) Raluy, E., Claver, C., Pàmies, O., and Diéguez, M. (2007) *Org. Lett.*, **9**, 49; (b) Raluy, E., Pàmies, O., and Diéguez, M. (2009) *Adv. Synth. Catal.*, **351**, 1648.

20 Mata, Y., Dièguez, M., Pàmies, O., and Claver, C. (2006) *Tetrahedron: Asymmetry*, **17**, 3282.

21 (a) Khiar, N., Araújo, C.S., Alvarez, E., and Fernández, I. (2003) *Tetrahedron Lett.*, **44**, 3401; (b) Khiar, N., Araújo, C.S., Suárez, B., and Fernández, I. (2006) *Eur. J. Org. Chem.*, 1685.

22 Khiar, N., Araújo, C.S., Suárez, B., Alvarez, E., and Fernández, I. (2004) *Chem. Commun.*, 714.

23 (a) Albinati, A., Pregosin, P.S., and Wick, K. (1996) *Organometallics*, **15**, 2419; (b) Barbaro, P., Currao, A., Herrmann, J., Nesper, R., Pregosin, P.S., and Salzmann, R. (1996) *Organometallics*, **15**, 1879.

24 (a) Khiar, N., Suàrez, B., and Fernández, I. (2006) *Inorg. Chim. Acta*, **359**, 3048; (b) Khiar, N., Navas, R., Álvarez, E., and Fernánadez, I. (2008) *ARKIVOC*, 211.

25 Nakano, H., Yokohama, J., Okuyama, Y., Fujita, R., and Hongo, H. (2003) *Tetrahedron: Asymmetry*, **14**, 2361.

26 (a) Guimet, E., Diéguez, M., Ruiz, A., and Claver, C. (2005) *Tetrahedron: Asymmetry*, **16**, 959; (b) Diéguez, M., Pàmies, O., and Claver, C. (2006) *J. Organomet. Chem.*, **691**, 2257.

27 (a) Khiar, N., Suárez, B., Stiller, M., Valdivia, V., and Fernández, I. (2005) *Phosphorus Sulfur Silicon*, **180**, 1253;

(b) Khiar, N., Suárez, B., Valdivia, V., and Fernández, I. (2005) *Synlett*, 2963.

28 Gläser, B. and Kunz, H. (1998) *Synlett*, 53.

29 (a) Yonehara, K., Hashizume, T., Mori, K., Ohe, K., and Uemura, S. (1999) *Chem. Commun.*, 415; (b) Yonehara, K., Hashizume, T., Mori, K., Ohe, K., and Uemura, S. (1999) *J. Org. Chem.*, **64**, 9374.

30 (a) Mata, Y., Diéguez, M., Pàmies, O., and Claver, C. (2005) *Adv. Synth. Catal.*, **347**, 1947; (b) Mata, Y., Pàmies, O., and Diéguez, M. (2009) *Adv. Synth. Catal.*, **351**, 3217.

31 (a) Pàmies, O., Diéguez, M., and Claver, C. (2005) *J. Am. Chem. Soc.*, **127**, 3646; (b) Diéguez, M. and Pàmies, O. (2008) *Chem. Eur. J.*, **14**, 3653.

32 Diéguez, M., Pàmies, O., Mata, Y., Teuma, E., Ribaudo, F., and van Leeuwen, P.W.N.M. (2008) *Adv. Synth. Catal.*, **350**, 2583.

33 (a) Borriello, C., Cucciolito, M.E., Panunzi, A., and Ruffo, F. (2003) *Inorg. Chim. Acta*, **353**, 238; (b) Brunner, H., Schönherr, M., and Zabel, M. (2003) *Tetrahedron: Asymmetry*, **14**, 1115; (c) Johannensen, S.A., Glegola, K., Sinou, D., Framery, E., and Skrydstrup, T. (2007) *Tetrahedron Lett.*, **48**, 3569; (d) Glegola, K., Johannensen, S.A., Thim, L., Goux-Henry, C., Skrydstrup, T., and Framery, E. (2008) *Tetrahedron Lett.*, **49**, 6635.

34 (a) Tollabi, M., Framery, E., Goux-Henry, C., and Sinou, D. (2003) *Tetrahedron: Asymmetry*, **14**, 3329; (b) Konovets, A., Glegola, K., Penciu, A., Framery, E., Jubault, P., Goux-Henry, C., Pietrusiewicz, K.M., Quirion, J.C., and Sinou, D. (2005) *Tetrahedron: Asymmetry*, **16**, 3183; (c) Glegola, K., Framery, E., Goux-Henry, C., Pietrusiewicz, K.M., and Sinou, D. (2007) *Tetrahedron*, **63**, 7133; (d) Bauer, T. and Bartoszewicz, A. (2007) *Pol. J. Chem.*, **81**, 2115.

35 Boog-Wick, K., Pregosin, P.S., and Trabesinger, G. (1998) *Organometallics*, **17**, 3254.

11
Carbohydrate-Derived Ligands in Asymmetric Heck Reactions

Montserrat Diéguez and Oscar Pàmies

11.1
Introduction

The asymmetric Mizoroki–Heck reaction, generally described as the palladium-mediated coupling of aryl or vinyl halides or triflates with alkenes in the presence of base, has become one of the most versatile methods for C–C bond formation (Scheme 11.1). This process has found extensive applications in asymmetric synthesis. Shibasaki and Overman have convincingly demonstrated the value of such a transformation in the synthesis of complex natural molecules [1].

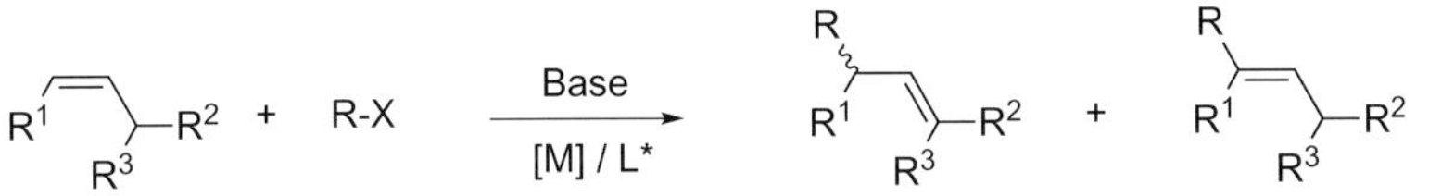

Scheme 11.1 Pd-catalyzed Heck reaction; X = halide or triflate.

The Heck reaction has been known to synthetic chemists since the late 1960s. However, reports of successful examples of the asymmetric Heck reaction were first published at the end of the 1980s. The bulk of the reported examples involve intramolecular reactions, which have the advantage of allowing easy control of alkene regiochemistry and geometry in the product [1g]. In contrast, successful intermolecular reactions have until very recently been limited to quite reactive substrates, principally O-,N-heterocycles, which again simplifies the question of alkene regiochemistry [1]. Nowadays several substrates have been applied in intermolecular asymmetric Heck reactions. Most of them are cyclic substrates, such as enol ethers, dihydropyrroles, dihydrodioxepines, and alkenes (Figure 11.1). Traditionally, 2,3-dihydrofuran has been the substrate of choice for testing a new ligand. With regard to the aryl or vinyl source, various triflate compounds have been applied. However, the most widely used is phenyl triflate. The base is also an

Carbohydrates – Tools for Stereoselective Synthesis, First Edition. Edited by Mike Martin Kwabena Boysen.
© 2013 Wiley-VCH Verlag GmbH & Co. KGaA. Published 2013 by Wiley-VCH Verlag GmbH & Co. KGaA.

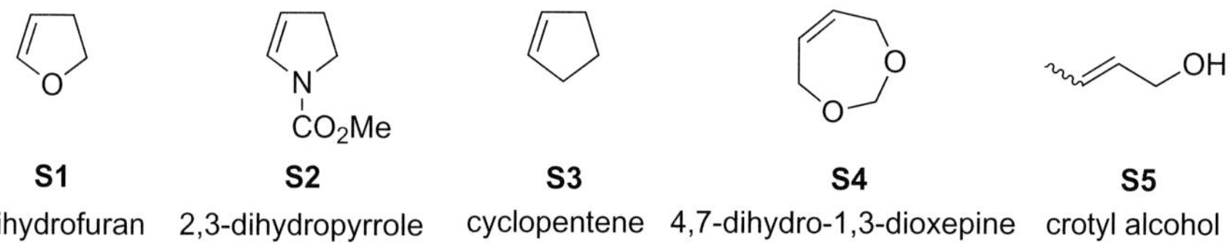

S1
2,3-dihydrofuran

S2
2,3-dihydropyrrole

S3
cyclopentene

S4
4,7-dihydro-1,3-dioxepine

S5
crotyl alcohol

Figure 11.1 The most common substrates used for the intermolecular asymmetric Heck reaction.

important parameter for high catalytic activity and enantioselectivity. A wide range of bases has been employed in this process, with N,N-diisopropylamine and proton sponge being the standard bases for testing new catalysts [1].

Notably, in the asymmetric intermolecular Heck reaction it is not only the enantioselectivity of the process that needs to be controlled, the regioselectivity is also a problem, because a mixture of regioisomers may be obtained. Consequently, for example, in the Heck reaction of 2,3-dihydrofuran **S1** with phenyl triflate a mixture of two products is obtained, namely, 2-phenyl-2,3-dihydrofuran (**1**) and the expected 2-phenyl-2,5-dihydrofuran (**2**, Scheme 11.2). The former is formed due to an isomerization process [1].

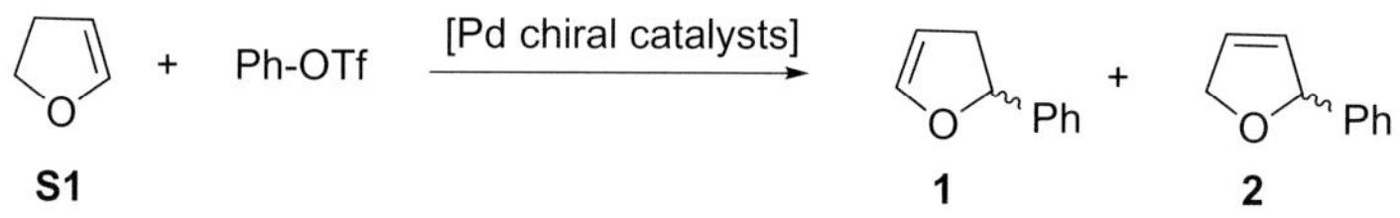

Scheme 11.2 Model Pd-catalyzed intermolecular Heck reaction.

Scheme 11.3 illustrates a proposed catalytic cycle for the phenylation of 2,3-dihydrofuran [1, 2]. The catalytic cycle starts with the oxidative addition of the organic triflate to Pd(0)-complex **3** to produce compound **4**. Since the triflate ligand in **4** is a good leaving group, coordination of 2,3-dihydrofuran on **4** induces dissociation of the triflate ligand to give the cationic phenylpalladium olefin species **5**, which has a 16-electron square-planar structure that is convenient for the subsequent enantioselective insertion of the olefin. The resulting alkyl-palladium(II) complex **6** undergoes β-hydride elimination leading to a hydrido-palladium olefin complex **7**. Dissociation of this π-complex furnishes **2** and a hydrido-palladium species **10**. Finally, the catalytic Pd(0) complex **3** is regenerated by reductive elimination of HOTf. Depending on the ligand, catalyst precursor, and reaction parameters, the palladium complex **7** can also undergo re-insertion of the hydride, which leads to the alkyl-palladium(II) complex **8**. β-Hydride elimination of **8** followed by dissociation of the resulting π-complex **9** leads to isomer **1** and hydride **10**. Reductive elimination of HOTf in **10** regenerates active species **3**.

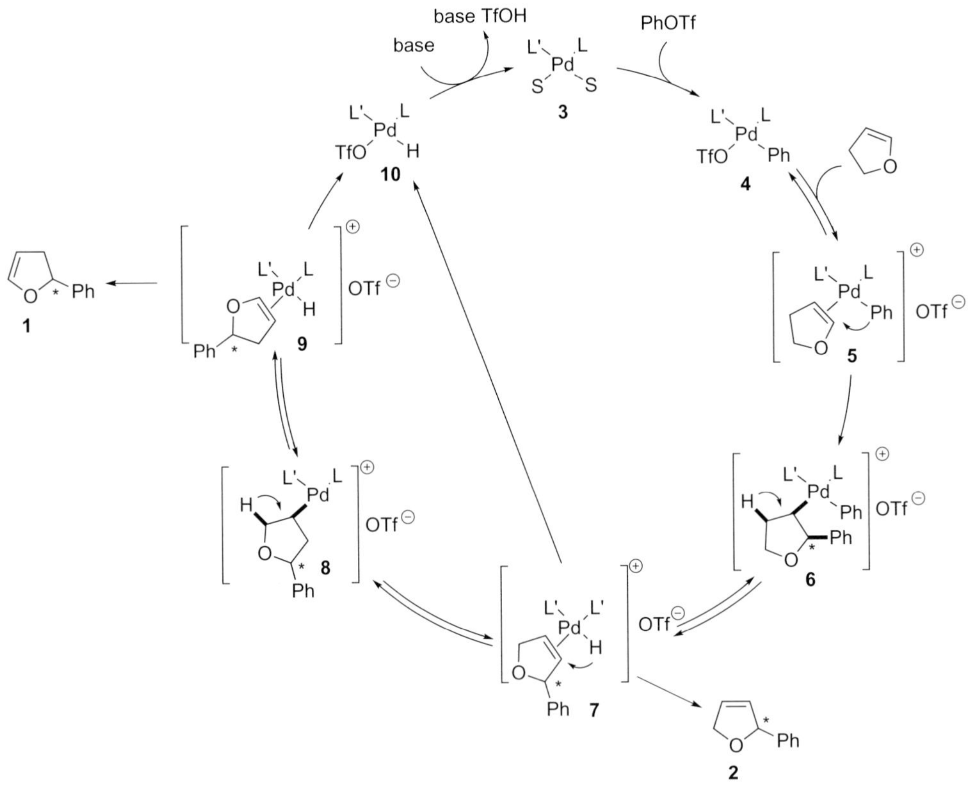

Scheme 11.3 Proposed mechanism for the Pd-catalyzed arylation of 2,3-dihydrofuran with phenyl triflate.

11.2 Ligands

In 1991, Ozawa and Hayashi reported the first example of the intermolecular version of the Heck reaction using 2,3-dihydrofuran and phenyl triflate (Scheme 11.2) [3]. Since then, this chemistry has been extensively studied using various chiral bidentate ligands. Diphosphines, which have played a key role in the success of the intramolecular version, were applied early on. Among these, Pd-BINAP was the first catalytic system that offered good regioselectivity in favor of product **1** and enantiocontrol [1]. In the last few years, a class of heterodonor ligands – phosphine–oxazolines – have emerged as suitable ligands for the intermolecular Heck reaction of several substrate types and triflate sources [4]. Two of the most representative examples of this type of ligands are the PHOX ligands developed by Pfaltz [4a,b] and coworkers (Figure 11.2) and the phosphine–oxazoline based on ketopinic acid developed by Gilbertson and coworkers [4e] (Figure 11.2). In contrast to Pd-BINAP systems, they offer preferentially isomer **2** in high enantioselectivities.

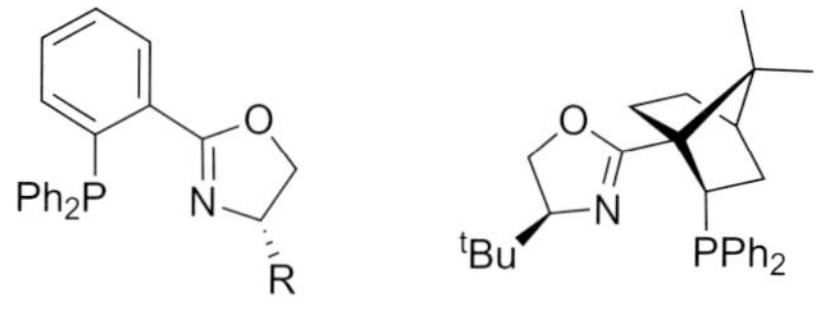

11 Carbohydrate-Derived Ligands in Asymmetric Heck Reactions

R= Ph, iPr, tBu, Bn
PHOX ligands   Gilbertson's Ligand

Figure 11.2 Phosphine–oxazoline ligand for asymmetric Pd-catalyzed Heck reactions.

Although carbohydrate-based ligands have been successfully used in other enantioselective reactions [5], there are only two reports of the highly enantioselective palladium-catalyzed asymmetric Heck reaction using this type of ligand [6, 7].

The first successful application of carbohydrate ligands in this process used the pyranoside phosphinite–oxazoline ligands **11** (Scheme 11.4), developed by Uemura and coworkers (Chapter 10, Scheme 10.13), in the Pd-catalyzed enantioselective arylation of 2,3-dihydrofuran (up to 96% ee, Scheme 11.4). The authors found an important oxazoline substituent effect on activity and enantioselectivity. Although there is no clear trend, activities and enantioselectivities were best when a benzyloxazoline substituent was present (Scheme 11.4). The reaction with *para*-tolyl triflate instead of phenyl triflate yielded the corresponding coupling product quantitatively but with moderate enantioselectivity (up to 78% ee). The reaction with *para*-methoxyphenyl triflate gave the product in good enantioselectivities (up to 90%) but in low yields. In contrast, no reaction took place when *para*-(trifluoromethyl)phenyl triflate was used. This set of ligands has also been applied in the phenylation of *trans*- and *cis*-crotyl alcohols with low enantioselectivity (up to 17% ee), representing the first example of the enantioselective intermolecular arylation of prochiral alkenes [6]. The authors also disclosed – by isolation of

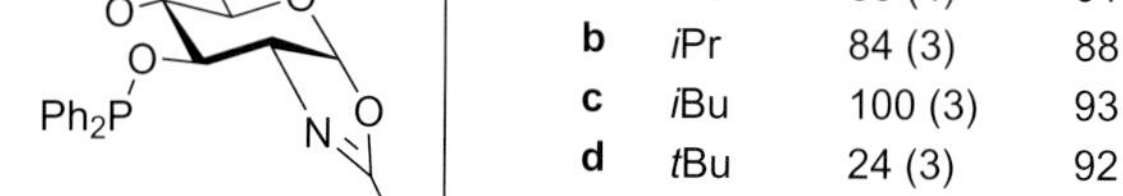

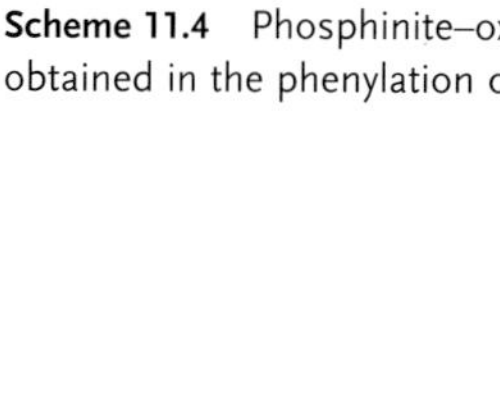

	R	%Yield (days)	%ee
a	Me	88 (4)	91
b	iPr	84 (3)	88
c	iBu	100 (3)	93
d	tBu	24 (3)	92
e	Ph	96 (2)	86
f	Bn	100 (1)	96

Scheme 11.4 Phosphinite–oxazoline ligands **11** and the yields and enantioselectivities obtained in the phenylation of **S1**.

the arylpalladium complex [(p-MeO$_2$CC$_6$H$_4$)PdI(**11a**)] and the stoichiometric reactions of the complex [PhPd(**11f**)]OTf (Scheme 11.3, **4**) with 2,3-dihydrofuran – mechanistic aspects of the Mizoroki–Heck-type reaction using P,N-ligands. They discovered that (i) the aryl moiety on the palladium is located *trans* to the nitrogen atom of the oxazoline moiety; this configuration might be responsible for an enantiofacial discrimination of 2,3-dihydrofuran to produce predominantly the (*R*)-enantiomer (Scheme 11.5); and (ii) deprotonation at the β-position with a base leading to an alkene(2-phenyl-2,5-dihydrofuran)palladium(0) complex (Scheme 11.3, **7**) is responsible for the selective formation of the product.

Scheme 11.5 Proposed origin of enantioselectivitiy using the chiral ligands **11a-f**.

Recently, replacement of the phosphinite group in ligands **11** by a phosphite moiety afforded phosphite–oxazoline ligands **12–15a–g** (Figure 11.3). These ligands, developed by Diéguez and coworkers (Chapter 10, Scheme 10.15), were

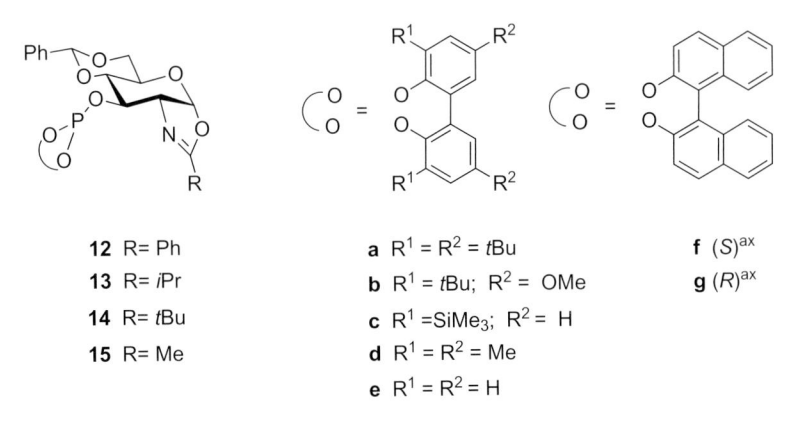

12 R= Ph
13 R= *i*Pr
14 R= *t*Bu
15 R= Me

a R^1 = R^2 = *t*Bu
b R^1 = *t*Bu; R^2 = OMe
c R^1 =SiMe$_3$; R^2 = H
d R^1 = R^2 = Me
e R^1 = R^2 = H

f (*S*)^{ax}
g (*R*)^{ax}

Figure 11.3 Phosphite–oxazoline ligands **12–15a–g**.

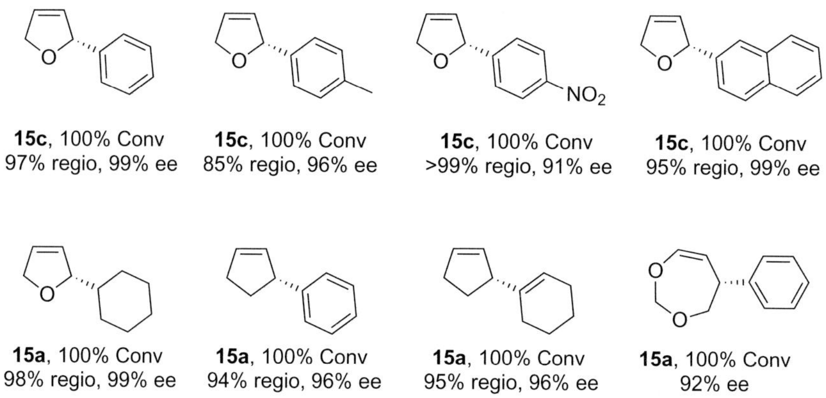

15c, 100% Conv
97% regio, 99% ee

15c, 100% Conv
85% regio, 96% ee

15c, 100% Conv
>99% regio, 91% ee

15c, 100% Conv
95% regio, 99% ee

15a, 100% Conv
98% regio, 99% ee

15a, 100% Conv
94% regio, 96% ee

15a, 100% Conv
95% regio, 96% ee

15a, 100% Conv
92% ee

Figure 11.4 Summary of the best results obtained in the Pd-catalyzed Heck reaction using ligands **12–15**.

successfully applied in the Pd-catalyzed asymmetric Heck reaction of several substrates and triflate sources. The results indicate that the degree of isomerization and the effectiveness in transferring the chiral information to the product and the activity can be tuned by choosing appropriate ligand components (phosphite and oxazoline substituents). The introduction of a biaryl phosphite moiety into the ligand design proved to be highly advantageous. Therefore, excellent activities (up to 100% conversion in 10 min), regio- (up to >99%) and enantioselectivities (up to 99% ee) were obtained for a wide range of substrates and triflate sources (Figure 11.4). The results also showed that the catalytic performance is highly influenced by the steric properties of the substituents on the oxazoline moiety, while the electronic properties have little effect. Bulky substituents in this position decreased activities and regio- and enantioselectivities (i.e., Ph ≈ Me > iPr > tBu). This is in contrast with the oxazoline-substituent effect observed for the vast majority of successful phosphine–oxazoline ligands, whose enantioselectivities are higher when bulky *tert*-butyl groups are present. The substituents on the biaryl phosphite were also found to have an important effect. Bulky substituents at the *ortho* position of the biaryl phosphite moiety are therefore needed for high activity and regio- and enantioselectivity.

11.3
Conclusions

The use of carbohydrate-based ligands in the asymmetric Heck reaction has been rare but the results obtained so far are among the best reported, especially those obtained with the glucosamine-based phosphite–oxazoline ligands. This together with the high modularity and availability of carbohydrates will lead to the design of new ligands in the near future.

Acknowledgments

We thank the Spanish Government (CTQ2010-15835, 2008PGIR/07 to O.P. and 2008PGIR/08 and ICREA Academia award to M.D. and to O.P.) and the Catalan Government (2009SGR116) for financial support.

References

1. For reviews see: (a) Tietze, L.T., Ila, H., and Bell, H.P. (2004) *Chem. Rev.*, **104**, 3453; (b) Dai, L.X., Tu, T., You, S.L., Deng, W.P., and Hou, X.L. (2003) *Acc. Chem. Res.*, **36**, 659; (c) Bolm, C., Hildebrand, J.P., Muñiz, K., and Hermanns, N. (2001) *Angew. Chem. Int. Ed.*, **44**, 3284; (d) Loiseleur, O., Hayashi, M., Keenan, M., Schemees, N., and Pfaltz, A. (1999) *J. Organomet. Chem.*, **576**, 16; (e) Beller, M., Riermeier, T.H., and Stark, G. (1998) in *Transition Metals for Organic Synthesis* (eds M. Beller and C. Bolm), Wiley-VCH Verlag GmbH, Weinheim, p. 208; (f) Dounay, A.B. and Overman, L.E. (2003) *Chem. Rev.*, **103**, 2945; (g) Jacobsen, E.N., Pfaltz, A., and Yamamoto, H. (1999) *Comprehensive Asymmetric Catalysis*, Springer, Berlin.
2. Ozawa, F., Kubo, A., Matsumoto, Y., and Hayashi, T. (1993) *Organometallics*, **12**, 4188.
3. Ozawa, F., Kubo, A., and Hayashi, T. (1991) *J. Am. Chem. Soc.*, **113**, 1417.
4. See for instance: (a) Loiseleur, O., Meier, P., and Pfaltz, A. (1996) *Angew. Chem. Int. Ed. Engl.*, **35**, 200; (b) Loiseleur, O., Hayashi, M., Schmees, N., and Pfaltz, A. (1997) *Synthesis*, 1338; (c) Tu, T., Hou, X.L., and Dai, L.X. (2003) *Org. Lett.*, **5**, 3651; (d) Gilbertson, S.R., Xie, D., and Fu, Z. (2001) *J. Org. Chem.*, **66**, 7240; (e) Gilbertson, S.R. and Fu, Z. (2001) *Org. Lett.*, **3**, 161; (f) Tu, T., Deng, W.P., Hou, X.L., Dai, L.X., and Dong, X.C. (2003) *Chem. Eur. J.*, **9**, 3073; (g) Gilberston, S.R., Genov, D.G., and Rheingold, A.L. (2000) *Org. Lett.*, **2**, 2885; (h) Hashimoto, Y., Horie, Y., Hayashi, M., and Saigo, K. (2000) *Tetrahedron: Asymmetry*, **11**, 2205; (i) Hou, X.L., Dong, D.X., and Yuan, K. (2004) *Tetrahedron: Asymmetry*, **15**, 2189; (j) Liu, D., Dai, Q., and Zhang, X. (2005) *Tetrahedron*, **61**, 6460; (k) Rubina, M., Sherrill, W.M., and Rubin, M. (2008) *Organometallics*, **27**, 6393; (l) Wu, W.-Q., Peng, Q., Dong, D.-X., Hou, X.-L., and Wu, Y.-D. (2008) *J. Am. Chem. Soc.*, **130**, 9717; (m) Mazuela, J., Pàmies, O., and Diéguez, M. (2010) *Chem. Eur. J.*, **16**, 3434.
5. See for example: (a) Diéguez, M., Pàmies, O., and Claver, C. (2004) *Chem. Rev.*, **104**, 3189; (b) Diéguez, M., Pàmies, O., Ruiz, A., Díaz, Y., Castillón, S., and Claver, C. (2004) *Coord. Chem. Rev.*, **248**, 2165; (c) Castillón, S., Díaz, Y., and Claver, C. (2005) *Chem. Soc. Rev.*, **34**, 702; (d) Diéguez, M., Claver, C., and Pàmies, O. (2007) *Eur. J. Org. Chem.*, 4621; (e) Boysen, M.M.K. (2007) *Chem. Eur. J.*, **13**, 8648; (f) Benessere, V., Del Litto, R., De Roma, A., and Ruffo, F. (2010) *Coord. Chem. Rev.*, **254**, 390.
6. Yonehara, K., Mori, K., Hashizume, T., Chung, K.G., Ohe, K., and Uemura, S. (2000) *J. Organomet. Chem.*, **603**, 40.
7. (a) Mata, Y., Diéguez, M., Pàmies, O., and Claver, C. (2005) *Org. Lett.*, **7**, 5597; (b) Mata, Y., Diéguez, M., and Pàmies, O. (2007) *Chem. Eur. J.*, **13**, 3296.

12
1,4-Addition of Nucleophiles to α,β-Unsaturated Carbonyl Compounds

Yolanda Díaz, M. Isabel Matheu, David Benito and Sergio Castillón

12.1
Copper-Catalyzed Reactions – Introduction

The conjugate addition reaction is a major synthetic transformation and has consequently been employed in numerous total syntheses [1]. Although several soft carbon nucleophiles such as malonates easily undergo this reaction, harder carbon nucleophiles such as classical organometallic reagents need the presence of a transition metal to avoid direct attack on the carbonyl group of the Michael acceptor. Traditionally, copper has found the broadest application, and various organocopper species have been widely used [2].

Four decades ago, it was already obvious that an asymmetric version of conjugate addition would be of great interest and, currently, the asymmetric conjugate addition of carbon nucleophiles to α,β-unsaturated carbonyl compounds is a useful method for the formation of single carbon–carbon bonds and the synthesis of chiral compounds [3]. There are several ways to tackle the problem: by functional group transformation, using chiral auxiliaries or reagents, or via the introduction of chiral ligands (Scheme 12.1). Until the mid-1990s, the stoichiometric approach (methods 2 and 3) was the most successful, with diastereoselectivities reaching >99% [3a, 4].

A catalytic approach can only be envisioned by method 4: hetero-cuprates in combination with external chiral ligands. Copper-catalyzed conjugate addition has its background in the first example described by Kharash, in 1941 [5]. Stoichiometric organocopper reagents are usually prepared by transmetalation of an organolithium or Grignard reagent. The catalytic process, however, is usually performed with Grignard reagents since organolithium reagents are too reactive towards the carbonyl functionality. Therefore, all of the enantioselective catalytic approaches were based on the use of Grignard reagents as primary organometallics. However, a prominent position in the rapid development of this process is occupied by the copper-catalyzed, ligand accelerated, 1,4-addition of diorganozinc reagents, especially $ZnEt_2$, which is a trend that was started by Alexakis [6, 7]. Trialkylaluminium

Carbohydrates – Tools for Stereoselective Synthesis, First Edition. Edited by Mike Martin Kwabena Boysen.
© 2013 Wiley-VCH Verlag GmbH & Co. KGaA. Published 2013 by Wiley-VCH Verlag GmbH & Co. KGaA.

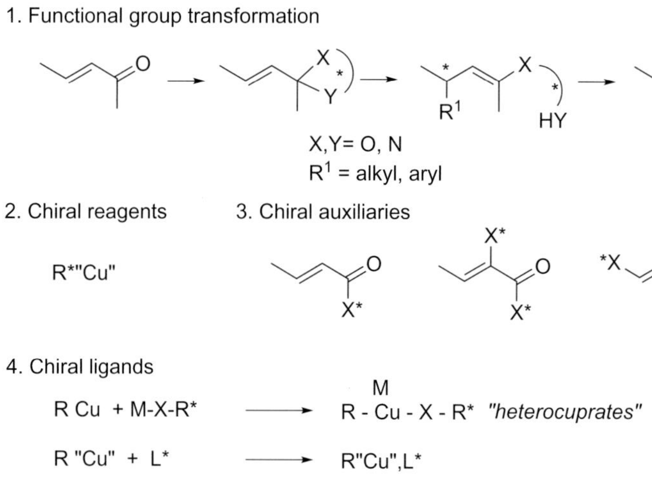

12 1,4-Addition of Nucleophiles to α,β-Unsaturated Carbonyl Compounds

1. Functional group transformation

X,Y= O, N
R^1 = alkyl, aryl

2. Chiral reagents

R*"Cu"

3. Chiral auxiliaries

4. Chiral ligands

R Cu + M-X-R* $\longrightarrow$ R - Cu - X - R* "heterocuprates"

R "Cu" + L* $\longrightarrow$ R"Cu",L*

Scheme 12.1 Strategies for asymmetric conjugate addition.

reagents have recently appeared as interesting alternatives to organozinc reagents since they are also readily available and offer additional hydro- and carboalumination possibilities for their preparation [8]. Additionally, organoaluminium reagents allow Cu-catalyzed 1,4-addition to very challenging substrates that are inert to organozinc methodologies [8f]. For example, these reagents provide a new efficient way to build chiral quaternary centers. Owing to their stronger Lewis acidity, a better activation of the substrates is achieved, overcoming the steric hindrance of β,β'-disubstituted enones.

The copper-catalyzed conjugate addition of dialkylzinc and trialkylaluminium species is believed to follow the generally accepted mechanism of stoichiometric organocopper reactions. Although no intermediate has been characterized, a tentative catalytic cycle has been proposed (Scheme 12.2) based on the catalysis data available for the stoichiometric organocuprate conjugate reaction [3, 9]. The proposed catalytic cycle postulates the reduction of the copper(II) salt to a Cu(I) species as a first step, followed by transmetalation between the stoichiometric organometallic compound and the Cu(I) species to give **A**. This bimetallic cluster, where copper and another metal (Zn or Al) are intimately associated, strongly coordinates to the oxygen atom of the enone by the most oxophilic metal (Zn or Al), activating it towards nucleophilic addition (Lewis acid effect) to give **B**. After or before coordination to the enone takes place, the copper species must be transformed into a higher order cuprate reagent by reaction with another equivalent of the organometallic compound for further reaction to take place. Formation of a Cu–olefin π-complex, involving d,π*-back bonding, oxidative addition to the β-carbon to form a d^8 Cu(III)-intermediate **E**, and ultimate reductive elimination yields the enolate and regenerates the resting state of the catalytic cycle.

The typical experiment in most studies is the reaction between diethylzinc and cyclic or acyclic enones, in the presence of 0.5–5 mol.% Cu(OTf)$_2$ and 1–10 mol.%

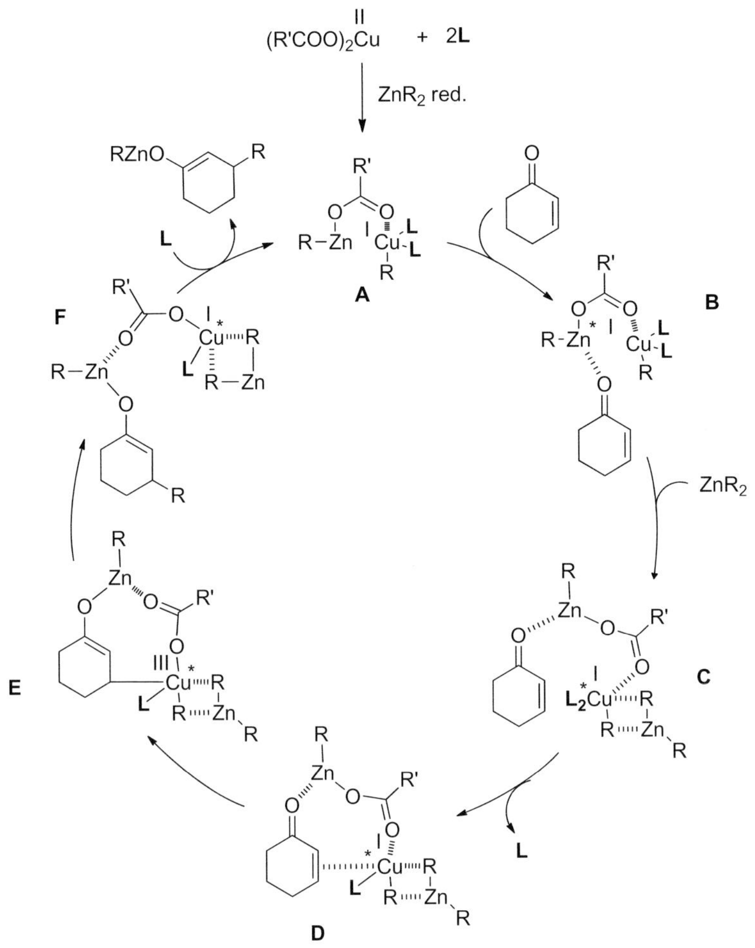

Scheme 12.2 Catalytic cycle proposed for the Cu-catalyzed conjugate addition of ZnR$_2$ or AlR$_3$ to α,β-unsaturated carbonyl compounds.

chiral ligand (Scheme 12.3). Concerning the solvent, the belief that a non-coordinating solvent like toluene is the most appropriate for the Cu-catalyzed conjugate addition of dialkylzinc species is in contrast with some results that show that several solvents such as Et$_2$O, THF, and EtOAc can be used as well, affording high degrees of enantioselectivity, although with slower reaction rates [9b].

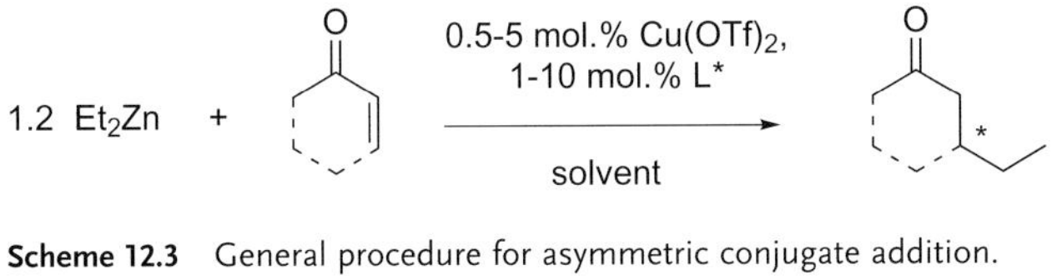

Scheme 12.3 General procedure for asymmetric conjugate addition.

Copper salt is essential for high catalytic activity and high enantioselectivity. Both copper(I) and copper(II) species have been tested. Copper(II) salts usually have the advantage of being cheaper and easier to handle. In this case, the reduction of the copper(II) species to the true catalytic copper(I) species must take place *in situ*. Although $Cu(OTf)_2$ has been used extensively in asymmetric addition reactions, leading in most cases to high degrees of enantioselectivities, it has been demonstrated that this Cu source can be replaced by much cheaper copper carboxylates, and that the Lewis acidity of the copper source, previously believed to play an important role, is not a significant factor [9b].

The chiral ligand is the centerpiece of this reaction, with the degree of enantioselectivity being entirely due to it [7]. P-donor ligands are among the most effective ligands in this reaction. They are of the phosphite and phosphoramidite type, most of them are monodentate, but some are bidentate. Aryl phosphines have been scarcely used in this reaction and only successfully when combined with other coordination sites. Phosphinite-based ligands have also been less studied. The most studied ligands are, by far, phosphites or phosphoramidites, where the phosphorus atoms are attached to three heteroatoms. In all cases, the phosphorus atom is incorporated in a ring formed from a diol. The chirality can be introduced through the diol moiety or by the exocyclic alcohol/amine, or by both. In the latter case, a matched or mismatched relationship may exist. Non-phosphorus ligands have been comparatively less used, although very high degrees of enantio-induction have also been obtained.

With regard to substrate, recently, conjugate additions of nucleophiles to cyclic enones have been studied extensively, as the products may be useful for the synthesis of biologically active compounds [3b, 3d, 10]. However, relatively few publications describing highly enantioselective additions of organometallics to linear aliphatic enones are found in the literature. Thus, for this class of substrate, the development of more active and enantioselective catalysts is still needed.

In contrast, one of the main problems revealed in these investigations is the dynamic behavior of equilibria between several species of organocopper compounds in solution. If the more reactive cuprates lead to a racemic product, a loss of enantioselectivity is unavoidable. It is, therefore, desirable to design and synthesize new chiral catalysts that can react rapidly with the substrate, and therefore suppress the undesired competing reactions. This fact justifies expanding the range of ligands used for Cu-catalyzed additions of organoaluminium reagents to cyclic and, more specifically, to linear substrates.

In this sense, using carbohydrate derivatives as starting materials for the synthesis of chiral ligands has several advantages: (i) the raw materials are of high optical purity and are readily available; (ii) their multifunctional property makes it possible to design an array of ligand structures through a series of modifications [11, 12].

This chapter covers the literature reports on enantioselective conjugate additions using carbohydrate derivatives as chiral ligands. The chapter is organized as follows. Section 12.2 presents the different ligands together with a brief outline of

12.2
Copper-Catalyzed Reactions – Ligands

This section gathers synthetic data of the carbohydrate-derived ligands that are divided according to the nature of the donor atoms present in their structures. Thus, bidentate P/P, P/N, P/S, S/O ligands and some monodentate P or S ligands will be considered. The syntheses are organized in turn according to the size of the ring of the carbohydrate backbone, that is, furanoside and pyranoside ligands. Other structurally closely related ligands with donor centers placed in peripheral imine or amide groups are considered alongside. The maximum ee values reached for each kind of ligand in the reaction with 2-cyclohexenone are shown in parentheses.

12.2.1
Bidentate P,P-Ligands

12.2.1.1 Xylo Ligands

Diphosphinite ligand **4** [13] and structurally related bulky diphosphite ligands **3a–c** [14] and **3d–f** [15] were efficiently synthesized from 1,2-isopropylidene-D-*xylo*-furanoside **2** [16], which is accessible from D-xylose **1**. Further reaction with either two equivalents of chloro(diphenyl)phosphine in THF in the presence of DMAP or two equivalents of the corresponding phosphorochloridite in the presence of pyridine afforded ligands **4** and **3a–f**, respectively (Scheme 12.4).

Mixed phosphine–phosphite ligands **8** and **9** [17], also with a xylofuranoside backbone, were synthesized starting from an oxetane intermediate **6** [18], which was obtained in turn from **5** via intramolecular displacement of the primary tosyl group at position 5 by the alkoxide formed under basic conditions. The ring opening of **6** with diphenylphosphide in DMF followed by reaction of the hydroxyphosphine formed with the corresponding phosphorochloridite afforded ligands **8** and **9** (Scheme 12.4).

Phosphite–phosphoramidite ligand **11** [19] was synthesized by reacting the monotosylate **5** with sodium azide to give the corresponding 5-azido derivative. Reduction with triphenylphosphine rendered the aminoalcohol **10**, which was reacted with two equivalents of the phosphorochloridite to obtain ligand **11**.

12.2.1.2 Ribo Ligands

To obtain the corresponding *ribo* structures an inversion of the configuration at C3 in the *xylo* backbone must be carried out. In this respect, oxidation of D-*xylo*-furanoside **2** with PCC and subsequent reduction with NaBH$_4$ afforded, after

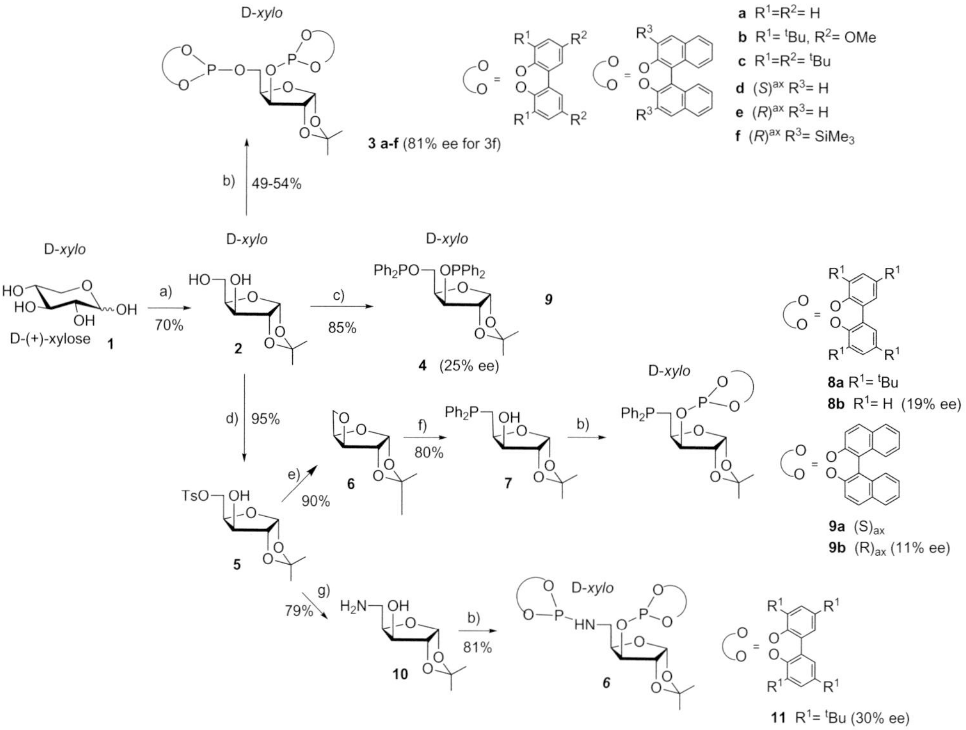

Scheme 12.4 Synthesis of bidentate P,P xylo ligands.

(a) CH$_3$COCH$_3$, H$_2$SO$_4$. (b) (RO)$_2$PCl, pyridine, toluene. (c) PPh$_2$Cl, pyridine, DMAP. (d) TsCl, NEt$_3$, CH$_2$Cl$_2$ (e) NaOH, EtOH, H$_2$O. (f) KPPh$_2$, DMF. (g) 1. NaN$_3$, DMF. 2. PPh$_3$, aqueous THF.

hydrolysis of the 5-O-benzoate group, the desired 1,2-isopropylidene-D-*ribo*-furanoside **13** [20], from which diphosphinite ligand **14** [13] and diphosphite ligands **15a–c** [14] were efficiently synthesized using the same methodology as for D-*xylo*-ligands (Scheme 12.5).

12.2.1.3 D-*Gluco*-, D-*Allo*-, L-*Talo*-, and D-*Galacto*-Ligands

A series of biphenol-based diphosphite ligands derived from D-glucose **16** was synthesized by Diéguez *et al.* using a procedure that involves the formation of epoxides in positions 5 and 6 [21, 22]. These ligands constitute three of the four diastereomers that can be obtained by varying the configuration of the C3 and C5 atoms of the sugar backbone, that is, the D-*gluco*- (**21**), D-*allo*- (**26**), L-*talo*- (**29**) products, in combination with several substituted bulky 2,2′-biphenyl moieties (Scheme 12.6). Thus, the D-*gluco* ligand **21** was obtained from the diisopropylidene-D-*gluco*-furanoside **17**. Acylation at position C3, hydrolysis of the terminal isopropylidene group, followed by monotosylation at the primary position, ring closure under basic conditions to give the 5,6-epoxide, and subsequent oxirane

(a) 1. BzCl, NEt₃, CH₂Cl₂. 2. PCC, NaOAc, CH₂Cl₂. (b) 1. NaBH₄, EtOH. 2. NaOMe, MeOH. (c) PPh₂Cl, pyridine, DMAP. (d) (RO)₂PCl, pyridine, toluene.

Scheme 12.5 Synthesis of bidentate P,P ribo ligands.

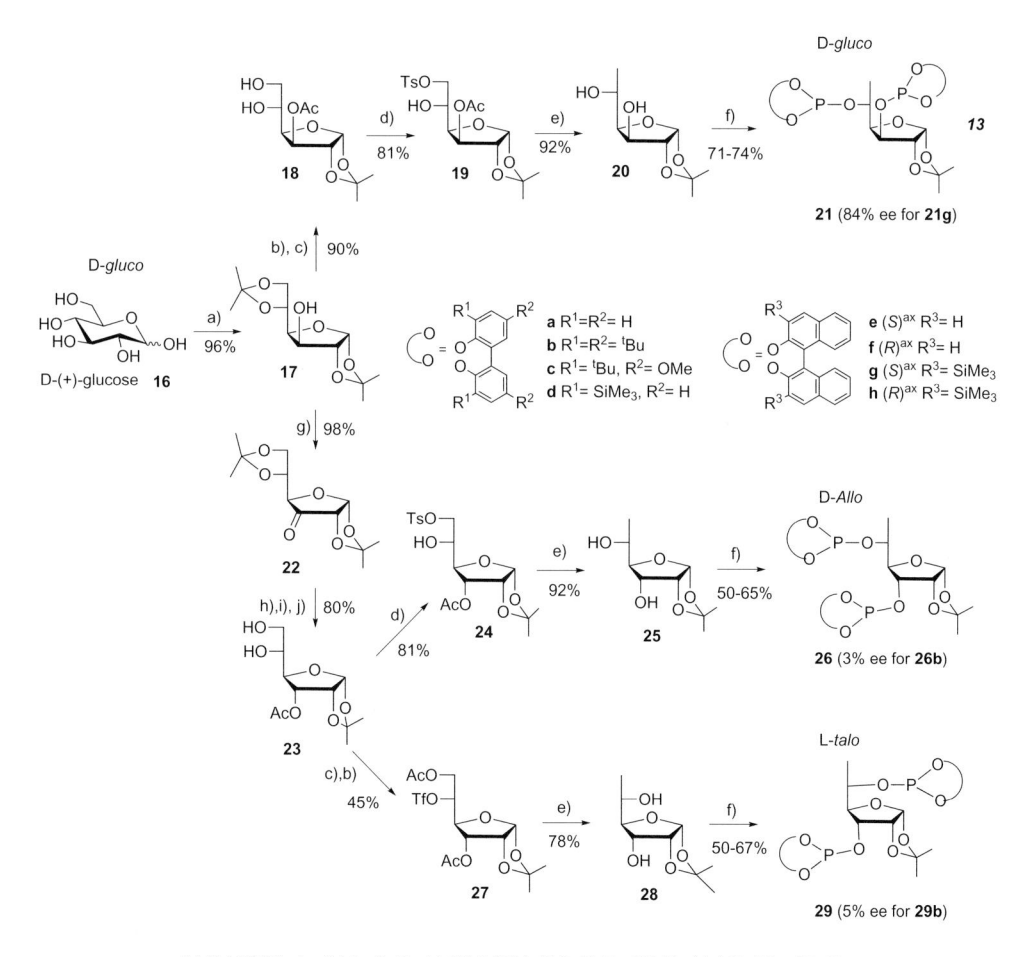

(a) CH₃COCH₃, I₂. (b) Ac₂O, Py. (c) CH₃COOH. (d) TsCl, Py, CH₂Cl₂. (e) 1.NaOMe, CH₂Cl₂. 2. LiAlH₄, THF. (f) (RO)₂PCl, Py, toluene. (g) PCC, NaOAc, CH₂Cl₂. (h) NaBH₄, EtOH, H₂O. (i) Ac₂O, pyridine. (j) CH₃COOH, H₂O.

Scheme 12.6 Synthesis of bidentate P,P-D-gluco-, D-allo-, and L-talo-ligands.

opening with LiAlH$_4$ rendered product **20**. The same methodology was applied to obtain ligands (**26**) and (**29**) but in this case prior inversion of the configuration at C3 was necessary, which was achieved by oxidation at position 3 and reduction of the carbonyl group with NaBH$_4$ to give, after protection, intermediate **23**. For the synthesis of product (**29**), an additional inversion at position 5 was carried out by introduction of a triflate leaving group at position C5 instead of position C6.

Diphosphite ligands **33a** (S^{ax}) and **33b** (R^{ax}), on the one hand, and **37a** (S^{ax}) and **37b** (R^{ax}), on the other hand based on pyranoside backbones of galactose and glucose, respectively, were synthesized stereospecifically by Chan et al. [23] in three steps from phenyl β-D-galactopyranoside **30** and phenyl β-D-glucopyranoside **34** (Scheme 12.7). Thus, tosylation of the starting materials produced the 6-tosyl derivatives in low to moderate yields. Treating the tosylated compounds with sodium hydroxide and subsequent reaction with a suitable phosphochloridite afforded the corresponding ligands **33a** (S^{ax}), **33b** (R^{ax}), **37a** (S^{ax}), and **37b** (R^{ax}). All of the ligands were stable during purification on silica gel under nitrogen atmosphere. The white solid ligands were also air-stable at room temperature.

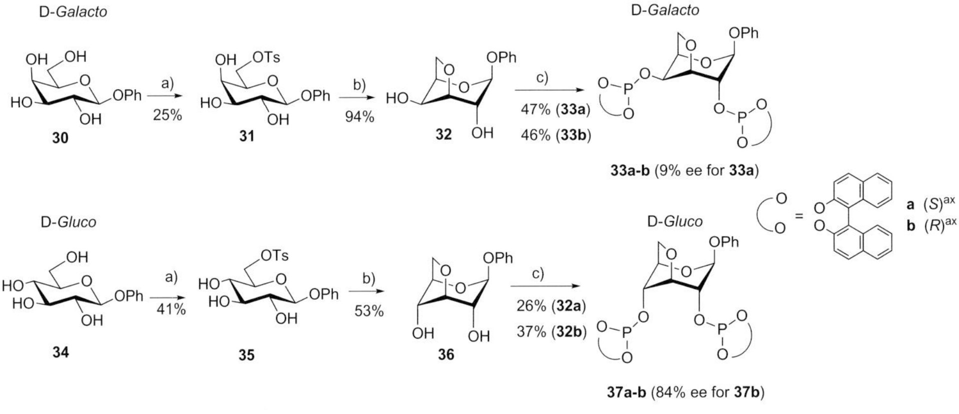

(a) TsCl, pyridine, CH$_2$Cl$_2$. (b) 1N NaOH, ethanol. (c) S or R-(1,1'-binaphthyl-2,2'-dioxy)-chlorophosphine, Tol, pyridine

Scheme 12.7 Synthesis of bidentate P,P-D-anhydro-D-*galacto*- and *gluco*-ligands.

Glucosamine was used as a starting material for the synthesis of ligand **42** [24] in a synthetic route that starts with *N*-acetylation, formation of the methyl glucoside **40** with methanol and acetyl chloride, and protection of position 4 and 6 as a benzylidene acetal. Hydrolysis of acetamide **40** furnished amine **41**, which was treated with various phosphorochloridites to afford phosphoramidite ligands **42a–c** (Scheme 12.8) [24, 25].

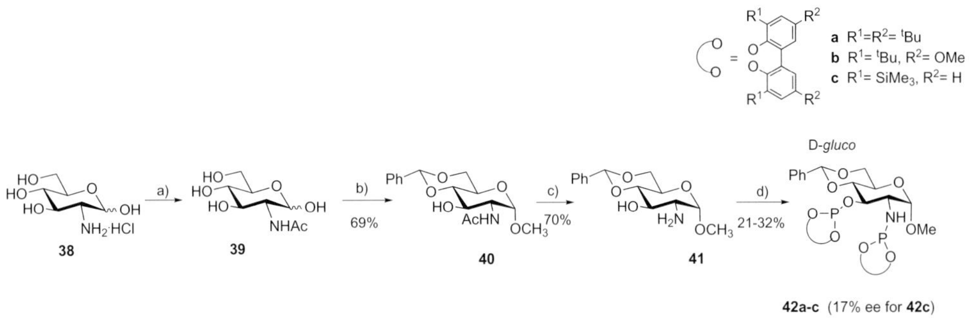

12.2 Copper-Catalyzed Reactions – Ligands

a $R^1=R^2={}^tBu$
b $R^1={}^tBu$, $R^2=OMe$
c $R^1=SiMe_3$, $R^2=H$

Scheme 12.8 Synthesis of bidentate P,P-D-*gluco*-ligands.

(a) $(CH_3CO)_2O$, NaOMe/MeOH, rt. (b) AcCl, MeOH then $PhCH(OMe)_2$, p-TsOH, DMF. (c) KOH (4 M) EtOH.
(d) Phosphorochloridite, Tol, pyridine.

12.2.2
Bidentate P,N-Ligands

12.2.2.1 *Xylo* and *Ribo* Ligands

Diéguez et al. [26] designed a series of amino–phosphite ligands with a furanoside backbone. Thus, the amino–phosphite D-*xylo*-**44** and D-*ribo*-**47** ligands were prepared in good yields (75–88%) by treating xylose-derived monotosylates **5** or **45** [27] with *tert*-butylamine, aniline, and isopropylamine in propan-2-ol at reflux followed by reaction with the corresponding phosphorochloridite in the presence of base (Scheme 12.9) [26].

12.2.2.2 D-*Gluco* Ligands

The sugar-based phosphite–oxazoline ligand libraries **50–54a–g** [24] are derived from D-glucosamine. They consist of a 4,6-O-protected glucopyranoside backbone with a phosphite moiety at the C3 position and a oxazoline ring fused to positions 1 and 2.

The sequence of ligand synthesis is illustrated in Scheme 12.10. Acylation of D-glucosamine hydrochloride **38** afforded the corresponding amides **39**. Subsequent protection of hydroxyl groups at the 4,6-positions with PhCHO and the 1,3-positions with Ac_2O gave the protected *N*-acyl-D-glucosamines **48**. They were subsequently converted into the corresponding oxazolines **49** in the presence of anhydrous $SnCl_4$. Deacetylation with K_2CO_3 and treatment with phosphorochloridite gave ligands **50–54** [24, 28].

12.2.3
Bidentate P,S- and O,S-Ligands

Thioether–alcohol ligands **56a–d** were synthesized from 1,2-*O*-isopropylidene-*xylo*-furanoside **2** by reaction with triflic anhydride and pyridine, followed by

262 | *12 1,4-Addition of Nucleophiles to α,β-Unsaturated Carbonyl Compounds*

(a) R^{1}NH$_2$, propan-2-ol. (b) (RO)$_2$PCl, pyridine, toluene. (c) TsCl, pyridine, CHCl$_3$.

Scheme 12.9 Synthesis of bidentate P,N-D-*xylo*- and D-*ribo*-ligands.

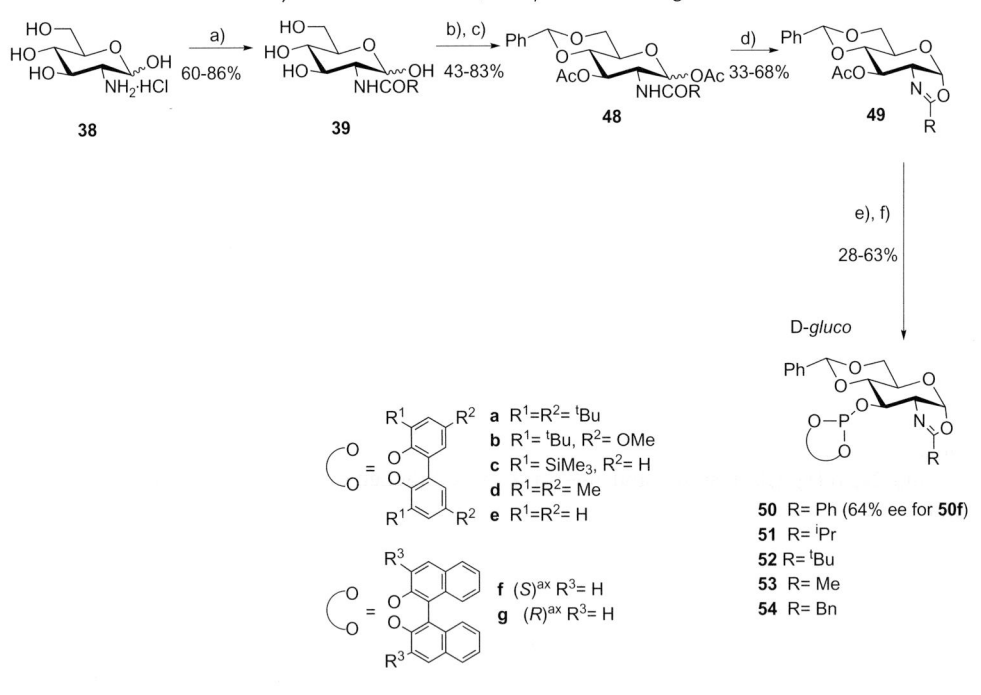

44a R^1= ^{t}Bu, R^2=R^3= ^{t}Bu
44b R^1= ^{t}Bu, R^2= ^{t}Bu, R^3= OMe
44c R^1= ^{i}Pr, R^2=R^3= ^{t}Bu
44d R^1= Ph, R^2=R^3= ^{t}Bu (63% ee)

47 R^1=Ph (8% ee)

a R^1=R^2= ^{t}Bu
b R^1= ^{t}Bu, R^2= OMe
c R^1= SiMe$_3$, R^2= H
d R^1=R^2= Me
e R^1=R^2= H

f (S)^{ax} R^3= H
g (R)^{ax} R^3= H

50 R= Ph (64% ee for **50f**)
51 R= ^{i}Pr
52 R= ^{t}Bu
53 R= Me
54 R= Bn

(a) (RCO)$_2$O, NaOMe/MeOH, rt or RCOCl, NaHCO$_3$ (aq). (b) PhCHO, ZnCl$_2$. (c) Ac$_2$O, pyridine. (d) SnCl$_4$, CH$_2$Cl$_2$.
(e) K$_2$CO$_3$, MeOH. (f) phosphorochloridite, Tol, pyridine.

Scheme 12.10 Synthesis of bidentate P,N-D-*gluco*-ligands.

displacement of the primary triflate **55** with the corresponding thiolate (Scheme 12.11) [29, 30]. Reaction of thioethers **56a–c** with chloro(diphenyl)phosphine in dry THF in the presence of pyridine and DMAP, or the biaryl-phosphorochloridite in the presence of pyridine, provided the desired thioether–phosphinite ligands **57a–c** [31] or the thioether–phosphite ligands **58a–e**, respectively [29, 32].

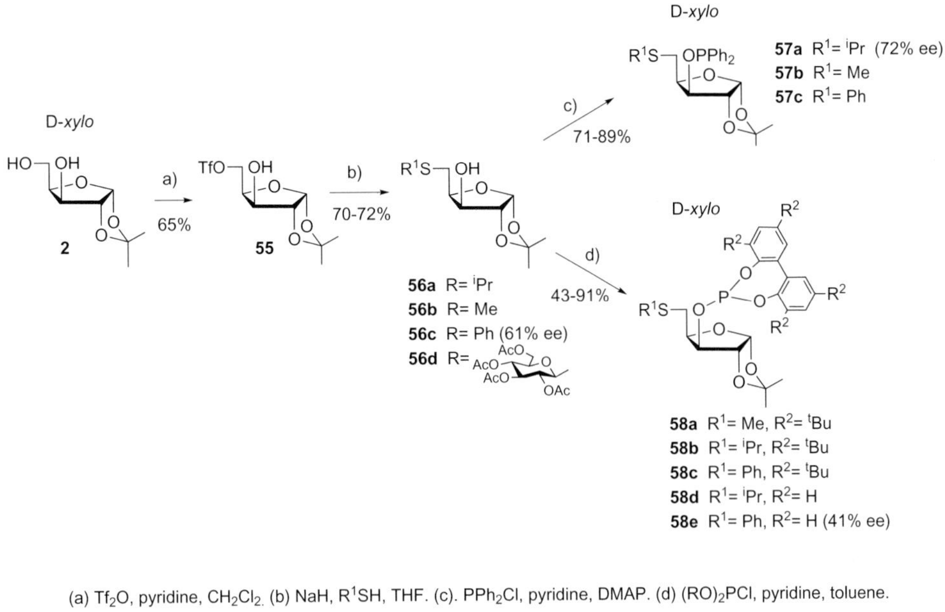

(a) Tf$_2$O, pyridine, CH$_2$Cl$_2$. (b) NaH, R^{1}SH, THF. (c). PPh$_2$Cl, pyridine, DMAP. (d) (RO)$_2$PCl, pyridine, toluene.

Scheme 12.11 Synthesis of bidentate P,S- and O,S-*xylo*-ligands.

12.2.4
Monodentate P or S-Ligands

The use of a highly modular sugar-based monophosphite ligand library in conjugate additions has been reported by Diéguez, Pàmies and Woodward [33] (ligands **59, 61, 66, 69, 71**) using previously described methodologies for their synthesis (see references therein). All ligands were synthesized by reaction of the corresponding sugar alcohols with PCl$_3$ and subsequent addition of the biaryl alcohols (a–f) in the presence of triethylamine. The sugar alcohols, in turn, were easily prepared on a large scale from D-(+)-glucose through the isopropylidene derivatives **17** and **22**, D-(+)-galactose **64**, and D-(–)-fructose **67** (Scheme 12.12). All ligands were stable during purification on neutral silica under an argon atmosphere and isolated in moderate yields as white solids.

Ligand **63** [35] was obtained by reaction of tetrakis(acetonitrile)copper(I) hexafluorophosphate [36] with the lithium salt of 1,2:5,6-di-*O*-isopropylidene-3-thio-α-D-glucofuranose [37]. This thiol was obtained by inversion at C3 of the alcohol **60** via triflate displacement by potassium thiocyanate and reduction with NaBH$_4$.

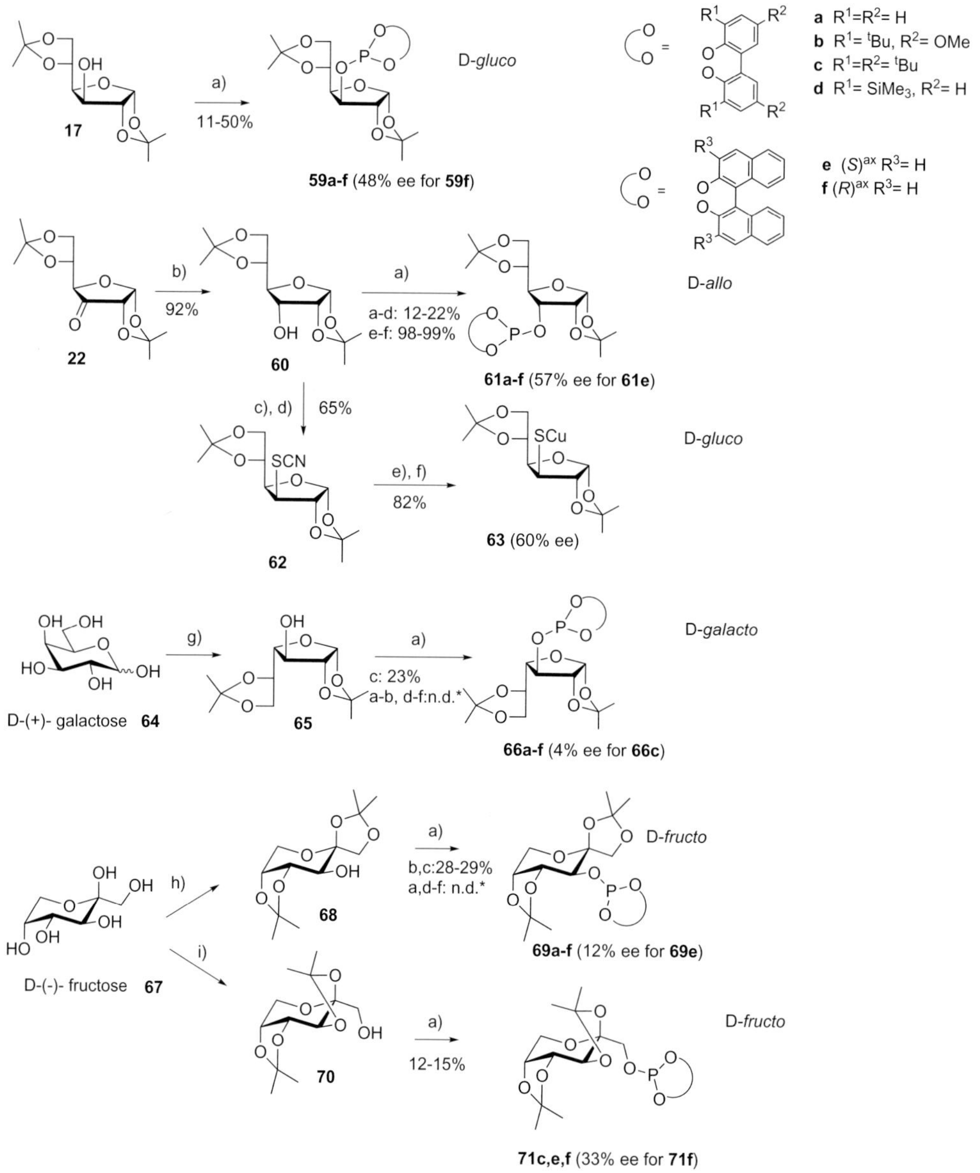

264 | *12 1,4-Addition of Nucleophiles to α,β-Unsaturated Carbonyl Compounds*

a $R^1=R^2=H$
b $R^1={}^tBu, R^2=OMe$
c $R^1=R^2={}^tBu$
d $R^1=SiMe_3, R^2=H$

e $(S)^{ax}\ R^3=H$
f $(R)^{ax}\ R^3=H$

(a) PCl$_3$, NEt$_3$, THF, biaryl alcohol (**a-f**) *n.d.: yields no described. (b) NaBH$_4$, EtOH. (c) Tf$_2$O, pyridine, CH$_2$Cl$_2$. (d) KSCN, CH$_3$CN. (e) NaBH$_4$, EtOH, H$_2$O. (f) BuLi, Et$_2$O, hexane. Cu(MeCN)$_4$PF$_6$. (g) DMF, acetone, Dowex H+. (h) HClO$_4$, dimethoxypropane, 0 °C. (i) HClO$_4$, dimethoxypropane, rt.

Scheme 12.12 Synthesis of monodentate P or S-ligands.

12.2.5
Miscellaneous Ligands with Peripheral Donor Centers. N,N-, N,O-, and P,N-Ligands

Ligands **76,78,79** [38] were obtained from D-ribose **72** by regioselective protection of the primary OH group with triisopropylsilyl chloride (TIPSCl), perbenzylation and removal of the silyl group to afford **74** (Scheme 12.12). Parikh–Doering oxidation of the alcohol afforded the aldehyde **75**, which was reacted with different functionalized aromatic amines to give iminic chiral ligands **76**. When the oxidation was performed with [bis(acetoxy)iodo]benzene (BAIB) and TEMPO, the corresponding acid **77** was obtained, which was employed for the synthesis of amidic chiral ligands **78** and **79** by reaction with different aromatic amines (Scheme 12.13).

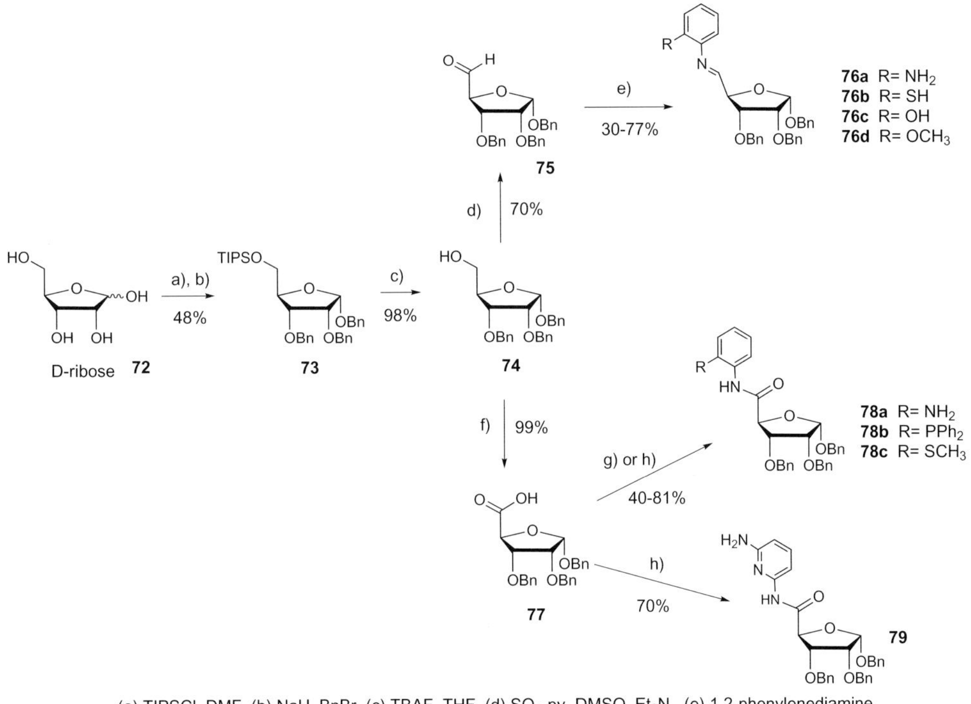

(a) TIPSCl, DMF. (b) NaH, BnBr. (c) TBAF, THF. (d) SO$_3$-py, DMSO, Et$_3$N. (e) 1,2-phenylenediamine (for **76a**), 2-aminobenzenethiol (for **76b**), 2-aminophenol (for **76c**) or 2-methoxyphenylamine (for **76d**), EtOH. (f) TEMPO, BAIB, CH$_3$CN/H$_2$O 1:1. (g) 1. EDAC, HOBT, DMF/CH$_2$Cl$_2$, 2. (i-Pr)$_2$NEt, CH$_2$Cl$_2$, 1,2-phenylenediamine (for **78a**). (h) 1. (COCl)$_2$, CH$_2$Cl$_2$ 2. Et$_3$N, THF, 2-diphenylphosphanyl-phenylamine (for **78b**), 2-methylsulfanyl-phenylamine (for **78c**), or 2,6-diaminopyridine (for **79**).

Scheme 12.13 Synthesis of miscellaneous ligands with peripheral donor centers N,N-, S,N-, N,O-, and P,N-ligands.

12.3
Enantioselective Copper-Catalyzed 1,4-Addition of Organometallics to α,β-Unsaturated Carbonyl Compounds

12.3.1
Substrate: 2-Cyclohexenone

2-Cyclohexenone has been the most widely studied substrate for copper-promoted asymmetric conjugate addition, being the substrate of choice for testing a new ligand. This enone is very reactive and has the advantage of being conformationally fixed. Thus, the problem of *s*-cis and *s*-trans conformational interconversion of acyclic substrates is avoided. Below, we present the results of conjugate additions of organometallics using this substrate. First, we cover the results using bidentate ligands with a furanoside backbone and, second, those with a pyranoside backbone. Finally, we review the catalytic data using monodentate ligands.

12.3.1.1 Bidentate Diphosphite and Diphosphinite Ligands

Bidentate Ligands with a Furanoside Backbone Copper-catalyzed asymmetric addition of alkyl nucleophiles to cyclohexenone using bulky diphosphite ligands derived from *xylo*- and *ribo*-furanose **3** and **15** proceeded with good conversions and almost complete regioselectivities (>95%) in the 1,4-product [14, 15]. Biphenyl-derived phosphite ligands **3a–c**, which only incorporate chirality at the sugar framework of the ligand, rendered moderate enantioselectivities, with ligand **3a**, which has a non-substituted biphenyl group, being the best (53% ee, Table 12.1, entry 1). Naphthyl-derived phosphite ligands **3d–f** (Table 12.1, entries 4–6), which led to more active catalysts, showed a different trend to those with biphenyl moieties. The more sterically demanding ligand **3f** led to a high degree of enantioselectivity (81% ee) that was the best of the series (Table 12.1, entry 6) and, furthermore, one of the best results using carbohydrate ligands. Results obtained from catalyst precursors **3d** and **3e** showed that the absolute configuration of the naphthyl moiety determines the activity and the sense of enantio-induction (Table 12.1, entry 4 versus entry 5).

Ribose-derived diphosphite ligands **15a–c**, in turn, gave slightly better results than those obtained with the xylose-biphenyl-derived phosphinite ligands **3a–c** in terms of activity, but the enantioselectivity was significantly lower, close to nil, showing the influence of the absolute configuration of C3 in this sugar framework.

With glucose-derived diphosphite ligands **21, 26,** and **29**, with biphenyl and binaphthyl moieties, good to excellent rates were observed [22]. The modular nature of these ligands allows a facile systematic variation of the configuration of the C3 and C5 stereogenic centers at the ligand as well as in the biaryl substituents; consequently, their influence on activity and enantioselectivity can be determined. Under the conditions that led to an optimum compromise between enantioselectivities and reaction rates for ligand **21a**, that is, a L/Cu ratio = 1, temperature of

0 °C, and dichloromethane as a solvent, the rest of the ligands were also tested. Systematic variation of the configuration at C3 and C5 of the sugar framework revealed that the enantioselectivity depended strongly on the absolute configuration at C3, and was much less affected by the configuration of the C5 stereocenter. The best enantioselectivities (20–84% ee) were obtained with ligands **21**, with an *(S)*-configuration at C3 and *(R)*-configuration at C5 (Table 12.1, entries 9–11), whereas ligands **26** and **29** led to nearly racemic mixtures. In particular, the best degree of enantio-induction (84% ee) was obtained with ligand **21g** as a catalyst precursor, with an *ortho*-trimethylsilyl substituted *(S)*-configured naphthyl moiety (Table 12.1, entry 11). This is one of the best results in Cu-catalyzed asymmetric addition to cyclohexenone using carbohydrate ligands.

Copper-catalyzed asymmetric addition of alkyl nucleophiles to cyclohexenone using diphosphinite ligands **4** and **14** as catalyst precursors led to moderate enantioselectivities (up to 32% ee), which did not substantially change with the nucleophilic species (Table 12.1, entries 7 and 8) [39]. Regioselectivities and activities were higher for the addition of triethylaluminium than with addition of diethylzinc.

Bidentate Ligands with a Pyranoside Diphosphite Backbone   For ligands **33** and **37**, proposed by Wang *et al.* [23], $Cu(OTf)_2$ was chosen as the Cu source for preparation of the active catalysts. In the first set of experiments, 2-cyclohexenone was treated with diethylzinc in the presence of $Cu(OTf)_2$ and ligands in toluene. Ligand **37b** was found to be more effective than ligands **33a**, **33b**, or **37a**, which gave substantially lower ee values (2–13% ee). These results clearly show the synergistic effects of the different chiral elements of ligands on the enantioselectivity of the reactions: one between the backbone stereocenters C2 and C4 and another between the stereocenters of the ligand backbone and the configuration of the binaphthyl phosphite moieties. The degree of enantioselectivity is controlled by all chiral elements, whereas the sense of the enantio-induction is mainly controlled by the configuration of the binaphthyl phosphite moieties.

When $(CuOTf)_2 \cdot C_6H_6$ was used as the copper source instead of $Cu(OTf)_2$, the yield and enantioselectivity increased. Moreover, a profound solvent effect on enantioselectivity was observed in this case. Ethereal solvents proved to be more effective than the non-coordinating toluene and dichloromethane. Thus, under the optimized condition reactions [$(CuOTf)_2 \cdot C_6H_6$, $ZnEt_2$ in THF] and using ligand **37b**, a high degree of enantioselectivity (84% ee) was observed (Table 12.1, entry 12). This is one of the best results in terms of yields and enantioselectivity reported in conjugate additions using carbohydrates as ligands.

12.3.1.2 Bidentate Ligands with Different Coordinating Functional Groups: P/P, P/N, S/O, and S/P

Mixed phosphine–phosphite ligands **8** and **9** were used in the asymmetric 1,4-addition of organometallics to cyclohexenone [32]. Although reaction rates with ligands **8** were good and chemoselectivities in 1,4-product were higher than 97%, enantioselectivities under optimized conditions [$Cu(OTf)_2$, CH_2Cl_2, 0 °C, L/Cu

Table 12.1 Representative results of conjugate addition of alkyl donors to 2-cyclohexenone.[a)f)]

Reaction scheme: 2-cyclohexenone + 1.4 equiv $MR^1{}_n$ → (with 1 mol% Cu source, 2 mol% ligand*, solvent) → 3-substituted cyclohexanone, R^1 = Et, Me

Entry	Ligand	Precursor	$MR^1{}_n$	Solvent
1	P,P donors (a–c)	3a, $R^1 = R^2 = H$	Cu(OTf)$_2$ [f)] (0.5 mol.%)	ZnEt$_2$	CH$_2$Cl$_2$
2		3b, $R^1 = {}^tBu$, $R^2 = OMe$	Cu(OTf)$_2$ [f)] (0.5 mol.%)	ZnEt$_2$	CH$_2$Cl$_2$
3		3c, $R^1 = R^2 = {}^tBu$	Cu(OTf)$_2$ [f)] (0.5 mol.%)	ZnEt$_2$	CH$_2$Cl$_2$
4		3d $(S)^{ax}$ $R^3 = H$	Cu(OTf)$_2$ [f)]	ZnEt$_2$	CH$_2$Cl$_2$
5		3e $(R)^{ax}$ $R^3 = H$	Cu(OTf)$_2$ [f)]	ZnEt$_2$	CH$_2$Cl$_2$
6	d–f	3f $(R)^{ax}$ $R^3 = SiMe_3$	Cu(OTf)$_2$ [f)]	ZnEt$_2$	CH$_2$Cl$_2$
7	Ph$_2$PO–OPPh$_2$ ligand	4	[CuMeCN)$_4$]BF$_4$ [f)]	AlEt$_3$	CH$_2$Cl$_2$
8	Ph$_2$PO / Ph$_2$PO ligand	14	[CuMeCN)$_4$]BF$_4$ [f)]	AlEt$_3$	CH$_2$Cl$_2$

Temperature (°C)	Conversion (%)[b] (t, min)	1,4-Selectivity[c] (%)	Yield[d] (%)	ee (%)[e]	Reference
0	6	>95	Nd	53	[14]
0	100	>95	Nd	22	[14]
0	75	>95	Nd	24	[14]
0	100	99	Nd	43 (S)	[15]
0	100	98	Nd	35 (R)	[15]
0	100	99	Nd	81 (R)	[15]
Room temp.	94	100	Nd	24 (S)	[38]
Room temp	97	100	Nd	32 (S)	[38]

(Continued)

Table 12.1 (Continued)

Entry	Ligand	Precursor	Cu source	MR^1_n	Solvent
9	(phosphite-sugar ligand)	21a, R^1 = R^2 = H	Cu(OTf)$_2$ [f]	ZnEt$_2$	CH$_2$Cl$_2$
10	(biphenyl phosphite)	21b, R^1 = R^2 = ^{t}Bu	Cu(OTf)$_2$ [f]	ZnEt$_2$	CH$_2$Cl$_2$
11	(binaphthyl phosphite)	21g, (S)^{ax} R^3 = SiMe$_3$	Cu(OTf)$_2$ [f]	ZnEt$_2$	CH$_2$Cl$_2$
12	(sugar-phosphite with OPh; binaphthyl phosphite a (S)^{ax}, b (R)^{ax})	37b (R)^{ax}	(CuOTf)$_2$·C$_6$H$_6$ (2 mol.%)	ZnEt$_2$ (2.4 equiv)	THF
13	(Ph-sugar phosphoramidite with NH, OMe; biphenyl phosphite)	42c, R^1 = SiMe$_3$, R^2 = H	Cu(OAc)$_2$ [g]	AlEt$_3$	Et$_2$O

Reaction scheme: cyclohexenone + 1.4 equiv MR^1_n → 3-substituted cyclohexanone (R^1 = Et, Me), with 1 mol% Cu source, 2 mol% ligand*, in Solvent.

Temperature (°C)	Conversion (%)[b] (t, min)	1,4-Selectivity[c] (%)	Yield[d] (%)	ee (%)[e]	Reference
0	75	75	Nd	60 (R)	[22]
0	100	99	Nd	25 (R)	[22]
0	98	98	Nd	84 (S)	[22]
−40	>99	No 1,2 (GC)	95	84 (R)	[23]
−30	89	Nd	22	17 (S)	[24]

(Continued)

Table 12.1 (Continued)

Entry	Ligand	Precursor	MR$^1{}_n$	Solvent
14	P,N donors	44d R^1 = Ph, R^2 = R^3 = ^{t}Bu	Cu(OTf)$_2$	ZnEt$_2$	CH$_2$Cl$_2$
15		50c R^1 = SiMe$_3$, R^2 = H	Cu(OAc)$_2$[g]	AlEt$_3$	Et$_2$O
16		50f (S)^{ax} R^3 = H	Cu(OAc)$_2$[g]	AlEt$_3$	Et$_2$O
17		56a R = ^{i}Pr	Cu(OTf)$_2$	ZnEt$_2$	CH$_2$Cl$_2$
18		56b R = Me	Cu(OTf)$_2$	ZnEt$_2$	CH$_2$Cl$_2$
19	S,O donors RS—OH	56c R = Ph	Cu(OTf)$_2$	ZnEt$_2$	CH$_2$Cl$_2$
20		56d R^1 = (AcO-glucosyl-OAc)	Cu(OTf)$_2$	ZnEt$_2$	CH$_2$Cl$_2$

Temperature (°C)	Conversion (%)[b] (t, min)	1,4-Selectivity[c] (%)	Yield[d] (%)	ee (%)[e]	Reference
0	90	No 1,2 (GC)	Nd	63 (S)	[26]
−30	99	Nd	50	64 (S)	[24]
−30	88	Nd	61	64 (S)	[24]
0	98	94	Nd	44 (S)	[30]
0	78	89	Nd	49 (S)	[30]
0	75	92	Nd	61 (S)	[30]
0	41	82	Nd	20 (S)	[30]

(Continued)

Table 12.1 (Continued)

Entry	Ligand	Precursor		MR^1_n	Solvent
21	P,S donors	57a R = ^{i}Pr	CuCN	ZnEt$_2$	CH$_2$Cl$_2$
22	RS—OPPh$_2$ (sugar-based)	57b R = Me	[CuMeCN)$_4$] BF$_4$	ZnEt$_2$	CH$_2$Cl$_2$
23		57c R = Ph	[CuMeCN)$_4$] BF$_4$	AlEt$_3$	CH$_2$Cl$_2$
24	R^{1}S—, biaryl phosphite with R^2 groups	58e R^1 = Ph, R^2 = H	Cu(OTf)$_2$[f]	ZnEt$_2$	CH$_2$Cl$_2$
25	P donors (sugar phosphite)	59b R^1 = ^{t}Bu, R^2 = OMe	Cu(OTf)$_2$[g]	AlMe$_3$	DME
26	bis-phosphite with R^1, R^2, R^3 (binaphthyl)	59f (R)^{ax} R^3 = H	Cu(OTf)$_2$[g]	AlMe$_3$	DME

Temperature (°C)	Conversion (%)[b] (t, min)	1,4-Selectivity[c] (%)	Yield[d] (%)	ee (%)[e]	Reference
Room temp.	100	100	Nd	64 (R)	[38]
Room temp.	100	100	Nd	53 (R)	[38]
0	100	100	Nd	48 (S)	[38]
0	64	>97	Nd	41 (R)	[32]
−30	99	Nd	61	37 (R)	[33]
−30	96	Nd	77	48 (R)	[33]

(Continued)

276 | *12 1,4-Addition of Nucleophiles to α,β-Unsaturated Carbonyl Compounds*

Table 12.1 (Continued)

Reaction scheme: cyclohexenone + 1.4 equiv MR^1_n → 3-substituted cyclohexanone (R^1 = Et, Me), with 1 mol% Cu source, 2 mol% ligand* in solvent.

Entry	Ligand	Precursor	MR^1_n	Solvent
27	(sugar-based phosphite ligand)	**61b** R^1 = ^{t}Bu, R^2 = OMe	Cu(OTf)$_2$[g]	AlMe$_3$	DME
28	(biphenyl/binaphthyl dioxy ligand)	**61e** $(S)^{ax}$ R^3 = H	Cu(OTf)$_2$[g]	AlMe$_3$	DME
29	(sugar-phosphite + biphenyl ligand)	**66c** R^1 = R^2 = ^{t}Bu	Cu(OTf)$_2$[g]	AlMe$_3$	DME
30	S donor (sugar-SCu ligand)	**63**	4 mol.% Cu source	BuMgBr	Et$_2$O

a) Typical reaction conditions: Cu precursor (1 mol.%), ligand (2 mol.%), MR^1_n (1.4 equiv) substrate (1 equiv) unless otherwise stated.
b) % Conversion determined by GC using undecane or dodecane as internal standard.
c) Regioselectivity determined by GC using undecane as internal standard.
d) Yield determined by CG using undecane or dodecane as internal standard, otherwise not determined (Nd).

Temperature (°C)	Conversion (%)[b] (t, min)	1,4-Selectivity[c] (%)	Yield[d] (%)	ee (%)[e]	Reference
−30	98	Nd	31	15 (S)	[33]
−30	99	Nd	55	57 (R)	[33]
−30	99	Nd	23	4 (R)	[33]
−78	Nd	>98	>90	60	[34]

e) Enantiomeric excess determined by GC or by ^{13}C NMR.
f) Ligand-to-Cu ratio = 1.
g) Ligand-to-Cu ratio = 4.

ratio = 1] were low (up to 19% ee). Replacement of a biphenyl moiety with an enantiomerically pure binaphthyl group at the phosphite moiety produced ligands **9** led to even lower enantioselectivities (up to 10% ee). The use of triethylaluminium as a nucleophilic species and [Cu(MeCN)$_4$]BF$_4$ as a copper source with ligands **8** and **9** led to similar activities and chemoselectivities in 1,4-products, but did not improve the enantioselectivities obtained with dialkylzinc.

Encouraged by the success of some bidentate P,N- and phosphoramidite ligands in conjugate addition, Diéguez *et al.* designed a series of phosphite–phosphoramidite and amino–phosphite ligands **11, 44, 47** with a furanoside backbone and tested them in this reaction [26].

The effect of an alkyl substituent at N on catalyst performance was investigated by comparing ligands **44a,c,** and **d** under the optimized reaction conditions [Cu(OTf)$_2$, L/Cu ratio = 1, $T = 0\,°C$, and dichloromethane]. The best conversions and enantioselectivities were obtained with the sterically less demanding phenyl derivative **44d**. Notably, the sense of enantio-induction was affected by the nature of the N-substituent. Ligands **44a,b** containing the bulky *tert*-butyl group at nitrogen gave the *(R)*-configured product, whereas the less sterically hindered ligands **44c** and **44d** gave the *(S)*-enantiomer preferentially. The effect of the ligand-to-copper ratio on the outcome of the reaction was also studied with ligand **44d**, with the highest reaction rates and enantioselectivities (90% conversion, 63% ee) being obtained with a ligand/Cu ratio = 2:1 (Table 12.1, entry 14).

Ligand **47**, whose configuration at C3 is opposite to that of ligand **44**, led to a similar conversion (84%) but the enantioselectivity dropped considerably [8% ee *(R)*]. Phosphoroamidite–phosphite ligand **11** afforded higher reaction rates, but moderate enantioselectivities [30% ee *(S)*].

A few P,N-heterodonor pyranoside ligands have been applied in copper-catalyzed conjugated addition to 2-cyclohexenone, but with moderate success. Phosphite–oxazoline ligands **50–54**, proposed by Woodward *et al.* [24], were designed to study the influence of systematically varying the electronic and steric properties of the oxazoline substituents and different substituents/configurations in the biaryl phosphite moiety. These authors also studied the performance of ligand **42** in the 1,4-addition of alkyl nucleophiles to cyclohexenone, where the oxazoline moiety present in ligands **50–54** was replaced by a phosphoroamidite.

Thus, phosphite–oxazoline **50–54a–g** and phosphite–phosphoroamidite **42a–c** ligands were tested [24] in the addition of triethylaluminium and diethylzinc by using Cu(OAc)$_2$ as a catalyst precursor. The results indicated that the enantioselectivity was mainly affected by the substituents/configuration at the biaryl phosphite moiety, while the oxazoline groups had little effect. The best results were obtained when using ligands **50c** and **50f** (Table 12.1, entries 15 and 16), which contain trimethylsilyl substituents at the *ortho* positions of the biphenyl phosphite moiety and an *(S)*-binaphthyl phosphite moiety, respectively (ee values up to 64%). In addition, replacement of the oxazoline by a phosphoroamidite moiety had a negative effect on the yield and enantioselectivity (Table 12.1, entry 15 versus 13).

Yields and enantioselectivities were lower when using diethylzinc in relation to triethylaluminium.

Chiral S,O ligands **56a–d** are active in the enantiomeric copper-catalyzed 1,4-addition of diethylzinc to cyclohexenone, and produce enantiomeric excesses of up to 61% (Table 12.1, entries 17–20) [30].

The effect of the thioether substituent on the outcome of the reaction was studied using the optimized reaction conditions for **56c** [Cu(OTf)$_2$, $T = 0\,°C$, dichloromethane]. Changing the thioether moiety produced an important effect on the rate and stereoselectivities. Conversions are higher for the catalyst precursor containing the electron-rich **56a** ligand (Table 12.1, entry 17), while better ee values were obtained with **56c** (Table 12.1, entry 19). The introduction of a sterically demanding group in the thioether moiety (**56d**) had a negative effect on enantioselectivity (Table 12.1, entry 20).

Structurally related thioether–phosphinite xylose-derived ligands **57** *(2)* [39] and thioether–phosphite ligands **58** [32] were also tested. Addition to cyclohexenone using the optimized conditions for **57a** (dichloromethane, L/Cu ratio = 2, and CuCN or [CuMeCN)$_4$]BF$_4$) led to complete conversions and 1,4-selectivities with moderate enantioselectivities (64% ee, Table 12.1, entry 21).

Substituents on sulfur affected the catalytic performance of ligands, the best one being the ligand with isopropyl substituent **57a** (64% ee, Table 12.1, entry 21), while the methyl and phenyl substituents showed lower enantioselectivities (Table 12.1, entries 22 and 23).

The effect of the temperature was also studied. Using ligand **57a** at 0 °C produced the 1,4-addition product with increased enantioselectivity to the detriment of conversion, which dropped substantially.

The performance of this series of ligands using triethylaluminium as a nucleophilic species was also explored. Under optimized conditions (dichloromethane, L/Cu ratio = 2, and Cu(OTf)$_2$ or [CuMeCN)$_4$]BF$_4$), 1,4-selectivities were complete, and the activities were higher than those for diethylzinc. In general, conversions were complete in 20 min as compared with ZnEt$_2$ (2 h) but, conversely, enantioselectivities were in general lower. The best ligand was in this case **57c**, with a phenyl substituent at sulfur (48% ee, Table 12.1, entry 23).

The presence of the thioether moiety instead of the phosphinite functionality at C5 led to lower catalyst activities, but higher chemoselectivities and enantioselectivities [compare ligands **4** with **57**, Table 12.1, entry 4 versus entries 21–23].

Changing a phosphinite moiety for a phosphite unit in the basic S,O-framework produced series of ligands **58** [32]. In all cases, regioselectivities in the 1,4-product were higher than 97% and no 1,2-product was observed by GC. In general, reaction rates were good for all ligands, being higher than those obtained with ligands **57**. By contrast, enantioselectivities under optimized conditions [Cu(OTf)$_2$, CH$_2$Cl$_2$, 0 °C, L/Cu ratio = 1] were in general lower that those obtained with ligands **57**, and modification of the sulfur substituent did not produce a significant effect, either on the rate or enantioselectivity, which ranged from 11 to 18% ee. The best enantioselectivity of this series of ligands was attained with *ortho*-phenyl substituted biphenyl-derived thioether–phosphinite **58e** (41% ee, Table 12.1, entry 24).

12.3.1.3 Monodentate P- or S-Ligands

In comparison with bidentate ligands, the use of monodentate ligands in the 1,4-addition of alkyl nucleophiles to cyclohexenone is rarer. In this respect, Diéguez, Pàmies, and Woodward synthesized phosphite ligands **59, 61, 66** with a furanoside framework, together with some pyranoside phosphite ligands **69** and **71** derived from fructose, and tested them in the asymmetric 1,4-conjugate addition of AlCl$_3$ to cyclohexenone [34]. The effect of several reaction parameters were studied using ligand **59f**. The best result was obtained using dimethoxyethane (DME) as solvent, Cu(OTf)$_2$ as the catalyst precursor, and a ligand-to-copper ratio of 4 at −30 °C (Table 12.1, entry 26, 77%, 48% ee). Under the optimized conditions, the remaining ligands, **59, 61, 66, 69**, were evaluated. The results indicated that selectivities are highly affected by the configuration of C4 of the carbohydrate backbone, the size of the ring of the sugar, and the cooperative effect between the configurations of C3 and that of the binaphthyl phosphite moiety, which resulted in a matched combination for ligand **61e** (Table 12.1, entry 28, 57% ee). In all cases, however, the formation of byproducts was observed.

In addition, biphenyl phosphite ligands **59a–d** were found to give *(R)*-configured addition products (see, for example, Table 12.1, entry 25), while biphenyl phosphite ligands **61a–d** (see, for example, Table 12.1, entry 27) were found to furnish the *(S)*-configured products. The authors attributed this opposite sense of enantioselectivity to the configuration of the biphenyl moieties of the ligands when coordinated to the copper species, with ligands **59a–d** adopting an *(R)*-configuration while ligands **61a–d** adopt an *(S)*-configuration.

Comparing the results using ligands **59** with **66**, which differ only in the configuration at C4, ligands **66** with an *(S)*-configuration at C4 were found to give lower enantioselectivities than ligands **59** with an opposite configuration at this position (Table 12.1 entry 29 versus 25 and 26). On the other hand, ligands **69** and **71**, which have a pyranoside backbone, provided lower yields and enantioselectivities than furanoside ligands (2–33% ee). In summary, the best results were obtained with ligand **61e**, which contains the best combination of the ligand parameters (ee values up to 57%; Table 12.1, entry 28).

In contrast, Spescha [35] introduced Cu-thiolate **63** for the conjugate addition of Grignard reagents to 2-cyclohexenone. The chemical yields and regioselectivities were in all the cases higher than 90% and 98%, respectively, and independent of the experimental conditions. Interestingly, the enantioselectivity was dependent on the halide of the Grignard reagent and reached a maximum of 60% when BuMgBr was used (Table 12.1, entry 30). The disadvantage of this reaction is its low application range; low amounts of the 1,4-product or enantioselectivities below 20% ee were obtained when applied to linear substrates.

12.3.1.4 Miscellaneous Ligands with Peripheral Donor Centers. N,N-, N,O-, and P,N-Ligands

Attempts at conjugate addition of Et$_2$Zn to cyclohexenone in the presence of amidic ligands **76**, **78**, and **79** with Cu(OTf)$_2$ as the catalyst gave modest results [38]. The formation of 3-ethylcyclohexanone was observed in high yields, but no significant enantiomeric excess (10–15%) was obtained. Although no more details about the obtained results were reported, Piancatelli and coworkers described a reduction of the reaction times by up to one-half only in the presence of the Cu(OTf)$_2$, showing the efficiency of the catalysis.

12.3.2
Substrate: 2-Cyclopentenone, 2-Cycloheptenone

Cyclopentenone is usually the most reactive substrate for conjugate addition, but the resulting enolate is reactive enough to undergo Michael addition to unreacted cyclopentenone, thus lowering the isolated yield of the reaction. Another problem with this substrate is the flatness of the molecule, which makes it less sensitive to the steric requirements of the chiral ligand. As a result, cyclopentenone generally affords lower ee values than cyclohexenone. Cycloheptenone behaves exactly as cyclohexenone, but usually provides higher ee values with the same ligands. These general trends, however, do not seem to apply to the addition of diethylzinc with catalytic systems developed with carbohydrate ligands **37b**. Thus, diphosphite ligand **37b** (Scheme 12.7) and Cu(OTf)$_2$ [23] (non-optimized copper source) showed erratic catalytic performance in terms of enantioselectivity related to the size of the ring: 35% ee for 2-cyclopentenone, 59% ee for 2-cyclohexenone, and 11% ee for 2-cycloheptenone (Table 12.2, entry 2), with very low yields for the cyclopentenone and cycloheptenone reactions (18% and 12%, respectively). The asymmetric conjugate addition of Et$_2$Zn to 2-cyclopentenone under optimized conditions (2 mol.% CuOTf·C$_6$H$_6$) proceeded with moderately good degrees of enantioselectivity although with a modest yield (36%, 64% ee, Table 12.2, entry 1).

12.3.3
Linear Substrates: *trans*-Non-3-en-2-one and *trans*-5-Methyl-3-hexen-2-one

Acyclic enones possessing only aliphatic substituents are a more demanding substrate class for asymmetric conjugated addition than 2-cyclohexenone. The high conformational flexibility of these substrates together with the presence of only subtle substrate–catalyst steric interactions makes the design of effective enantioselective systems challenging. For this type of substrate, the use of organoaluminium reagents has been very useful for obtaining high levels of enantioselectivity.

Thus, phosphite–phosphoramidite ligands **42b,c** and phosphite–oxazolines **50a–g, 51–54a** have been also tested with linear ketones [24]. The effect of the second donor group in the framework of the pyranoside ligand was explored in the asymmetric 1,4-conjugate addition of AlMe$_3$ to *trans*-non-3-en-2-one.

Table 12.2 Representative results of conjugate addition of diethylzinc to 2-cyclopentenone and 2-cycloheptenone.

Entry	Substrate	Ligand	Precursor	Yield (%)[a]	ee (%)[b]	Reference
1	(cyclopentenone)	(phosphite, OPh) 37b (R)^{ax}	(CuOTf)$_2$·C$_6$H$_6$	36	64 (R)	[23]
2	(cycloheptenone)	a (S)^{ax} b (R)^{ax}	Cu(OTf)$_2$	12	11 (R)	[23]

a) Yield determined by CG using dodecane as internal standard.
b) Enantiomeric excess determined by GC.

Replacement of the phosphoramidite moiety with an oxazoline ring led to slightly lower conversion but slightly improved enantioselectivity in the 1,4-addition product (Table 12.3, entry 1 versus 2). Using phosphite–oxazoline **50a–g**, yields and enantioselectivities were affected by the substituents at the biaryl phosphite moiety. Thus, the best yields (up to 85%) and enantioselectivities (up to 62%) were obtained with ligand **50f**, which contains sterically demanding substituents in the biaryl phosphite moiety and phenyl oxazoline group (Table 12.3, entry 4).

The effect of several reaction parameters, such as catalyst precursor {i.e., Cu(OTf)$_2$, CuTC, Cu(OAc)$_2$, CuI, and [Cu(MeCN)$_4$]BF$_4$}, solvent (i.e., diethyl ether, CH$_2$Cl$_2$, THF, and tert-butyl methyl ether) and ligand-to copper ratio was also studied. Interestingly, in contrast to the conjugate addition to 2-cyclohexenone, an important positive effect using [Cu(MeCN)$_4$]BF$_4$ as a catalyst precursor on enantioselectivity was found. The enantioselectivity increased from 51% to 78% ee (Table 12.3, entry 3). The rest of the reaction parameters did not improve the catalytic performance.

For *trans*-5-methyl-3-hexen-2-one, yields and enantioselectivities were also affected by the substituents/configurations at the biaryl phosphite moiety and by the oxazoline groups. Again, the best result in terms of yields and

12.3 Enantioselective Copper-Catalyzed | 283

Table 12.3 Representative results of conjugate addition of alkyl donors to *trans*-3-nonen-2-one.[a]

Reaction: enone + 1.4 equiv AlMe$_3$, 1 mol% Cu precursor, 4 mol% ligand*, Et$_2$O or THF → product

Entry	Ligand	Precursor	Conversion (%)[b]	1,4-Selectivity[c]	Yield[d] (%)	ee (%)[e]	Reference
1	(carbohydrate phosphite-phosphoramidite, OMe)	42b, R^1 = ^{t}Bu, R^2 = OMe	Cu(OTf)$_2$, 80	Nd	32	26 (S)	[24]
2	(carbohydrate phosphite, Ph)	50b, R^1 = ^{t}Bu, R^2 = OMe	Cu(OTf)$_2$, 60	Nd	32	32 (R)	[24]
3	(biphenyl phosphite)	50c, R^1 = SiMe$_3$, R^2 = H	[Cu(MeCN)$_4$]BF$_4$, 94	Nd	48	78 (R)	[24]
4	(binaphthyl phosphite, R^3)	50f, (S)^{ax}, R^3 = H	Cu(OTf)$_2$, 92	Nd	85	62 (R)	[24]
5	(RS—OH sugar ligand, R = Ph)	56c, R = Ph	[Cu(MeCN)$_4$]BF$_4$[f] 10 mol.%, 40	80	Nd	22 (R)	[30]

Table 12.3 (Continued)

Entry	Ligand	Precursor	Conversion (%)[b]	1,4-Selectivity[c]	Yield[d] (%)	ee (%)[e]	Reference
6	69f (R)[ax] R³ = H	Cu(OTf)₂	91	Nd	66	52 (R)	[33]

a) Typical reaction conditions: Cu precursor (1 mol.%), ligand (4 mol.%), AlMe₃ (1.4 equiv), and substrate (1 equiv) unless otherwise stated.
b) % Conversion determined by GC using undecane as internal standard.
c) Regioselectivity determined by GC using undecane as internal standard.
d) Yield determined by CG using undecane as internal standard, otherwise not determined (Nd).
e) Enantiomeric excess determined by GC.
f) Ligand-to-Cu ratio = 2.

enantioselectivities (90%, 80% ee, Table 12.4, entry 3) were obtained with ligand **50f**, which contains sterically encumbered biaryl phosphite moieties and a phenyl oxazoline group. Replacement of the oxazoline with a phosphoroamidite moiety had a negative effect on the enantioselectivity (Table 12.4, entry 1 versus 2). In contrast to the conjugate addition to *trans*-3-nonen-2-one, yields and enantioselectivities did not improve on using different catalyst precursors, solvents, and ligand-to-copper ratios.

Chiral S,O ligands **56a–d** have also been applied to the addition of trimethylaluminium to *trans*-non-3-en-2-one. The best results were obtained using [CuMeCN)₄]BF₄ instead of Cu(OTf)₂ as a Cu source, and THF as a solvent. Conversions ranged between moderate and good (40–73%), and the enantioselectivities were between low and moderate (up to 34%). Performing the reaction at −20 °C led to the optimum enantioselectivities (22% ee at best, Table 12.3, entry 5), with only a slight decrease of conversion with respect to higher temperatures. Lower temperatures (−40 °C) led to higher enantioselectivities (34% ee), but the

Table 12.4 Representative results of conjugate addition of alkyl donors to *trans*-5-methyl-3-hexen-2-one.

Entry	Ligand	Conversion (%)[a]	Yield (%)[b]	ee[c] (%)	Reference
1	42c, R¹ = SiMe₃, R² = H	93	88	8 (S)	[25]
2	50c, R¹ = SiMe₃, R² = H	94	94	68 (R)	[25]
3	50f, (S)^{ax} R³ = H	96	90	80 (R)	[25]
4	69f, (R)^{ax} R³ = H	97	97	52 (S)	[33]

a) % Conversion determined by GC using undecane as internal standard.
b) Yield determined by CG using undecane as internal standard.
c) Enantiomeric excess determined by GC.

conversion dropped drastically (10%). As previously observed for 2-cyclohexenone, the introduction of sterically demanding groups in the thioether moiety had a negative effect on selectivity, while at low temperatures selectivities were better than at 0 °C. The best enantioselectivities were obtained with the catalyst precursor containing **56c** (Table 12.3, entry 5).

The use of monodentate ligands **59, 61, 66, 69**, and **71 a–f** in the addition of triethylaluminium to *trans*-3-nonen-2-one and *trans*-5-methyl-3-hexen-2-one has also been studied [34]. The results with these ligands indicated that the optimum trade-off between yields and enantioselectivity for *trans*-3-nonen-2-one was obtained when diethyl ether was used as a solvent, the ligand-to-copper ratio was 4, and $Cu(OTf)_2$ was used as a catalyst precursor.

Under optimized conditions, unlike cyclic substrates, pyranoside ligands **69 *(31)*** provided better enantioselectivities than the corresponding furanoside ligands **59 *(28)*.** The best result (ee values up to 52%) was obtained with ligand **69f *(31f)***, which contains the best combination of the ligand parameters (Table 12.3, entry 6).

Similar trends were observed for *trans*-5-methyl-3-hexen-2-one. Thus, the best enantioselectivity (ee values up to 52%) was obtained with ligand **69f *(31f)***, which contains an *R* binaphthyl phosphite moiety attached to the pyranoside backbone (Table 12.4, entry 4).

12.4
Rhodium-Catalyzed Reactions

While copper-catalyzed 1,4-additions of diorganozinc or trialkylaluminium reagents to unsaturated carbonyl compounds have been known for several decades, the rhodium-catalyzed conjugate addition of boronic acids to enones (Hayashi–Miyaura reaction) has only quite recently emerged as a powerful tool for stereoselective C–C bond formation [40, 41]. Along with this transformation chiral olefins have evolved as a new class of stereodirecting ligands [42]. These chiral olefin ligands are either dienes (the first examples were designed independently by Hayashi [43] and Carreira [44]) or structures containing one olefin moiety in combination with a another donor center, mainly phosphorus (first reported by Grützmacher [45] and later also by Hayashi [46]) and are generally superior to traditional diphosphorus donor ligands in rhodium-catalyzed 1,4-additions.

Boysen *et al.* have designed a family of novel chiral diene and olefin–phosphinite hybrid ligands derived from D-glucose (Scheme 12.14) and D-arabinose [47]. The olefin–phosphinite hybrid based on glucose led to excellent levels of stereoselectivity in 1,4-additions of boronic acids to enones (99% ee from cyclohexenone and cyclopentenone, Scheme 12.15), whereas the dienes only led to modest enantioselectivity.

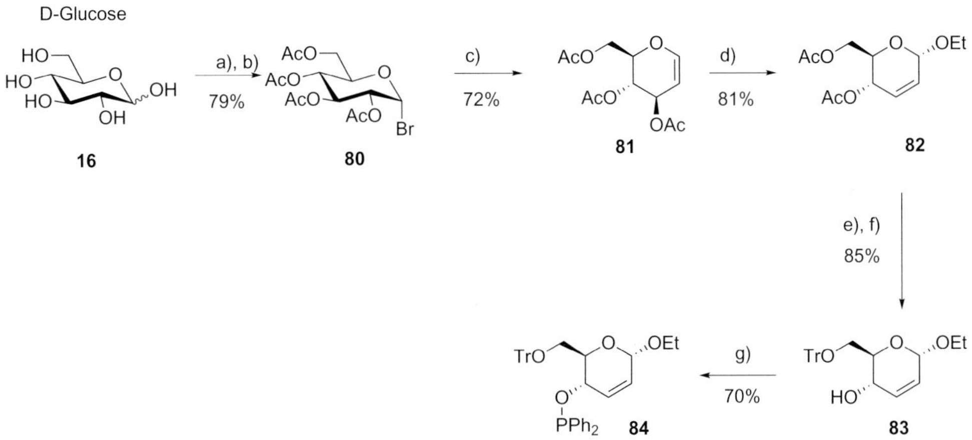

Scheme 12.14 Synthesis of a *gluco*-configured olefin–phosphinite ligand.

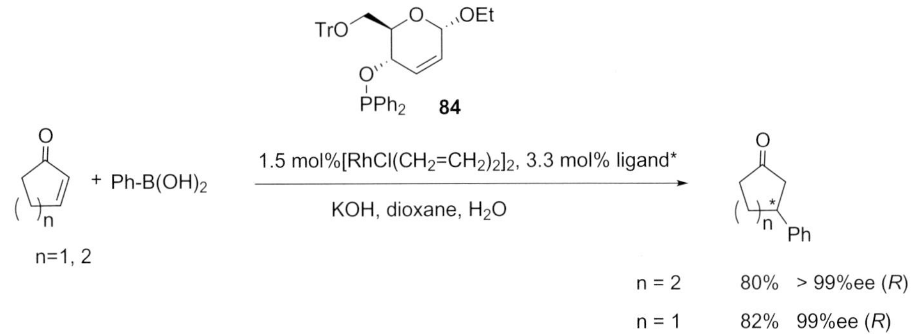

(a) Ac$_2$O, NaOAc. (b) HBr, HOAc. (c) Zn, CuSO$_4$, HOAc. (d) BF$_3$ * Et$_2$O, CH$_2$Cl$_2$. (e) MeOH, H$_2$O, Et$_3$N. (f) TrCl, pyridine. (g) Ph$_2$PCl, Et$_3$N, THF.

Scheme 12.15 Rhodium-catalyzed addition of phenylboronic acid to enones.

n = 2   80%   > 99%ee (R)
n = 1   82%   99%ee (R)

12.5 Conclusions

The use of carbohydrate derivatives as chiral ligands has had a modest impact on copper-catalyzed conjugated additions of organometallic reagents to α,β-unsaturated carbonyl compounds compared to the extensive use of these ligands in other catalytic asymmetric reactions. Furthermore, the a priori advantageous modular nature of carbohydrate-derived ligands have not afforded improved results in the asymmetric conjugate addition with regard to other kinds of ligands that have led to very high degrees of enantioselectivity.

Most catalytic data collected use 2-cyclohexenone as a substrate for conjugate addition due to its high reactivity, which makes it ideal for probing the efficiency of new catalytic systems. As far as this enone is concerned, carbohydrate-derived ligands have shown in general to be fairly active and 1,4-regioselective, with no 1,2-addition product being detected, but have not led to high degrees of enantio-induction, giving ee values lower than 60%. Most of the experimental data are for the addition of diethylzinc as an alkyl donor. Exceptionally good enantioselectivities (81–84% ee) have been obtained using $Cu(OTf)_2$ and bidentate diphosphite ligands **3f, 21g**, and **37b** with axially chiral naphthyl moieties as a diol substituent at phosphorus with both furanoside and pyranoside rings in the sugar backbone. Notably, the donor atoms in these ligands are in a *cis*-relationship and form a seven-membered chelate.

The conjugate addition to more demanding substrates such as cyclopentenone or linear carbonyl compounds using carbohydrate derivatives as chiral ligands has received considerably less attention. The few data available, however, indicate that the efficiency of these ligands is only very moderate and, in general, low yields and enantioselectivities of 1,4-addition products are obtained. For the Cu-catalyzed asymmetric addition to cyclopentenone, the catalytic system formed by the conformationally restricted naphthyl diphosphite ligand **37b**/$(CuOTf)_2 \cdot C_6H_6$ has led to moderately good degrees of enantio-induction (64% ee), although with low yields of the 1,4-product. For the 1,4-addition to *trans*-3-nonen-2-one, the organometallic reagent most used is the more reactive trimethylaluminium, and, in some cases, the use of $[CuMeCN)_4]BF_4$ as a copper source has found advantage over the classic $Cu(OTf)_2$. The best catalytic systems reported in terms of enantioselectivity for this linear substrate, **50c**/$[CuMeCN)_4]BF_4$ and **50f**/$(CuOTf)_2$, involve conformationally restricted binaphthyl phosphite–oxazoline glucoside ligands, with ee values of 78% ee and 62% ee, respectively, although the latter gives moderate yields.

In contrast, outstanding enantioselectivities (99% ee from cyclohexenone and cyclopentenone) have been obtained in the rhodium-catalyzed addition of boronic acid to enones using D-glucose-derived olefin–phosphinite hybrids as chiral ligands.

The catalytic asymmetric conjugate addition has made spectacular advances in the last few years, and an impressive number of catalytic systems with non-carbohydrate ligands has been designed. The reaction, however, still suffers from high substrate specificity, and therefore intensive research in this area is still necessary. In this sense, the high variability of carbohydrate scaffolds together with the modular installation of different donor atoms and/or auxiliary units in their structure might well allow for the design of highly efficient catalysts of general application to most substrates.

References

1 (a) Perlmutter, P. (1992) *Conjugate Addition Reaction in Organic Synthesis*, Tetrahedron Organic Chemistry Series, no. 9, Pergamon Press, Oxford; (b)

Yamamoto, Y. (1995) *Methods Org. Chem. (Houben-Weyl) (Stereoselective Synthesis)*, **4**, 2041–2057.

2 (a) Kozlowski, J.A. (1991) in *Comprehensive Organic Synthesis*, vol. 4 (eds B.M. Trost and I. Fleming), Pergamon Press, Oxford, pp. 169–198; (b) Taylor, R.J.K. (1985) *Synthesis*, 364–392; (c) Posner, G.H. (1974) *Org. React.*, **22**, 253–400; (d) Posner, G.H. (1972) *Org. React.*, **19**, 1–113; (e) Lipshutz, B.H. and Sengupta, S. (1992) *Org. React.*, **41**, 135–631.

3 For reviews on asymmetric conjugate addition: see: (a) Rossiter, B.E. and Swingle, N.M. (1992) *Chem. Rev.*, **92**, 771–808; (b) Sibi, M.P. and Manyem, S. (2000) *Tetrahedron*, **56**, 8033–8061; (c) Krause, N. and Hoffmann-Röder, A.A. (2001) *Synthesis*, 171–196; (d) Alexakis, A. and Benhaim, C. (2002) *Eur. J. Org. Chem.*, 3221–3236; (e) Fleming, F.F. and Wang, Q.Z. (2003) *Chem. Rev.*, **103**, 2035–2078; (f) Ballini, R., Bosica, G., Fiorini, D., Palmieri, A., and Petrini, M. (2005) *Chem. Rev.*, **105**, 933–971; (g) Feringa, B.L., Naasz, R., Imbos, R., and Arnold, A.A. (2002) *Copper-Catalyzed Enantioselective Conjugate Addition Reactions of Organozinc Reagents in Modern Organocopper Chemistry* (ed. N. Krause), Wiley-VCH Verlag GmbH, Weinheim, p. 224; (h) Christoffers, J., Koripelly, G., Rosiak, A., and Rössle, M. (2007) *Synthesis*, **9**, 1279–1300.

4 (a) Alexakis, A. (1994) in *Organocopper Reagents, A Practical Approach* (ed. R.J.K. Taylor), ch. 8. Oxford University Press, Oxford, pp. 159–183; (b) Noyori, R. (1994) *Asymmetric Catalysis in Organic Synthesis*, John Wiley & Sons, Inc., New York; (c) Ojima, I. (1993) *Catalytic Asymmetric Synthesis*, VCH, Weinheim; (d) Krause, N. and Gerold, A. (1997) *Angew. Chem. Int. Ed. Engl.*, **36**, 186–204.

5 Kharash, M.S. and Tawney, P.O. (1941) *J. Am. Chem. Soc.*, **63**, 2308–2316.

6 Alexakis, A., Frutos, J.C., and Mangeney, P. (1993) *Tetrahedron: Asymmetry*, **4**, 2427–2430.

7 Alexakis, A., Bäckvall, J.E., Krause, N., Pàmies, O., and Diéguez, M. (2008) *Chem. Rev.*, **108**, 2796–2823.

8 See for instance: (a) Takemoto, Y., Kuraoka, S., Humaue, N., Aoe, K., Hiramatsu, H., and Iwata, C. (1996) *Tetrahedron*, **52**, 14177–14188; (b) Chataigner, I., Gennari, C., Ongeri, S., Piarulli, U., and Ceccarelli, S. (2001) *Chem. Eur. J.*, **7**, 2628–2634; (c) d'Augustin, M., Palais, L., and Alexakis, A. (2005) *Angew. Chem. Int. Ed.*, **44**, 1376–1378; (d) Su, L., Li, X., Chan, W.L., Jia, X., and Chan, A.S.C. (2003) *Tetrahedron: Asymmetry*, **14**, 1865–1869; (e) Liang, L. and Chan, A.S.C. (2002) *Tetrahedron: Asymmetry*, **13**, 1393–1396; (f) Alexakis, A., Albrow, V., Biswas, K., d'Augustin, M., Prieto, O., and Woodward, S. (2005) *Chem. Commun.*, 2843–2845.

9 (a) Mandoli, A., Arnold, L.A., de Vries, A.H.M., Salvadori, P., and Feringa, B.L. (2001) *Tetrahedron: Asymmetry*, **12**, 1929–1937; (b) Alexakis, A., Benhaim, C., Rosset, S., and Humam, M. (2002) *J. Am. Chem. Soc.*, **124**, 5262–5263.

10 For some examples, see: (a) Krauss, I.J. and Leighton, J.L. (2003) *Org. Lett.*, **5**, 3201–3203; (b) Kwak, Y.-S. and Corey, E.J. (2004) *Org. Lett.*, **6**, 3385–3388; (c) Lopez, F., Harutyunyan, S.R., Minnaard, A.J., and Feringa, B.L. (2004) *J. Am. Chem. Soc.*, **126**, 12784–12785; (d) Lee, K.-S., Brown, M.K., Hird, A.W., and Hoveyda, A.H. (2006) *J. Am. Chem. Soc.*, **128**, 7182–7184.

11 (a) Becher, J., Seidel, I., Plass, W., and Klemm, D. (2006) *Tetrahedron*, **62**, 5675–5681; (b) Diéguez, M., Pàmies, O., and Claver, C. (2006) *J. Organomet. Chem.*, **691**, 2257–2262.

12 Diéguez, M., Pàmies, O., and Claver, C. (2004) *Chem. Rev.*, **104**, 3189–3216.

13 Guimet, E., Diéguez, M., Ruiz, A., and Claver, C. (2004) *Tetrahedron: Asymmetry*, **15**, 2247–2251.

14 Pàmies, O., Net, G., Ruiz, A., and Claver, C. (1999) *Tetrahedron: Asymmetry*, **10**, 2007–2014.

15 Pàmies, O., Diéguez, M., Net, G., Ruiz, A., and Claver, C. (2000) *Tetrahedron: Asymmetry*, **11**, 4377–4383.

16 (a) Kartha, K.P.R. (1986) *Tetrahedron Lett.*, **27**, 3415–3416; (b) Vogel, A.I. (1994) *Vogel's Textbook of Practical Organic Chemistry*, 5th edn, Longman, New York.

17 Pàmies, O., Diéguez, M., Net, G., Ruiz, A., and Claver, C. (2000) *Chem. Commun.*, 2383–2384.

18 Goueth, P.Y., Fauvin, M.A., Mashoudi, M., Ramiz, A., Ronco, G.L., and Villa, P.J. (1994) *J. Nat.*, **6**, 3–6.

19 Diéguez, M., Ruiz, A., and Claver, C. (2001) *Tetrahedron: Asymmetry*, **12**, 2827–2834.

20 (a) Levene, P.A. and Raymond, A.L. (1933) *J. Biol. Chem.*, **102**, 317–330; (b) Hollenberg, D.H., Klein, D.H., and Fox, J.J. (1978) *Carbohydr. Res.*, **67**, 491–494; (c) Ritzmann, G., Klein, R.S., Hollenberg, D.H., and Fox, J.J. (1975) *Carbohydr. Res.*, **39**, 227–236; (d) Tsutsumi, H., Kawai, Y., and Ishido, Y. (1979) *Carbohydr. Res.*, **73**, 293–296.

21 (a) Diéguez, M., Pàmies, O., Ruiz, A., Castillón, S., and Claver, C. (2001) *Chem. Eur. J.*, **7**, 3086–3094; (b) Schmidt, O.T. (1963) in *Methods in Carbohydrate Chemistry*, vol. 2 (eds R.L. Whistler and M.L. Wolfrom), Academic Press, New York, p. 318.

22 Diéguez, M., Ruiz, A., and Claver, C. (2001) *Tetrahedron: Asymmetry*, **12**, 2895–2900.

23 Wang, L., Li, Y.-M., Yip, C.-W., Qiu, L., Zhou, Z., and Chan, A.S.C. (2004) *Adv. Synth. Catal.*, **346**, 947–953.

24 Mata, Y., Diéguez, M., Pàmies, O., Biswas, K., and Woodward, S. (2007) *Tetrahedron: Asymmetry*, **18**, 1613–1617.

25 Emmerson, D.P.G., Hems, W.P., and Davis, B.G. (2005) *Tetrahedron: Asymmetry*, **16**, 213–221.

26 Diéguez, M., Ruiz, A., and Claver, C. (2001) *Tetrahedron: Asymmetry*, **12**, 2861–2866.

27 (a) Lu, Y. and Just, G. (2001) *Tetrahedron*, **57**, 1677; (b) Kiss, J., D'souza, R., Koeveringe, J.A., and Arnold, W. (1982) *Helv. Chim. Acta*, **65**, 1522–1537.

28 (a) Mata, Y., Diéguez, M., Pàmies, O., and Claver, C. (2005) *Adv. Synth. Catal.*, **347**, 1943–1947; (b) Yonehara, K., Hashizume, T., Mori, K., Ohe, K., and Uemura, S. (1999) *J. Org. Chem.*, **64**, 9374–9380.

29 Pàmies, O., Diéguez, M., Net, G., Ruiz, A., and Claver, C. (2000) *Organometallics*, **19**, 1488–1496.

30 Pàmies, O., Net, G., Ruiz, A., Claver, C., and Woodward, S. (2000) *Tetrahedron: Asymmetry*, **11**, 871–877.

31 Guimet, E., Diéguez, M., Ruiz, A., and Claver, C. (2005) *Tetrahedron: Asymmetry*, **16**, 959–963.

32 Diéguez, M., Pàmies, O., Net, G., Ruiz, A., and Claver, C. (2002) *J. Mol. Catal. A*, **185**, 11–16.

33 Mata, Y., Diéguez, M., Pàmies, O., and Woodward, S. (2007) *J. Organometal. Chem.*, **692**, 4315–4320.

34 Spescha, M. and Ribs, G. (1993) *Helv. Chim. Acta*, **76**, 1219–1230.

35 Kubas, G.J., Monzyk, B., and Crumbliss, A.L. (1979) *Inorg. Synth.*, **19**, 90–92.

36 Spescha, M. (1993) *Helv. Chim. Acta*, **76**, 1832–1846.

37 Pigro, M.C., Angiuoni, G., and Piancatelli, G. (2002) *Tetrahedron*, **58**, 5459–5466.

38 Guimet, E., Diéguez, M., Ruiz, A., and Claver, C. (2005) *Tetrahedron: Asymmetry*, **16**, 2161–2165.

39 Takaya, Y., Ogasawara, M., Hayashi, T., Sakai, M., and Miyaura, N. (1998) *J. Am. Chem. Soc.*, **120**, 5579–5580.

40 For reviews see: (a) Hayashi, T. and Yamasaki, K. (2003) *Chem. Rev.*, **103**, 2829–2844; (b) Hayashi, T. (2004) *Pure Appl. Chem.*, **76**, 465–476.

41 (a) Glorius, F. (2004) *Angew. Chem. Int. Ed.*, **43**, 3364; (b) Defieber, C., Grützmacher, H., and Carreira, E.M. (2008) *Angew. Chem. Int. Ed.*, **47**, 4482.

42 (a) Hayashi, T., Ueyama, K., Tokunaga, N., and Yoshida, K. (2003) *J. Am. Chem. Soc.*, **125**, 11508; (b) Berthon-Gelloz, G. and Hayashi, T. (2006) *J. Org. Chem.*, **71**, 8957.

43 (a) Defieber, C., Paquin, J.-F., Serna, S., and Carreira, E.M. (2004) *Org. Lett.*, **6**, 3873; (b) Fischer, C., Defieber, C., Suzuki, T., and Carreira, E.M. (2004) *J. Am. Chem. Soc.*, **126**, 1628.

44 (a) Deblon, S., Grützmacher, H., and Schönberg, H. (2003) WO 03/048175 A1, ; (b) Maire, P., Maire, P., Deblon, S., Breher, F., Geier, J., Böhler, C., Rüegger, H., Schönberg, H., and Grützmacher, H. (2004) *Chem. Eur. J.*, **10**, 4198; (c) Piras, E., Läng, F., Rüegger, H., Stein, D.,

Wörle, M., and Grützmacher, H. (2006) *Chem. Eur. J.*, **12**, 5849.

45 (a) Shintani, R., Duan, W.-L., Nagano, T., Okada, A., and Hayashi, T. (2005) *Angew. Chem. Int. Ed.*, **44**, 4611; (b) Duan, W.-L., Iwamura, H., Shintani, R., and Hayashi, T. (2007) *J. Am. Chem. Soc.*, **129**, 2130.

46 (a) Minuth, T. and Boysen, M.M.K. (2009) *Org. Lett.*, **11**, 4212–4215; (b) Grugel, H., Minuth, T., and Boysen, M.M.K. (2010) *Synthesis*, 3248–3258.

13
1,2-Addition of Nucleophiles to Carbonyl Compounds

M. Isabel Matheu, Yolanda Díaz, Patricia Marcé and Sergio Castillón

13.1
Introduction

The asymmetric metal-catalyzed addition of different reagents to carbonyl compounds is a useful and efficient procedure for synthesizing chiral secondary and tertiary alcohols. Similarly, the addition to imines affords chiral amines. Various catalysts and ligands have been explored for each process, and in some of them sugar ligands have been tested [1]. In general, the catalytic asymmetric addition to ketones and ketimines, which gives tertiary alcohols and amines [2], is more challenging than the addition to aldehydes and aldimines, which gives secondary alcohols and amines. Many catalytic systems are initially explored for the catalyzed addition of organometallic compounds to aldehydes and ketimines. Scheme 13.1 summarizes the processes of additions to carbonyl compounds that involve carbohydrate-derived ligands. The most widely studied process is addition of dialkylzinc reagents to aldehydes (via route a, R^2 = H, in Scheme 13.1) [3], which has been used as a reference for testing new ligands. This process can be catalyzed by

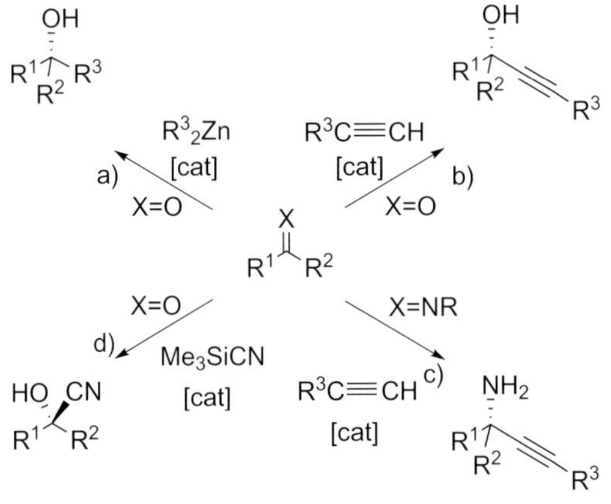

Scheme 13.1 Carbonyl addition processes involving carbohydrate-derived chiral ligands.

Carbohydrates – Tools for Stereoselective Synthesis, First Edition. Edited by Mike Martin Kwabena Boysen.
© 2013 Wiley-VCH Verlag GmbH & Co. KGaA. Published 2013 by Wiley-VCH Verlag GmbH & Co. KGaA.

aminoalcohols in the presence or absence of additional metal compounds. Propargylic alcohols and amines are interesting intermediates for synthesizing natural products, and the asymmetric synthesis of these compounds by the transition-metal-catalyzed addition of acetylenes to aldehydes [4] and imines [5] has been studied more recently (via routes b and c, Scheme 13.1). Finally, the asymmetric synthesis of cyanohydrins and aminonitriles are also processes of interest since they can give further access to α-hydroxy acids and α-amino acids [6] (via route d, Scheme 13.1).

13.2
Addition of Organometallic Reagents to Aldehydes

13.2.1
Addition of Diethylzinc to Aldehydes. N,O Donor Ligands (Aminoalcohols)

The addition of diethylzinc to aldehydes has been the most widely studied 1,2-addition process. The discovery by Oguni et al. [7] that (S)-leucinol catalyzed the addition of diethylzinc to benzaldehyde in 49% ee prompted research in this field that grew enormously during the 1990s. Subsequently, Noyori discovered that aminoalcohol **1** provided up to 95% ee in a similar reaction (Scheme 13.2) [8]. Since then, chiral aminoalcohols have been the most widely explored ligands for this process, although amino thiols, diols, disulfides, and other ligands have also

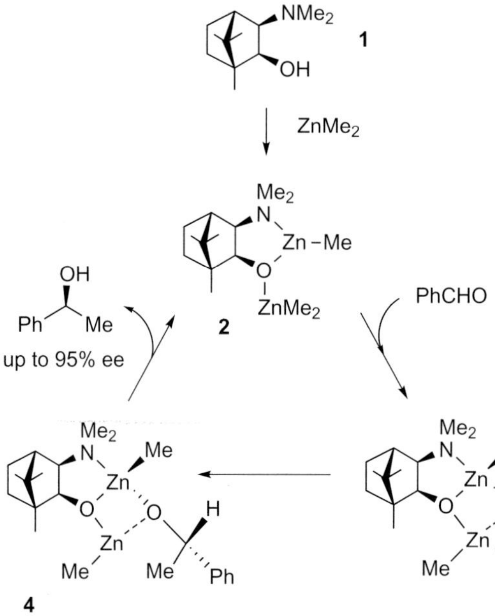

Scheme 13.2 Addition of ZnMe$_2$ to benzaldehyde catalyzed by Noyori's aminoalcohol **1**.

Figure 13.1 Possible dimeric species from ZnMe$_2$/**1**.

been used. Zinc reagents bearing other organic residues such as methyl, *i*-propyl, vinyl, alkynyl, and aryl have also been tested.

The active species in the catalysis is complex **2**, which has two zinc atoms coordinated to ligand **1** (Scheme 13.2). Aldehyde coordinates to zinc *anti* to the chiral ligand to afford **3**, and transfer of the methyl group to the aldehyde then takes place from the *Si* face in **3** to generate **4**, from which the methylzinc alcoholate is shifted by reaction with dimethylzinc, regenerating the catalyst. This reaction shows a nonlinear effect in the enantioselectivity because of the formation of dimers and because the homodimer **5a** is less stable than heterodimer **5b** (Figure 13.1) and, consequently, dissociates more easily to give rise to the active catalytic species (Scheme 13.2).

Cho and Kim [9] synthesized a series of 5-deoxy-5-amino-1,2-O-isopropylidene-D-xylo-furanosides (**8**, Scheme 13.3), which have different substituents on the nitrogen, in a straightforward manner by reacting the corresponding amines with

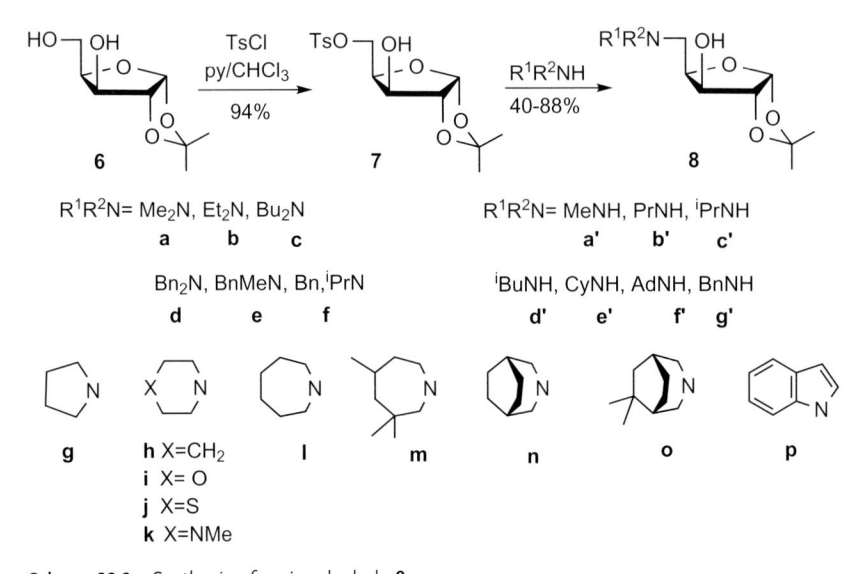

Scheme 13.3 Synthesis of aminoalcohols **8**.

5-O-tosyl-1,2-O-isopropylidene-D-xylo-furanoside (**7**). These ligands were tested in the addition of diethylzinc to benzaldehyde (Scheme 13.4). The enantioselectivities were dependent on the structure of the chiral ligand and the following general trends were observed: (i) catalysts with ligands **8** bearing tertiary amines provided better results than those with secondary amines (**8a′–g′**); (ii) ligands **8g–j**, with pyrrolidino, piperidino, morpholino, and thiomorpholino groups, respectively, afforded the highest enantioselectivities; and (iii) ligands bearing seven-membered rings (**8l,m**) provided inferior results to those obtained with five- or six-membered heterocyclic amine moieties (**8g–j**).

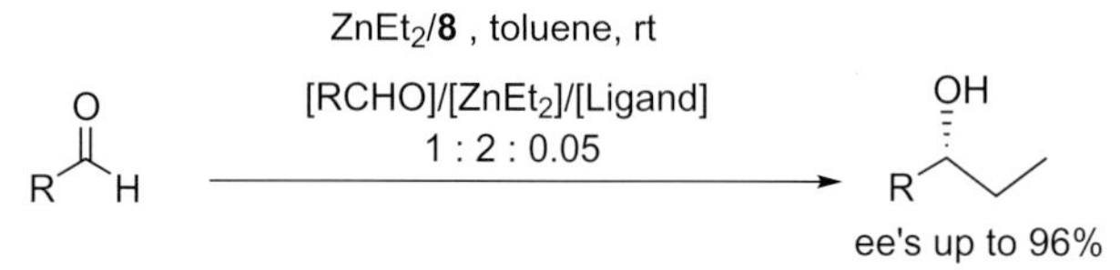

$$\text{RCHO} \xrightarrow[\text{[RCHO]/[ZnEt}_2\text{]/[Ligand]} = 1:2:0.05]{\text{ZnEt}_2/\mathbf{8}, \text{ toluene, rt}} \text{R-CH(OH)-Et}$$

ee's up to 96%

	8g	8h	8i	8j	8k	8l	8m	8n	8o	8p	8a′	8b′	8c′	8d′	8e′	8f′	8g′
R=Ph, yield (%)	90	92	**90**	**88**	99	98	98	90	90	88	50	78	78	83	83	95	95
ee (%)	90	87	**96**	**94**	78	76	62	76	80	50	7	69	58	77	77	72	82

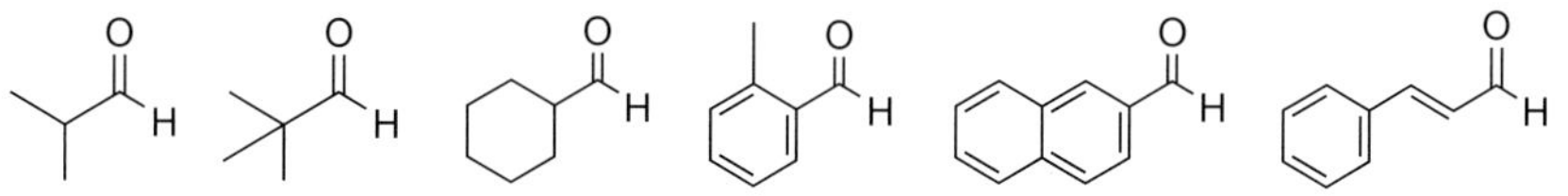

	iPr-CHO	tBu-CHO	Cy-CHO	o-tolyl-CHO	2-naphthyl-CHO	PhCH=CH-CHO
8i/8j yield (%)	93/92	86/85	88/85	88/–	90/92	88/75
ee (%)	93/92	93/76	96/85	89/–	86/61	42/18

Scheme 13.4 Addition of diethylzinc to aldehydes catalyzed by aminoalcohols **8i, 8j**.

The best results were obtained by performing the reaction in toluene at 25 °C using ZnEt$_2$ (2 mol equiv.) and ligand **8i** (5%). Under these conditions, (R)-1-phenylpropan-1-ol was obtained in 90% yield and 96% ee (Scheme 13.4). Different aliphatic and aromatic aldehydes were reacted with ZnEt$_2$ in the presence of ligands **8g–j** to give the corresponding alcohols with enantioselectivities of up to 96%. In general, ligand **8i** gave better ee values with aromatic and sterically hindered aliphatic aldehydes, and ligand **8j** worked better with unhindered aliphatic aldehydes (Scheme 13.4).

Ligand **8i** was found to be highly effective for the isopropylation of both aliphatic and aromatic aldehydes, leading to the corresponding alcohols in very high enantiomeric excesses (94–98%) [10]. In the case of α,β-unsaturated aldehydes, the reaction provided low enantioselectivity.

Glucosamine is a promising candidate for the preparation of aminoalcohols ligands because of its availability and easy modification. Bauer synthesized aminoalcohols **9** and **10** (Scheme 13.5), from methyl 4,6-O-benzylidene-α- and benzyl β-D-glucosamine by reaction with tosyl chloride [11].

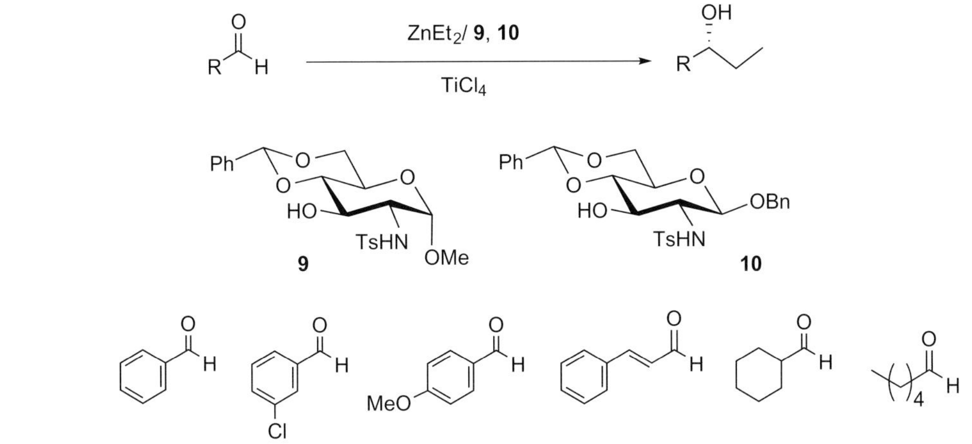

13.2 Addition of Organometallic Reagents to Aldehydes | 297

9/10	yield(%)	99/98	85	95	94	73	80
	ee (%)	97/10	90	89	86	39	88

Scheme 13.5 Addition of diethylzinc to aldehydes catalyzed by aminoalcohols **9** and **10**.

While the previous studies were carried out with Lewis bases as catalyst, the diethylzinc addition with **9** and **10** was carried out using a Lewis acid catalyst. Thus, when benzaldehyde (1 mmol) was reacted with diethylzinc (3 mmol) at room temperature in the presence titanium tetrachloride (1.4 mmol) and **9** (7 mol.%), *(R)*-1-phenyl-propan-1-ol was obtained in 99% yield and 97% ee (Scheme 13.5). The reaction was still highly enantioselective when the amount of ligand was reduced to 1.8 mol.%, but the conversion dropped. The influence of the configuration at the anomeric center was studied by carrying out the reaction in the presence of ligand **10** (0.1 mmol). The yield was quantitative, but the enantioselectivity was very low (10%); the results are not fully comparable since the anomeric substituent is different. Other aromatic aldehydes and hexanal also reacted with diethylzinc under similar conditions in the presence of TiCl$_4$/**9** to provide ee values of up to 90%.

Davis carried out a systematic study preparing four diastereomeric ligand families (**14–17**, Scheme 13.6) that result from varying the configurations at positions 1, 2, and 3, and introducing different substituents on the nitrogen [12]. The ligands were easily prepared from glucosamine. Notably, the reaction of **14a** with H$_2$O$_2$/Na$_2$WO$_4$ [13], followed by reduction with LiAlH$_4$, afforded **17a** in a simple and fast procedure for inverting the amine configuration. The following conclusions were derived from this study (Table 13.1): (i) The stereoselectivity varied according to the ligand configuration in the order α-D-*gluco* (**14**) > β-D-*gluco* (**15**) > α-D-*allo* (**16**) > α-D-*manno* (**17**). (ii) The presence of a tertiary amine at C2 caused a reversal from *(S)* to *(R)* in enantioselectivity [e.g., tertiary amine ligand **14e** the *(S)* enantiomer in 56% ee and secondary amine ligand **14d** preferentially gave the *(R)* enantiomer in 30% ee], suggesting different modes of asymmetric induction. (iii) Bulkier NR^{1}R^2 groups also influenced the enantioselectivity, in the

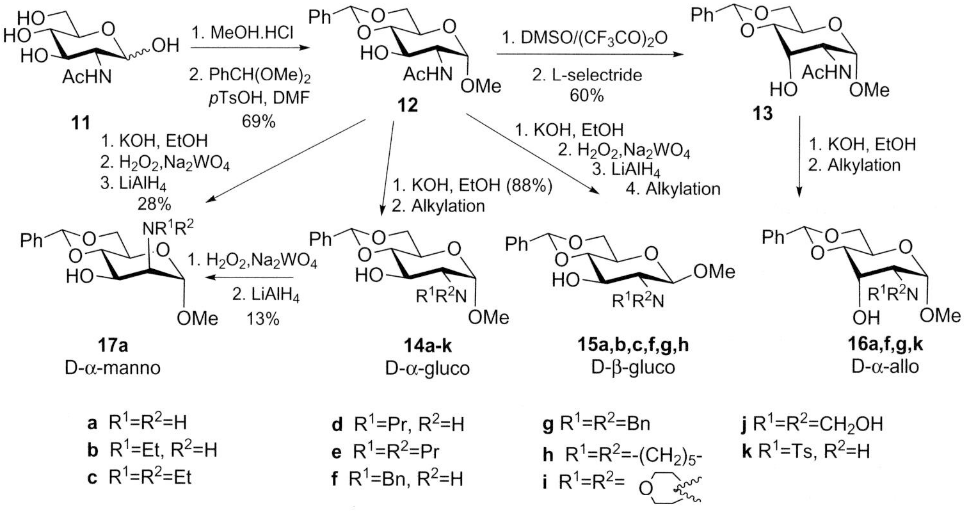

Scheme 13.6 Synthesis of aminoalcohols **14–17**.

Table 13.1 Addition of diethylzinc to benzaldehyde catalyzed by aminoalcohols **14–17**.[a]

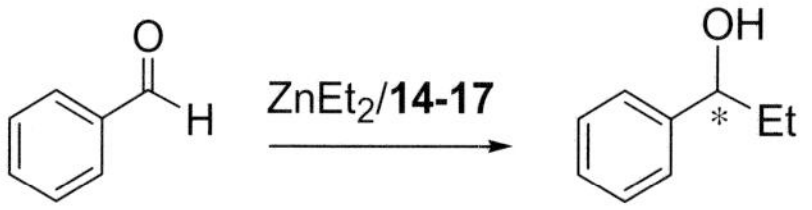

Entry	Ligand	R^1/R^2	Yield (%)	ee (%) (configuration.)	Entry	Ligand	R^1/R^2	Yield (%)	ee (%)[b] (configuration)
1	14a	H/H	66	63 (S)	11	14k	H/Ts	55	19 (S)
2	14b	Et/H	68	20 (R)	12	15a	H/H	68	46 (S)
3	14c	Et/Et	91	53 (S)	13	15b	Et/H	81	17 (R)
4	14d	H/Pr	92	30 (R)	14	15c	Et/Et	93	26 (S)
5	14e	Pr/Pr	96	56 (S)	15	15f	H/Bn	74	14 (S)
6	14f	H/Bn	61	35 (R)	16	15g	Bn/Bn	71	32 (S)
7	14g	Bn/Bn	71	38 (S)	17	15h	$(CH_2)_5$	77	28 (S)
8	14h	$(CH_2)_5$	90	58 (S)	18	15i	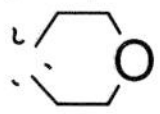 morpholine	69	28 (S)
9	14i	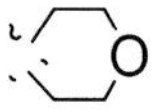 morpholine	86	65 (S)	19	16a	H/H	94	32 (S)
10	14j	$(CH_2OH)_2$	64	12 (R)	20	17a	H/H	88	21 (S)

a) Conditions: 10 mol.% ligand, toluene, room temp.
b) Enantiomeric excess in % and configuration of the major isomer.

order morpholino (**i**) > pyrrolidino (**h**) > diisopropylamino (**e**) > diethylamino (**c**) > dibenzylamino (**g**), ranging between 65% and 38%. The results were better when the amine was part of a morpholine ring. Basicity seems to have a strong influence on the reaction, since the presence of Ts as substituent on the amino groups decreased the reaction rate and provided a low ee. This result is in contrast with previous work in which ligand **9** (identical to **14k**) provided high conversion and enantioselectivity in the addition of diethylzinc to benzaldehyde in the presence of titanium Lewis acids.

Thus, the best ligand for this reaction in terms of enantioselectivity was **14i** with an α-D-*gluco* configuration and a 2-morpholino residue. This study clearly shows the possibilities of tuning carbohydrate ligands through a systematic variation of configuration and substituents at the different positions of the carbohydrate, as well as the remarkable influence of substituents neighboring the coordination sites on the enantioselectivity.

Ligands **19–27** [14, 15] (Schemes 13.7 and 13.8) constitute a family of structurally related ligands that contain pyridine or quinoline moieties incorporated in a fructopyranose or glucofuranose backbone, whose main structural features are as

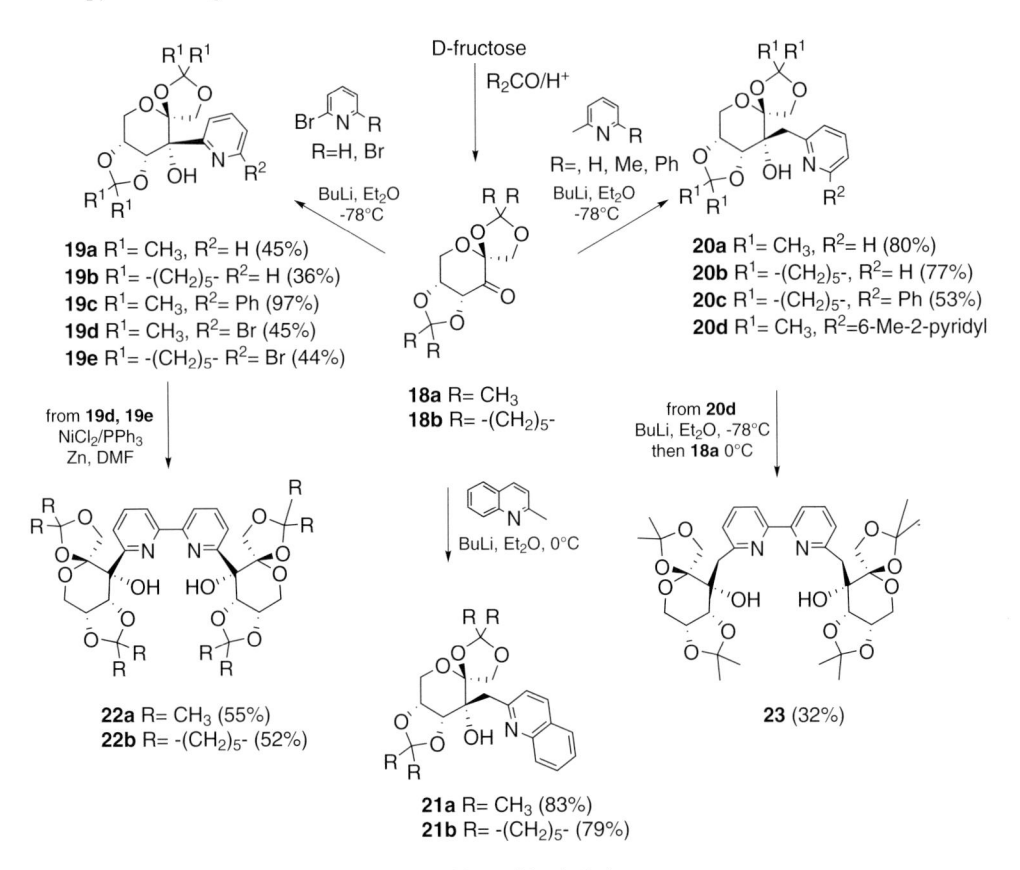

Scheme 13.7 Synthesis of pyridyl-, quinolyl-, and bipyridyl-/alcohols **19–23**.

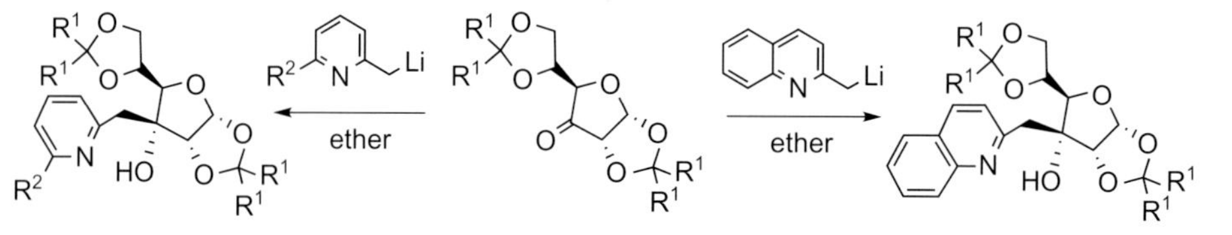

Scheme 13.8 Synthesis of pyridyl- and quinolyl-alcohols **25–27**.

25a R^1= Me, R^2= H (51%)
25b R^1= -(CH$_2$)$_5$-, R^2= H (74%)
26a R^1= Me, R^2= Me (54%)
26b R^1= -(CH$_2$)$_5$-, R^2= Me (78%)

24a R^1= Me
24b R^1= -(CH$_2$)$_5$-

27a R^1= Me (57%)
27b R^1= -(CH$_2$)$_5$- (64%)

follows (i) chirality is provided by the fructopyranose and glucofuranose derivatives (Figure 13.2), (ii) all of them are pyridyl alcohols with a tertiary alcohol function (iii) chelates with different ring size with the metal can be formed (**19** versus **20, 21**), (iv) the pyridyl unit can be modified by introducing substituents at position 6 (**19c, 20c**) or quinoline may be used instead of pyridine (**21** and **27**), and (v) tetradentate ligands **22** and **23** are dimers of **19a,b** and **20a**, respectively (Scheme 13.7).

The synthesis of these ligands is summarized in Schemes 13.7 and 13.8. They were obtained by the addition of different organolithium compounds to ketones **18a,b** and **24** derived from fructopyranose and glucofuranose. The starting materials **18a,b** and **24** can be obtained in one and two steps, from fructose (by reaction with acetone/H$^+$) and glucose (by reaction with acetone/H$^+$ and oxidation with PCC), respectively. Ligands **22a,b** were prepared by homocoupling of bromopyridines **19d,e**, catalyzed by nickel. Ligand **23** was prepared in 32% overall yield from 6,6'-dimethyl-2,2'-bipyridine by lithiation and addition to **18a** to give **20d**, and a second lithiation and addition to **18a**. Similarly, ligands **25** and **26** were prepared from ketones **24** by addition of the corresponding organolithium pyridyl and quinolyl derivatives.

Table 13.2 shows the results obtained with ligands **19–23** and **25–27** in the addition of ZnEt$_2$ to benzaldehyde. From these results, it can be deduced that, in the family of ligands **19–23**, ligands **20** and **21** (entries 4–8), which form a six-membered chelate ring, provide higher yields and enantioselectivities than ligands **19** (entries 1–3), which form a five-membered chelate ring. This tendency was also observed for ligands **22a** and **23** (entries 9, 11), which have similar structure but contain different backbone length. The different behavior of these ligands was explained by conformational analysis of their ethylzinc amino-alkoxide complex formed *in situ*.

It is not easy to explain the lower ee value obtained for ligand **22b** and the fact that the sense of asymmetric induction is opposite to that of all of the other ligands. Based on related work [16], it is proposed that the C_2 symmetric bipyridyl alcohols **22** and **23** may act as bidentate ligands.

13.2 Addition of Organometallic Reagents to Aldehydes

Table 13.2 Addition of diethylzinc to benzaldehyde catalyzed by aminoalcohols **19–23** and **25–27**.[a]

PhCHO + ZnEt$_2$ /**19-23, 25-27** → PhCH(OH)Et

Entry	Ligand (R)	Yield (%)	ee (%) (configuration)	Entry	Ligand (R)	Yield (%)	ee (%) (configuration)
1	19a (R = Me)	76	21 (R)	12	25a (R = Me)	92	73 (S)
2	19b [R = (CH$_2$)$_5$]	65	25 (R)	13	25b [R = (CH$_2$)$_5$]	99	81 (S)
3	19c (R = Me)	50	7 (R)	14	26a (R = Me)	99	87 (S)
4	20a (R = Me)	94	79 (R)	15	26b (R = (CH$_2$)$_5$)	99	89 (S)
5	20b [R = (CH$_2$)$_5$]	96	82 (R)	16	27b (R = (CH$_2$)$_5$)	99	79 (S)
6	20c (R = Me)	94	75 (R)	17[b]	26b (R = (CH$_2$)$_5$)	99	90 (S)
7	21a (R = Me)	99	80 (R)	18[c]	26b (R = (CH$_2$)$_5$)	99	92 (S)
8	21b [R = (CH$_2$)$_5$]	99	83 (R)	19	27a (R = Me)	99	76 (S)
9	22a (R = Me)	65	15 (R)				
10	22b [R = (CH$_2$)$_5$]	51	15 (S)				
11	23	97	73 (R)				

a) Conditions: benzaldehyde/Et$_2$Zn/ligand = 1.0/2.2/0.05, toluene, 0 °C, 12 h, (Et$_2$Zn 1 M in hexanes).
b) −10 °C.
c) −10 °C (Et$_2$Zn 1.1 M in toluene).

Ligands **25–27** provided enantioselectivities ranging from 73 to 89% (entries 12–19), which are slightly higher than those provided by the related ligands **20** and **21**. More significantly, the configuration of the main enantiomer was *(S)* in all cases, confirming their behavior as pseudo-enantiomers of ligands **19–23** (Scheme 13.7). Stereoselectivity was strongly influenced by the substitution of the pyridine ring and the protecting groups on the D-glucose skeleton of ligands **25–27**. Thus, the methyl group at 6-position of the pyridine ring is key for obtaining high ee. The ee could be improved up to 92% by decreasing the temperature to −10 °C (Table 13.2, entries 18 and 19). The solvent had a marginal effect on yield and enantioselectivity.

D-Fructo ligands **20b** and **21b** and D-gluco ligands **26a,b**, which provided the best results, were also tested on a series of representative aromatic and aliphatic aldehydes (Scheme 13.9). Aromatic aldehydes gave the diethyl addition products with good ee values, and no obvious electronic or steric effects of the substituents on the aromatic residue were observed. Especially relevant is the result obtained with 2-*O*-Me-benzaldehyde. However, with the sterically demanding cyclohexylcarbaldehyde, the enantioselectivity was moderate with ligands **20b** and **21b**, and good with ligand **26b**, while with cinnamaldehyde, all ligands provided low enantioselectivity. In all cases, ligand **26b** afforded the opposite enantiomer to the one obtained with ligands **20b** and **21b**.

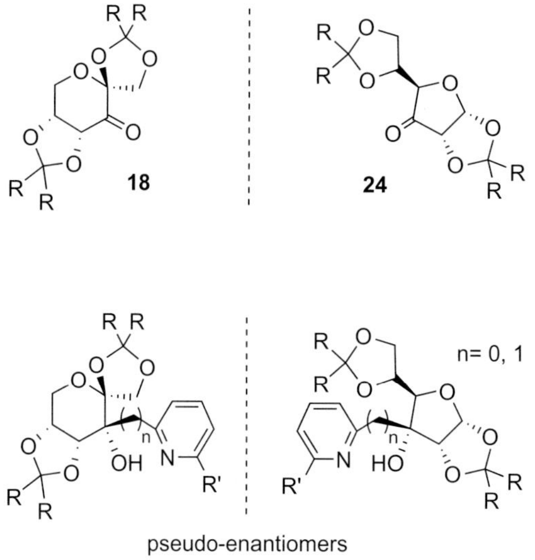

13 1,2-Addition of Nucleophiles to Carbonyl Compounds

Figure 13.2 Pseudo-enantiomeric structural relationship between ketones **18** and **24** and their derivatives.

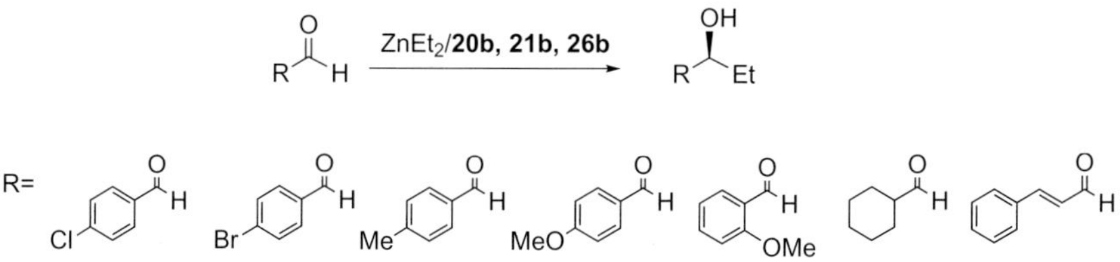

Scheme 13.9 Addition of diethylzinc to aldehydes catalyzed by pyridyl-alcohols **20b**, **26b**, and quinolyl-alcohol **21b**.

R=	4-Cl-C6H4	4-Br-C6H4	4-Me-C6H4	4-MeO-C6H4	2-MeO-C6H4	cyclohexyl	cinnamyl	
Yield (%), ee (%) **20b**	99, 81	98, 80	99, 78	99, 80	99, 81	78, 50	85, 34	R
21b	99, 80	97, 79	96, 80	95, 83	99, 82	75, 41	85, 36	
26b	99, 92	99, 90	93, 92	81, 88	99, 94	78, 90	87, 35	S

13.2.2
Addition of Trialkylaluminium Compounds to Aldehydes. P,P and P,O-Donor Ligands

More recently, Woodward and coworkers reported the asymmetric addition of trialkylaluminium compounds to aldehydes employing nickel as a catalyst with phosphoramidite ligands [17]. Trialkylaluminium compounds can be easily obtained from aluminium hydride and olefins. Later on, this reaction was explored using chelate ligands such as phosphite–oxazoline (**30–33**) [18] and phosphite–phosphoramidite (**35**) ligands derived from glucosamine (Scheme 13.10) [19], and monophosphite ligands (**41–45**) derived from glucose [20], galactose, and fructose (Scheme 13.11).

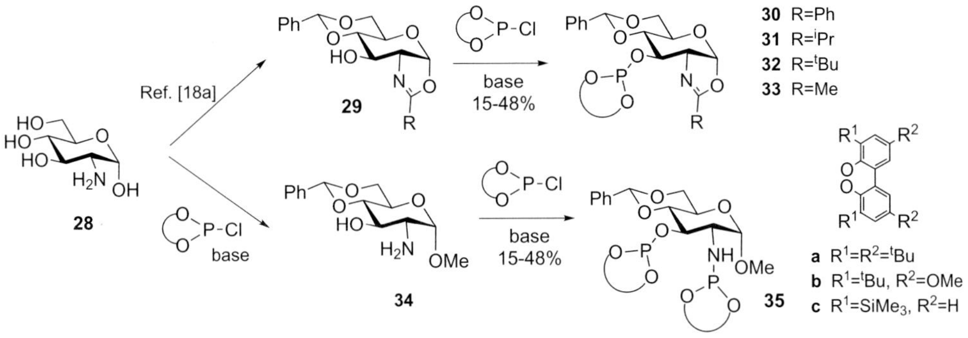

13.2 Addition of Organometallic Reagents to Aldehydes | 303

Scheme 13.10 Synthesis of oxazoline–phosphite ligands **30–33**, and phosphite–phosphoramidite ligands **35**.

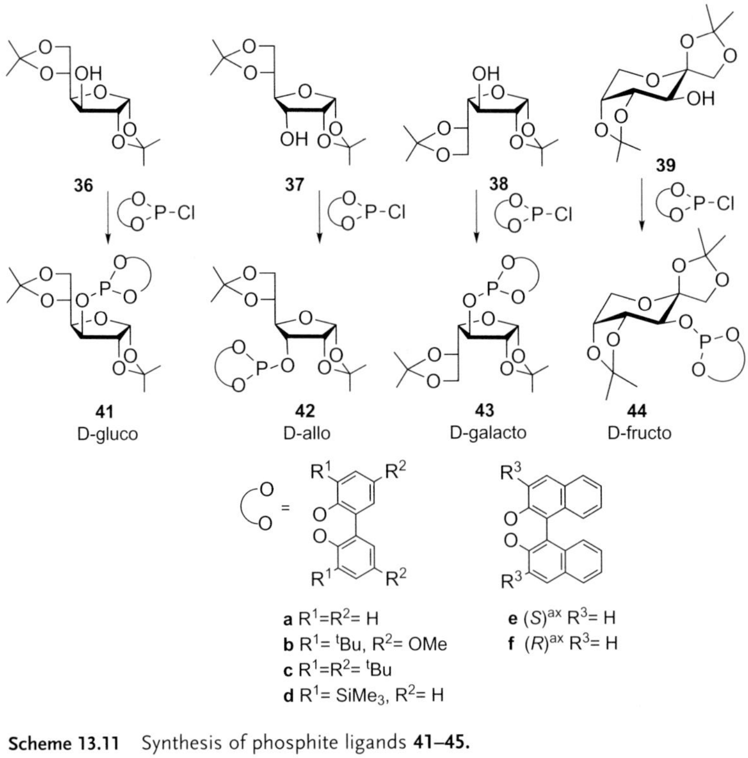

30 R=Ph
31 R=^{i}Pr
32 R=^{t}Bu
33 R=Me

a R^1=R^2=^{t}Bu
b R^1=^{t}Bu, R^2=OMe
c R^1=SiMe$_3$, R^2=H

Scheme 13.11 Synthesis of phosphite ligands **41–45**.

a R^1=R^2= H
b R^1= ^{t}Bu, R^2= OMe
c R^1=R^2= ^{t}Bu
d R^1= SiMe$_3$, R^2= H

e (S)^{ax} R^3= H
f (R)^{ax} R^3= H

Ligands **30–34** and **35** provided moderate to low conversions and enantioselectivities in the Ni-catalyzed asymmetric addition of trimethylaluminium to benzaldehyde and other benchmark aldehydes. The best enantioselectivity (59%) was obtained in the addition of AlMe$_3$ to benzaldehyde in the presence of ligand **32a**, but the conversion was very low.

Phosphites **41–45** were selected to explore the influence of different structural parameters such as configuration of the C3 stereocenter bearing the phosphite group (**41, 42**), configuration at C4 (**41, 43**), ring size (**42, 44**), and the presence of the phosphite moiety on a primary carbon (**45**) (Scheme 13.11). The diol moiety in the phosphite function can also be systematically modified by introducing substituents in the bisphenol unit (**a–d**) or by using chiral bisnaphthol (**e, f**). In the latter case, the matched–mismatched effect between bisnaphthol and the chiral carbohydrate backbone must be explored. Ligands **41–45** were prepared, in yields lower than 12–30% (**a–d**) and 75–98% (**e, f**), by reaction of the corresponding alcohols with the phosphochloridites, which in turn were obtained by reaction of bisphenol and bisnaphthol derivatives with PCl_3 in the presence of Et_3N. Alcohol precursors than **41** and **43** were obtained in one step from glucose and galactose, respectively, by reaction with acetone. Alcohol precursors of **44** and **45** were both obtained from D-fructose by reaction with acetone in acid medium under kinetic and thermodynamic conditions, respectively. Finally, **42** was obtained from 1,2:5,6-di-*O*-isopropylidene-D-glucofuranose by oxidation with PCC to give the ketone, followed by reduction with $NaBH_4$.

An initial screening of ligands **41–45** was carried out using $AlMe_3$ and benzaldehyde (R = H) (Table 13.3). The results indicated that the catalytic performance (activities and enantioselectivities) is highly affected by the configuration at C3 and C4, and the size of the ring of the sugar backbone, with the best enantioselectivity obtained with ligand **41**. The substituents of the biaryl moieties exert a remarkable influence (entries 1–6, Table 13.3). Substituents at the *ortho* positions of the biaryl phosphite moiety affected the yield, while enantioselectivities were mainly affected by substituents at the *para* positions. Bulky *ortho* substituents are required for high yields, while a *tert*-butyl group at the *para* positions provides the best ee values (entry 3). Thus, the best trade-off between yield and enantioselectivity was obtained using ligand **41c**, and the best results were obtained with a ligand/metal ratio 1:1 and by reducing the reaction time to 1 h (entry 11). The use of binaphthyl phosphite derivatives provided very low conversions and low ee values (entries 5, 6).

The presence of the bisphenol moiety **c** in ligands **42–45** also afforded the best results, but ees were lower than those obtained with **41c** (entries 7–10 versus 3). The best enantioselectivity was achieved with *p*-methoxybenzaldehyde (94% ee, entry 13), but the ee is hardly affected by the presence of the electron-withdrawing or electron-donating groups in the substrate (entries 14, 15). However, substrates substituted at *meta* and *ortho* positions afforded lower enantioselectivities (entries 16, 17), probably due to steric factors. With more flexible substrates, such as cinnamaldehyde and 3-phenylpropanal, the ee was only 25%.

The reaction of other aldehydes bearing different substituents, and other aluminium reagents such as $AlEt_3$ and DABAL-Me_3, an air-stable aluminium reagent, was also explored. The results of using triethylaluminium (entries 12, 15) indicated that the catalytic performance followed the same trend as for the trimethylaluminium addition. The use of DABAL-Me_3 showed similar trends, although yields were lower than in the trimethylaluminium addition. Ligand **41c** was also the best ligand in this case.

Table 13.3 Addition of AlR$_3$ to aldehydes catalyzed by [Ni(acac)$_2$]/**41–45**.[a]

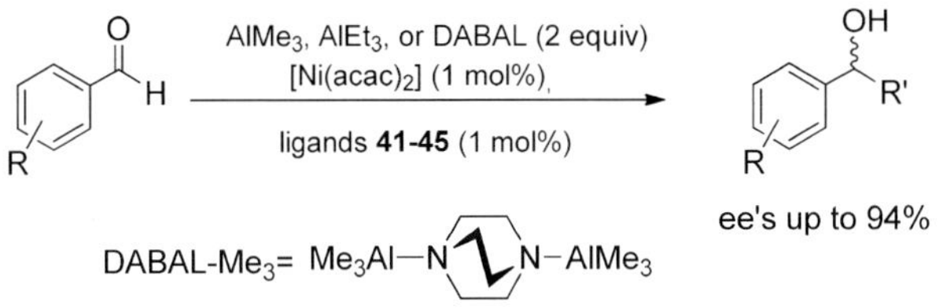

$$\text{R-C}_6\text{H}_4\text{-CHO} \xrightarrow[\text{ligands } \mathbf{41\text{-}45} \text{ (1 mol\%)}]{\text{AlMe}_3, \text{AlEt}_3, \text{ or DABAL (2 equiv)} \\ [\text{Ni(acac)}_2] \text{ (1 mol\%)}} \text{R-C}_6\text{H}_4\text{-CH(OH)R'}$$

ee's up to 94%

DABAL-Me$_3$ = Me$_3$Al–N(CH$_2$CH$_2$)$_2$N–AlMe$_3$

Entry	Substrate (R)	AlR$_3$	Ligand	L*/Ni	Yield (%)	ee (%)
1	H	AlMe$_3$	41a	2	15	27 (R)
2	H	AlMe$_3$	41b	2	80	82 (S)
3	H	AlMe$_3$	41c	2	85	89 (S)
4	H	AlMe$_3$	41d	2	87	52 (S)
5	H	AlMe$_3$	41e	2	11	41 (R)
6	H	AlMe$_3$	41f	2	12	10 (R)
7	H	AlMe$_3$	42c	2	100	44 (R)
8	H	AlMe$_3$	43c	2	60	70 (R)
9	H	AlMe$_3$	44c	2	64	52 (S)
10	H	AlMe$_3$	45c	2	52	36 (S)
11	H	AlMe$_3$	41c	1	100	90 (S)
12	H	AlEt$_3$	41c	1	100	88 (S)
13	4-OMe	AlMe$_3$	41c	1	53	94 (S)
14	4-CF$_3$	AlMe$_3$	41c	1	95	93 (S)
15	4-CF$_3$	AlEt$_3$	41c	1	96	94 (S)
16	2-Cl	AlMe$_3$	41c	1	85	41 (S)
17	3-Cl	AlMe$_3$	41c	1	73	74 (R)
18	4-Cl	AlMe$_3$	41c	1	82	91 (S)
19	4-Cl	AlEt$_3$	41c	1	83	90 (S)
20	4-Cl	DABAL-Me$_3$	41c	1	64	91 (S)

a) Conditions: substrate (0.25 mmol), AlR$_3$ (0.50 mmol), [Ni(acac)$_2$] (1 mol.%), THF (2 ml), $T = -20\,°\text{C}$.

13.2.3
Zinc-Triflate-Catalyzed Addition of Alkynes to Aldehydes and Imines. N,O Ligands

The addition of acetylene to carbonyl compounds to provide propargyl alcohols has been rarely studied and there are few methods that give high yields and enantioselectivities. Among the more relevant catalytic systems are chiral oxazaborolidines [21] and the catalytic systems Zn(OTf)$_2$/**46** [22] or **47** (Figure 13.3) [23].

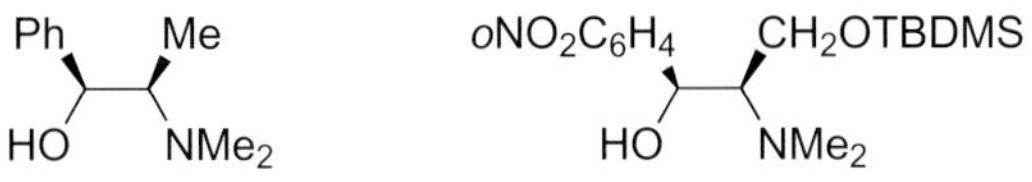

Ph–CH(OH)–CH(Me)–NMe$_2$
46

oNO$_2$C$_6$H$_4$–CH(OH)–CH(CH$_2$OTBDMS)–NMe$_2$
47

Figure 13.3 Aminoalcohols **46** and **47**.

Table 13.4 Addition of phenylacetylene to aldehydes catalyzed by Zn(OTf)$_2$/**14l–17l**.[a]

$$R\text{-CHO} + Ph\text{-}\!\!\!\equiv\!\!\!\text{-H} \xrightarrow[\text{Toluene, Et}_3\text{N, rt}]{\text{Zn(OTf)}_2,\ \mathbf{11l\text{-}14l}} R\text{-CH(OH)-C}\!\!\equiv\!\!\text{C-Ph}$$

ee's up to 99%

14l D-α-gluco; **15l** D-β-gluco; **16l** D-α-allo; **17l** D-α-manno

Entry	R	Ligand	Yield (%)	ee (%) (configuration)	Entry	R	Ligand	Yield (%)	ee (%) (configuration)
1	Cy	14l	95	97 (R)	6	c-Pr	15l	71	94 (R)
2	Cy	15l	94	99 (R)	7	t-Bu	15l	62	97 (R)
3	Cy	16l	92	22 (R)	8	Ph(CH$_2$)$_2$	15l	15	98 (R)
4	Cy	17l	8	35 (R)	9	p-MePh	15l	38	97 (R)
5	i-Pr	15l	70	97 (R)	10	p-ClPh	15l	36	98 (R)

a) Conditions: ratio Zn(OTf)$_2$/ligand/Et$_3$N/acetylene/aldehyde = 1.1 : 1.2 : 1.2 : 1.2 : 1.0, toluene, 40 °C.

Since remarkable results in the addition of acetylenes to aldehydes were obtained using carbohydrate-derived aminoalcohols, the library of ligands **14–17** (see Scheme 13.6) was also assayed in this reaction (Table 13.4) [24]. Ligands **14c,e,j,k,l** and **14l–17l** were selected for this screening. It was observed that cyclic amine substituents at C2 (**15l–17l**) gave enantioselectivities and yields higher than those of acyclic ones, and α- (**14l**) and in particular β-gluco ligands (**15l**) gave selectivities higher than those of α-allo or α-manno derivatives (entries 1–4, Table 13.4). Thus, the addition of phenylacetylene to different aldehydes in the presence of Zn(OTf)$_2$ afforded excellent ee values, between 94% and 99%, when ligand **15l** was employed (entries 5–10, Table 13.4).

Finally, other acetylenes such as 4-phenylbut-1-yne, 2-methylbut-3-yn-2-ol, but-3-yn-1-ol, and (triethylsilyl)acetylene were also successfully added to cyclohexylcarbaldehyde using the catalytic system Zn(OTf)$_2$/**15l** (Table 13.5).

Since the first report by Li [25] on the asymmetric addition of alkynes to imines [5] catalyzed by a Cu(I)/pybox catalytic system, this transformation has attracted the interest of researchers due to the synthetic applications of the propargylamines obtained.

The ligand *gluco*Pybox (**50**) [26], which is related to traditional pybox ligands and incorporates two glucose units, was recently prepared from glucosamine

13.2 Addition of Organometallic Reagents to Aldehydes

Table 13.5 Addition of acetylenes to cyclohexylcarbaldehyde catalyzed by Zn(OTf)$_2$/15I.[a]

$$\text{Cy-CHO} + \text{R}\equiv\text{H} \xrightarrow[\text{Toluene, Et}_3\text{N, rt}]{\text{Zn(OTf)}_2,\ 15\text{I}} \text{Cy-CH(OH)-C}\equiv\text{C-R}$$

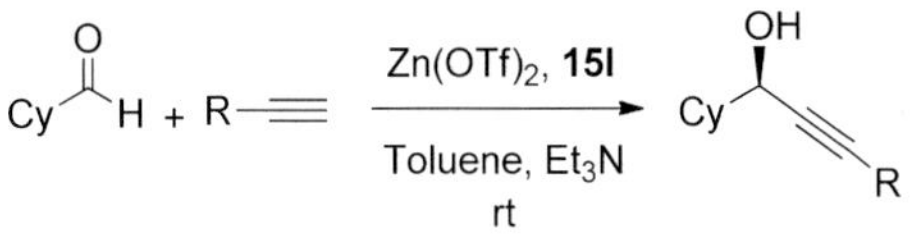

Entry	R	Yield (%)	ee (%) (configuration)
1	Ph(CH$_2$)$_2$	89	95 (R)
2	Me$_2$COH	71	99 (R)
3	Me$_2$COTMS	68	81 (R)
4	HO(CH$_2$)$_2$	90	98 (R)
5	Et$_3$Si	42	98 (R)

a) Conditions: similar to Table 13.4, T = 50 °C.

hydrochloride by reaction with 2,6-dipicolinic acid to give **49**, followed by treatment with neat AcCl and base (Scheme 13.12).

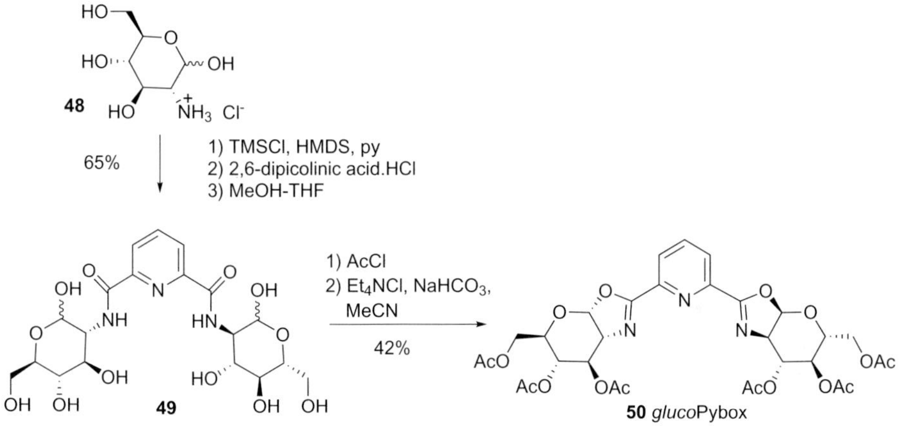

Scheme 13.12 Synthesis of *gluco*Pybox ligand **50**.

The catalytic system CuOTf/*gluco*Pybox afforded an excellent 99% ee and 69% yield in the addition of phenylacetylene to imines (Table 13.6, entry 1). The catalytic system appears to be very sensitive to the substituents R^1–R^3. Thus, for R^1 or R^2 = 4-MeOPh, the ee decreased to 80 and 90%, respectively (entries 2 and 3). The ee was also 90% when (trimethylsilyl)acetylene was used (R^3 = TMS), although the yield dropped significantly (entry 4). However, when substituents of the amino and acetylene moieties were simultaneously modified racemic product was obtained (entry 5). The ee values did not exceeded 75% when substituted benzaldehyde or aliphatic aldehydes were used (entry 6).

Table 13.6 Enantioselective alkynylation of imines with Cu/glucoPybox (**50**).

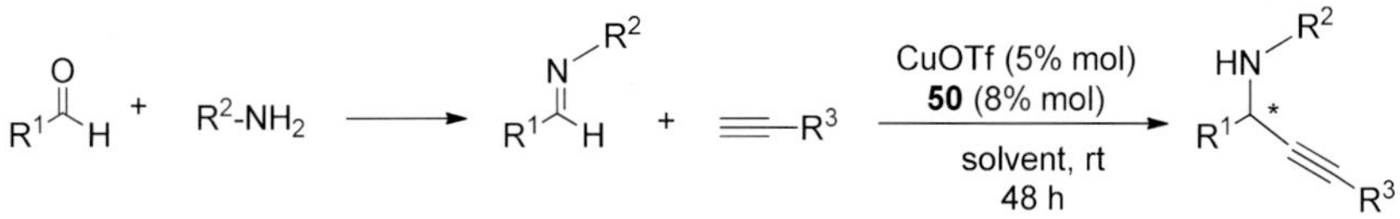

Entry	R^1	R^2	R^3	Yield (%)	ee (%)
1	Ph	Ph	Ph	69	99 (R)
2	4-MeOPh	Ph	Ph	33	80 (S)
3	Ph	4-MeOPh	Ph	38	90 (S)
4	Ph	Ph	TMS	21	90 (S)
5	Ph	4-MeOPh	TMS	29	Racemic
6	i-Pr	Ph	Ph	57	75 (R)

13.3
Addition of Trimethylsilyl Cyanide to Ketones. P=O,O Ligands

Shibasaki has developed a new type of Lewis acid–Lewis base bifunctional catalyst that has been successfully applied in asymmetric cyanosilylation of ketones [27]. The most successful catalyst was the aluminium derivative **51**, where Al and the oxygen atom of the phosphine oxide work cooperatively as Lewis acid and Lewis base to activate the aldehyde and TMSCN, respectively. In a search for more general and reactive catalysts, glucose derivative **52** was selected as a catalyst since it presented structural features similar to **51** (Figure 13.4) [28].

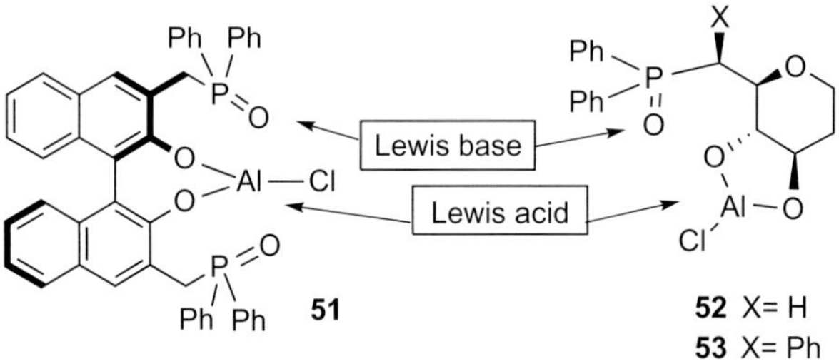

Figure 13.4 Structural relationship between bifunctional catalysts **51–53**.

The authors hypothesized that restriction of the conformational freedom of the ligand would improve the enantioselectivity. With this goal, ligand **53** was prepared from glucal **54** following the reaction sequence shown in Scheme 13.13. After reduction of the double bond in **54**, a deprotection–protection process afforded **55** that was then transformed into the ketone **56**. Further, **56** was subjected to a reaction sequence involving ketone reduction, hydroxyl activation, substitution with Ph$_2$PK, oxidation, removal of the MOM groups and oxidation to

13.3 Addition of Trimethylsilyl Cyanide to Ketones. P=O,O Ligands | 309

Scheme 13.13 Synthesis of bifunctional catalysts **52** and **53**.

give **59**. The reaction of **59** with Me$_2$AlCl in basic medium yielded **53**. Compound **52** can be prepared from **55** by a synthetic pathway similar to that used in the transformation of **57** into **53**. Ligands **60–63** (Figure 13.5) were prepared in a similar way starting from the appropriate glycals.

Complexes **52**, **53**, and **60–63** were prepared to study the influence of the stereochemistry at positions 3 and 4 of the sugar ring on the enantioselectivity. Initially, the addition of TMSCN to benzaldehyde was studied, and ligand **52** provided the best results, although the enantioselectivity was moderate (99%, 46% ee).

Cyanosilylation of benzaldehyde using catalyst **53** provided 96% yield and 80% ee. This catalyst afforded similar enantioselectivities in the cyanosilylation of aldehydes such as cinnamaldehyde or heptanal. However, it was only moderately efficient in the cyanosilylation of ketones.

In a search for a more efficient catalysts able to provide high ee in the cyanosilylation of ketones, ligand **52** was again modified, by anchoring a catechol moiety to the 3-OH, with the aim of providing an additional coordinating atom (Scheme 13.14) [29]. Thus, ligand **68a** was prepared by the reaction of **65** with alcohol **64** to

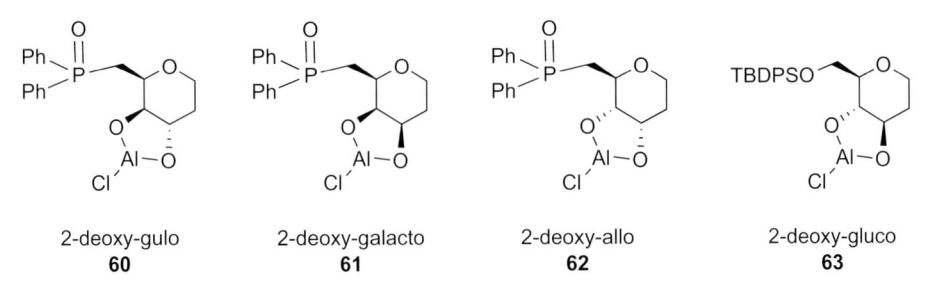

Figure 13.5 Bifunctional catalysts **60–63**.

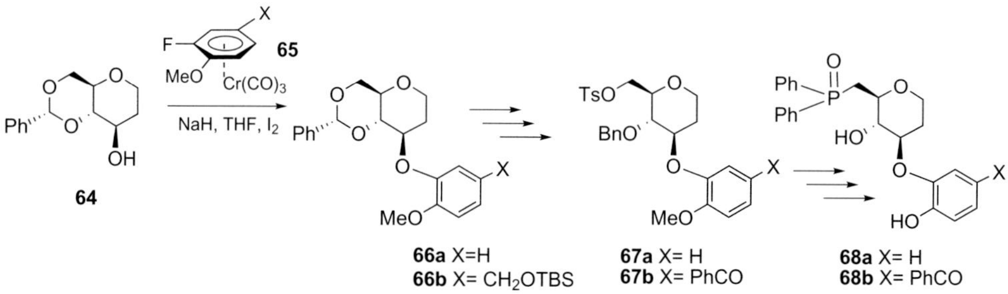

Scheme 13.14 Synthesis of ligands **68**.

afford **66a**. Selective deprotection of the benzylidene acetal using DIBAL-H and tosylation to afford **67a**, substitution with Ph$_2$PK, oxidation and deprotection of the hydroxyl groups then provided **68a**. Ligand **68a** was coordinated to different metals (Al, Yb, Zr, and Ti) and the best results in the cyanosilylation of acetophenone were obtained with the titanium derivative.

Scheme 13.15 shows the results obtained in the cyanosilylation of different ketones under the best reactions conditions (10 mol.% of catalyst loading, CH$_2$Cl$_2$, −40 °C); enantioselectivities up to 95% ee were obtained. This was the first example of a general cyanosilylation of ketones.

Yield (%),ee (%)							
68a	80, 92	82, 95	72, 69	89, 91	72, 91	86, 90	86, 76
68b	76, 97		90, 84	90, 92		91, 93	71, 86

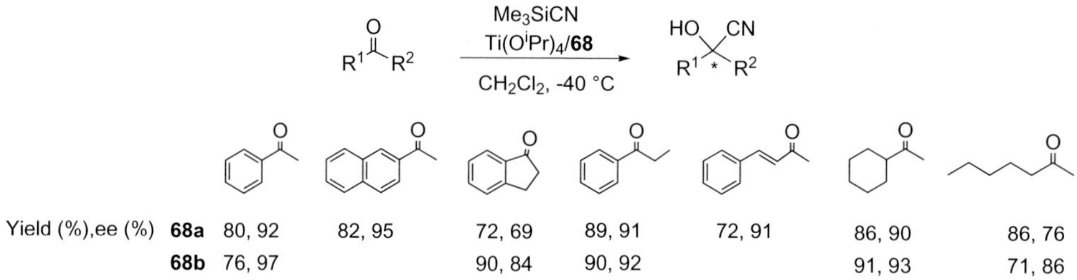

Scheme 13.15 Addition of Me$_3$SiCN to ketones catalyzed by Ti/**68**.

Mechanistic studies carried out by isotopic labeling and kinetic studies allowed the proposal of transition state model **69** for the titanium/**68a**-catalyzed cyanosilylation of ketones (Figure 13.6). In this transition state model, the titanium cyanide would act only as a Lewis acid, and not as a CN source.

With the aim of improving some moderate results obtained with catalyst Ti/**68a** (Scheme 13.15) and reducing the high catalyst loading required (10 mol.%), ligand **68a** was modified by adding a benzoyl group at the *para* position of the phenol (**68b**) [29b]. With this modification, it was intended to strengthen the Ti–phenoxide bond, giving rise to a more stable catalyst and to sterically shield site B (Figure 13.6), forcing coordination of the ketone at site A. In general, the enantioselectivities were improved, particularly on those substrates that provided poor results with ligand **68a**. Moreover, catalysts loadings of 1–2.5 mol.% were enough to achieve similar conversions in similar reaction times as for ligand **68a**.

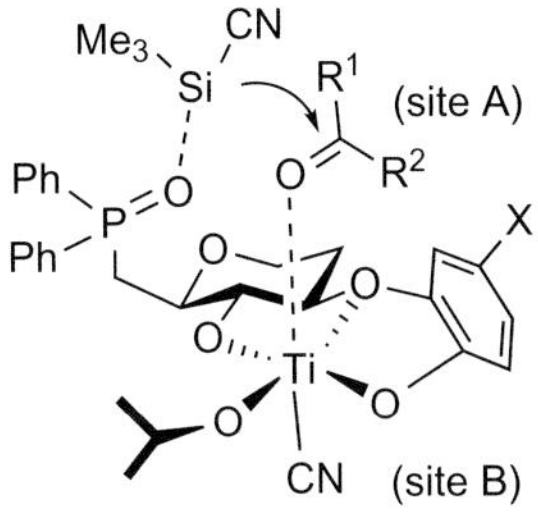

Figure 13.6 Working model for the titanium/**68a**-catalyzed cyanosilylation of ketones.

13.4 Conclusion

Carbohydrate-derived ligands with a diversity of coordinating heteroatoms have been successfully used in organometallic additions to carbonyl compounds. In the addition of ZnEt$_2$ to aldehydes, aminoalcohols have provided ee values of up to 96% (**8i**). The Lewis acid [Ti(Oi-Pr)$_4$] catalyzed addition of ZnEt$_2$/aminoalcohol (**9**) to aldehydes gives improved yields and ee values. Aminoalcohols **15** proved to be also very efficient ligands in the Zn(OTf)$_2$-catalyzed addition of alkynes to aldehydes, affording ee values up to 99%. An oxazoline ring can be easily integrated in a pyranose ring starting from amino-carbohydrates such as glucosamine. Thus, oxazoline–phosphite ligands (**30–33**) were used in the nickel-catalyzed addition of AlR$_3$ reagents to aldehydes, although conversions and ees were moderate. However, in a nice extension of these ligands, pyridine–bisoxazoline ligands (**50**, *gluco*Pybox), an analog of pybox, were prepared and used in the copper-catalyzed addition of alkynes to imines, affording ee values of up to 99%. Monophosphites also provided excellent ees in the nickel-catalyzed addition of AlMe$_3$ to aldehydes.

Ligands **68a,b** led to an innovative new class of bifunctional Lewis acid–Lewis base catalytic system that allowed the efficient titanium-catalyzed addition of Me$_3$SiCN to ketones, providing ee values of up to 97%.

As expected, ee values are very dependent on the configuration of the carbon bearing the coordinating heteroatoms. Carbohydrate chemistry provides easy access to ligands with all possible configurations at those positions, allowing easy optimization of results, as illustrated in the two processes studied with ligands **14–17**. Complexes **52**, **53**, **60–63**, and Al,Y,Ti/**68a,b** also constitute a very nice example of the optimization of a catalytic process while taking advantage of the possibilities of modifying carbohydrates. The configuration of positions neighboring the coordinating heteroatoms also has a strong influence on the enantioselectivity (**9** versus **10**). Finally, the possibility of obtaining both enantiomers in a catalytic process using D-carbohydrates is illustrated by the realization that ketones **18** and **24** are pseudo-enantiomers. From these ketones, ligands **20**, **25**, and **26** were prepared and used as ligands in the addition of diethylzinc to aldehydes, effectively delivering the opposite enantiomers.

References

1 (a) Boysen, M.M.K. (2007) *Chem. Eur. J.*, **13**, 8648–8659; (b) Diéguez, M., Claver, C., and Pàmies, O. (2007) *Eur. J. Org. Chem.*, 4621–4634; (c) Castillón, S., Claver, C., and Díaz, Y. (2005) *Chem. Soc. Rev.*, **34**, 702–713; (d) Diéguez, M., Pàmies, O., Ruiz, A., Díaz, Y., Castillón, S., and Claver, C. (2004) *Coord. Chem. Rev.*, **248**, 2165–2192; (e) Diéguez, M., Pàmies, O., and Claver, C. (2004) *Chem. Rev.*, **104**, 3189–3215.

2 (a) Shibasaki, M. and Kanai, M. (2008) *Chem. Rev.*, **108**, 2853–2873; (b) Riant, O. and Hannedouche, J. (2007) *Org. Biomol. Chem.*, **5**, 873–888.

3 (a) Pu, L., and Yu, H.B. (2001) *Chem. Rev.*, **101**, 757–824; (b) Soai, K. and Niwa, S. (1992) *Chem. Rev.*, **92**, 833–856; (c) Noyori, R. and Kitamura, M. (1991) *Angew. Chem. Int. Ed. Engl.*, **30**, 49–69.

4 Lu, G., Li, Y.-M., Li, X.-S., and Chan, A.S.C. (2005) *Coord. Chem. Rev.*, **249**, 1736–1744.

5 Zani, L. and Bolm, C. (2006) *Chem. Commun.*, 4263–4275.

6 Gröger, H. (2003) *Chem. Rev.*, **103**, 2795–2827.

7 (a) Oguni, N., Omi, T., Yamamoto, Y., and Nakamura, A. (1983) *Chem. Lett.*, 841–842; (b) Oguni, N. and Omi, T. (1984) *Tetrahedron Lett.*, **25**, 2823–2824.

8 Kitamura, M., Suga, S., Kawai, K., and Noyori, R. (1986) *J. Am. Chem. Soc.*, **108**, 6071–6072.

9 Cho, B.T. and Kim, N.J. (1996) *J. Chem. Soc., Perkin Trans. 1*, 2901–2907.

10 Yang, W.Y. and Cho, B.T. (2000) *Tetrahedron: Asymmetry*, **11**, 2947–2953.

11 Bauer, T., Tarasiuk, J., and Pasniczek, K. (2002) *Tetrahedron: Asymmetry*, **13**, 77–82.

12 Emmerson, D.P., Villard, R., Mugnaini, C., Batsanov, A., Howard, J.A.K., Hems, W.P., Tooze, T.P., and Davis, B.G. (2003) *Org. Biomol. Chem.*, **1**, 3826–3838.

13 Kahr, K. and Berther, C. (1960) *Chem. Ber.*, **93**, 132–136.

14 Huang, H., Chen, H., Hu, X., Bai, C., and Zheng, Z. (2003) *Tetrahedron: Asymmetry*, **14**, 297–304.

15 Huang, H., Zheng, Z., Chen, H., Hu, X., Bai, C., and Wang, J. (2003) *Tetrahedron: Asymmetry*, **14**, 1285–1289.

16 (a) Collomb, P. and von Zelewsky, A. (1998) *Tetrahedron: Asymmetry*, **3**, 3911–3917; (b) Kwong, H.L. and Lee, W.S. (1999) *Tetrahedron: Asymmetry*, **10**, 3791–3801; (c) Goanvic, D.L., Holler, M., and Pale, P. (2002) *Tetrahedron: Asymmetry*, **13**, 119–121.

17 Biswas, K., Prieto, O., Goldsmith, P.J., and Woodward, S. (2005) *Angew. Chem. Int. Ed.*, **44**, 2232–2234.

18 (a) Mata, Y., Diéguez, M., Pàmies, O., and Claver, C. (2005) *Adv. Synth. Catal.*, **347**, 1943–1947; (b) Mata, Y., Diéguez, M., and Pàmies, O. (2007) *Chem. Eur. J.*, **13**, 3296–3304.

19 Mata, Y., Diéguez, M., Pàmies, O., and Woodward, S. (2008) *Inorg. Chim. Acta*, **361**, 1381–1384.

20 Mata, Y., Diéguez, M., Pàmies, O., and Woodward, S. (2006) *J. Org. Chem.*, **71**, 8159–8165.

21 Corey, E.J. and Cimprich, K.A. (1994) *J. Am. Chem. Soc.*, **116**, 3151–3152.

22 (a) Frantz, D.E., Fässler, R., and Carreira, E.M. (2000) *J. Am. Chem. Soc.*, **122**, 1806–1807; (b) Frantz, D.E., Fässler, R., Tomooka, C.S., and Carreira, E.M. (2000) *Acc. Chem. Res.*, **33**, 373–381.

23 (a) Jiang, B., Chen, Z., and Xiong, W. (2002) *Chem. Commun.*, 1524–1525; (b) Yamashita, M., Yamada, K., and Tomioka, K. (2005) *Adv. Synth. Catal.*, **347**, 1649–1652.

24 Emmerson, D.P., Hems, W.P., and Davis, B.G. (2006) *Org. Lett.*, **8**, 207–210.

25 (a) Wei, C. and Li, C.-J. (2002) *J. Am. Chem. Soc.*, **124**, 5638–5639; (b) Wei, C., Mague, J.T., and Li, C.-J. (2004) *Proc. Natl. Acad. Sci. USA*, **101**, 5749–5754.

26 Irmak, M. and Boysen, M.M.K. (2008) *Adv. Synth. Catal.*, **350**, 403–405.

27 Hamashima, Y., Sawada, D., Kanai, M., and Shibasaki, M. (1999) *J. Am. Chem. Soc.*, **121**, 2641–2642.

28 Kanai, M., Hamashima, Y., and Shibasaki, M. (2000) *Tetrahedron Lett.*, **41**, 2405–2409.

29 (a) Hamashima, Y., Kanai, M., and Shibasaki, M. (2000) *J. Am. Chem. Soc.*, **122**, 7412–7413; (b) Hamashima, Y., Kanai, M., and Shibasaki, M. (2001) *Tetrahedron Lett.*, **42**, 691–694.

14
Cyclopropanation
Mike M.K. Boysen

Asymmetric cyclopropanation [1] is not only an important method for setting up chiral three-membered rings, which occur as motifs in many natural products [2] but, due to the inherent reactivity of the cyclopropyl ring, the products may also be transformed into other useful chiral building blocks. Chiral bis(oxazolines) (box) and pyridyl bis(oxazolines) (pybox) [3] are among the most useful ligands for transition-metal-catalyzed asymmetric cyclopropanation of alkenes with diazo compounds via carbene complexes. The first highly stereoselective examples of intermolecular cyclopropanations with copper(I) box complexes were reported independently by Evans [4] and Masamune [5] and have inspired the design of many new ligands.

A successful family of carbohydrate-derived box ligands was reported by the Boysen group. In the first approach, D-glucosamine (**1**) was transformed into box ligand **5** [6] in four steps (Scheme 14.1). As the 3-position of the pyranose subunits in **5** is closest to the coordinating nitrogen atoms, this position was envisaged to have a major impact on stereoselectivity. Therefore ligands **9a–f** with 3-*O*-acyl, -alkyl, or -silyl groups of varying steric demand and type were prepared from 4,6-*O*-benzylidene protected thioglucoside **6** as key intermediate [7]. The cyclization to the desired 3-*O*-R^1 *gluco*Box ligands (**12a–f**) was effected using NIS (Scheme 14.1).

Ligands **5** and **9a–f** were subsequently tested in the asymmetric cyclopropanation of styrene (**10**) with ethyl diazoacetate (**11**), yielding the diastereomeric products **12** *trans* and **12** *cis* (Table 14.1). The per-*O*-acetylated ligand **5** gave the products **12** in moderate overall yield, a *trans/cis* ratio of 70:30 and 82% ee for both diastereomers (entry 1). The series of selectively 3-*O*-functionalized ligands **9a–f** led to some important findings: As expected, the size of the 3-*O*-substituent had a fundamental impact on stereoselectivity and 3-*O*-acetylated ligand **9a** gave a greatly improved result with 90% yield and 93% ee for the major trans-diastereomer of **12** (entry 2). However, an increase of the steric demand of the 3-*O*-acyl residues (**9b,c**, R^1 = Bz, Piv) led do a drop in enantioselectivity (64% ee and 62% ee respectively), while yield and diastereoselectivity were less affected (entries 3, 4); therefore, improved stereoselectivity was expected for ligand **9d** with

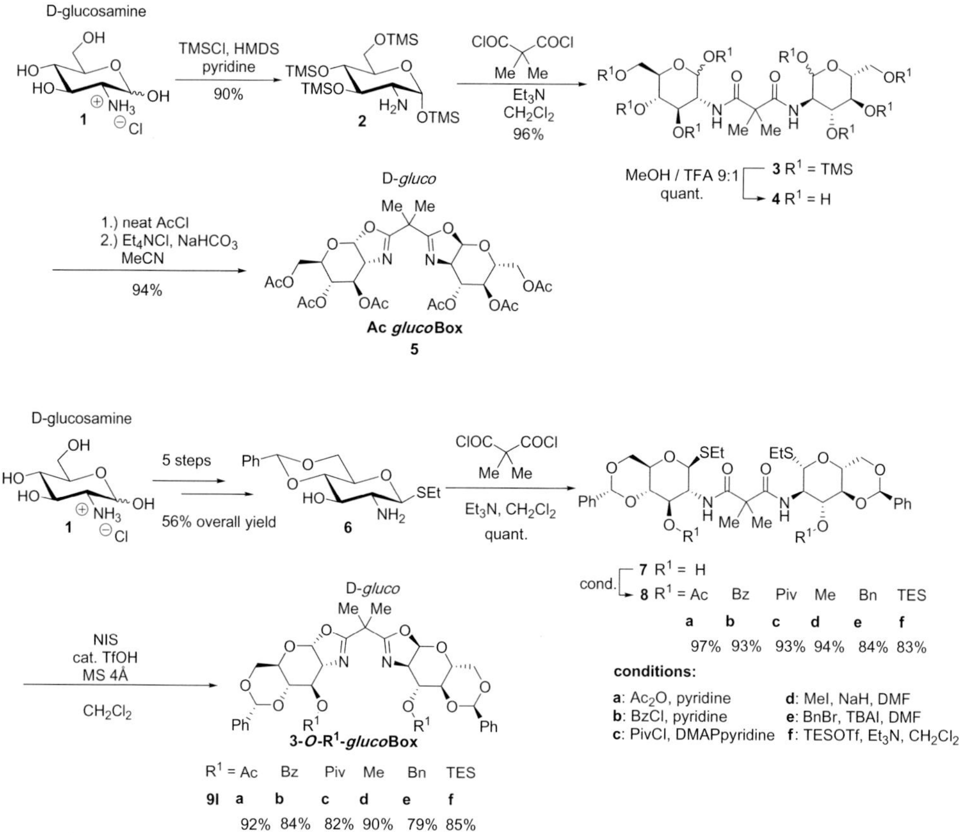

Scheme 14.1 Preparation bis(oxazoline) of Ac glucoBox (**5**) and 3-O-R¹-glucoBox (**9**) from D-glucosamine (**1**).

small 3-O-methyl ether substituents. Surprisingly, this derivative gave only 73% ee for **12** trans (entry 5) and an increase in steric size for 3-O-alkyl and silyl residues resulted in less than 50% ee (**9e,f**, R^1 = Bn, TES). When the 3-O-position was left unprotected, the stereoselectivity also remained modest (Table 14.1, entry 8). These findings show that the efficiency of the chirality transfer for ligands **9a–g**, does not only depend on the steric demand of the 3-O residue but also on its nature and, obviously, a 3-O-acyl group is a prerequisite for high stereoselectivity.

Apart from these steric considerations, a conformational effect appears to be in operation as well. The main difference between ligand per-O-acetylated **5** and 3-O-acetylated **9a** is the absence and presence of the cyclic 4,6-O-benzylidene protective group, which has fundamental consequences for the pyranoside conformation: Unprotected D-glucosamine typically adopts a $^4C_1$ chair conformation (Figure 14.1, structure **19**) while the pyranoside unit is significantly distorted in per-O-acetylated oxazoline **20** and exists in a $^OS_1$ twist conformation [8]. In the presence

14 Cyclopropanation

Table 14.1 Asymmetric cyclopropanation of styrene (**10**) with ethyl diazoacetate (**10**) using carbohydrate bis(oxazoline)s **5**, **9**, **13**, and **14–18**.

Entry	Ligand			Combined yield (%)	Ratio trans/cis	ee trans (%)	ee cis (%)
1	D-gluco (structure with AcO, AcO, OAc)		**5**	60	70:30	82	82
2	D-gluco (structure with Ph, R¹)	R¹ = Ac	**9a**	90	79:21	93	82
3		R¹ = Bz	**9b**	74	69:31	64	36
4		R¹ = Piv	**9c**	90	64:36	62	49
5		R¹ = Me	**9d**	87	77:23	73	39
6		R¹ = Bn	**9e**	95	62:38	48	28
7		R¹ = TES	**9f**	71	64:36	33	7
8		R¹ = H	**9g**	87	68:32	51	44
9		R¹ = formyl	**9h**	95	71:29	95	94
10	D-allo (structure with Ph, R¹)	R¹ = Ac	**13a**	75	66:34	Racemate	Racemate
11		R¹ = Bz	**13b**	70	67:33	21	7
12		R¹ = Me	**13d**	98	68:32	14	12
13		R¹ = Bn	**13e**	56	66:34	22	15
14		R¹ = formyl	**13h**	75	64:36	Racemate	Racemate
15	3-deoxy-D-gluco (structure with Ph)		**14**	86	69:31	Racemate	Racemate
16	D-allo (structure with AcO, AcO, OAc)		**15**	79	70:30	71	87
17	3-deoxy-D-gluco (structure with AcO, AcO)		**16**	75	74:26	78	72

Reaction scheme: Ph-CH=CH₂ (**10**) + N₂=CH-CO₂Et (**11**) → ligand (1.1 mol%), CuOTf·0.5 C₆H₆ (1 mol%), CH₂Cl₂ → Ph-cyclopropane-CO₂Et (**12** trans) + Ph-cyclopropane-CO₂Et (**12** cis)

Table 14.1 (Continued)

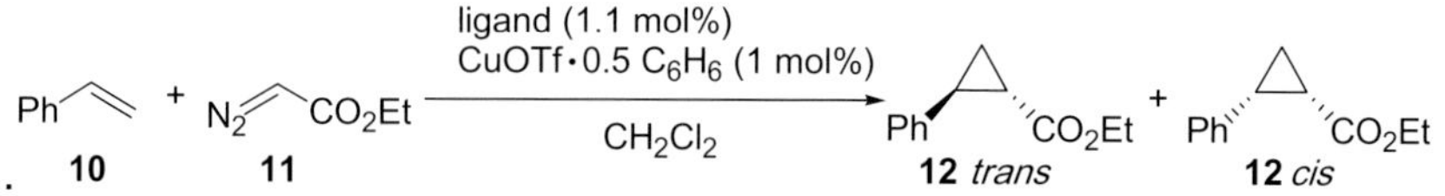

Ph—= 10 + N₂=CO₂Et 11 →[ligand (1.1 mol%) CuOTf·0.5 C₆H₆ (1 mol%)][CH₂Cl₂] Ph—△—CO₂Et **12 trans** + Ph—△—CO₂Et **12 cis**

Entry	Ligand		Combined yield (%)	Ratio trans/cis	ee trans (%)	ee cis (%)
18	D-gluco 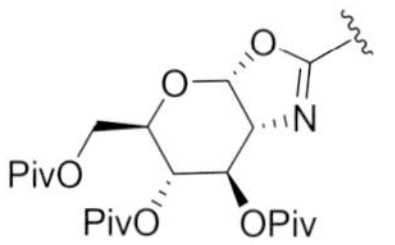 (PivO, PivO, OPiv substituted oxazoline)	17	86	63:37	84	94
19	D-gluco 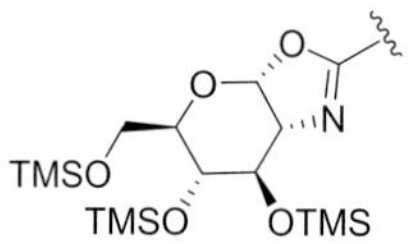 (TMSO, TMSO, OTMS substituted oxazoline)	18	56	71:29	68	58

of a cyclic 4,6-O-benzylidene group, the situation is different: Introduction of the cyclic acetal into glucosamine results in compound **21**, containing a rigid trans-decalin like structure. Oxazolines with this rigid bicyclic motif (structure **22**) deviate significantly less from the $^4C_1$ conformation; distortion occurs mainly in the front part of the pyranose (C1–C3), while the rear part is fixed in a chair

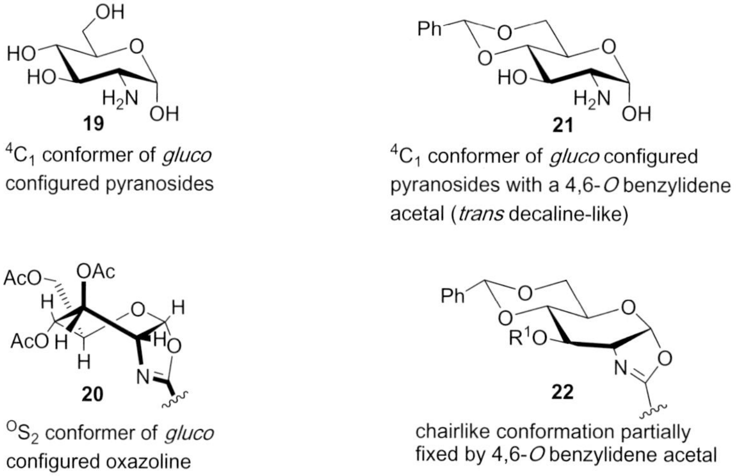

19
$^4C_1$ conformer of *gluco* configured pyranosides

21
$^4C_1$ conformer of *gluco* configured pyranosides with a 4,6-O benzylidene acetal (*trans* decaline-like)

20
$^OS_2$ conformer of *gluco* configured oxazoline

22
chairlike conformation partially fixed by 4,6-O benzylidene acetal

Figure 14.1 Pyranoside conformations in free monosaccharides and oxazolines in the absence and presence of cyclic 4,6-O-benzylidene protection.

conformation, a finding that is supported by ^{1}H NMR data [7b, 9]. This difference in pyranose geometry appears to be directly responsible for the improved stereoselectivity of 93% ee observed for **9a**, as opposed to 82% ee obtained with per-acetylated **5**. As both the cyclic 4,6-O-protective group and a 3-O-acyl residue were prerequisites for high levels of stereoinduction, 3-O-formyl ligand **9h** was prepared, giving the best results in this series with 95% yield and >90% ee for both diastereomers of **12** (Table 14.1, entry 9) [9]. Thus, the carbohydrate box ligands were successfully optimized via simple variation of the protective groups, exploiting steric, electronic, and conformational effects.

To study the influence of the stereocenter at position 3 of the pyranoside unit, *allo*-configured ligands **13** [9] (and G. Özüduru and M.M.K. Boysen, unpublished results) and 3-deoxygenated ligand **14** [9] with cyclic benzylidene protection were prepared and tested (Table 14.1, entries 10–15). None of these ligands achieved high levels of stereoinduction but, interestingly, the 3-O-acetylated and formylated *allo*-configured derivatives **13a** and **13h** gave only racemic mixtures of **12**, while 3-O-benzoyl ligand **13b** and the 3-O-alkylated derivatives **13d** and **13e** did at least produce enantiomeric excesses in the range of 15–20%. The 3-deoxygenated structure **14** gave only a racemate, which clearly shows the importance of both presence and configuration of the stereocenter at pyranose position 3. Surprisingly, *allo*- and 3-deoxy ligands **15** and **16**, missing the cyclic benzylidene acetal, gave substantial levels stereoinduction (entries 16, 17; 71% ee and 78% ee for **12** *trans* respectively), highlighting once again the importance of the overall pyranose conformation for the stereoinduction process. The results obtained with the *gluco*-configured, per-O-pivalated and per-silylated ligands **18** and **17** on the other hand confirmed the importance of acyl-based protective groups for high stereoselectivity, with pivalate ligand **17** giving 84% ee for the major product **11** *trans* and 94% ee for minor diastereomer **12** *cis*, while per-O-TMS modified ligands **18** only gave enantioselectivities of around 60% ee (Table 14.1, entries 18 and 19) [7b].

Apart from carbohydrate box-ligands a carbohydrate-derived pyridyl bis(thiazoline) ligand was explored in asymmetric cyclopropanation by the same authors; however only modest yields and stereoselectivities were obtained [10].

Optimized ligand 3-O-formyl *gluco*Box (**9h**) was evaluated in the asymmetric copper-catalyzed cyclopropanation of non-activated aliphatic alkenes [11], which still remain challenging substrates for this reaction [12]. The reaction of 1-octene (**23a**) and 1-nonene (**23b**) with ethyl diazoacetate (**11**) yielded the cyclopropanation products **24a,b** *trans* and **24a,b** *cis* in combined yield of 75–80% and stereoselectivities above 90% ee (Scheme 14.2). These results are among the best reported for these non-activated olefin substrates. Cyclopropanation product **24b** *trans* was subsequently employed as starting material for the stereoselective synthesis of the (+)-grenadamide (*ent*-**25**), the enantiomer of the natural product (–)-grenadamide (Scheme 14.2).

In summary, the ease of optimization and the high stereoselectivities obtained with aliphatic alkenes demonstrate the synthetic usefulness of carbohydrate box ligands, making them an interesting addition to the toolbox for asymmetric catalysis.

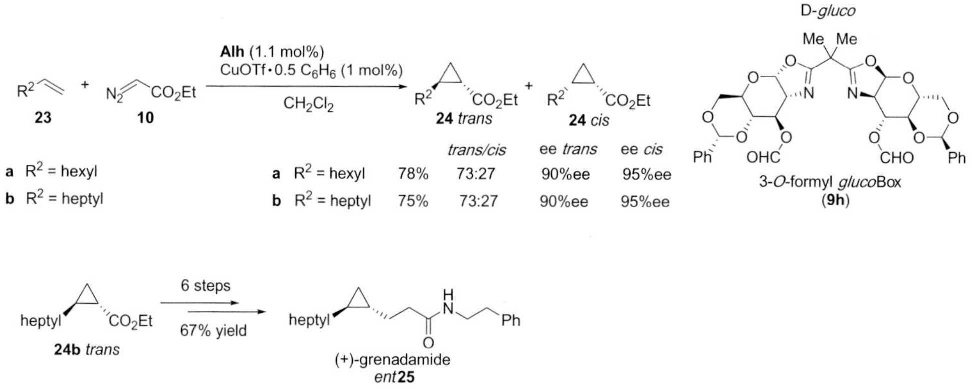

Scheme 14.2 Asymmetric cyclopropanation of non-activated aliphatic alkenes (**23a,b**) with ethyl diazoacetate (**11**) in the presence of optimized carbohydrate bis(oxazoline) 3-O-formyl glucoBox (**9h**). Stereoselective synthesis of (+)-grenadamide (*ent*-**25**).

References

1 For reviews see: (a) Pellissier, H. (2008) *Tetrahedron*, **64**, 7041; (b) Lebel, H., Marcoux, J.-F., Molinaro, C., and Charette, A.B. (2003) *Chem. Rev.*, **103**, 977; (c) Doyle, M.P. and Forbes, D.C. (1998) *Chem. Rev.*, **98**, 911; (d) Doyle, M.P. and Protopopova, M.N. (1998) *Tetrahedron*, **54**, 7919.

2 (a) Taylor, R.E., Engelhardt, F.C., and Schmitt, M.J. (2003) *Tetrahedron*, **59**, 5623; (b) Donaldson, W.A. (2001) *Tetrahedron*, **57**, 8589; (c) Wessjohann, L.A., Brandt, W., and Thiemann, T. (2003) *Chem. Rev.*, **103**, 1625.

3 (a) Desimoni, G., Faita, G., and Jørgensen, K.A. (2006) *Chem. Rev.*, **106**, 3561; (b) McManus, H.A. and Guiry, P.J. (2004) *Chem. Rev.*, **104**, 4151; (c) Desimoni, G., Faita, G., and Quadrelli, P. (2003) *Chem. Rev.*, **103**, 315; (d) Ghosh, A.K., Mathivanan, P., and Capiello, J. (1998) *Tetrahedron: Asymmetry*, **9**, 1.

4 (a) Evans, D.A., Woerpel, K.A., Hinman, M.M., and Faul, M.M. (1991) *J. Am. Chem. Soc.*, **113**, 726; (b) Loewenthal, R.E., Abiko, A., and Masamune, S. (1990) *Tetrahedron Lett.*, **31**, 6005.

5 Loewenthal, R.E., Abiko, A., and Masamune, S. (1990) *Tetrahedron Lett.*, **31**, 6005.

6 Irmak, M., Groschner, A., and Boysen, M.M.K. (2007) *Chem. Commun.*, 177.

7 (a) Minuth, T. and Boysen, M.M.K. (2008) *Synlett*, 1483; (b) Minuth, T., Irmak, M., Groschner, A., Lehnert, T., and Boysen, M.M.K. (2009) *Eur. J. Org. Chem.*, 997.

8 (a) Nashed, M.A., Slife, C.W., Kiso, M., and Anderson, L. (1980) *Carbohydr. Res.*, **82**, 237; (b) Foces-Foces, C., Cano, F.H., Bernabe, M., Penades, S., and Martin-Lomas, M. (1984) *Carbohydr. Res.*, **135**, 1.

9 Minuth, T. and Boysen, M.M.K. (2010) *Beilstein J. Org. Chem.*, **6** (23), doi: 10.3762/bjoc.6.23

10 Irmak, M., Lehnert, T., and Boysen, M.M.K. (2007) *Tetrahedron Lett.*, **48**, 7890.

11 Minuth, T. and Boysen, M.M.K. (2010) *Synthesis*, 2799.

12 For successful examples see: (a) with planar chiral ligands: Lo, M.M.-C. and Fu, G.C. (1998) *J. Am. Chem. Soc.*, **120**, 10270; (b) with boron-bridged bis(oxazolines): Mazet, C., Köhler, V., and Pfaltz, A. (2005) *Angew. Chem. Int. Ed.*, **44**, 4888.

Part IV
Carbohydrate Organocatalysts

15
Oxidations

O. Andrea Wong, Brian Nettles and Yian Shi

15.1
Oxidations

15.1.1
Epoxidations with Ketone Catalysts and Oxone

Olefins can be efficiently epoxidized by dioxiranes generated *in situ* from ketones and Oxone ($2KHSO_5 \cdot KHSO_4 \cdot K_2SO_4$) [1–6]. In theory, only a catalytic amount of ketone is required because the ketone can be regenerated upon epoxidation (Scheme 15.1). It can be anticipated that using a chiral ketone catalyst would provide an enantioselective epoxidation. In reality, it is extremely challenging to develop an effective chiral ketone catalyst for epoxidation. The reactivity and enantioselectivity of a chiral ketone catalyst is highly sensitive to the steric and electronic environment around the reacting carbonyl group. It proved to be difficult to obtain a proper balance between steric and electronic effects to achieve the desired enantioselectivity and reactivity. In 1984, Curci and coworkers first reported that up to 12.5% ee was obtained for the epoxidation of *trans*-β-methylstyrene with (+)-isopinocamphone as promoter [7]. During the last two decades, various types of chiral ketone catalysts have been investigated for asymmetric epoxidation [8–15], such as C_2-symmetric binaphthyl-based [13, 16–21], biphenyl-based [8, 22], and related ketones [23–26], ammonium ketones [8, 22, 27, 28], carbocyclic ketones [29–41], and bicyclo[3.2.1]octan-3-ones [42–45].

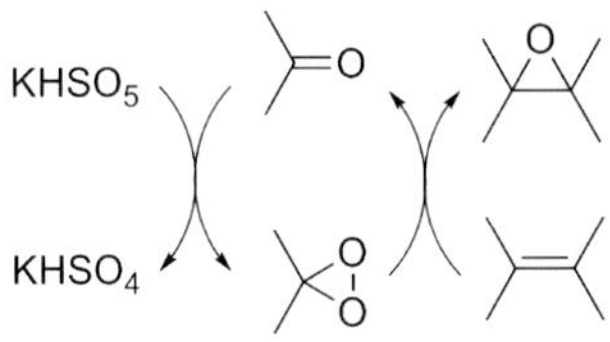

Scheme 15.1 Catalytic cycle of ketone-mediated epoxidations.

Carbohydrates are readily available in enantiopure form. They and their derivatives have been widely utilized in asymmetric synthesis as chiral starting materials, chiral auxiliaries, chiral reagents, and chiral ligands for metal catalysts. In the past decade, carbohydrate-based ketones [9–12, 14, 15] have been shown to be highly effective organocatalysts for asymmetric epoxidation of olefins with very broad substrate scope. The hydroxyl groups can be used as handles for chiral controlling elements and as precursors to carbonyl groups. In 1996, D-fructose-derived ketone **1** (Scheme 15.2) was reported to be a highly effective epoxidation catalyst for a wide variety of *trans-* and trisubstituted olefins [46, 47]. Ketone **1** can be readily synthesized in large quantities from D-fructose by ketalization with acetone and oxidation of the resulting alcohol (Scheme 15.2) [47–49]. Efforts have also been made to develop a synthesis of L-fructose from L-sorbose to allow easy access to the enantiomer of ketone **1** (*ent-***1**, Scheme 15.3) [50, 51]. Alcohol **2**, obtained from the ketalization of L-sorbose with 2,2-dimethoxypropane and SnCl$_2$, is converted into mesylate **3**, which undergoes a series of transformations in one pot to furnish L-fructose. *Ent-***1** can be readily synthesized from L-fructose via ketalization and oxidation.

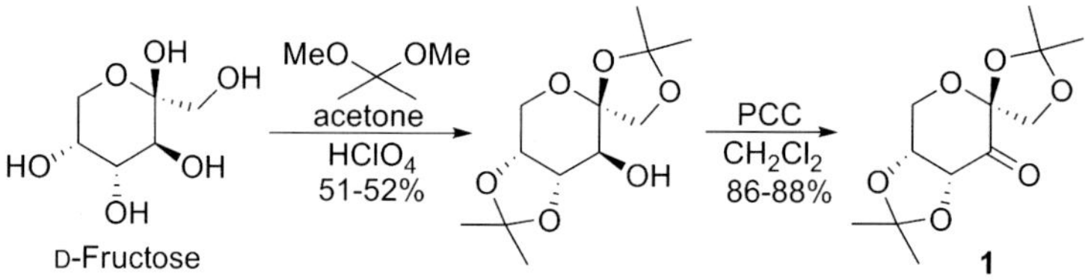

Scheme 15.2 Synthesis of ketone **1**.

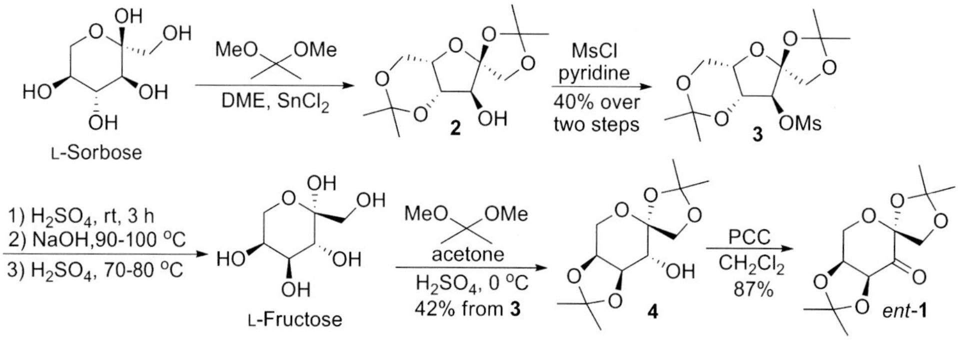

Scheme 15.3 Synthesis of *ent-***1**.

A proposed catalytic cycle for ketone **1**-mediated epoxidation is outlined in Scheme 15.4. Ketone-mediated epoxidations are often sensitive to the reaction pH and are usually carried out at pH 7–8 since Oxone autodecomposes readily at high pH [52, 53]. However, ketone **1** was found to rapidly decompose in the presence of Oxone at pH 7–8, presumably via Baeyer–Villiger oxidation to form lactones **8**

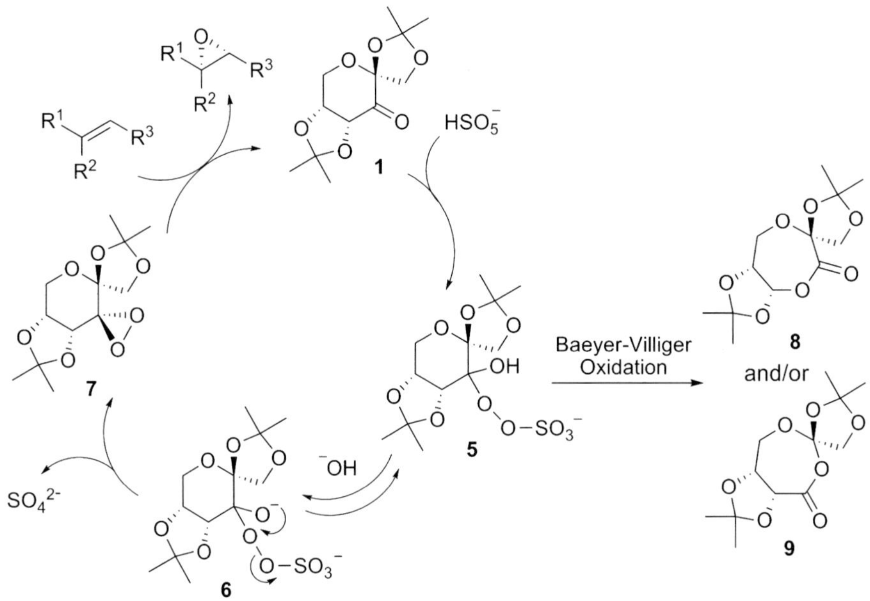

Scheme 15.4 Catalytic cycle of epoxidation mediated by ketone **1**.

and/or **9**. The epoxidation was found to be much more efficient at higher pH, possibly by suppressing the competing Baeyer–Villiger oxidation and favoring the formation of intermediate **6** and dioxirane **7** [54, 55]. Under these reaction conditions, ketone **1** appears to be reactive enough that the autodecomposition of Oxone at high pH is presumably overridden.

Ketone **1** provides high levels of enantioselectivity for the epoxidation of a wide range of 1,2-*trans*- and trisubstituted olefins (Figure 15.1) [46, 47, 56]. Conjugated dienes and enynes [57–59] (Figure 15.2), enol esters [60–63], and 2,2-disubstituted vinyl silanes [64] are also suitable substrates (Figure 15.3). Optically active α-hydroxyl and α-acyloxy ketones can be obtained by stereoselective rearrangement of the chiral enol ester epoxides. The choice of Lewis acid determines whether the rearrangement would lead to retention or inversion in configuration (an example is shown in Scheme 15.5) [61, 62]. Additionally, the chiral epoxides

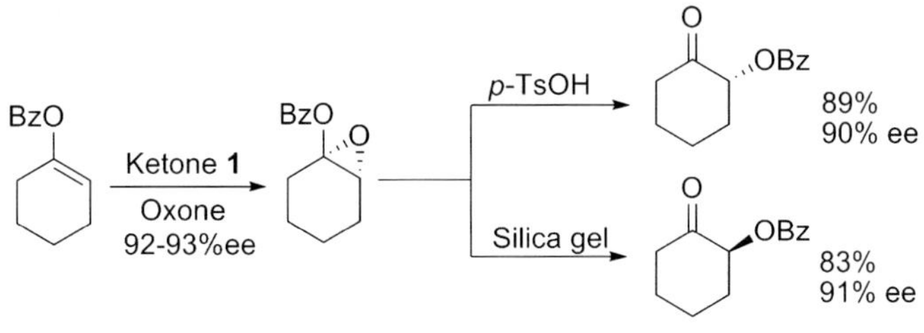

Scheme 15.5 Rearrangement of chiral enol ester epoxides.

Figure 15.1 Examples of epoxidation of *trans*- and trisubstituted olefins with ketone **1**.

Figure 15.2 Examples of epoxidation of conjugated dienes and enynes with ketone **1**.

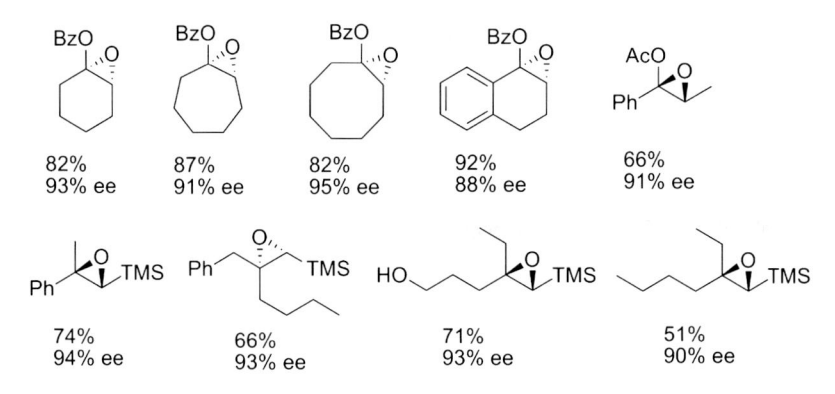

Figure 15.3 Examples of epoxidation of enol esters and 2,2-disubstituted vinyl silanes with ketone **1**.

of 2,2-disubstituted vinyl silanes can undergo desilylation to obtain 1,1-disubstituted terminal epoxides [64].

Two extreme epoxidation transition states—spiro and planar—have been proposed for the epoxidation with dioxiranes (Figure 15.4). Both experimental and computational studies suggest that the spiro transition state is energetically favored due to a stabilizing secondary orbital interaction between the nonbonding orbital of the dioxirane oxygen and the π^* orbital of the olefin [3, 4, 17, 46, 47, 65–72]. Studies have also shown that spiro transition state **A** is favored for the epoxidations of *trans-* and trisubstituted olefins with ketone **1** based on the obtained epoxide configurations (Figure 15.5). The stereochemical outcome of the epoxidation reveals that planar transition state **B** is competing and that the competition is dependent upon the steric and electronic nature of the olefin substituents [46, 47]. Generally speaking, conjugating aromatic rings, alkenes, and alkynes to the reacting double bond enhances the aforementioned secondary orbital interaction, thus favoring spiro **A** over planar **B**, which leads to an increase in the enantioselectivity of the resulting epoxide. A small R^1 group and/or a large R^3 group is also beneficial for the enantioselectivity by sterically favoring spiro **A** and/or disfavoring planar **B**. The transition state model is validated by the results of kinetic resolution of

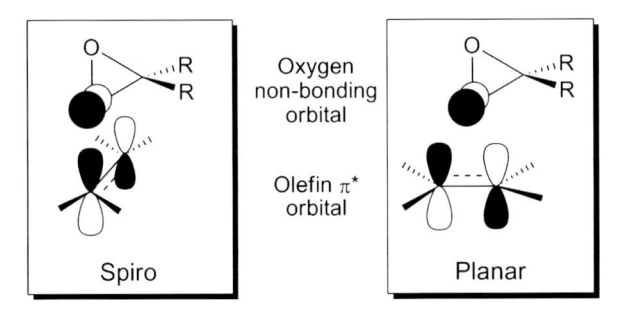

Figure 15.4 Spiro and planar transition states for the dioxirane epoxidation of olefins.

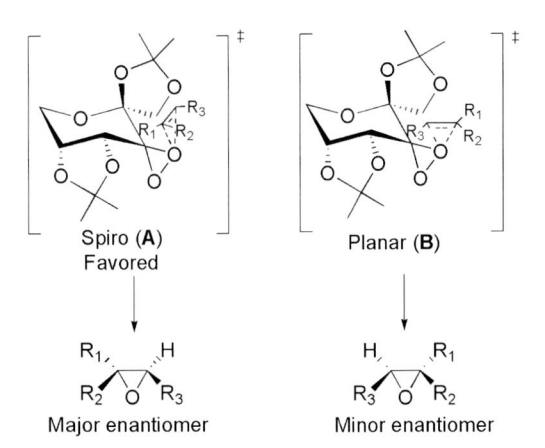

Figure 15.5 Competing spiro and planar transition states for the epoxidation with ketone **1**.

1,6- and 1,3-disubstituted cyclohexanes [73] and desymmetrization of cyclohexane derivatives [74] with ketone **1**.

While Oxone has been commonly used in ketone-mediated epoxidations, hydrogen peroxide can also be employed as oxidant in combination with a nitrile for epoxidation with ketone **1** [75–78]. Peroxyimidic acid **10**, the reaction product of H_2O_2 and nitriles, is likely to be the active oxidant (Scheme 15.6). Mixed organic solvents, such as $MeCN/EtOH/CH_2Cl_2$, are used in these epoxidations to improve conversions for substrates with poor solubilities. Epoxidation results obtained using the $H_2O_2/MeCN$ protocol are comparable to that of using Oxone. In addition, no slow addition of the oxidant is necessary and the amount of salt involved is greatly reduced.

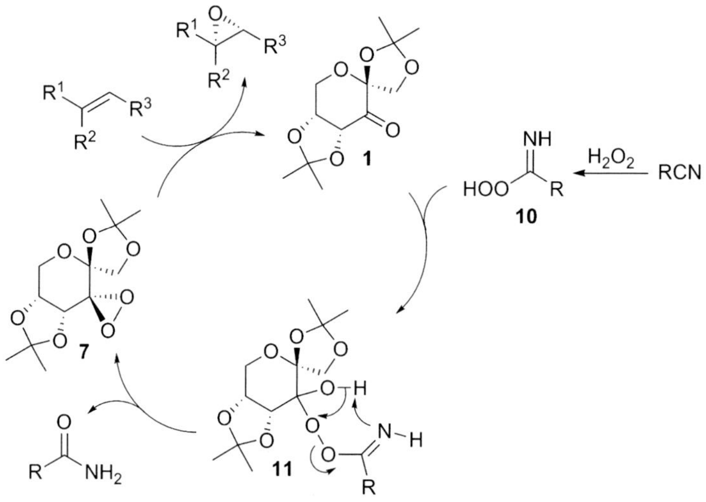

Scheme 15.6 Catalytic cycle of ketone **1**-mediated epoxidation using H_2O_2.

Ketone **1** is effective for a wide variety of *trans*- and trisubstituted olefins and is readily accessible. Consequently, the epoxidation using ketone **1** has been applied to the synthesis of complex chiral molecules by various researchers [15]. For example, recently, Morimoto and coworkers reported that diepoxy-alcohol **13** is obtained in >6:1 dr from the epoxidation of diene **12** in the first asymmetric total synthesis of (+)-enshuol to confirm its structure (Scheme 15.7) [79, 80]. Bis-epoxide **13** cyclized to the desired ring system **14** in the presence of an acid catalyst. Subsequently, diene **15** underwent a selective epoxidation with ketone *ent*-**1** to give monoepoxide **16** in >8:1 dr with the terminal alkene still intact. Monoepoxide **16** was then transformed into (+)-enshuol (**17**) in four steps. In 2007, McDonald and coworkers employed ketone **1** in the synthesis of the enantiomers of marine natural products nakorone and abudinol B via tandem epoxide cyclizations [81]. Bis-epoxide **19** was obtained from the selective epoxidation of triene-yne **18** on the two more electron-rich alkenes (Scheme 15.8). The electron-deficient alkene next

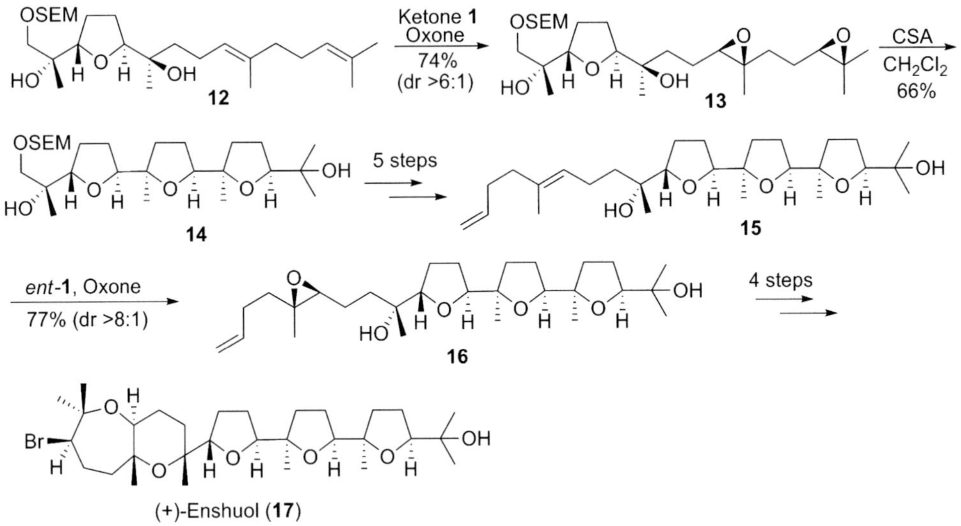

Scheme 15.7 Synthesis of (+)-enshuol (17).

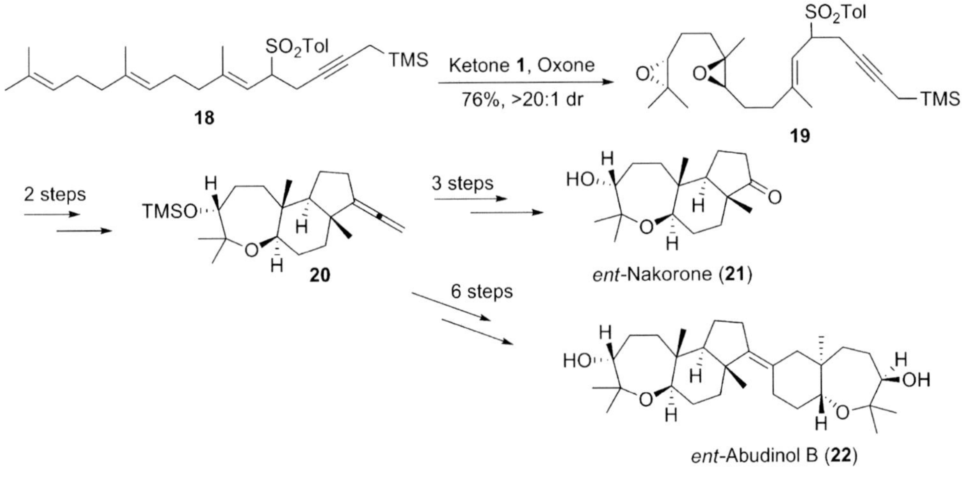

Scheme 15.8 Synthesis of *ent*-nakorone (21) and *ent*-abudinol B (22).

to the electron-withdrawing sulfone group was left intact. Both *ent*-nakorone and *ent*-abudinol B can be obtained from intermediate **20** after a series of transformations. Also in 2007, in their studies on Lewis-acid-mediated tandem oxacyclizations of polyepoxides, McDonald and coworkers reported that polyepoxides such as **25**, obtained from the epoxidation of **24** with ketone **1**, can undergo *endo*-mode oxacyclization under the influence of TBSOTf and 2,6-di-*tert*-butyl-4-methyl pyridine

(DTBMP) to form fused bis(pyran) product **26** (Scheme 15.9). On the other hand, when TMSOTf is used the *exo*-mode of oxacyclization takes place to furnish product **27** [82].

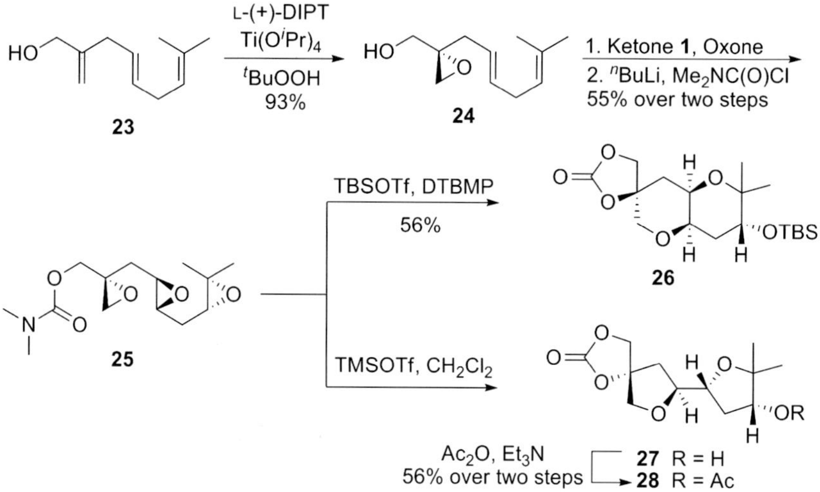

Scheme 15.9 Divergent oxacyclization of tris-epoxide **25**.

In 2000, Corey and coworkers reported an enantioselective total synthesis of glabrescol (**32**) and its structure elucidation (Scheme 15.10) [83, 84]. Tetraene **29** was epoxidized with ketone **1** to give tetraepoxide **30**, which underwent acid-catalyzed cyclization to form tetraol **31**. Glabrescol (**32**) was obtained in two steps from tetraol **31**. In 2007, Jamison and coworkers reported on epoxide-opening cascades promoted by water at 70 °C [85]. Under the reaction conditions, formation of desired tetrahydropyran is favored over the formation of tetrahydrofuran. The

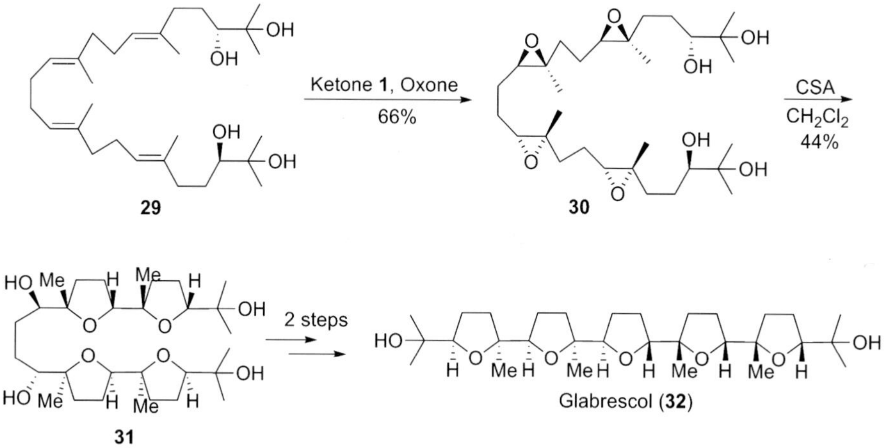

Scheme 15.10 Synthesis of glabrescol (**32**).

polyepoxides that were used in this study could be obtained from the epoxidation of the corresponding polyene with ketone **1** (an example is shown in Scheme 15.11).[1)]

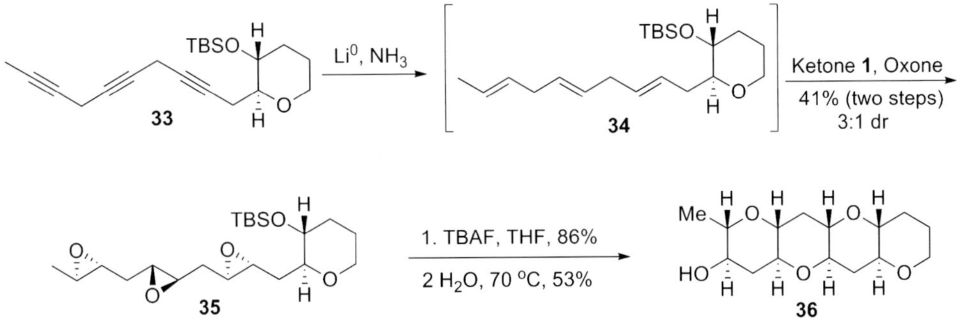

Scheme 15.11 Synthesis of tetrahydropyran ring system **36**.

Replacing the 4,5-acetonide ketal of ketone **1** with more electron-withdrawing oxazolidinone or diacetates led to more robust ketones **37** and **38**, presumably (at least partially) due to the reduction of Baeyer–Villiger decomposition (Figures 15.6 and 15.7) [86, 87]. When 5 mol.% (1 mol.% in some cases) of ketone **37** was used, the epoxidation results obtained were comparable to that of when 20–30 mol.% of ketone **1** was employed (Figure 15.6) [86]. Diacetate ketone **38** was found to be an effective catalyst for the epoxidation of α,β-unsaturated esters, which are electron-deficient substrates and react sluggishly when ketone **1** is used (Figure 15.7) [87].

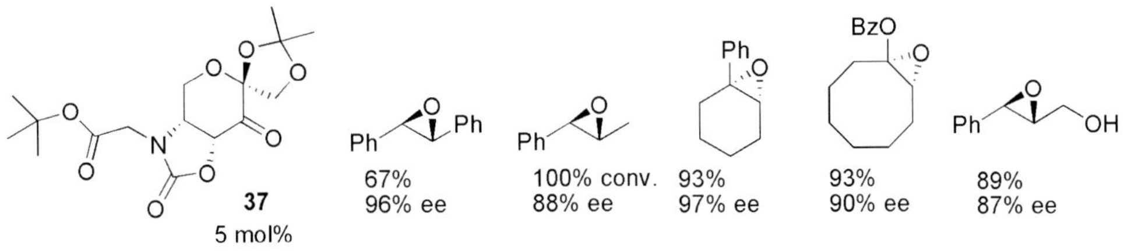

Figure 15.6 Examples of epoxidation of olefins with oxazolidinone ketone **37**.

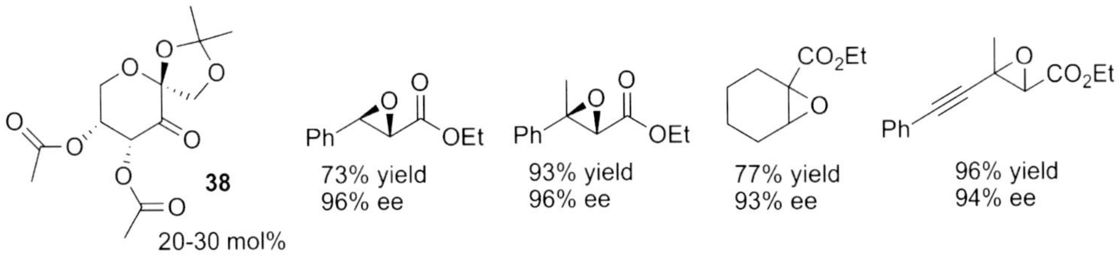

Figure 15.7 Examples of epoxidation of α,β-unsaturated esters with diacetate ketone **38**.

1) For more synthetic applications using ketone **1**, see Reference [15] and references therein.

To further probe the structural requirements for an efficient ketone catalyst, a series of carbohydrate-derived ketones were investigated for asymmetric epoxidation [88]. First, various ketones with a D-fructose scaffold were prepared. Ketones **40**, with different fused ketal and spiro ketal moieties, were synthesized in a manner similar to ketone **1** (Scheme 15.12). Ketones **44**, having different fused ketal groups, were synthesized by selective deketalization of alcohol **41**, reketalization with different ketones, and oxidation of the resulting alcohols (Scheme 15.13). Ketones **47** were prepared by treating D-fructose with chloroethanol and HCl to give tetraol **45**, followed by ring closure, ketalization, and oxidation of the resulting alcohols (Scheme 15.14). Ketones **48–50** were also prepared from tetraol **45** (Scheme 15.14). Ketones **52** were synthesized from D-arabinose by glycosylation with alcohols, ketalization, and oxidation (Scheme 15.15). Ketone **54** was obtained from the oxidation of D-mannose-derived alcohol **53** (Scheme 15.16). To compare furanose to pyranose derived structures, ketone **55a** was prepared from L-sorbose by ketalization and oxidation, and ketone **55b** was obtained from oxidation of diacetone-D-glucose (Scheme 15.17) [89, 90]. Finally, 4-fluorinated ketone **56a** was prepared from ketone **1**, and **56b**, a carbocyclic analog of **1**, was obtained from quinic acid (Figure 15.8) [37].

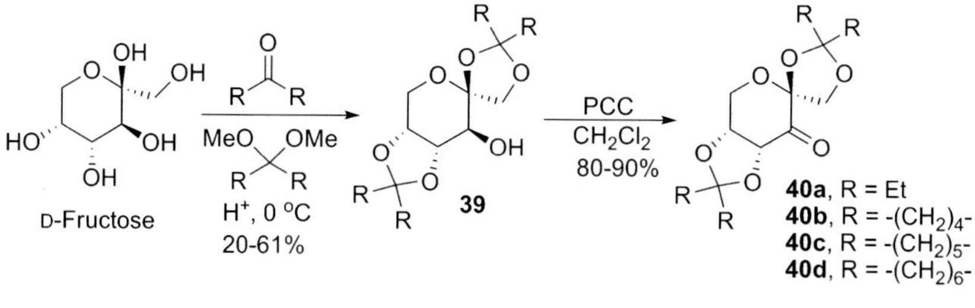

Scheme 15.12 Syntheses of D-fructose-based ketones **40**.

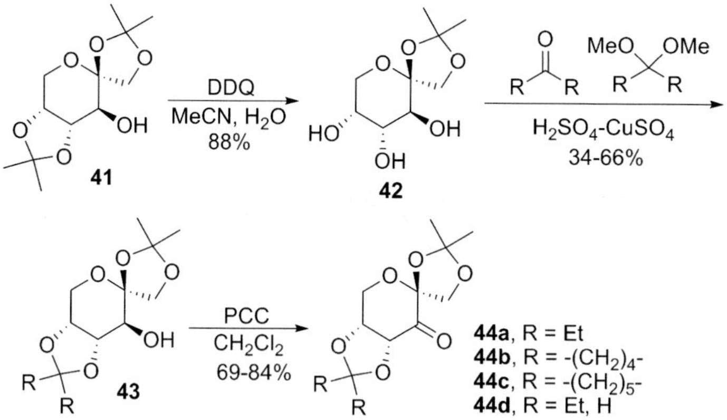

Scheme 15.13 Syntheses of D-fructose-based ketones **44**.

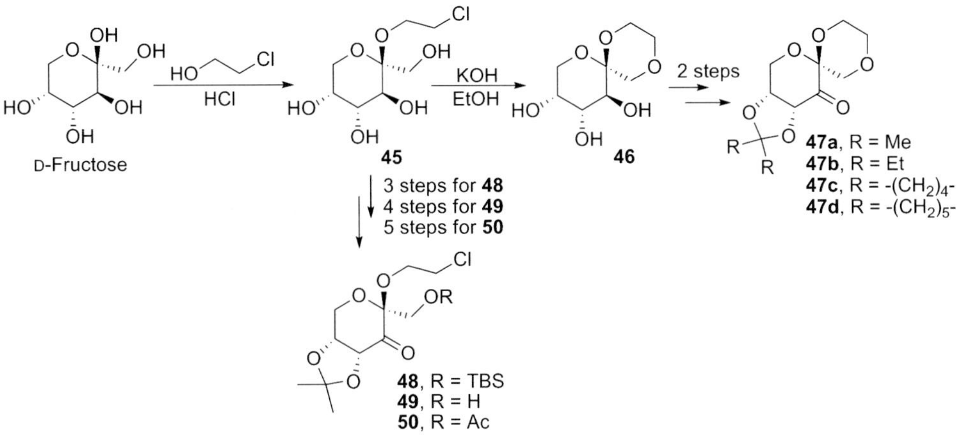

15.1 Oxidations | 331

Scheme 15.14 Syntheses of D-fructose-based ketones **47–50**.

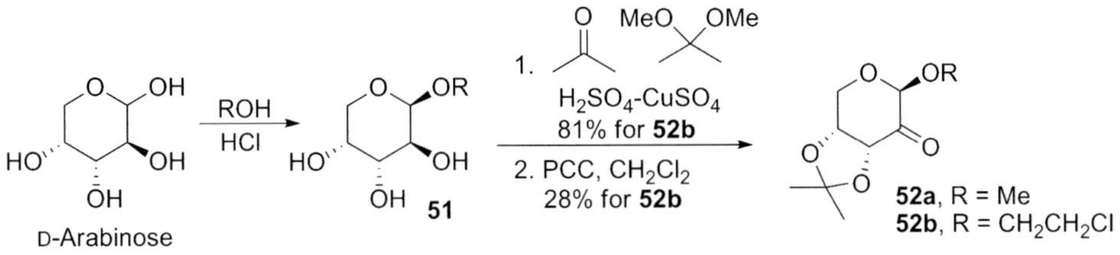

47a, R = Me
47b, R = Et
47c, R = -(CH$_2$)$_4$-
47d, R = -(CH$_2$)$_5$-

48, R = TBS
49, R = H
50, R = Ac

Scheme 15.15 Syntheses of D-arabinose-based ketones **52**.

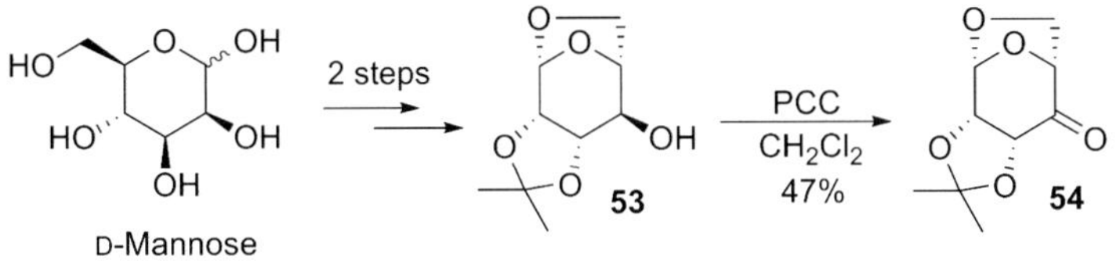

52a, R = Me
52b, R = CH$_2$CH$_2$Cl

Scheme 15.16 Synthesis of D-mannose-based ketone **54**.

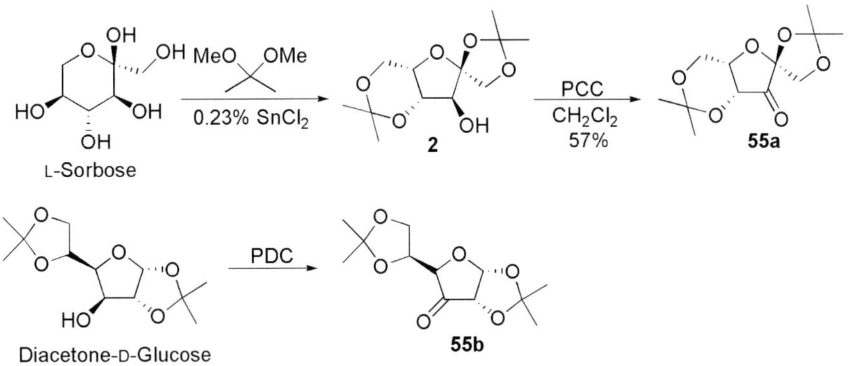

Scheme 15.17 Syntheses of furanose ketones **55**.

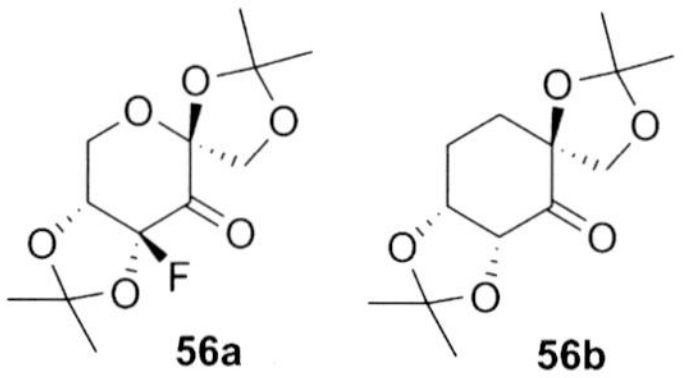

Figure 15.8 Ketones **56**.

Table 15.1 summarizes the selected epoxidation results for the ketones **1, 40, 44, 47–50, 52,** and **54–56**. In terms of both reactivity and enantioselectivity, ketone **1** gave overall the best results while some other ketones might give slightly higher ees in certain cases. Ketones with smaller groups on the spiro ketal moiety generally provide higher reactivity and selectivity (Table 15.1, entry 1 versus 2–5). The size on the fused ketal group may affect the conformation of the ketone catalyst, thus influencing its reactivity (Table 15.1, entry 1 versus 6–9). The five-membered spiro ketal group in ketone **1** is more effective than the six-membered dioxane rings in ketones **47** and the acyclic groups in ketones **48–50** and **52** (Table 15.1, entry 1 versus 10–18). Bicyclic ketone **54** provided low conversions and ees for the olefins tested (Table 15.1, entry 19). Ketones **55**, derived from L-sorbose and D-glucose, are less effective catalysts likely due to their more facile Baeyer–Villiger oxidation (Table 15.1, entries 20 and 21). α-Fluorine-bearing ketone **56a** gave both low reactivity and enantioselectivity for the olefins tested (Table 15.1, entry 22). The carbocyclic analog of ketone **1** (**56b**), in which the pyranose oxygen has been replaced with a less electronegative CH_2 group, causes a reduction in reactivity (Table 15.1, entry 23) [37]. The X-ray structures of ketones **1** and **56b** also indicate that the conformations of the two ketones are somewhat different. The conformation of ketone **56b** may disfavor the major spiro transition state, thus causing a decrease in conversion and enantioselectivity.

With the ketone catalysts discussed above, high enantioselectivity was obtained only for *trans*- and trisubstituted olefins. It is envisioned that changing the spiro ketal structure in ketone **1** to an oxazolidinone may change the interaction between the olefin and the catalyst. Such a structure can be conveniently prepared from glucose via an Amadori rearrangement. In 2000, glucose-derived ketone **60** was reported to be an effective catalyst for *cis*- and some terminal olefins (Figure 15.9) [91–93]. No isomerization was observed in the epoxidation of *cis*-olefin, meaning only *cis*-epoxide was obtained from *cis*-olefin. Ketone **60** can be obtained in six steps on a multigram scale without extensive chromatographic purification (Scheme 15.18) [94]. Alcohol **57**, obtained from Amadori rearrangement of D-glucose and dibenzyl amine, was converted into alcohol **58** by ketalization. Alcohol **58** was then transformed into ketone **59** by hydrogenation to remove the benzyl groups, reaction with phosgene to form the oxazolidinone, and oxidation with pyridinium dichromate (PDC). The Boc group was added in high yield by treatment of ketone **59** with $(Boc)_2O$ and a catalytic amount of DMAP.

Table 15.1 Epoxidation of olefins with ketones **1, 40, 44, 47–50, 52,** and **54–56**.[a)]

Entry	Ketone	Ph-CH=CH-Ph Conversion (ee) (%)	Ph-C(Me)=CH-Ph Conversion (ee) (%)	Ph-CH=CH-Me Conversion (ee) (%)
1	**1**	75 (97)	100 (72)	93 (92)
2	**40a**, R = Et	16 (96)	40 (57)	32 (86)
3	**40b**, R = -(CH$_2$)$_4$-	57 (99)	100 (68)	89 (94)
4	**40c**, R = -(CH$_2$)$_5$-	41 (98)	80 (58)	51 (87)
5	**40d**, R = -(CH$_2$)$_6$-	30 (98)	37 (66)	37 (91)
6	**44a**, R = Et	38 (94)	98 (57)	91 (93)
7	**44b**, R = -(CH$_2$)$_4$-	52 (98)	100 (69)	95 (93)
8	**44c**, R = -(CH$_2$)$_5$-	59 (92)	100 (63)	100 (89)
9	**44d**, R = Et, H	38 (96)	100 (65)	79 (91)
10	**47a**, R = Me	34 (90)	65 (84)	44 (61)
11	**47b**, R = Et	25 (85)	63 (76)	33 (61)
12	**47c**, R = -(CH$_2$)$_4$-	34 (91)	81 (76)	36 (61)
13	**47d**, R = -(CH$_2$)$_5$-	35 (78)	80 (74)	52 (52)
14	**48**, R = TBS	0	8 (17)	10 (40)
15	**49**, R = H	2 (not determined)	6 (68)	8 (65)
16	**50**, R = Ac	6 (96)	11 (67)	5 (66)
17	**52a**, R = Me	10 (88)	17 (30)	15 (59)
18	**52b**, R = CH$_2$CH$_2$Cl	10 (90)	18 (29)	16 (67)
19	**54**	27 (74)	5 (35)	5 (41)

(Continued)

Table 15.1 (Continued)

Entry	Ketone	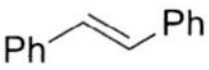 Ph–CH=CH–Ph Conversion (ee) (%)	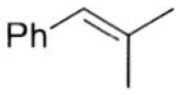 Ph–C(CH3)=CH2 Conversion (ee) (%)	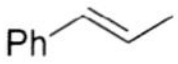 Ph–CH=CH2 Conversion (ee) (%)
20	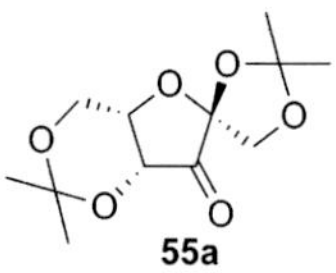 55a	14 (75)	57 (20) (S)	41 (62)
21	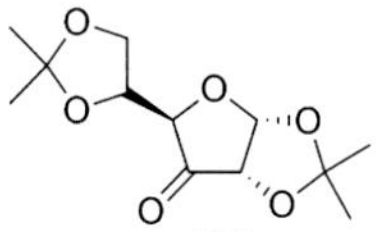 55b	—	—	4 (23) (S,S)
22	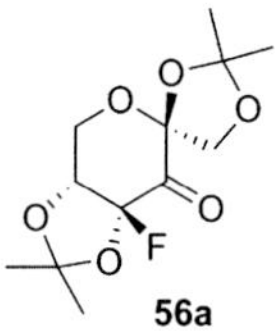 56a	3 (11)	23 (12)	11 (5)
23	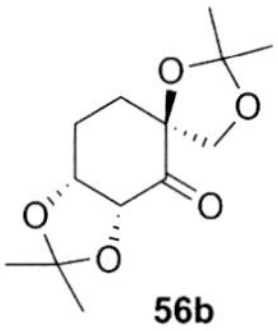 56b	10 (88)	—	61 (87)

a) The epoxides have the (R,R) or (R) configurations unless otherwise noted.

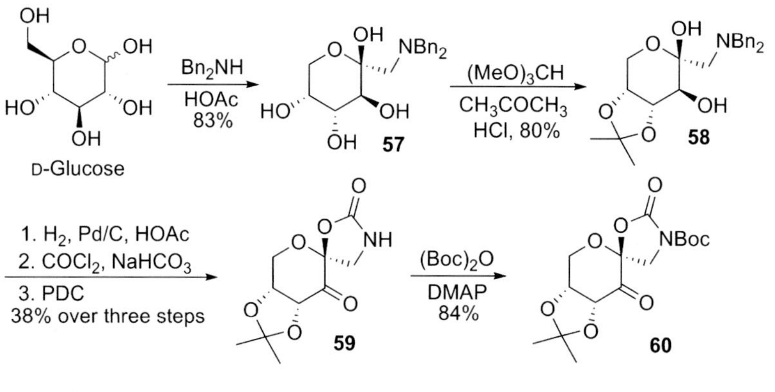

Scheme 15.18 Synthesis of ketone **60**.

15.1 Oxidations

Ph (87%, 91% ee)	(88%, 83% ee)	(61%, 97% ee)	NC- (61%, 91% ee)	
Ph (100% conv., 81% ee)	(86%, 84% ee)	Cl- (74%, 83% ee)	F- (87%, 82% ee)	

Figure 15.9 Examples of epoxidation of *cis*- and terminal olefins with D-glucose-derived ketone **60**.

Spiro (**C**) Favored Spiro (**D**)

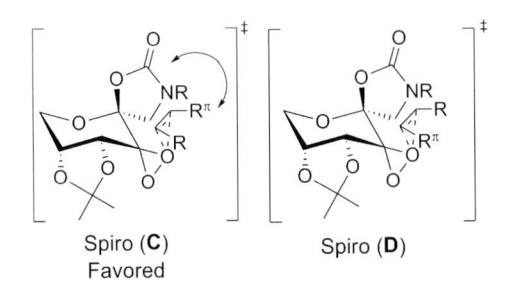

Figure 15.10 Spiro transition states for the epoxidation with ketone **60**.

Determination of the absolute configurations of several epoxide products revealed that the olefin substituents with π-system prefer to be proximal to the oxazolidinone moiety of the catalyst. There seems to be an attractive interaction between the R_π substitution on the olefin and the oxazolidinone of the ketone catalyst (spiro **C** favored over spiro **D**, Figure 15.10). The epoxidation of 1-phenylcyclohexene with ketone **1** and **60** also validate this hypothesis. While the (R,R) enantiomer was obtained in 98% ee with ketone **1** via favored spiro **E** transition state, the opposite absolute configuration was obtained in 39% ee with ketone **60** (Figure 15.11) [93, 95]. This result indicated that planar **H** became the major transition state for the epoxidation with ketone **60**, presumably due to the attractive interaction between the oxazolidinone moiety of the catalyst and the phenyl group of the substrate [93]. Interestingly, the epoxidation of 1-phenylcyclohexene with carbocyclic analog of ketone **60** (**61**) (Figure 15.11) gave the (R,R) epoxide in 40% ee, suggesting that spiro **I** became the major transition state again in this case (Figure 15.11) [95]. X-ray studies showed that ketones **60** and **61** have very similar conformations in the solid state. It is likely that the pyranose oxygen in ketone **60** influences the epoxidation transition state via electronic effects rather than conformational effects. The replacement of the pyranose oxygen with a carbon atom causes an increase in energy of the nonbonding orbital of the dioxirane oxygen,

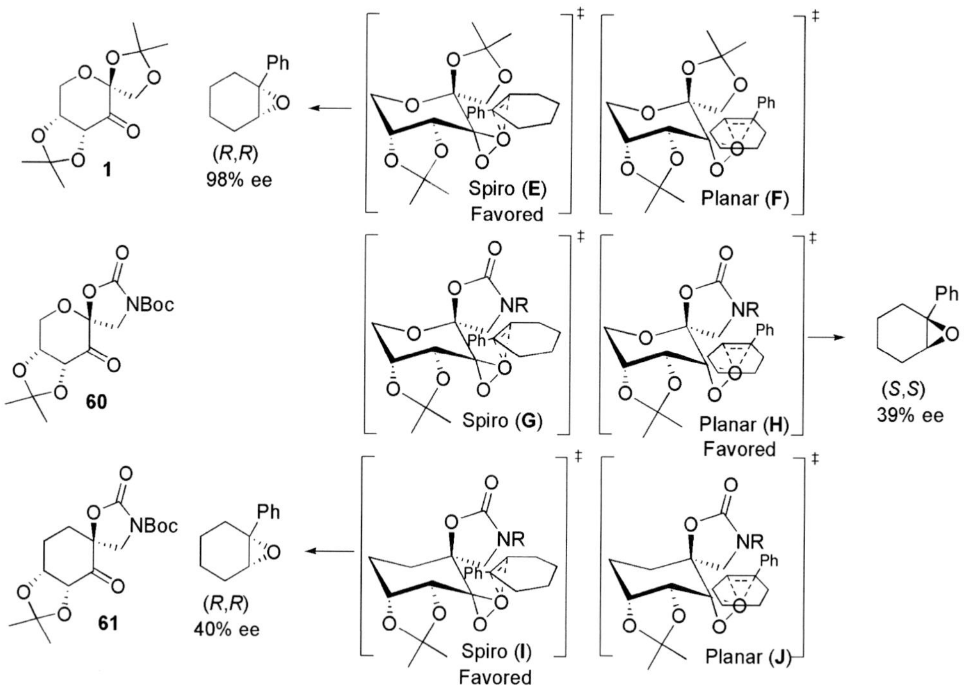

Figure 15.11 Competing transition states for the epoxidation of 1-phenylcyclohexene with ketones **1**, **60**, and **61**.

thus enhancing the secondary orbital interactions between the nonbonding orbital of the dioxirane and the π* orbital of the olefin (Figure 15.4), and consequently favoring spiro **I** over planar **J** (Figure 15.11). The results obtained from ketones **60** and **61** demonstrate that the substituents on the ketones have delicate influences on the catalytic properties of ketones.

Studies have shown that the spiro rings in the aforementioned ketones are important for stereodifferentiation in epoxidation reactions. For a more detailed investigation, a series of ketones (**63–67**) with different spiro ring substitution patterns were subjected to epoxidation studies (Scheme 15.19) [96]. Ketones **63–67** can all be synthesized from D-fructose via the common intermediate **62**. Ketone **63** is structurally similar to ketone **1**. The only difference between these catalysts is the replacement of one of the oxygens in the spiro ketal ring in **1** with a carbon atom. Ketone **64** differs from ketone **63** by the replacement of the methyl groups with two hydrogens. Ketones **65–67** contain spiro lactone moieties in place of the spiro ketal in ketone **63**. Epoxidation results indicated that replacement of one of the oxygens in the spiro ring with a carbon atom leads to no significant changes in enantioselectivity (ketone **1** versus **63**) (Figure 15.1 and Table 15.2, entry 1). However, the replacement of methyl groups with smaller substituents (ketone **63** versus **64**) decreased the enantioselectivity of the epoxidation of *trans*-olefins,

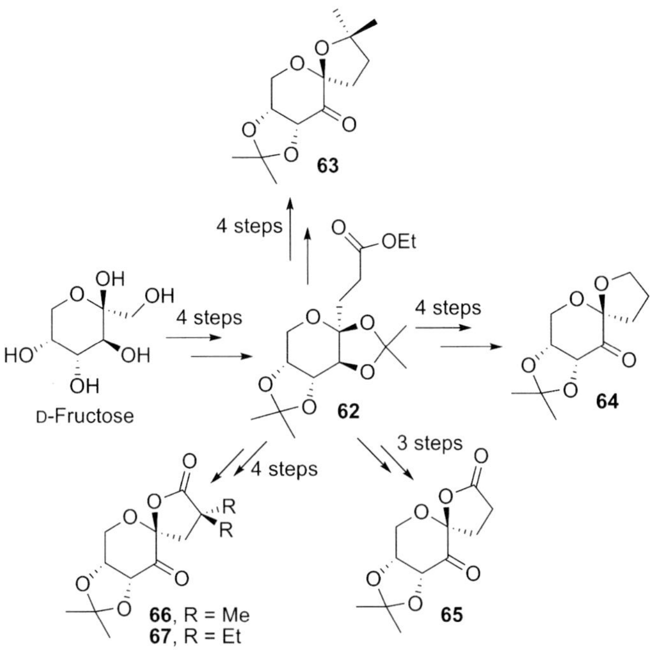

Scheme 15.19 Syntheses of ketones **63–67**.

presumably due to increased competition from disfavored transition states (Table 15.2, entries 3 and 4). Higher enantioselectivity was obtained with lactone-containing ketones (**65–67**) for the epoxidation of *cis*-olefins compared to that of the spiro ketal-containing ketones (**63, 64**) (Table 15.2, entries 5–7), suggesting that the carbonyl group of the lactone-containing ketones (or oxazolidinone-containing for ketone **60**) contributes to the attractive interactions between the catalyst and the R_π group of the olefin. Furthermore, nonbonding interactions such as van der Waals forces and/or hydrophobic interactions between the substrate and the catalyst may also contribute to the stereodifferentiation in the epoxidation.

Ketones **68** were investigated to further understand the effect of N-substituents on the epoxidation using **60** (Figure 15.12) [97, 98]. Among these ketones, **68c** and **68d** can be practically useful since they give high ees for various olefins and can be prepared from inexpensive anilines in large quantities in four steps [99]. As an example, synthesis of **68c** is shown in Scheme 15.20.

Various substituted β-methyl styrenes can be epoxidized in high ee and conversion with ketone **68c** (Figure 15.13) [100]. The substituents on the styrene phenyl ring, from electron-withdrawing to electron-donating groups, increase the enantioselectivity of the resulting epoxide. These results indicate that the substituents on the phenyl group further enhance the attractive interactions in spiro **K** (Figure 15.14). A series of cyclic analog of *cis*-β-methylstyrene (6- and 8-substituted 2,2-dimethylchromenes) was studied to limit the reacting approaches in order to

Table 15.2 Epoxidation of olefins with ketones **63–67**.

Entry	Ketone	Ph⁀/ Conversion (ee) (%)	Ph⁀\ Conversion (ee) (%)
1	**1**	93 (92)	(39)
2	**60**	91[a] (77)	65[a] (94)
3	**63**	76 (96)	87 (12)
4	**64**	91 (76)	100 (45)
5	**65**	66 (73)	55 (61)
6	**66**	76 (83)	89 (70)
7	**67**	100 (80)	100 (68)

a) Isolated yield.

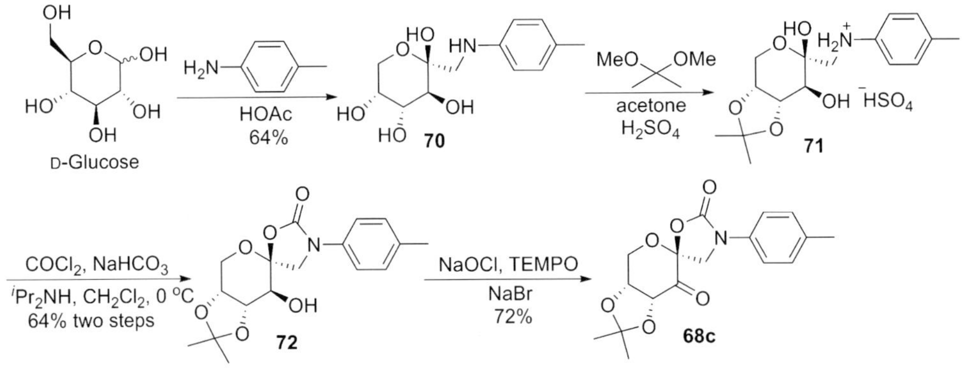
D-Glucose
HOAc
64%
70
MeO OMe
acetone
H_2SO_4
71
$COCl_2$, $NaHCO_3$
$^iPr_2NH$, CH_2Cl_2, 0 °C
64% two steps
72
NaOCl, TEMPO
NaBr
72%
68c

Scheme 15.20 Synthesis of ketone **68c.**

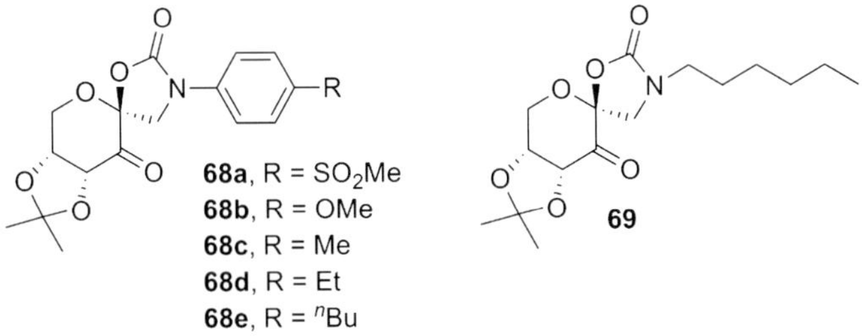
68a, R = SO_2Me
68b, R = OMe
68c, R = Me
68d, R = Et
68e, R = ^nBu
69

Figure 15.12 Glucose-derived ketone **68** and **69.**

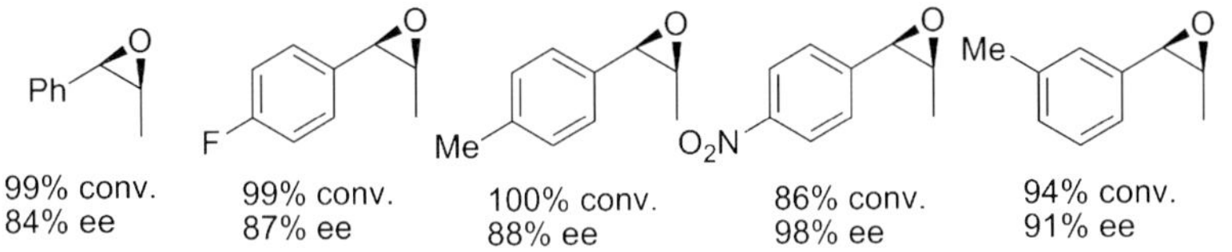
99% conv.
84% ee
99% conv.
87% ee
100% conv.
88% ee
86% conv.
98% ee
94% conv.
91% ee

Figure 15.13 Examples of epoxidation of substituted *cis*-β-methylstyrenes with ketone **68c.**

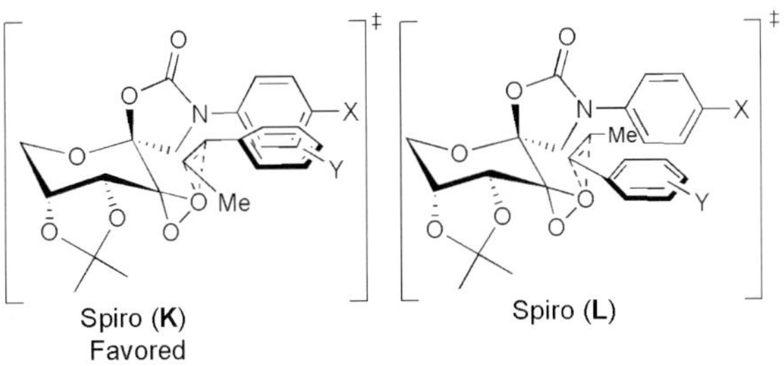
Spiro (**K**)
Favored
Spiro (**L**)

Figure 15.14 Spiro transition states for the epoxidation of *cis*-β-methylstyrenes with ketones **68.**

further examine the interaction between the substituent of the substrate and the catalyst [101]. For the epoxidation of 6-substituted chromenes with ketone **68d**, the ees increased with all the substituents tested (Figure 15.15). However, for the epoxidation of 8-substituted chromenes the ees increase with electron-withdrawing groups and decreased with electron-donating groups. The change in enantioselectivity is likely due to electronic influences of the substituents at the 8-position. In addition to electronic effects, the substituents at the 6-position can provide beneficial attractive interactions between the substrate and the catalyst (spiro **M**, Figure 15.16). Such interaction is not geometrically feasible in the favored transi-

Figure 15.15 Examples of epoxidation of 6- or 8-substituted chromenes with ketone **68d**.

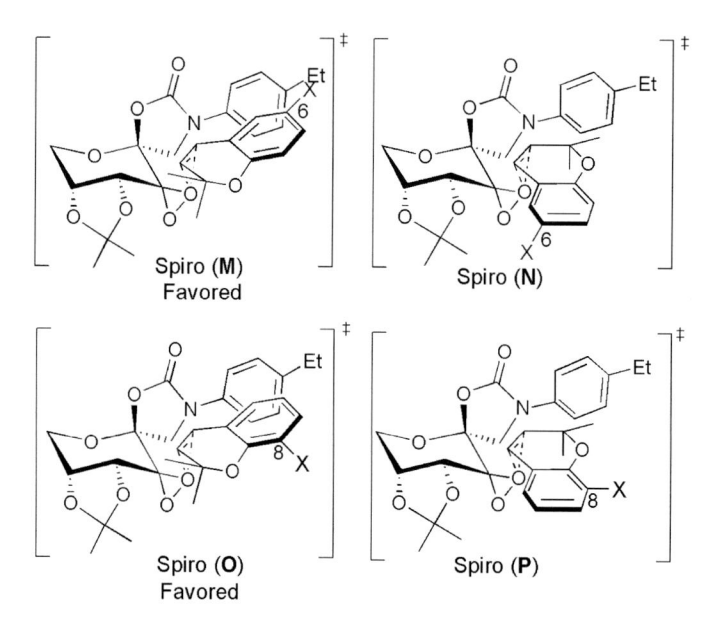

Figure 15.16 Spiro transition states for the epoxidation of 6- and 8-substituted 2,2-dimethylchromenes with ketone **68d**.

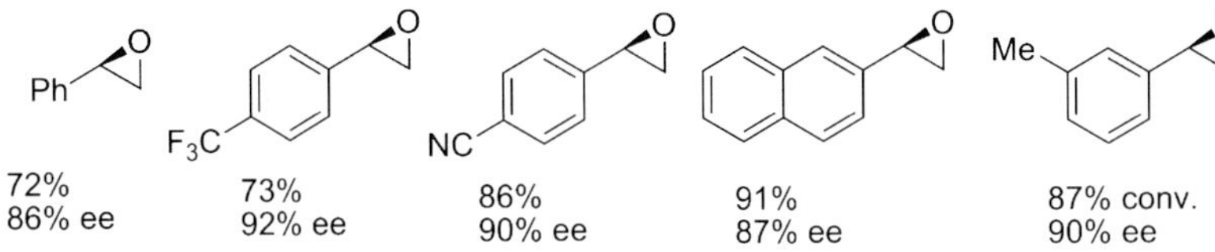

Figure 15.17 Examples of epoxidation of styrenes with ketone **68d**.

tion state for the epoxidation of 8-substituted chromenes (spiro O, Figure 15.16). An N-alkyl substituted ketone (**69**) was also employed in the epoxidation of 6- and 8-substituted 2,2-dimethylchromenes (Figure 15.12). Ketone **69** gave similar epoxidation results to those of ketone **68d**, which suggests that van der Waals forces and/or hydrophobic effects might play important roles in the beneficial interactions between the substrate and the catalyst. Notably, 6-methyl-2H-chromene was also epoxidized in good ee (Figure 15.15). Styrenes with various substituents can also be epoxidized with ketone **68d** in high yields and enantioselectivities (Figure 15.17) [98].

Trisubstituted benzylidenecyclobutanes were epoxidized with ketone **68c** in high yields and ees. The resulting benzylidenecyclobutane oxides were converted into 2-aryl-cyclopentanones using Et$_2$AlCl or LiI with either inversion or net retention of configuration (Scheme 15.21) [102]. When this epoxidation–rearrangement process was extended to tetrasubstituted cyclobutylidenes, optically active 2-alkyl-2-aryl-cyclopentanones can be obtained in good ees, allowing the generation of all-carbon quaternary chiral centers (Figure 15.18) [103]. The epoxidation of benzylidenecyclobutane likely proceeds mainly via spiro transition state **Q** (Figure 15.19). Furthermore, γ-butyrolactones can be obtained by epoxidation of benzylidenecyclopropanes, subsequent epoxide rearrangement, and Baeyer–Villiger oxidation (Scheme 15.22) [104]. By suppressing Baeyer–Villiger rearrangement with more catalyst and less Oxone, chiral cyclobutanones can also be obtained (Figure 15.20).

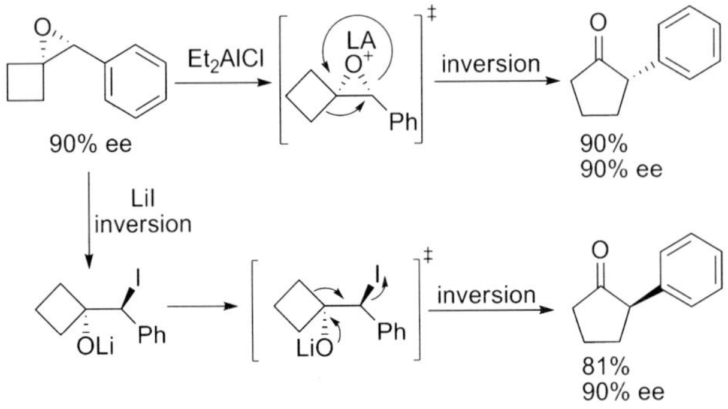

Scheme 15.21 Synthesis of 2-aryl-cyclopentanone.

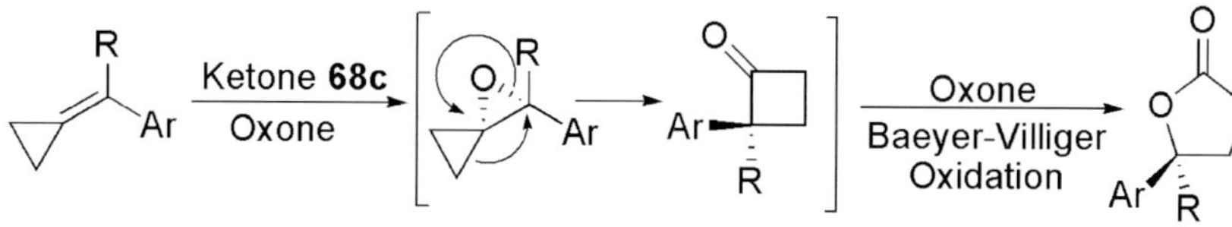
R
Ar
Ketone **68c**
Oxone
O
R
Ar
O
Ar
R
Oxone
Baeyer-Villiger
Oxidation
O
O
Ar
R

Scheme 15.22 Synthesis of γ-aryl-γ-butyrolactone.

Conjugated *cis*-dienes [105] and *cis*-enynes [106] were also found to be effective substrates (Figure 15.21). No isomerization was observed during epoxidation, giving only *cis*-epoxide for *cis*-olefin. Alkene and alkyne appear to be effective directing groups to favor the desired transition state (spiro **S** and **U**, Figure 15.22). The epoxidation with ketones **68** has also been studied using H_2O_2 as oxidant [107].

Epoxidation using carbohydrate-derived ketones was also subsequently investigated by other groups. In 2002, Shing and coworkers reported D-glucose-derived ulose catalyst **73** (Figure 15.23). *trans*-Stilbene oxide could be obtained in up to

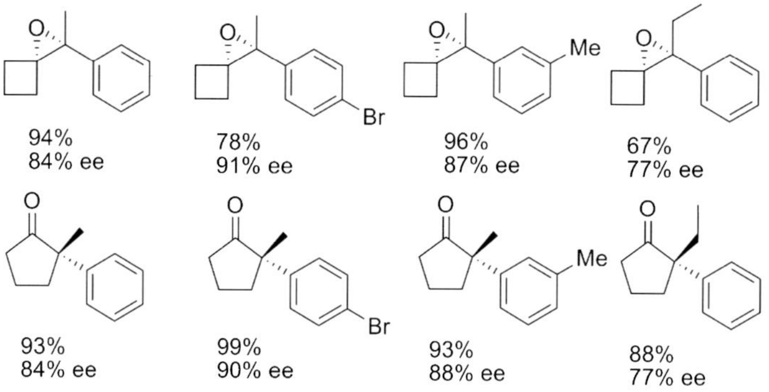
94%
84% ee
78%
91% ee
Br
96%
87% ee
Me
67%
77% ee
93%
84% ee
99%
90% ee
Br
93%
88% ee
Me
88%
77% ee

Figure 15.18 Examples of epoxidation and rearrangement of tetrasubstituted cyclobutylidenes with ketone **68c.**

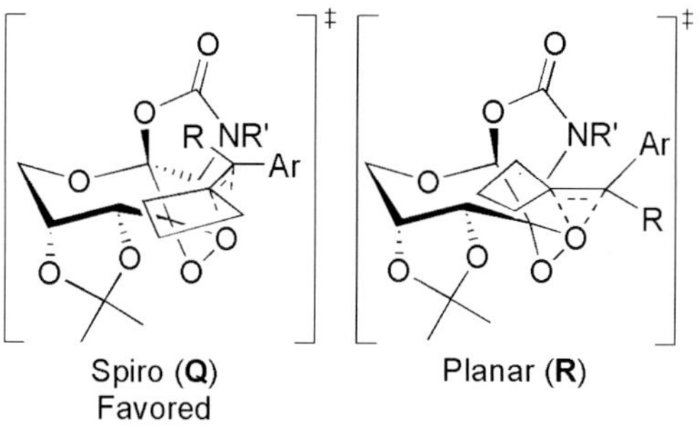
R NR'
Ar
NR'
Ar
R
Spiro (**Q**)
Favored
Planar (**R**)

Figure 15.19 Competing transition states for the epoxidation of benzylidene-cyclobutanes with ketones **68.**

15.1 Oxidations

Figure 15.20 Examples of γ-aryl-γ-butyrolactones and a cyclobutanone synthesized with ketone **68c**.

68%, 90% ee
52%, 91% ee
48%, 91% ee
54%, 86% ee
54%, 87% ee
68%, 90% ee

66% (**68e**), 85% ee
64% (**68e**), 94% ee
74% (**68e**), 94% ee
61% (**68c**), 93% ee

78% (**68d**), 93% ee
84% (**68d**), 90% ee
76% (**68c**), 93% ee
66% (**68d**), 97% ee

Figure 15.21 Examples of epoxidation of conjugated dienes and enynes with ketones **68**.

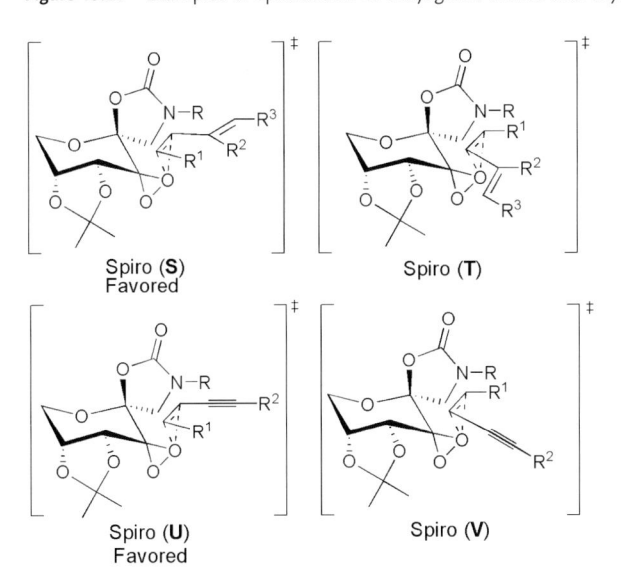

Spiro (**S**) Favored

Spiro (**T**)

Spiro (**U**) Favored

Spiro (**V**)

Figure 15.22 Spiro transition states for the epoxidation of conjugated dienes and enynes with ketones **68**.

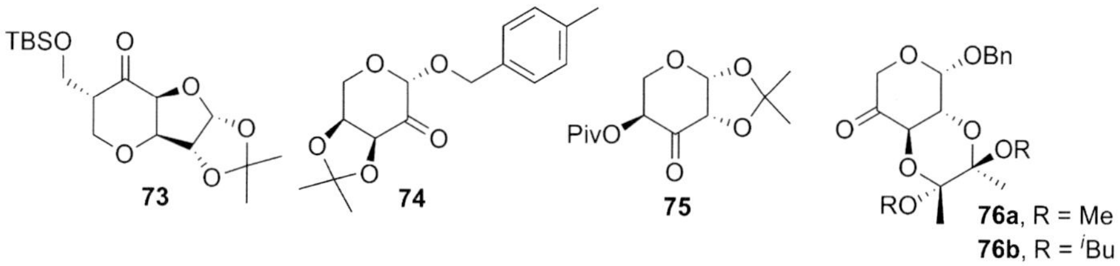

Figure 15.23 Ketones **73–76** by Shing et al.

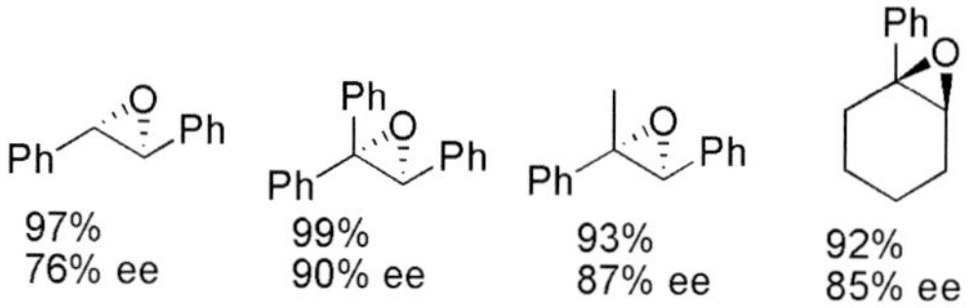

Figure 15.24 Examples of epoxidation with ketone **76b**.

63% yield and 65% ee using this ketone [108]. Later, Shing and coworkers also reported L-arabinose-derived uloses **74** and **75** [109]. Ulose **74** gave 90% ee for the epoxidation of *trans*-stilbene; however, the yield was unsatisfactory due to the decomposition of the catalyst during epoxidation. To overcome this problem, more sterically demanding α-pivalate ulose **75** was studied (Figure 15.23). The yield of the epoxidation was dramatically improved and up to 68% ee was obtained for phenylstilbene oxide. Another series of L-arabinose-derived uloses containing tunable steric sensors (**76**) were also reported by Shing and coworkers (Figure 15.23) [110, 111]. Enantioselectivity increased with increasing R group size. Good yield and enantioselectivities were achieved with various *trans*- and trisubstituted olefins with ketone **76b** (Figure 15.24). Epoxide **78**, synthesized from olefin **77** using ketone **76a** in 68% ee, was transformed into taxol side-chain **79** in five steps (Scheme 15.23) [112].

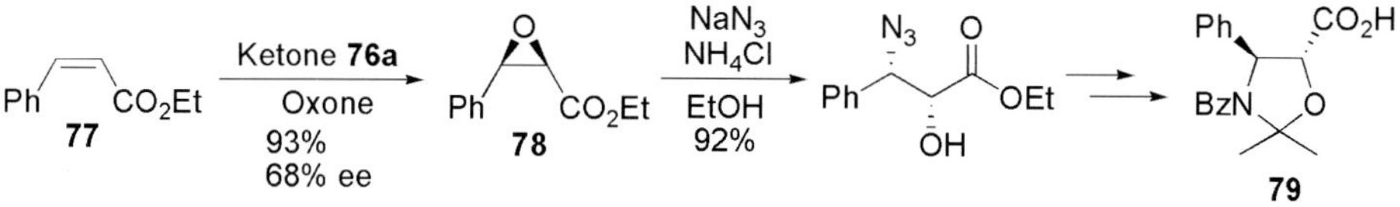

Scheme 15.23 Synthesis of taxol side chain **79**.

In 2003, Zhao and coworkers described compounds **80–82** as epoxidation catalysts for several unfunctionalized olefins (Figure 15.25) [113]. Up to 94% ee was obtained for the epoxidation of *trans*-stilbene with aldehyde **80**. Also in 2003, Wong and coworkers described a β-cyclodextrin-derived ketoester **83** for the epoxidation of several unfunctionalized olefins (Figure 15.26) [114]. Up to 40% ee was obtained for *p*-chlorostyrene with ketoester **83**.

Figure 15.25 Fructose-derived catalysts **80–82** by Zhao et al.

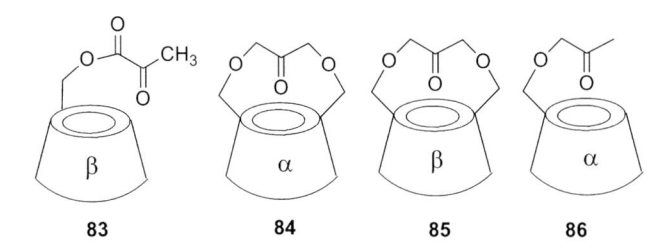

Figure 15.26 α- and β-Cyclodextrin-based ketones **83–86** by Wong and Bols.

In 2004, Bols and coworkers synthesized a series of cyclodextrin derivatives from both α- and β-cyclodextrin and employed them as epoxidation catalysts (**84–86**) (Figure 15.26) [115]. Catalysts **84** and **85** contain a 1,3-dialkoxyacetone bridge. During epoxidation with these catalysts, a considerable amount of diols was formed from the olefins and up to 12% ee could be obtained for styrene oxide with catalyst **84**.

15.1.2
Oxidation of Miscellaneous Substrates

Besides epoxidation, other types of oxidations have also been investigated by carbohydrate-derived catalyst. In 1998, Adam and coworkers described an enantioselective oxidation of *vic*-diols to the corresponding α-hydroxy ketones using fructose-derived ketone **1** [116]. Optically active α-hydroxy ketones can be accessed through desymmetrization of *meso-vic*-diols (Scheme 15.24) and kinetic resolution of racemic *vic*-diols (Schemes 15.25 and 15.26). By using the appropriate diastereomer of *vic*-diol, both enantiomers of the resulting α-hydroxy ketone can be achieved. Electronic effects on the enantioselectivity of these C–H oxidations have

Ar = Ph, 89% conv. 45% ee (**1**)
Ar = *p*-OMePh, 95% conv. 24% ee (**1**)
Ar = *p*-FPh, 95% conv. 58% ee (**1**)
Ar = Ph, 80% yield 87% ee (**60**)
Ar = *p*-FPh, 65% yield 87% ee (**60**)

Scheme 15.24 Examples of desymmetrization of *meso-vic*-diols with ketone **1** or **60**.

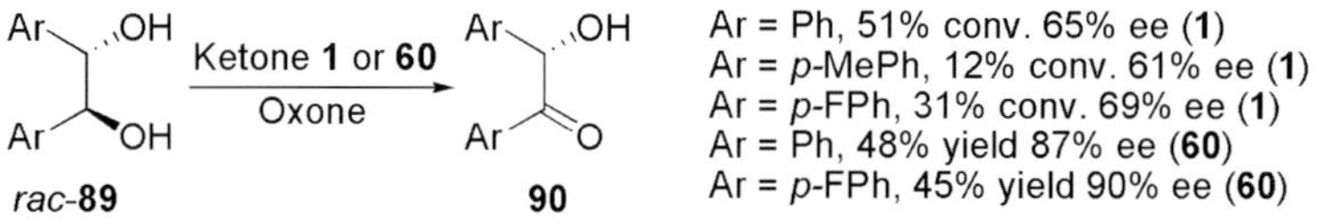

Scheme 15.25 Examples of kinetic resolution of rac-vic-diols with ketone 1 or 60.

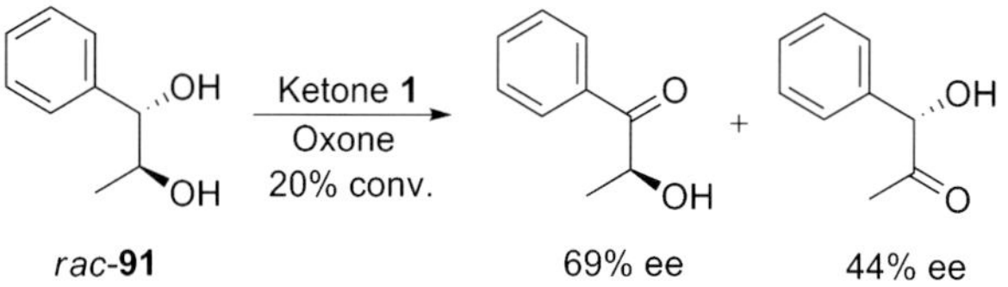

Ar = Ph, 51% conv. 65% ee (1)
Ar = p-MePh, 12% conv. 61% ee (1)
Ar = p-FPh, 31% conv. 69% ee (1)
Ar = Ph, 48% yield 87% ee (60)
Ar = p-FPh, 45% yield 90% ee (60)

Scheme 15.26 Kinetic resolution of unsymmetric vic-diol with ketone 1.

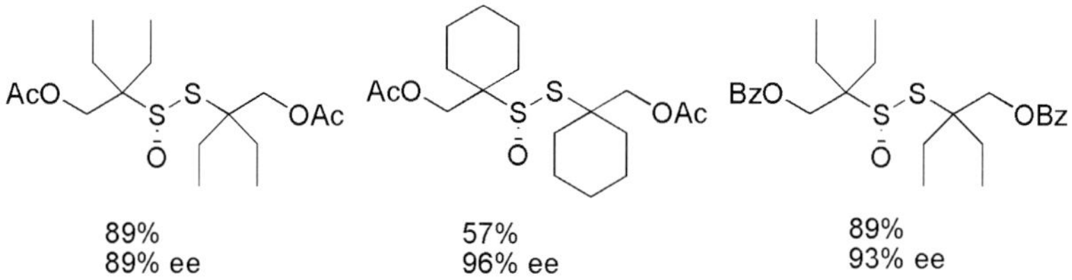

Figure 15.27 Examples of disulfide oxidation with ketone 1.

been observed [117]. Electron-withdrawing substituents on the phenyl ring of the substrate increase the enantioselectivity of the reaction (Schemes 15.24 and 15.25). C–H oxidation of vic-diols was also studied using ketone **60** [118]. Good yields and ees were obtained for various meso- (Figure 15.24) and racemic vic-diols (Figure 15.25).

The dioxirane derived from ketone **1** was also used in the synthesis of chiral sulfoxides [119]. Up to 26% ee was obtained for methyl p-methylphenyl sulfoxide. In 2005, Colonna and coworkers described the oxidation of disulfides with ketone **1** and Oxone [120]. tert-Butyl disulfide was oxidized to the corresponding sulfoxide in 75% ee. More recently, the oxidation of functionalized sterically hindered disulfides with ketone **1** was reported by Khiar and coworkers [121]. Various disulfides can be oxidized in good yields and ees (Figure 15.27).

15.2
Conclusion

In summary, enantiopure carbohydrates are commercially available in large quantities. The hydroxyl groups of carbohydrates allow for versatility in catalyst synthesis and they also provide an accessible avenue to dioxirane precursors. In the past decade, significant progress has been made in the epoxidation of *trans*-, trisubstituted, *cis*-, certain terminal, and tetrasubstituted olefins. For example, fructose-derived ketone **1** has shown a broad substrate scope for *trans* and trisubstituted olefins, many of which remain challenging for metal-based systems. The epoxidation with ketone **1** is practical, general, predictable, and scalable; it has become a widely used method in organic synthesis. Studies also show that the structural requirement for an effective catalyst is very stringent. A proper balance between sterics and electronics appears to be crucial for the development of more effective catalysts. Ketones **1** and **60** were also employed in the study of asymmetric C–H oxidation and sulfur oxidation. Good enantioselectivities have been achieved for the oxidation of disulfides. Carbohydrate-based ketone catalysts have already been demonstrated to be highly useful in oxidations and their synthetic potential warrants further exploration.

Fructose-derived ketone **1** is an early example of a highly effective organic catalyst. The level of practicality and generality exhibited by ketone **1** had not been previously achieved by organic-based catalyst systems [122]. This system exemplified the potential of small organic molecules as practical catalysts for asymmetric reactions. Asymmetric catalysis with small organic molecules, often referred to as organic catalysis [123, 124] or organocatalysis [124], has rapidly emerged as a dynamic field in modern organic chemistry.

References

1 Murray, R.W. (1989) *Chem. Rev.*, **89**, 1187–1201.
2 Adam, W., Curci, R., and Edwards, J.O. (1989) *Acc. Chem. Res.*, **22**, 205–211.
3 Curci, R., Dinoi, A., and Rubino, M.F. (1995) *Pure Appl. Chem.*, **67**, 811–822.
4 Adam, W. and Smerz, A.K. (1996) *Bull. Soc. Chim. Belg.*, **105**, 581–599.
5 Adam, W., Saha-Möller, C.R., and Ganeshpure, P.A. (2001) *Chem. Rev.*, **101**, 3499–3548.
6 Adam, W., Saha-Möller, C.R., and Zhao, C.-G. (2002) *Org. React.*, **61**, 219–516.
7 Curci, R., Fiorentino, M., and Serio, M.R. (1984) *J. Chem. Soc., Chem. Commun.*, 155–156.
8 Denmark, S.E. and Wu, Z. (1999) *Synlett*, 847–859.
9 Frohn, M. and Shi, Y. (2000) *Synthesis*, 1979–2000.
10 Shi, Y. (2002) *J. Synth. Org. Chem. Jpn.*, **60**, 342–349.
11 Shi, Y. (2004) *Modern Oxidation Methods* (ed. J.-E. Bäckvall), Wiley-VCH Verlag GmbH, Weinheim, ch. 3.
12 Shi, Y. (2004) *Acc. Chem. Res.*, **37**, 488–496.
13 Yang, D. (2004) *Acc. Chem. Res.*, **37**, 497–505.
14 Shi, Y. (2006) *Handbook of Chiral Chemicals* (ed. D. Ager), CRC Press, Taylor & Francis Group, Boca Raton, ch. 10.
15 Wong, O.A. and Shi, Y. (2008) *Chem. Rev.*, **108**, 3958–3987.
16 Yang, D., Yip, Y.-C., Tang, M.-W., Wong, M.-K., Zheng, J.-H., and Cheung, K.-K. (1996) *J. Am. Chem. Soc.*, **118**, 491–492.
17 Yang, D., Wang, X.-C., Wong, M.-K., Yip, Y.-C., and Tang, M.-W. (1996) *J. Am. Chem. Soc.*, **118**, 11311–11312.
18 Yang, D., Wong, M.-K., Yip, Y.-C., Wang, X.-C., Tang, M.-W., Zheng, J.-H., and Cheung, K.-K. (1998) *J. Am. Chem. Soc.*, **120**, 5943–5952.
19 Song, C.E., Kim, Y.H., Lee, K.C., Lee, S.-G., and Jin, B.W. (1997) *Tetrahedron: Asymmetry*, **8**, 2921–2926.

20 Matsumoto, K. and Tomioka, K. (2001) *Chem. Pharm. Bull.*, **49**, 1653–1657.
21 Stearman, C.J. and Behar, V. (2002) *Tetrahedron Lett.*, **43**, 1943–1946.
22 Denmark, S.E. and Matsuhashi, H. (2002) *J. Org. Chem.*, **67**, 3479–3486.
23 Adam, W. and Zhao, C.-G. (1997) *Tetrahedron: Asymmetry*, **8**, 3995–3998.
24 Carnell, A.J., Johnstone, R.A.W., Parsy, C.C., and Sanderson, W.R. (1999) *Tetrahedron Lett.*, **40**, 8029–8032.
25 Matsumoto, K. and Tomioka, K. (2001) *Heterocycles*, **54**, 615–617.
26 Matsumoto, K. and Tomioka, K. (2002) *Tetrahedron Lett.*, **43**, 631–633.
27 Denmark, S.E., Forbes, D.C., Hays, D.S., DePue, J.S., and Wilde, R.G. (1995) *J. Org. Chem.*, **60**, 1391–1407.
28 Denmark, S.E., Wu, Z., Crudden, C.M., and Matsuhashi, H. (1997) *J. Org. Chem.*, **62**, 8288–8289.
29 Brown, D.S., Marples, B.A., Smith, P., and Walton, L. (1995) *Tetrahedron*, **51**, 3587–3606.
30 Wang, Z.-X. and Shi, Y. (1997) *J. Org. Chem.*, **62**, 8622–8623.
31 Armstrong, A. and Hayter, B.R. (1997) *Tetrahedron: Asymmetry*, **8**, 1677–1684.
32 Yang, D., Yip, Y.-C., Chen, J., and Cheung, K.-K. (1998) *J. Am. Chem. Soc.*, **120**, 7659–7660.
33 Wang, Z.-X., Miller, S.M., Anderson, O.P., and Shi, Y. (1999) *J. Org. Chem.*, **64**, 6443–6458.
34 Adam, W., Saha-Möller, C.R., and Zhao, C.-G. (1999) *Tetrahedron: Asymmetry*, **10**, 2749–2755.
35 Solladié-Cavallo, A. and Bouérat, L. (2000) *Org. Lett.*, **2**, 3531–3534.
36 Bortolini, O., Fogagnolo, M., Fantin, G., Maietti, S., and Medici, A. (2001) *Tetrahedron: Asymmetry*, **12**, 1113–1115.
37 Wang, Z.-X., Miller, S.M., Anderson, O.P., and Shi, Y. (2001) *J. Org. Chem.*, **66**, 521–530.
38 Solladié-Cavallo, A., Bouérat, L., and Jierry, L. (2001) *Eur. J. Org. Chem.*, 4557–4560.
39 Bortolini, O., Fantin, G., Fogagnolo, M., Forlani, R., Maietti, S., and Pedrini, P. (2002) *J. Org. Chem.*, **67**, 5802–5806.
40 Solladié-Cavallo, A., Jierry, L., Norouzi-Arasi, H., and Tahmassebi, D. (2004) *J. Fluorine Chem.*, **125**, 1371–1377.
41 Bortolini, O., Fantin, G., Fogagnolo, M., and Mari, L. (2006) *Tetrahedron*, **62**, 4482–4490.
42 Armstrong, A. and Hayter, B.R. (1998) *Chem. Commun.*, 621–622.
43 Armstrong, A., Ahmed, G., Dominguez-Fernandez, B., Hayter, B.R., and Wailes, J.S. (2002) *J. Org. Chem.*, **67**, 8610–8617.
44 Armstrong, A., Dominguez-Fernandez, B., and Tsuchiya, T. (2006) *Tetrahedron*, **62**, 6614–6620.
45 Klein, S. and Roberts, S.M. (2002) *J. Chem. Soc., Perkin Trans. 1*, 2686–2691.
46 Tu, Y., Wang, Z.-X., and Shi, Y. (1996) *J. Am. Chem. Soc.*, **118**, 9806–9807.
47 Wang, Z.-X., Tu, Y., Frohn, M., Zhang, J.-R., and Shi, Y. (1997) *J. Am. Chem. Soc.*, **119**, 11224–11235.
48 Mio, S., Kumagawa, Y., and Sugai, S. (1991) *Tetrahedron*, **47**, 2133–2144.
49 Tu, Y., Frohn, M., Wang, Z.-X., and Shi, Y. (2003) *Org. Synth.*, **80**, 1–8.
50 Chen, C.-C. and Whistler, R.L. (1988) *Carbohydr. Res.*, **175**, 265–271.
51 Zhao, M.-X. and Shi, Y. (2006) *J. Org. Chem.*, **71**, 5377–5379.
52 Ball, D.L. and Edwards, J.O. (1956) *J. Am. Chem. Soc.*, **78**, 1125–1129.
53 Montgomery, R.E. (1974) *J. Am. Chem. Soc.*, **96**, 7820–7821.
54 Wang, Z.-X., Tu, Y., Frohn, M., and Shi, Y. (1997) *J. Org. Chem.*, **62**, 2328–2329.
55 Frohn, M., Wang, Z.-X., and Shi, Y. (1998) *J. Org. Chem.*, **63**, 6425–6426.
56 Wang, Z.-X. and Shi, Y. (1998) *J. Org. Chem.*, **63**, 3099–3104.
57 Frohn, M., Dalkiewicz, M., Tu, Y., Wang, Z.-X., and Shi, Y. (1998) *J. Org. Chem.*, **63**, 2948–2953.
58 Cao, G.-A., Wang, Z.-X., Tu, Y., and Shi, Y. (1998) *Tetrahedron Lett.*, **39**, 4425–4428.
59 Wang, Z.-X., Cao, G.-A., and Shi, Y. (1999) *J. Org. Chem.*, **64**, 7646–7650.
60 Zhu, Y., Tu, Y., Yu, H., and Shi, Y. (1998) *Tetrahedron Lett.*, **39**, 7819–7822.
61 Zhu, Y., Manske, K.J., and Shi, Y. (1999) *J. Am. Chem. Soc.*, **121**, 4080–4081.
62 Feng, X., Shu, L., and Shi, Y. (1999) *J. Am. Chem. Soc.*, **121**, 11002–11003.
63 Zhu, Y., Shu, L., Tu, Y., and Shi, Y. (2001) *J. Org. Chem.*, **66**, 1818–1826.

64 Warren, J.D. and Shi, Y. (1999) *J. Org. Chem.*, **64**, 7675–7677.
65 Baumstark, A.L. and McCloskey, C.J. (1987) *Tetrahedron Lett.*, **28**, 3311–3314.
66 Baumstark, A.L. and Vasquez, P.C. (1988) *J. Org. Chem.*, **53**, 3437–3439.
67 Bach, R.D., Andrés, J.L., Owensby, A.L., Schlegel, H.B., and McDouall, J.J.W. (1992) *J. Am. Chem. Soc.*, **114**, 7207–7217.
68 Houk, K.N., Liu, J., DeMello, N.C., and Condroski, K.R. (1997) *J. Am. Chem. Soc.*, **119**, 10147–10152.
69 Jenson, C., Liu, J., Houk, K.N., and Jorgensen, W.L. (1997) *J. Am. Chem. Soc.*, **119**, 12982–12983.
70 Armstrong, A., Washington, I., and Houk, K.N. (2000) *J. Am. Chem. Soc.*, **122**, 6297–6298.
71 Deubel, D.V. (2001) *J. Org. Chem.*, **66**, 3790–3796.
72 Singleton, D.A. and Wang, Z. (2005) *J. Am. Chem. Soc.*, **127**, 6679–6685.
73 Frohn, M., Zhou, X., Zhang, J.-R., Tang, Y., and Shi, Y. (1999) *J. Am. Chem. Soc.*, **121**, 7718–7719.
74 Lorenz, J.C., Frohn, M., Zhou, X., Zhang, J.-R., Tang, Y., Burke, C., and Shi, Y. (2005) *J. Org. Chem.*, **70**, 2904–2911.
75 Shu, L. and Shi, Y. (1999) *Tetrahedron Lett.*, **40**, 8721–8724.
76 Shu, L. and Shi, Y. (2000) *J. Org. Chem.*, **65**, 8807–8810.
77 Shu, L. and Shi, Y. (2001) *Tetrahedron*, **57**, 5213–5218.
78 Wang, Z.-X., Shu, L., Frohn, M., Tu, Y., and Shi, Y. (2003) *Org. Synth.*, **80**, 9–17.
79 Morimoto, Y., Yata, H., and Nishikawa, Y. (2007) *Angew. Chem. Int. Ed.*, **46**, 6481–6484.
80 Morimoto, Y. (2008) *Org. Biomol. Chem.*, **6**, 1709–1719.
81 Tong, R., Valentine, J.C., McDonald, F.E., Cao, R., Fang, X., and Hardcastle, K.I. (2007) *J. Am. Chem. Soc.*, **129**, 1050–1051.
82 Tong, R., McDonald, F.E., Fang, X., and Hardcastle, K.I. (2007) *Synthesis*, 2337–2342.
83 Xiong, Z. and Corey, E.J. (2000) *J. Am. Chem. Soc.*, **122**, 4831–4832.
84 Xiong, Z. and Corey, E.J. (2000) *J. Am. Chem. Soc.*, **122**, 9328–9329.
85 Vilotijevic, I. and Jamison, T.F. (2007) *Science*, **317**, 1189–1192.
86 Tian, H., She, X., and Shi, Y. (2001) *Org. Lett.*, **3**, 715–718.
87 Wu, X.-Y., She, X., and Shi, Y. (2002) *J. Am. Chem. Soc.*, **124**, 8792–8793.
88 Tu, Y., Wang, Z.-X., Frohn, M., He, M., Yu, H., Tang, Y., and Shi, Y. (1998) *J. Org. Chem.*, **63**, 8475–8485.
89 Andersson, F. and Samuelsson, B. (1984) *Carbohydr. Res.*, **129**, C1–C3.
90 Mazur, A., Tropp, B.E., and Engel, R. (1984) *Tetrahedron*, **40**, 3949–3956.
91 Tian, H., She, X., Shu, L., Yu, H., and Shi, Y. (2000) *J. Am. Chem. Soc.*, **122**, 11551–11552.
92 Tian, H., She, X., Xu, J., and Shi, Y. (2001) *Org. Lett.*, **3**, 1929–1931.
93 Tian, H., She, X., Yu, H., Shu, L., and Shi, Y. (2002) *J. Org. Chem.*, **67**, 2435–2446.
94 Shu, L., Shen, Y.-M., Burke, C., Goeddel, D., and Shi, Y. (2003) *J. Org. Chem.*, **68**, 4963–4965.
95 Hickey, M., Goeddel, D., Crane, Z., and Shi, Y. (2004) *Proc. Natl. Acad. Sci. USA*, **101**, 5794–5798.
96 Crane, Z., Goeddel, D., Gan, Y., and Shi, Y. (2005) *Tetrahedron*, **61**, 6409–6417.
97 Shu, L., Wang, P., Gan, Y., and Shi, Y. (2003) *Org. Lett.*, **5**, 293–296.
98 Goeddel, D., Shu, L., Yuan, Y., Wong, O.A., Wang, B., and Shi, Y. (2006) *J. Org. Chem.*, **71**, 1715–1717.
99 Zhao, M.-X., Goeddel, D., Li, K., and Shi, Y. (2006) *Tetrahedron*, **62**, 8064–8068.
100 Shu, L. and Shi, Y. (2004) *Tetrahedron Lett.*, **45**, 8115–8117.
101 Wong, O.A. and Shi, Y. (2006) *J. Org. Chem.*, **71**, 3973–3976.
102 Shen, Y.-M., Wang, B., and Shi, Y. (2006) *Angew. Chem. Int. Ed.*, **45**, 1429–1432.
103 Shen, Y.-M., Wang, B., and Shi, Y. (2006) *Tetrahedron Lett.*, **47**, 5455–5458.
104 Wang, B., Shen, Y.-M., and Shi, Y. (2006) *J. Org. Chem.*, **71**, 9519–9521.
105 Burke, C.P. and Shi, Y. (2006) *Angew. Chem. Int. Ed.*, **45**, 4475–4478.
106 Burke, C.P. and Shi, Y. (2007) *J. Org. Chem.*, **72**, 4093–4097.

107 Burke, C.P., Shu, L., and Shi, Y. (2007) *J. Org. Chem.*, **72**, 6320–6323.

108 Shing, T.K.M. and Leung, G.Y.C. (2002) *Tetrahedron*, **58**, 7545–7552.

109 Shing, T.K.M., Leung, Y.C., and Yeung, K.W. (2003) *Tetrahedron*, **59**, 2159–2168.

110 Shing, T.K.M., Leung, G.Y.C., and Yeung, K.W. (2003) *Tetrahedron Lett.*, **44**, 9225–9228.

111 Shing, T.K.M., Leung, G.Y.C., and Luk, T. (2005) *J. Org. Chem.*, **70**, 7279–7289.

112 Shing, T.K.M., Luk, T., and Lee, C.M. (2006) *Tetrahedron*, **62**, 6621–6629.

113 Bez, G. and Zhao, C.-G. (2003) *Tetrahedron Lett.*, **44**, 7403–7406.

114 Chan, W.-K., Yu, W.-Y., Che, C.-M., and Wong, M.-K. (2003) *J. Org. Chem.*, **68**, 6576–6582.

115 Rousseau, C., Christensen, B., Petersen, T.E., and Bols, M. (2004) *Org. Biomol. Chem.*, **2**, 3476–3482.

116 Adam, W., Saha-Möller, C.R., and Zhao, C.-G. (1998) *Tetrahedron: Asymmetry*, **9**, 4117–4122.

117 Adam, W., Saha-Möller, C.R., and Zhao, C.-G. (1999) *J. Org. Chem.*, **64**, 7492–7497.

118 Jakka, K. and Zhao, C.-G. (2006) *Org. Lett.*, **8**, 3013–3015.

119 Dieva, S.A., Eliseenkova, R.M., Efremov, Y.Y., Sharafutdinova, D.R., and Bredikhin, A.A. (2006) *Russ. J. Org. Chem.*, **42**, 12–16.

120 Colonna, S., Pironti, V., Drabowicz, J., Brebion, F., Fensterbank, L., and Malacria, M. (2005) *Eur. J. Org. Chem.*, 1727–1730.

121 Khiar, N., Mallouk, S., Valdivia, V., Bougrin, K., Soufiaoui, M., and Fernández, I. (2007) *Org. Lett.*, **9**, 1255–1258.

122 Dalko, P.I. and Moisan, L. (2001) *Angew. Chem. Int. Ed.*, **40**, 3726–3748.

123 Kagan, H.B. (1999) *Comprehensive Asymmetric Catalysis* (eds E.N. Jacobsen, A. Pfaltz, and H. Yamamoto), Springer-Verlag, Berlin Heidelberg, ch. 2.

124 Ahrendt, K.A., Borths, C.J., and MacMillan, D.W.C. (2000) *J. Am. Chem. Soc.*, **122**, 4243–4244.

16
Enantioselective Addition Reactions Catalyzed by Carbohydrate-Derived Organocatalysts

Jun-An Ma and Guang-Wu Zhang

16.1
Introduction

Enantioselective organocatalysis is regarded as a complementary protocol to the long-established metal-complex and enzyme-catalyzed procedures. A century ago, Bredig and Fiske reported the use of natural alkaloids to catalyze the synthesis of cyanohydrins [1]. About 60 years later, two industrial groups independently reported the now well-known intramolecular aldol reaction catalyzed by L-proline [2]. Even though after this pioneering work there were several reports on small organic molecules promoting synthetic transformations without any assistance by metal compounds [3], for many years organocatalysis received very little attention. Only since the beginning of the new century are we witnessing a remarkable renewal of interest in enantioselective organocatalysis [4].

In this upcoming area of chemistry, a lot of synthetic tools are based on well-known compounds of the *chiral pool*, such as amino acids, alkaloids, and tartaric acid. Carbohydrates, which are inexpensive and readily available natural materials, have only recently also been employed as chiral backbones of organocatalysts [5]. For example, in the 1990s Shi and coworkers successfully applied the fructose-derived ketones as chiral organocatalysts in asymmetric epoxidation reactions of olefins [6] (cf. also Chapter 15); and Roberts' group described carbohydrate-derived thiols as chiral organocatalysts for the enantioselective radical additions (see below). In the following sections, the focus will be on several enantioselective addition reactions to C=N, C=C, and C=O bonds, mediated by carbohydrate-derived organocatalysts.

16.2
Strecker, Mannich, and Nitro-Mannich Reactions

1,2-Additions of carbon nucleophiles to C=N bonds such as Strecker [7] and Mannich [8] reactions are very important and fundamental C–C bond-forming reactions in organic synthesis. The adducts can be readily be converted into valuable nitrogen-containing building blocks, such as α- or β-amino acids, 1,2-diamines, and 2-amino alcohols. Therefore, these transformations have received much attention over the past decades, and considerable effort has been directed towards the development of enantioselective versions, catalyzed by transition-metal complexes [9]. In recent years enantioselective, organocatalytic Strecker and Mannich reactions have become a new focus of interest for organic chemists [10].

16.2.1
Strecker Reactions

The Strecker reaction constitutes one of the easiest and most efficient methods for the preparation of α-amino acids. Over past decades, many efficient chiral catalysts have been applied in asymmetric Strecker reactions for the production of optically active α-amino acids [10a–d]. Jacobsen and coworkers found that the combination of chiral 1,2-*trans*-diaminocyclohexane with urea or thiourea functions as hydrogen bridge-forming groups affords excellent organocatalysis to promote the hydrocyanation of imines in high enantiomeric excess [3e, 4b, 11]. Based on this concept, Kunz and coworkers developed a series of new interesting organocatalysts by replacing the chiral 1,2-*trans*-diaminocyclohexane moiety with readily available D-glucosamine **1** [12]. In principle, there are four possible ways of attaching a Schiff base and a urea-derived side chain to the carbohydrate backbone (Scheme 16.1: structures **A–D**). The synthesis of organocatalyst **2** is shown in Scheme 16.1. Acetylated β-D-glucosamine was prepared from D-glucosamine **1** in three steps. The allyloxycarbonyl (Aloc) group was introduced to furnish allyl carbamate, which was treated with trimethylsilyl azide to give glycosyl azide. Formation of the urea linkage with the L-*tert*-leucine benzylamide was achieved by a Staudinger–*aza*-Wittig-type reaction to give the *N*-glycosylurea derivative. Subsequently, deprotection in the presence of a palladium(0)-complex, and condensation with a salicylaldehyde furnished catalyst **2**. Other imine–urea catalysts were prepared in a similar way. When these derivatives were employed in enantioselective Strecker reactions, it was found that only the β-configured compound **2** with the urea moiety linked to the anomeric center and the Schiff base at the 2-amino function exhibited efficient reactivity and enantioselectivity. Other combinations or modified imine–urea catalysts displayed low enantioselectivities (<36% ee). For many aromatic aldimines formed from substituted benzaldehydes and allyl or benzyl amine, good to high enantiomeric excesses were observed in the presence of **2**; only the *p*-nitro aldimine gave low enantioselectivity due to racemization of the resulting *p*-nitrophenyl-glycinonitrile. On the other hand, aliphatic aldimines, ketimines, and the reactive furfuryl derivatives only gave moderate enantioselectivities (Table 16.1).

Shortly afterwards, the Kunz group introduced another type of imine as organocatalyst for the enantioselective Strecker reaction. These were prepared from a tetra-*O*-pivalated D-galactosyl-amine and planar-chiral [2, 2]paracyclophane derivatives (Scheme 16.2) [13]. When employed in Strecker reactions, *N*-galactosyl

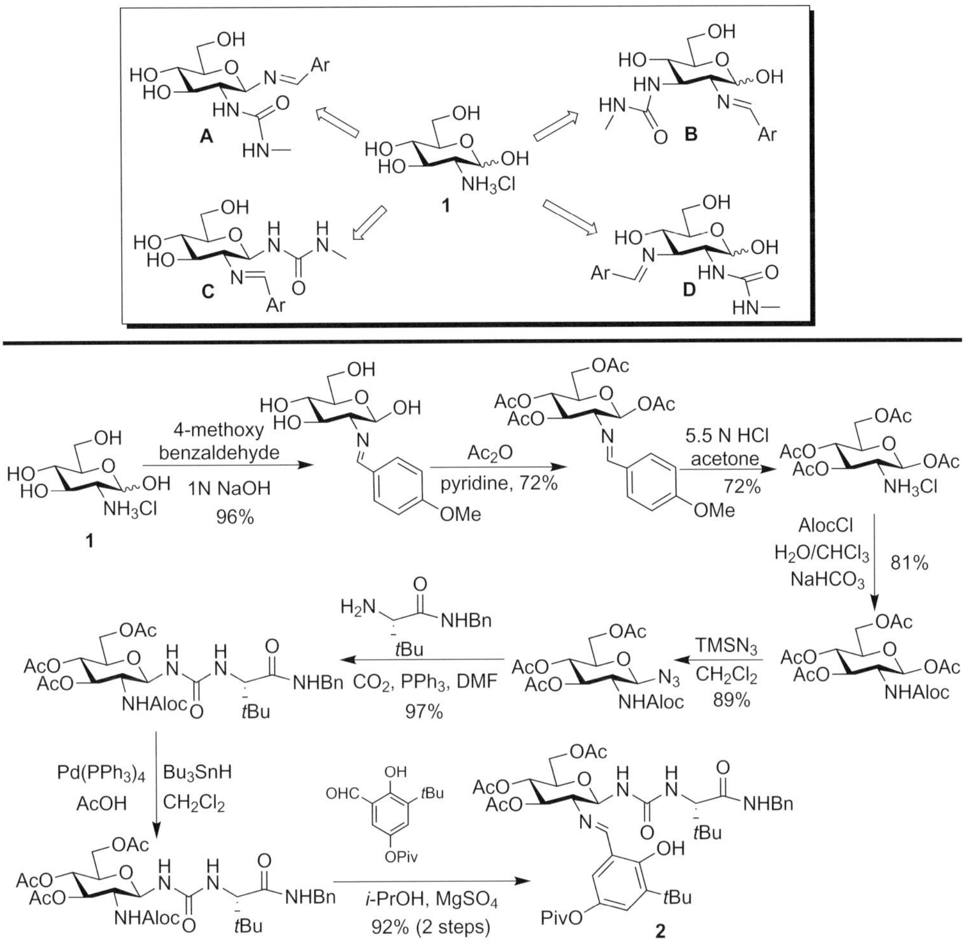

Scheme 16.1 Four possible Jacobsen-type organocatalysts and synthesis of **2** from glucosamine **1**.

aldimine **4**, bearing a methoxycarbonyl group on the paracyclophane moiety, proved to be more efficient than compound **3** without the additional ester group. Optimal conditions were achieved when organocatalyst **4**, TMSCN, and methanol (for *in situ* generation of HCN) were dissolved in toluene beforehand, and the substrates were added at −50 °C. Imines of aliphatic and aromatic aldehydes react in high yields and with excellent enantioselectivities to give the *(S)*-amino nitriles (Table 16.2).

The effect of compounds **3** and **4** is very surprising as they contain neither a hydrogen-bond-donor structure nor a Brønsted acid group. Their catalytic activities should arise from the Lewis basic center cooperatively generated by the imine nitrogen and the carbonyl oxygen of the 2-pivaloyl group. This basic center could

Table 16.1 Organocatalyzed enantioselective Strecker reaction of various imines.

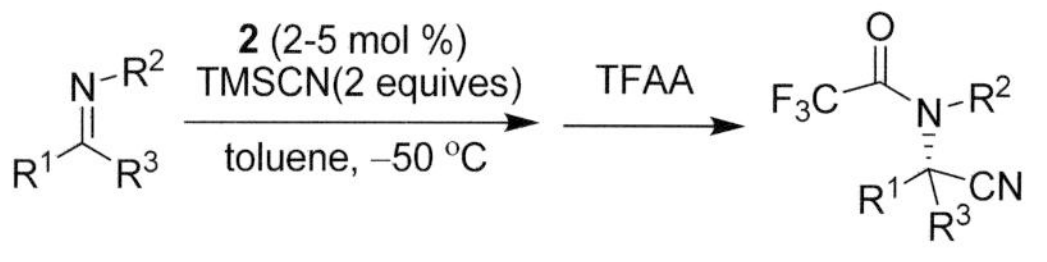

R¹	R²	R³	Yield (%)	ee (%)
2-MeC$_6$H$_4$	Allyl	H	81	69
3-MeC$_6$H$_4$	Allyl	H	93	78
4-MeC$_6$H$_4$	Allyl	H	83	75
2-MeOC$_6$H$_4$	Allyl	H	98	64
3-MeOC$_6$H$_4$	Allyl	H	96	66
4-MeOC$_6$H$_4$	Allyl	H	79	82
2-BrC$_6$H$_4$	Allyl	H	69	72
3-BrC$_6$H$_4$	Allyl	H	83	74
4-BrC$_6$H$_4$	Allyl	H	77	84
4-t-BuC$_6$H$_4$	Allyl	H	82	84
2-NO$_2$C$_6$H$_4$	Allyl	H	67	0
2-Furfuryl	Allyl	H	87	50
1-Naphthyl	Allyl	H	78	80
2-Naphthyl	Allyl	H	73	76
Cyclohexyl	Benzyl	H	60	47
6,7-Dimethoxyisoquinoline		H	97	86
Ph	4-BrC$_6$H$_4$CH$_2$	Me	63	50
Ph	allyl	H	72	84

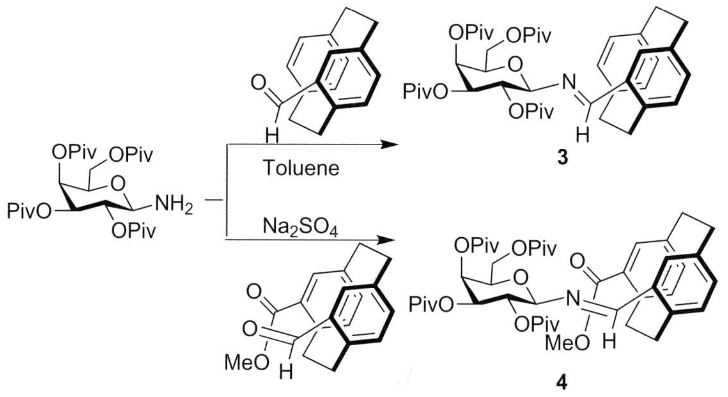

Scheme 16.2 Synthesis of imino organocatalysts from D-galactosyl-amine and paracyclophane derivatives.

Table 16.2 N-Galactosyl aldimine (4)-catalyzed enantioselective Strecker reactions with aldimines.

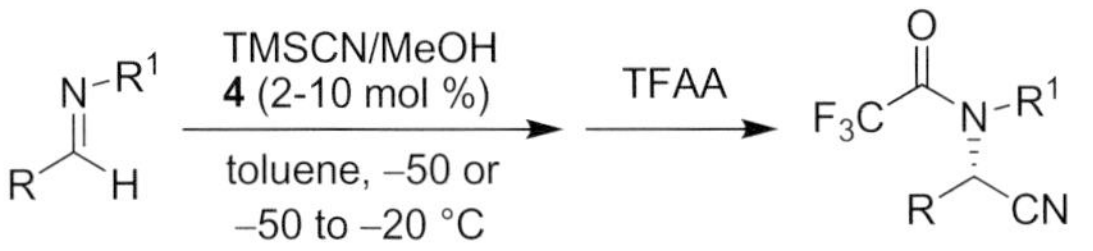

R	R¹	Yield (%)	ee (%)
Ph	Allyl	55	71
4-MeC$_6$H$_4$	Allyl	68	67[a]
i-Pr	Allyl	89	96
c-Hex	Benzyl	87	88
Isoamyl	Benzyl	84	99
4-MeOC$_6$H$_4$	Benzyl	87	82

a) Catalyst 3 was used.

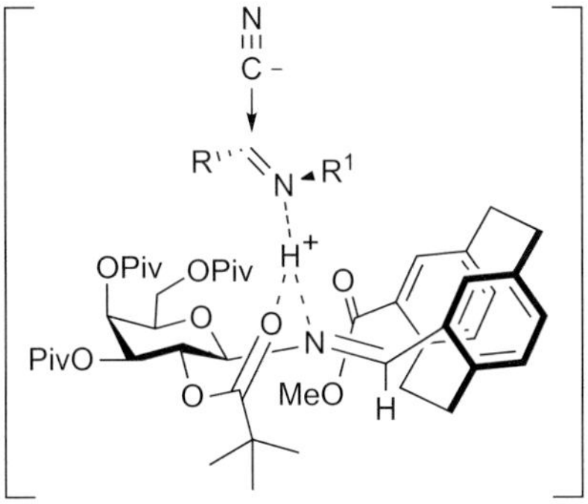

Figure 16.1 Proposed transition state of the Strecker reaction catalyzed by galactosyl imine **4**.

trap the proton from the weak acid HCN to form the protonated catalysts, which then accounts for the activation of the imine substrate and is also responsible for the observed stereoselectivity by positioning the substrate in a favorable geometry for the attack of the nucleophile. An S-shaped transition state model has been postulated to explain the selective formation of the (S)-amino nitriles (Figure 16.1).

16.2.2
Mannich Reactions

Like α-amino acids generated by the Strecker reaction, β-amino acid derivatives, which are accessible via Mannich reactions, are important building blocks in synthetic chemistry [14]. Kunz's group employed D-glucosamine-derived urea/salen organocatalyst **2** in the enantioselective Mannich reaction of N-Boc-aldimine

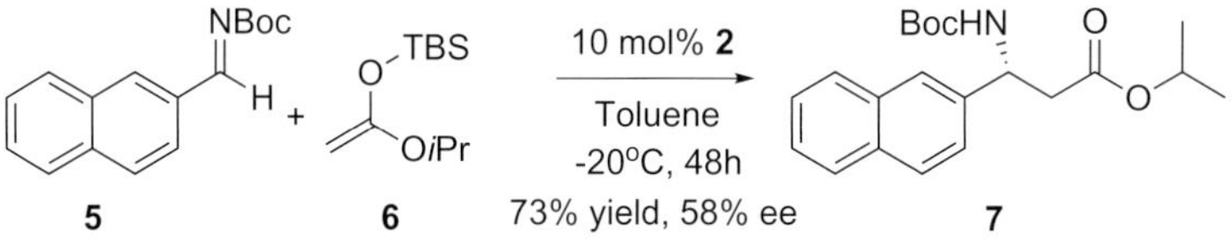

Scheme 16.3 Mannich reaction catalyzed by urea/salen carbohydrate organocatalyst **2**.

5 with silyl ketene acetal 6 (Scheme 16.3) [12]. The corresponding β-amino acid ester 7 was obtained in good yield, albeit with moderate enantioselectivity.

16.2.3
Nitro-Mannich Reactions

Aspiring to imitate enzymatic synergistic cooperation, chemists have succeeded in developing many kinds of bifunctional catalysts for asymmetric synthesis [4e, 10c, 15, 16]. One site of the bifunctional catalyst could involve a Lewis acid moiety or a hydrogen-bond donor that coordinates electrophilic substrates while another site activates nucleophilic reagents. The catalyst positions the two reaction partners in close proximity and with the correct relative geometry, thus facilitating the reaction in a synergistic manner. The first (thio)urea-based bifunctional catalyst was reported Takemoto and coworkers. They found that the tertiary amine and thiourea were indispensable for the reaction efficiencies and selectivities [17a]. Subsequently, the design of bifunctional (thio)urea catalysts has received considerable attention [17b–f].

In 2006, Ma's group developed a new class of chiral bifunctional amine–thiourea catalysts based on the monosaccharide D-glucose, and the disaccharides maltose and lactose [18]. Starting from commercially available α-D-glucopyranose, glycosyl isothiocyanate 8 was readily prepared via acetylation, bromination, and nucleophilic substitution [19]. Consequently, addition of $(1S,2S)$-N^1,N^1-dimethylcyclohexane-1,2-diamine 9 to isothiocyanate afforded the desired bifunctional thiourea catalyst 10a, with the glycosyl scaffold and the tertiary amine group, in 55% yield (Scheme 16.4). Using the same procedure, other thiourea catalysts 10b–h were also synthesized in moderate to good yields.

The catalytic asymmetric nitro-Mannich reaction is also a useful process for the synthesis of versatile β-nitroamine derivatives [20, 21]. Ma's group employed the tertiary amino–thioureas 10a–h as bifunctional organocatalysts in asymmetric nitro-Mannich reaction. Most of these thiourea compounds were proven to be effective organocatalysts for the enantioselective nitro-Mannich reaction of benzalimine with nitromethane (Table 16.3) [22]. Notably, both the carbohydrate structure and the configuration of the 1,2-diaminocyclohexane moiety influence the enantioselectivity of this reaction. It was found that the combination of a β-D-glucopyranose and a *(R,R)*-configured 1,2-diaminocyclohexane unit lead to a matched combination of the two chiral subunits, giving a high ee. In addition, the stereoselectivity was found to be partially dependent on the alkyl substituents of

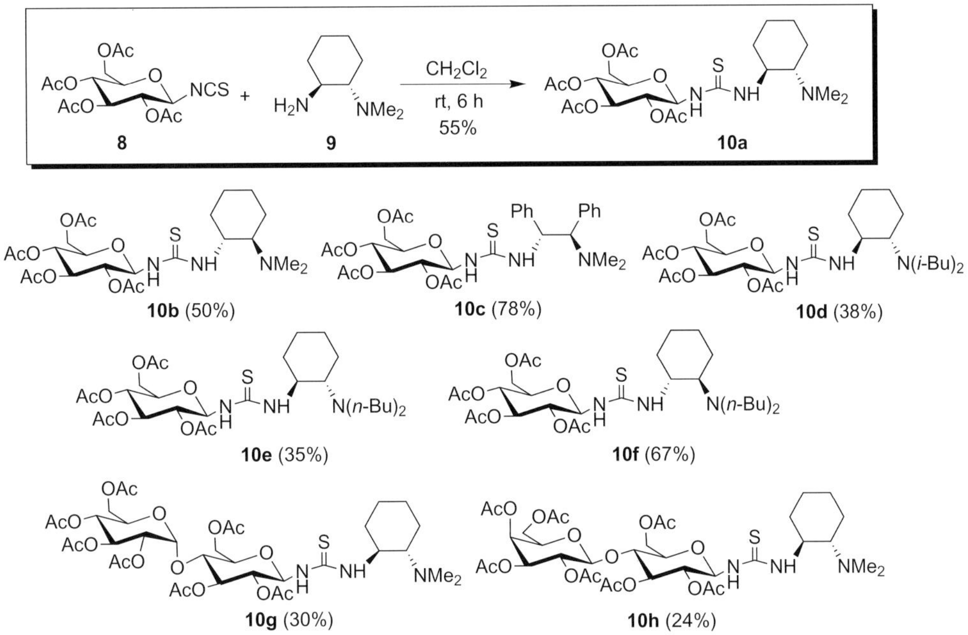

Scheme 16.4 Synthesis of carbohydrate-derived bifunctional tertiary amino–thioureas.

Table 16.3 Catalytic activity of carbohydrate-derived bifunctional tertiary amino–thioureas.

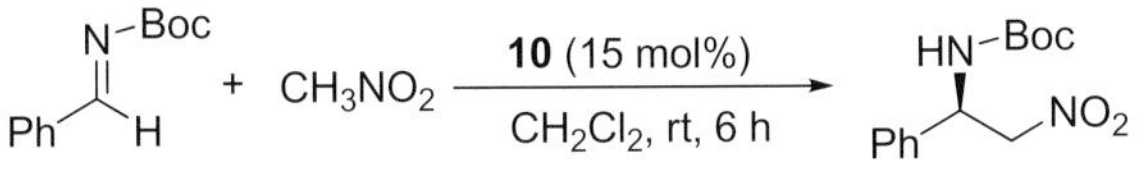

Catalyst	Yield (%)	ee (%)
10a	80	84
10b	82	91
10c	78	89
10d	<5	0
10e	78	89
10f	91	98
10g	76	88
10h	80	85

the tertiary amine moiety. N,N-Di-n-butyl substituted organocatalyst **10f** gave the best result (98% ee) at room temperature. At the same time, Zhou and coworkers reported their results of the nitro-Mannich reaction [23]. In the presence of the catalyst **10b**, a series of aromatic imines, bearing electron-rich, electron-neutral, or electron-deficient groups, reacted with nitromethane to afford the corresponding products in high yields and enantioselectivities (Table 16.4).

Table 16.4 Bifunctional amine–thiourea **10b**-catalyzed nitro-Mannich reaction.

$$\text{Ar-CH=N-Boc} + CH_3NO_2 \xrightarrow[\text{CH}_2\text{Cl}_2,\ -78\ °C]{15\ \text{mol\%}\ \mathbf{10b}} \text{Ar-CH(NHBoc)-CH}_2\text{NO}_2$$

Ar	Yield (%)	ee (%)
Ph	86	>99
4-MeOC$_6$H$_4$	94	94
4-MeC$_6$H$_4$	93	83
4-ClC$_6$H$_4$	93	>99
3-FC$_6$H$_4$	87	>99
4-FC$_6$H$_4$	91	>99
2-ClC$_6$H$_4$	85	92
1-Naphthyl	95	>99
2-Furyl	86	90
4-F$_3$CC$_6$H$_4$	85	94
2-F$_3$CC$_6$H$_4$	84	96

16.3
Michael Additions

The conjugate addition of nucleophiles to electron-deficient olefins is an important tool for the construction of highly functionalized carbon skeletons. Among the variants of these Michael additions, the direct catalytic addition of carbonyl compounds to α-β unsaturated nitroolefins is one of the most facile routes to produce useful building blocks in an atom-economical manner [24]. In the last decade, several small organic molecules have been developed as efficient organocatalysts for asymmetric conjugate additions of aldehydes and ketones as well as more reactive nucleophiles, such as malonic esters, ketoesters, 1,3-diketones, and 1,3-dinitriles to nitroolefins [25]. Very recently, carbohydrate derivatives have also been elaborated into successful organocatalysts for several Michael addition reactions.

16.3.1
Bifunctional Primary Amino–Thiourea Catalysts

In 2006, Ma's group described enantioselective conjugate addition of aromatic methyl ketones to β-*trans*-nitrostyrenes by using chiral bifunctional primary amine–thioureas [26]. These organocatalysts were readily prepared from commercially available carbohydrates and 1,2-cyclohexanediamine (Scheme 16.5). In the presence of optimized organocatalyst **11b**, good to excellent enantioselectivities were achieved for various aromatic nitroolefins and substituted acetophenones. High stereoselectivity was also observed with alkyl-substituted nitroolefins, albeit in low yield (Table 16.5).

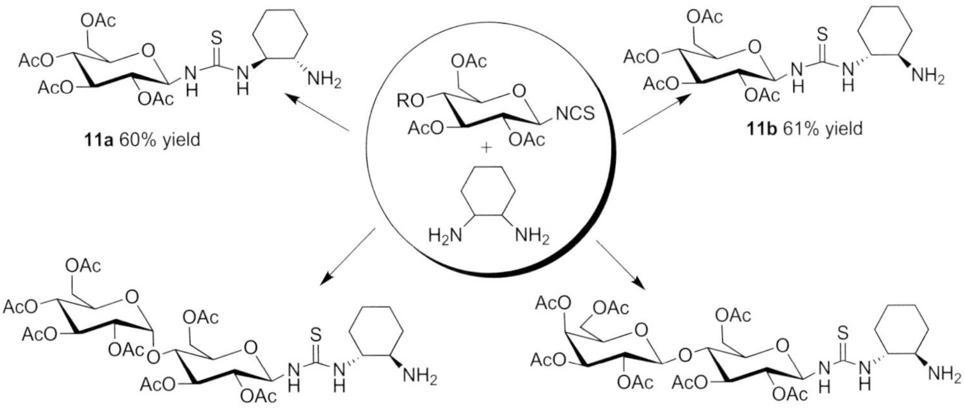

Scheme 16.5 Synthesis of saccharide-derived bifunctional primary amine–thioureas.

Table 16.5 Enantioselective addition of aromatic ketones to nitroolefins.

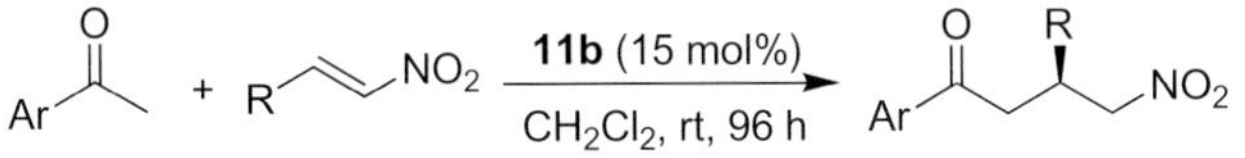

$$\text{Ar}\overset{O}{\underset{}{\diagup\hspace{-0.5em}\diagdown}} + R\diagdown\hspace{-0.5em}\diagup NO_2 \xrightarrow[\text{CH}_2\text{Cl}_2,\text{ rt, 96 h}]{\text{11b (15 mol\%)}} \text{Ar}\overset{O}{\underset{}{\diagup\hspace{-0.5em}\diagdown}}\underset{R}{\diagdown\hspace{-0.5em}\diagup}NO_2$$

Ar	R	Yield (%)	ee (%)
Ph	Ph	72	97
Ph	4-MeC$_6$H$_4$	89	97
Ph	4-MeOC$_6$H$_4$	78	96
Ph	4-ClC$_6$H$_4$	65	95
Ph	2-BrC$_6$H$_4$	83	97
Ph	2-ClC$_6$H$_4$	99	98
Ph	2-Naphthyl	74	94
Ph	2-Furyl	99	98
Ph	Et	20	94
4-MeC$_6$H$_4$	Ph	84	96
4-MeOC$_6$H$_4$	Ph	64	97
2-ClC$_6$H$_4$	Ph	92	96
2-Naphthyl	Ph	46	95

Subsequently, the same group developed enantioselective organocatalytic conjugate addition of a series of ketones to α,β-γ,δ-unsaturated nitro compounds using modified bifunctional primary amine–thiourea catalysts [27]. Interestingly, in the presence of catalyst **11e**, 1,4-addition of aromatic ketones to α,β-γ,δ-unsaturated nitro compounds proceeded well, whereas bifunctional amine–thiourea **11f** promoted 1,4-addition of both cyclic and acyclic aliphatic ketones to α,β-γ,δ-unsaturated nitro compounds (Table 16.6).

Table 16.6 Enantioselective addition of ketones to nitrodienes and nitro-eneyne.

Ar^1—CH=CH—CH=CH—NO_2 + CH_3-CO-Ar^2 →[11e (15 mol%), PhCO$_2$H (5 mol%), CH$_2$Cl$_2$, rt, 6 days] Ar^1-CH=CH-CH(CH_2NO_2)-CH$_2$-CO-Ar^2

	Ar^1	Ar^2	Yield (%)	ee (%)
	Ph	Ph	78	97
	Ph	4-MeOC$_6$H$_4$	77	97
	Ph	4-MeC$_6$H$_4$	90	96
	Ph	4-BrC$_6$H$_4$	80	96
	Ph	4-ClC$_6$H$_4$	72	94
	Ph	4-FC$_6$H$_4$	83	96
	Ph	2-Naphthyl	70	97
	Ph	3,4-OCH$_2$OC$_6$H$_4$	89	98
	Ph	3-Py	79	94
	Ph	2-Py	90	84
	Ph	2-Furyl	58	95
	Ph	2-Thienyl	56	85
	2-NO$_2$C$_6$H$_4$	Ph	94	96
	4-MeOC$_6$H$_4$	Ph	48	99

Ar^1—CH=CH—CH=CH—NO_2 + cyclic ketone →[11f (15 mol%), PhCO$_2$H (5 mol%), H$_2$O (2.0 equiv), toluene, rt] product

Ar^1	Ketone	Yield (%)	syn/anti	ee (%) (syn)
Ph	Cyclohexanone	94	70/30	95
Ph	Cyclopentanone	97	63/37	97
Ph	Dihydro-2H-pyran-4(3H)-one	98	50/50	99
Ph	Dihydro-2H-thiopyran-4(3H)-one	92	55/45	88
Ph	Acetone	75	—	95
Ph	4-Methylpentan-2-one	44	—	93
2-NO$_2$C$_6$H$_4$	Cyclohexanone	85	99/1	96
2-NO$_2$C$_6$H$_4$	Acetone	78	—	84
4-MeOC$_6$H$_4$	Cyclohexanone	98	66/34	91
4-MeOC$_6$H$_4$	Acetone	70	—	91

Table 16.6 (Continued)

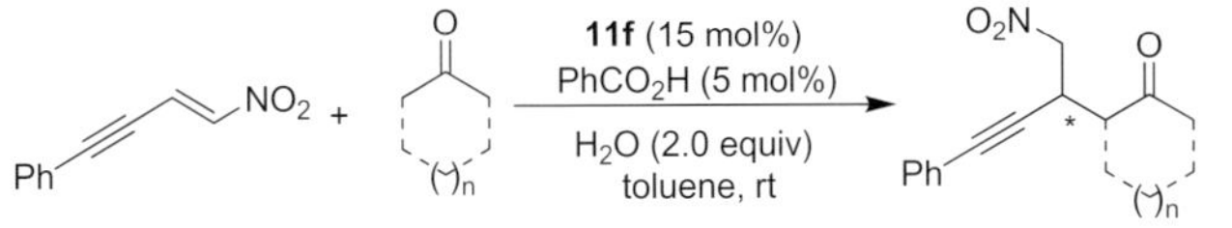

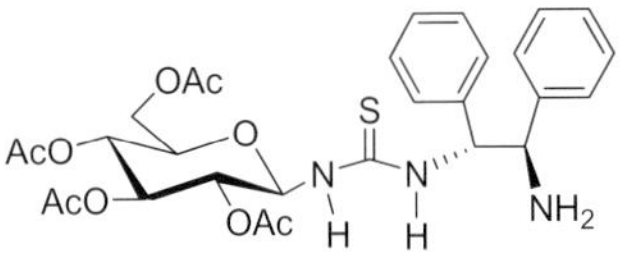

Ketone	Yield (%)	syn/anti	ee (%) (syn)
Acetophenone	85	—	96
Cyclohexanone	89	71/29	95
Acetone	85	—	94

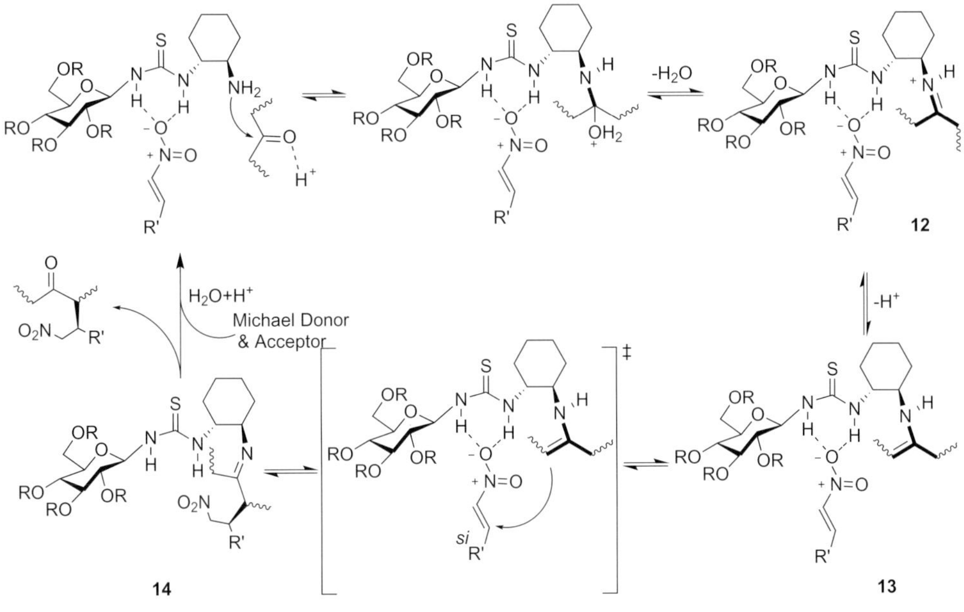

11f

Figure 16.2 Plausible mechanism for bifunctional amine–thiourea catalysis.

A bifunctional catalytic mechanism was suggested in which the thiourea moiety interacts through hydrogen bonding with the nitro group of the α-β unsaturated nitroolefin enhancing its electrophilicity while the neighboring primary amine activates ketones by formation of an enamine intermediate **13** which is promoted by acid additives. Subsequently, the C-nucleophile attacks the *Si*-face of the β-carbon at the nitroolefin to give the imine intermediate **14**. Finally, the catalyst is regenerated through hydrolysis of the imine **14** by water (Figure 16.2).

16.3.2
Bifunctional Secondary Amino–Thiourea Catalysts

Chiral bifunctional thioureas bearing secondary amino groups were also used as efficient organocatalysts for the asymmetric Michael addition of ketones to nitroolefins [28]. Recently, Zhou and coworkers developed two pyrrolidine–thioureas **15** derived from (S)- or (R)-1-Boc2-(aminomethyl)pyrrolidine and glucosyl isothiocyanate (Scheme 16.6) [29]. Owing to slow decomposition on storage, these bifunctional secondary amino–thioureas were converted into the corresponding trifluoroacetates and the active organocatalysts were generated *in situ* by addition of organic base. Pyrrolidine–thiourea **15b** proved to be matched combination of carbohydrate and pyrrolidine for the asymmetric addition of cyclohexanone to nitroolefins. A series of adducts **16** were obtained in good to high yields with

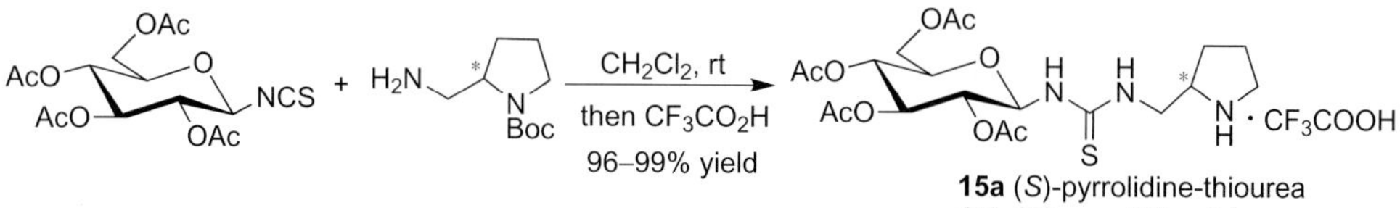

Scheme 16.6 Synthesis of saccharide-derived bifunctional secondary amino–thioureas.

Table 16.7 Asymmetric addition of cyclohexanone to nitroolefins catalyzed by bifunctional secondary amino–thiourea **15a**.

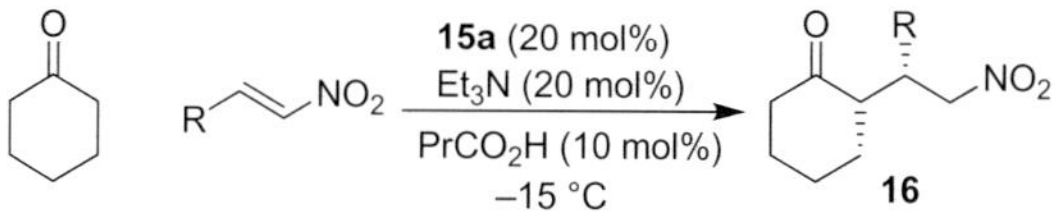

R	Yield (%)	syn/anti	ee (%)
Ph	98	99/1	95
2-CF$_3$C$_6$H$_4$	76	97/3	97
3-CF$_3$C$_6$H$_4$	95	99/1	93
4-CF$_3$C$_6$H$_4$	71	97/3	95
4-ClC$_6$H$_4$	86	98/2	94
4-FC$_6$H$_4$	78	99/1	95
2,4-Cl$_2$C$_6$H$_3$	87	99/1	94
2-NO$_2$-5-ClC$_6$H$_3$	98	98/2	97
2-MeOC$_6$H$_4$	82	98/2	94
4-MeOC$_6$H$_4$	96	98/2	92
4-MeC$_6$H$_4$	81	98/2	93
2,4-(MeO)$_2$C$_6$H$_3$	90	97/3	93
4-Benzyloxy	99	97/3	92
Benzo[d][1,3]dioxol-5-yl	99	99/1	96
1-Naphthyl	88	99/1	94
2-Furyl	81	92/8	92
(E)-Cinnamyl	87	95/5	94
Phenylethyl	73	76/24	93

high diastereo- and enantioselectivities (Table 16.7). However, the asymmetric addition of other ketones, such as acetone and cyclopentanone, gave the corresponding adducts in moderate yields with moderate to good stereoselectivities.

16.3.3
Bifunctional Tertiary Amino–Thiourea Catalysts

Since the pioneering research of Takemoto's group [17a–d], several bifunctional tertiary amine–thioureas have been developed and applied in various asymmetric Michael additions [10d, 17e–f]. In these reactions, a tertiary amine plays the role of chiral base, resulting in an ion pair formed by the corresponding chiral ammonium cation and the enolate. The latter is the nucleophilic species involved in this type of addition reaction.

In 2008, Ma and coworkers employed the saccharide-derived tertiary amine–thioureas **10** as chiral bifunctional catalysts in the direct Michael addition reactions of malonates to nitroolefins [18b]. In this reaction, nitroolefin and malonate could be concertedly activated by the thiourea–tertiary amine catalyst. Excellent yields and enantioselectivities were obtained for malonate and various nitroolefins with catalyst **10a** or **10e** (Table 16.8). Furthermore, β-ketoesters were also used as

Table 16.8 Enantioselective Michael addition reaction of malonate to nitroolefins with catalysts **10a** and **10e** (values in parentheses obtained with catalyst **10e**).

R	Yield (%)	ee (%)
Ph	99 (99)	94 (99)
4-BrC$_6$H$_4$	99 (99)	94 (90)
2-BrC$_6$H$_4$	99 (99)	97 (98)
4-ClC$_6$H$_4$	98 (99)	93 (90)
2-ClC$_6$H$_4$	99 (99)	95 (97)
4-FC$_6$H$_4$	97 (99)	92 (92)
4-MeOC$_6$H$_4$	99	95
4-MeC$_6$H$_4$	99	92
2-Naphthyl	99	96
3-PhOC$_6$H$_4$	99	91
2-Furyl	99	93
4-NO$_2$C$_6$H$_4$	99 (99)	90 (91)
PhCH=CH	92	92
2-NO$_2$PhCH=CH	86	92
4-MeOPhCH=CH	91	90
MeCH$_2$CH$_2$	97	90
Me$_2$CH$_2$CH$_2$	89	90

Table 16.9 Bifunctional tertiary amino–thiourea promoted asymmetric Michael addition of acetylacetone to nitroolefins.

R	Yield (%)	ee (%)
Ph	93	96
4-MeC$_6$H$_4$	80	89
4-BnOC$_6$H$_4$	96	85
4-MeOC$_6$H$_4$	99	88
3-MeOC$_6$H$_4$	99	92
2-MeOC$_6$H$_4$	89	92
2,4-(MeO)$_2$C$_6$H$_3$	99	84
4-CF$_3$C$_6$H$_4$	76	91
3-CF$_3$C$_6$H$_4$	82	89
2-CF$_3$C$_6$H$_4$	81	95
4-ClC$_6$H$_4$	91	88
1-Naphthyl	97	95
2-Furyl	99	94
5-Me-2-furyl	99	85
(E)-PhCH=CH-	85	81

nucleophiles to afford the Michael adducts in high yield with good diastereo- and enantioselectivity.

At the same time, Zhou and coworkers reported the enantioselective Michael addition of acetylacetone to nitroolefins by using saccharide-derived bifunctional tertiary amine–thioureas [30]. In terms of enantioselectivity, catalyst **10b** derived from D-glucopyranose and N-[(1R,2R)-2-aminocyclohexyl]-N,N-dimethylamine gave the best results (Table 16.9). Various nitroolefins containing aryl and heteroaryl groups gave the desired adducts **17** in high yields (76–99%) and enantioselectivities (80–96% ee). However, the reaction of aliphatic nitroolefins and nitro-substituted aryl nitroolefins did not proceed at all.

16.4
Miscellaneous Reactions

Besides the reactions described in the previous sections, carbohydrate-derived organocatalysts have also been applied to other asymmetric transformations such as enantioselective radical reactions [31] and aldol reactions [4a, 32]. These reactions extended the application of different carbohydrate-derived catalysts in asymmetric synthetic chemistry.

16.4.1
Radical-Chain Reactions

Roberts and coworkers examined the carbohydrate-derived thiols **18a–h** as organocatalysts for the enantioselective radical addition of triphenyl silane to methylene-lactones [33]. These carbohydrate-derived organocatalysts are readily accessible or easily prepared from commercially available materials. For example, treatment of acetylated α-D-mannopyranose with HBr/HOAc in CH_2Cl_2 gave 2,3,4,6-tetra-O-acetyl-α-D-mannopyranosyl bromide, which was reacted with thiourea to afford mainly the S-α-mannosylthiopseudouronium salt. The α-mannose thiol **18c** was the major products after hydrolysis of this salt (Scheme 16.7). In addition, a small amount of the β-thiol **18d** was also obtained via crystallization from ethanol. In the radical addition reaction, acetylated β-mannopyranose thiol **18d** gave significantly higher enantioselectivity than β-glucopyranose and β-galactopyranose thiols (**18a** and **18b**) (Table 16.10). The α-mannose thiol **18c** was ineffective at transferring its chirality to the hydrosilylation products. When the acetyl residues in **18d** were replaced with several more bulky acyl groups (**18e**), no increase in the ee of the adducts **19** was found. Even with D-glucosamine-derived thiol compounds (**18f**), the obtained ee was also poor. Notably, when the relatively rigid mannofuranose thiols **18g** and **18h** were employed, the sense of asymmetric induction was reversed, albeit only moderate stereoselectivities were observed.

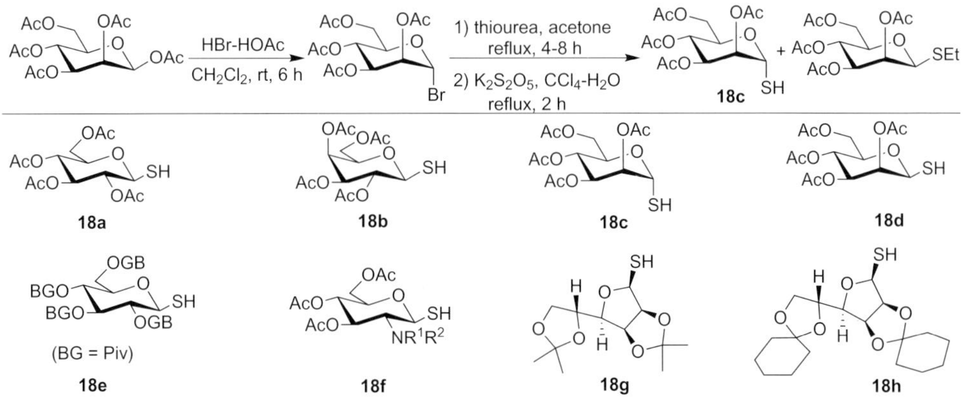

Scheme 16.7 Carbohydrate-derived thiols **18** and synthesis of **18c**.

For these addition reactions, the thiols act as protic polarity-reversal organocatalysts [34] to promote the radical-chain hydrosilylation of alkenes, via the propagation cycle as shown in Figure 16.3.

Table 16.10 Enantioselective radical-chain reactions catalyzed by carbohydrate-derived thiols.

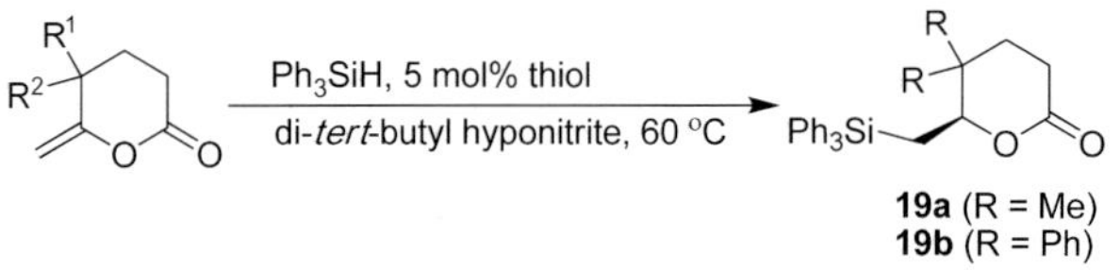

Catalyst	Solvent	Product	Yield (%)	ee (%)
18a	Hexane	19a	72	50
18a	Hexane/dioxane (5:1)	19b	93	87
18b	Hexane	19a	79	40
18b	Hexane/dioxane (5:1)	19b	96	84
18c	Hexane	19a	79	3
18c	Hexane/dioxane (5:1)	19b	92	5
18d	Hexane	19a	84	76
18d	Hexane/dioxane (5:1)	19b	90	95
18e ($R^1 = R^2 =$ Piv)	Hexane	19a	77	44
18e ($R^1 = R^2 =$ Piv)	Hexane/dioxane (5:1)	19b	95	87
18f ($R^1 =$ H, $R^2 =$ Ac)	Hexane/dioxane (7:1)	19a	67	25
18f ($R^1,R^2 =$ phthalimide)	Dioxane	19a	95	5
18g	Hexane/dioxane (4:1)	19b	82	53 (S)
18g	Hexane/dioxane (4:1)	19b	74	58 (S)

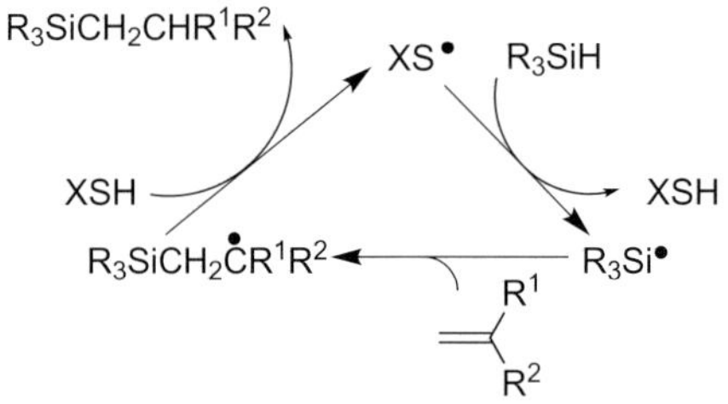

Figure 16.3 Propagation cycle of thiol-catalyzed radical-chain reactions.

Non-racemic products were also obtained in the thiol-catalyzed, silane-mediated free-radical reductive alkylation of methylene-lactones. In this reaction the adduct **20** was obtained with 54% ee using β-glucose thiol **18a** as catalyst and in 53% ee with the β-mannose thiol **18d** (Scheme 16.8) [35].

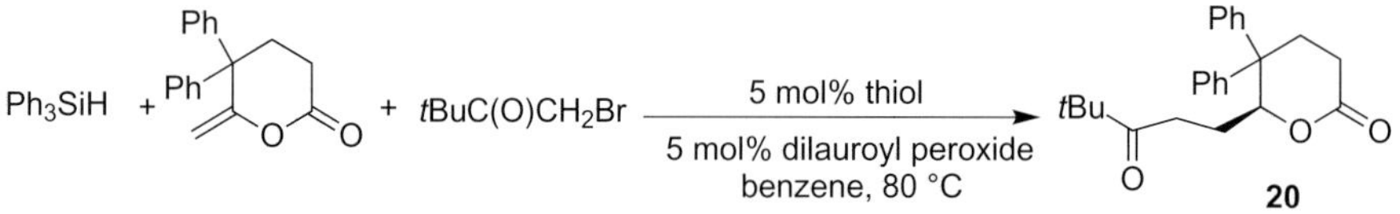

thiol **18a**: 70% yield, 54% ee
thiol **18d**: 74% yield, 53% ee

Scheme 16.8 Carbohydrate-derived thiol-catalyzed reductive alkylation of methylene-lactone.

16.4.2 Aldol Reactions

Since List, Lerner, and Barbas reported the highly stereoselective proline-promoted aldol reaction in 2000 [4a], organocatalytic aldol condensations have become standard model reactions for numerous new organocatalysts. Usually, these aldol condensations were carried out in organic media. To investigate the aldol reaction in the presence of water [36], Machinami's group integrated a carbohydrate auxiliary into proline-based catalysts to provide a hydrophilic active site (Scheme 16.9) [37]. For the model reaction of acetone with 4-nitrobenzadehyde, aminoacylamidoglucosides **21a–23a** gave the *(R)*-aldol with moderate enantioselectivities in the presence of water. The stereoinduction was reversed with aminoacyloxy-glucosides **21b–23b.** In addition, the amount of water influenced the reaction yields and enantioselectivity. Possibly, the hydroxyl group on the glucoside moiety of **21a** and **21b** stabilizes the transition state of the aldol reaction since the hydroxyl group of sugar is considered to compete with water molecules for its coordination to organic molecules.

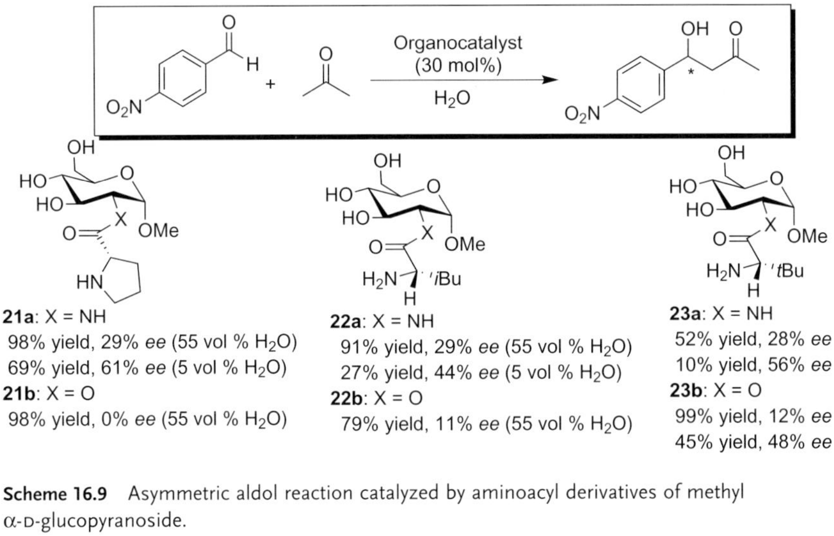

21a: X = NH
98% yield, 29% *ee* (55 vol % H_2O)
69% yield, 61% *ee* (5 vol % H_2O)
21b: X = O
98% yield, 0% *ee* (55 vol % H_2O)

22a: X = NH
91% yield, 29% *ee* (55 vol % H_2O)
27% yield, 44% *ee* (5 vol % H_2O)
22b: X = O
79% yield, 11% *ee* (55 vol % H_2O)

23a: X = NH
52% yield, 28% *ee* (55 vol % H_2O)
10% yield, 56% *ee* (5 vol % H_2O)
23b: X = O
99% yield, 12% *ee* (55 vol % H_2O)
45% yield, 48% *ee* (5 vol % H_2O)

Scheme 16.9 Asymmetric aldol reaction catalyzed by aminoacyl derivatives of methyl α-D-glucopyranoside.

16.5 Outlook

In summary, several interesting examples of carbohydrate-derived organocatalysts have appeared in recent times, leading to efficient stereochemical control in asymmetric reactions. However, the exploration of carbohydrate-based organocatalysts is just at its very start and, therefore, new and exciting examples of these tools will no doubt emerge in the future.

Acknowledgment

We are grateful for the National Natural Science Foundation of China (No. 20972110 and 21172170) for generous financial support of our research.

References

1 Bredig, G. and Fiske, P.S. (1912) *Biochem. Z.*, **46**, 7.
2 (a) Eder, U., Sauer, G., and Weichert, R. (1971) *Angew. Chem. Int. Ed.*, **10**, 496; (b) Hajos, Z.G. and Parrish, D.R. (1974) *J. Org. Chem.*, **39**, 1615.
3 (a) O'Donnell, M.J., Bennett, W.D., and Wu, S.D. (1989) *J. Am. Chem. Soc.*, **111**, 2353; (b) Tu, Y., Wang, Z.X., and Shi, Y. (1996) *J. Am. Chem. Soc.*, **118**, 9806; (c) Corey, E.J., Xu, F., and Noe, M.C. (1997) *J. Am. Chem. Soc.*, **119**, 12414; (d) Yang, D., Wong, M.K., Wang, X.C., and Tang, Y.C. (1998) *J. Am. Chem. Soc.*, **120**, 6611; (e) Sigman, M.S. and Jacobsen, E.N. (1998) *J. Am. Chem. Soc.*, **120**, 4901; (f) Corey, E.J. and Grogan, M.J. (1999) *Org. Lett.*, **1**, 157; (g) Denmark, S.E. and Wu, Z.C. (1999) *Synlett*, 847.
4 (a) List, B., Lerner, R.A., and Barbas, C.F., III (2000) *J. Am. Chem. Soc.*, **122**, 2395; (b) Sigman, M.S., Vachal, P., and Jacobsen, E.N. (2000) *Angew. Chem. Int. Ed.*, **39**, 1279; for reviews, see: (c) Dalko, P.L. and Moisan, L. (2004) *Angew. Chem. Int. Ed.*, **43**, 5138; (d) Berkessel, A. and Gröger, H. (2004) *Asymmetric Organocatalysis*, Wiley-VCH Verlag GmbH, Weinheim; (e) Seayad, J. and List, B. (2005) *Org. Biomol. Chem.*, **3**, 719; (f) Dalko, P.I. (ed.) (2007) *Enantioselective Organocatalysis*, Wiley-VCH Verlag GmbH, Weinheim.
5 Boysen, M.M.K. (2007) *Chem. Eur. J.*, **13**, 8648.
6 For reviews on organocatalytic oxidations: (a) Shi, Y. (2004) *Acc. Chem. Res.*, **37**, 488; (b) Wong, O.D. and Shi, Y. (2008) *Chem. Rev.*, **108**, 3958.
7 Strecker, A. (1850) *Ann. Chem. Pharm.*, **75**, 27.
8 Mannich, C. and Krosche, W. (1912) *Arch. Pharm.*, **250**, 647.
9 Jacobsen, E.N., Pfaltz, A., and Yamamoto, H. (eds) (1999) *Comprehensive Asymmetric Catalysis*, Springer, Berlin.
10 For reviews on enantioselective Strecker reactions, see: (a) Gröger, H. (2003) *Chem. Rev.*, **103**, 2795; (b) Vilaivan, T., Bhanthumnavin, W., and Sritana-Anant, Y. (2005) *Curr. Org. Chem.*, **9**, 1315; (c) Taylor, M.S. and Jacobsen, E.N. (2006) *Angew. Chem. Int. Ed.*, **45**, 1520; (d) Doyle, A.G. and Jacobsen, E.N. (2007) *Chem. Rev.*, **107**, 5713; for reviews on enantioselective Mannich reactions, see: (e) Notz, W., Tanaka, F., and Barbas, C.F., III (2004) *Acc. Chem. Res.*, **37**, 580; (f) Verkade, J.M.M., van Hemert, L.J.C., Quaedflieg, P.J.L.M., and Rutjes, F.P.J.T. (2008) *Chem. Soc. Rev.*, **37**, 29.
11 (a) Vachal, P. and Jacobsen, E.N. (2002) *J. Am. Chem. Soc.*, **124**, 10012; (b) Wenzel, A.G. and Jacobsen, E.N. (2002) *J. Am. Chem. Soc.*, **124**, 12964; (c) Joly, G.D. and Jacobsen, E.N. (2004) *J. Am. Chem. Soc.*, **126**, 4102; (d) Taylor, M.S. and Jacobsen, E.N. (2004) *J. Am. Chem. Soc.*, **126**, 10558; (e) Yoon, T.P. and Jacobsen, E.N. (2005) *Angew. Chem. Int. Ed.*, **44**, 466; (f) Fuerst, D.E. and Jacobsen, E.N. (2005) *J. Am. Chem. Soc.*, **127**, 8964.
12 Becker, C., Hoben, C., and Kunz, H. (2007) *Adv. Synth. Catal.*, **349**, 417.
13 Negru, M., Schollmeyer, D., and Kunz, H. (2007) *Angew. Chem. Int. Ed.*, **46**, 9339.
14 For reviews on enantioselective synthesis of β-amino acids, see: (a) Ma, J.-A. (2003) *Angew. Chem. Int. Ed.*, **42**, 4290; (b) Miller, J.A. and Nguyen, S.T. (2005) *Mini-Rev. Org. Chem.*, **2**, 39; (c) Weiner, B., Szymański, W., Janssen, D.B., Minnaard, A.J., and Feringa, B.L. (2010) *Chem. Soc. Rev.*, **39**, 1656.

15 For reviews on metallic bifunctional catalysts, see: (a) Gröger, H. (2001) *Chem. -Eur. J.*, **7**, 5246; (b) Shibasaki, M. and Yoshikawa, N. (2002) *Chem. Rev.*, **102**, 2187; (c) Ma, J.-A. and Cahard, D. (2004) *Angew. Chem.*, **116**, 4666; (2004) *Angew. Chem. Int. Ed.*, **43**, 4566; (d) France, S., Weatherwax, A., Taggi, A.E., and Lectka, T. (2004) *Acc. Chem. Rev.*, **37**, 592.

16 For reviews on bifunctional organic catalysts, see: (a) Berkessel, A. and Gröger, H. (2004) *Asymmetric Organocatalysis: From Biomimetic Concepts to Applications in Asymmetric Synthesis*, Wiley-VCH Verlag GmbH, Weinheim; (b) Pihko, P.M. (2004) *Angew. Chem.*, **116**, 2110; (2004) *Angew. Chem. Int. Ed.*, **43**, 2062; (c) special issue on organocatalysis, Houk, K.N., and List, B. (eds) (2004) *Acc. Chem. Res.*, **37**, 487; (d) Marcelli, T., van Maarseveen, J.H., and Hiemstra, H. (2006) *Angew. Chem. Int. Ed.*, **45**, 7496.

17 (a) Okino, T., Hoashi, Y., and Takemoto, Y. (2003) *J. Am. Chem. Soc.*, **125**, 12672; (b) Okino, T., Nakamura, S., Furukawa, T., and Takemoto, Y. (2004) *Org. Lett.*, **6**, 625; (c) Okino, T., Hoashi, Y., Furukawa, T., Xu, X., and Takemoto, Y. (2005) *J. Am. Chem. Soc.*, **127**, 119; (d) Hoashi, Y., Okino, T., and Takemoto, Y. (2005) *Angew. Chem. Int. Ed.*, **44**, 4032; for reviews on bifunctional thiourea-based organocatalysis, see: (e) Takemoto, Y. (2005) *Org. Biomol. Chem.*, **3**, 4299; (f) Connon, S.J. (2006) *Chem. -Eur. J.*, **12**, 5418.

18 (a) Ma, J.-A., Liu, K., Cui, H.-F., Nie, J., Dong, K.-Y., and Li, X.-J. (20 Dec 2006) Preparation and application of chiral bifunctional amine-thiourea catalysts based on saccharides, Faming Zhuanli, Shenqing Gongkai Shuomingshu, CN1974009; (b) Li, X.-J., Liu, K., Ma, H., Nie, J., and Ma, J.-A. (2008) *Synlett*, 3242.

19 (a) Lindhorst, T. and Kieburg, C. (1995) *Synthesis*, 1228; (b) Selkti, M., Kassab, R., Lopez, H.P., Villain, F., and de Rango, C.J. (1999) *Carbohydr. Chem.*, **18**, 1019.

20 Westermann, B. (2003) *Angew. Chem. Int. Ed.*, **42**, 151.

21 Ting, A. and Schaus, S.E. (2007) *Eur. J. Org. Chem.*, 5797.

22 Li, X.-J. (2010) Design and Application of Chiral Organocatalysts in Catalytic Asymmetric Reactions, PhD thesis, Tianjin University.

23 Wang, C., Zhou, Z., and Tang, C. (2008) *Org. Lett.*, **10**, 1707.

24 Enders, D. and Chow, S. (2006) *Eur. J. Org. Chem.*, 4578.

25 For reviews on organocatalytic Michael addition reactions, see: (a) Tsogoeva, S.B. (2007) *Eur. J. Org. Chem.*, 1701; (b) Enders, D., Grondal, C., and Hüttl, M.R.M. (2007) *Angew. Chem. Int. Ed.*, **46**, 1570.

26 Liu, K., Cui, H.-F., Nie, J., Dong, K.-Y., Li, X.-J., and Ma, J.-A. (2007) *Org. Lett.*, **9**, 923.

27 Ma, H., Liu, K., Zhang, F.-G., Zhu, C.-L., Nie, J., and Ma, J.-A. (2010) *J. Org. Chem.*, **75**, 1402.

28 (a) Cao, Y.-J., Lai, Y.-Y., Wang, X., Li, Y.-J., and Xiao, W.-J. (2007) *Tetrahedron Lett.*, **48**, 21; (b) Cao, Y.-J., Lu, H.-H., Lai, Y.-Y., Lu, L.-Q., and Xiao, W.-J. (2006) *Synthesis*, 3795; (c) Cao, C.-L., Ye, M.-C., Sun, X.-L., and Tang, Y. (2006) *Org. Lett.*, **8**, 2901.

29 Lu, A., Gao, P., Wu, Y., Wang, Y., Zhou, Z., and Tang, C. (2009) *Org. Biomol. Chem.*, **7**, 3141.

30 Gao, P., Wang, C., Wu, Y., Zhou, Z., and Tang, C. (2008) *Eur. J. Org. Chem.*, 4563.

31 For a review on radicals in asymmetric organocatalysis, see: Bertelsen, S., Nielsen, M., and Jørgensen, K.A. (2007) *Angew. Chem. Int. Ed.*, **46**, 7356.

32 For reviews on organocatalytic asymmetric aldol reactions, see: (a) Saito, S. and Yamamoto, H. (2004) *Acc. Chem. Res.*, **37**, 570; (b) Guillena, G., Nájera, C., and Ramón, D.J. (2007) *Tetrahedron: Asymmetry*, **18**, 2249.

33 (a) Haque, M.B. and Roberts, B.P. (1996) *Tetrahedron Lett.*, **37**, 9123; (b) Haque, M.B., Roberts, B.P., and Tocher, D.A. (1998) *J. Chem. Soc., Perkin Trans. 1*, 2881; (c) Cai, Y., Roberts, B.P., and Tocher, D.A. (2002) *J. Chem. Soc., Perkin Trans. 1*, 1376.

34 Roberts, B.P. (1999) *Chem. Soc. Rev.*, **28**, 25.

35 Dang, H.-S., Elsegood, M.R.J., Kim, K.-M., and Roberts, B.P. (1999) *J. Chem. Soc., Perkin Trans. 1*, 2061.

36 For a very interesting discussion of enantioselective organocatalysis "in water" or "in the presence of water", see: (a) Brogan, A.P., Dickerson, T.J., and Janda, K.D. (2006) *Angew. Chem. Int. Ed.*, **45**, 8100; (b) Hayashi, Y. (2006) *Angew. Chem. Int. Ed.*, **45**, 8103; (c) Blackmond, D.G., Armstrong, A., Coombe, V., and Wells, A. (2007) *Angew. Chem. Int. Ed.*, **46**, 3798.

37 Tsutsui, A., Takeda, H., Kimura, M., Fujimoto, T., and Machinami, T. (2007) *Tetrahedron Lett.*, **48**, 5213.

Index

References to figures are given in *italic* type. References to tables are given in **bold** type.

a
ent-adubinol B   326, *327*
acetylacetone   364, **364**
acivicin   98
acrylic esters and amides
– 1,4 nucleophilic addition   27–40
– – 1,2:5,6-di-*O*-isopropylidene-α-D-allo-hexofuranose auxiliary   39, 40
– – hexapyranoside templates   30, 31
– – methyl 6-deoxy-2,3, di-*O*-(*t*-butyldimethylsilyl)-α-D-glucopyranoside auxiliary for organometallic addition   33–35
– – natural products   35, 36
– – organometallics to α,β unsaturated esters   32
– – radical species   36, 37
– 1,4- nucleophilic addition, methyl 6-deoxy-2,3, di-*O*-(*t*-butyldimethylsilyl)-α-D-glucopyranoside auxiliary for organometallic addition   32–35
– Diels-Alder reactions   65, 66, 69
acylated 2,5-dihydropyrrole   195
N-acyloxazolidindones   27–29
aldehydes
– 1,2-nucleophilic addition   137, 138, **139**, **140**
– – allyltitanium reagents   149–151
– – allylsilicon reagents   151–153
– – diethylzinc   294–302
– – trialkylaluminiums   302–304
– – zinc triflate-catalysed   305–307
– amino acid   6
– Strecker reaction   3
aldimines   6, 7
aldol condensation   56–59, 367
aldol reactions   47, 48, 367
– alkylation   48–55

– – 6-deoxy-3,4-di-*O*-(*tert*-butyldimethylsilyl)-α-D-glucopyranoside auxiliary   53, *54*, 59
– – α-hydroxycarboxylic acids   54, 55, *54*
– – oxazolidindone auxiliary   51, 52
– enolate formation, titanium Lewis acids   143–149
alkenylidene acetals   *109*, 110, 114–116
– epoxidation   114–116, *116*
alkylation   48–55
allenyl ethers   80, **81**, **82**
– Diels-Alder reactions   80–83, **81**, **82**
allose derivatives   23, 72, 133, 144, 258–260
– furanoside   *225*
– hydroformylation   187
– nucleophilic substitution   23, *24*
allyl acetate   200
allyl bromide   137
allyl reagents
– silicon   151–153
– titanium   149–151
allylation   13, 14
– benzaldehydes   137–139
allylic glycosides   107
– cyclopropanation   108–110
– epoxidation   114, *115*
allylic substitution   217, *218*
– Tsuji-Trost, *see* Tsuji-Trost reaction
aluminohydrides   128–130
amides   **115**
– Tsuji-Trost reactions   240
– Ugi reaction   8–12
– *see also* acrylic esters and amides
amines
– allylation   13, 14
– as Michael addition catalysts   361
α-amino acids

– galactosyl 8, 9
– Strecker reaction 3–8
β-amino acids 355, 356
– via Mannich reactions 15–18
amino acids, via Ugi reaction 12, 13
amino alcohols 296–302, *298*
amino thioureas
– primary 358–362
– secondary 362, 363
arabinose derivatives 5, 69
– hydrogenation 179, 180, *180*
– Ugi reaction 10
aryl methyl ketones, hydrosilylation *213*, *214*
arylation 246
azides, Ugi reaction 10
azo compounds 92, 93

b

benzaldehyde 300, **301**
benzomorphane derivatives 43–45, *44*
4-benzyl-5,6-dehydropiperidin-2-ones 43, 44
benzylidenecyclobutanes, epoxidation 341
BINAP (2,2′-bis(diphenylphosphino)-1,1′-binaphthyl) 219
BINAPHOS 183
BINOL (1,1′-Bi-2-naphthol) 170, 187
bis(oxazolines) 306, 307, *307*, 314, 317
borohydrides 129–134
(+)-bostrycin 84
butanoic acids 33, 34
tert-butyl 2-(aminomethyl)pyrrolidine-1-carboxylate 362, 363
γ-butyrolactones 341, *343*

c

carbonyl compounds
– 1,2- nucleophilic addition 135–140, 266–268, 293, 294
– – bidentate phosphorus donors 257–260
– – bidentate phosphorus-nitrogen donors 260
– – bidentate phosphorus-sulfur donors 260–263
– – copper catalysed 253–257
– – cyclopentenone and cycloheptenone substrates 281
– – diethylzinc to aldehydes 294–302
– – ketones 308–310
– – ligands with peripheral donor centers 265
– – organometallic ligands 266–280
– – rhodium-catalysed 286, 287
– – stoichiometric reagents 149–153
– – trialkylaluminiums to aldehydes 302–304
– 1,4- nucleophilic addition 253–257
– – monodentate P- and S- ligands 280
– – rhodium-catalysed 286, 287
carboxylic esters 22, 23
α-chloro carboxylic esters 22–24
α-chloronitroso compounds 75–80, **76**, **77**
N-chlorosuccinimide (NCS) 50, *51*
chromenes 340, 341, *340*
copper salts 256
crotyl alcohol 246
cyanosilylation 310
cycloadditions
– [2 + 2] 101–104
– Diels-Alder, *see* Diels-Alder reactions
– 1,3- dipolar 93–101
– – dipolarophiles attached to auxiliary 101
– – using oximes 97
cyclodextrins 344, 345
– phosphine 174, 175, *175*
2-cycloheptenone 281, **282**
cyclohexanone 362
2-cyclohexenone 266, **268–276**, 280, 281
cyclopentadienes 72
– Diels-Alder reaction 70
cyclopentene 246, *246*
2-cyclopentenone 281, **282**
cyclopropanation 107–113, 313–317
– chiral phosphines 120–123
– chiral sulfur and phosphorus center construction 117–120
– diazoacetates 112
– epoxidation 113–116
– Simmons-Smith 107, *109*
– styrenes 313, **315**, **316**

d

DAG-methodology 118–122
decahydroquinolines *41*, 42
dehydropiperidones 18, *18*
demethoxydaunomycinone 86, 84
6-deoxy-3,4-di-*O*-(*tert*-butyldimethylsilyl)-α-D-glucopyranosides 53, 54, 58, 59, *59*
diacetone-D-glucose derivatives (1,2:5,6-di-*O*-isopropylidene-alpha-D-glucofuranose derivatives) 56, *56*, 58, 59, *59*, 118, 120
– chiral phosphines 120–122
dialkylzincs 254
diazo compounds 110–113, *112*
diazoacetates, cyclopropanation *112*
2,6-di-*tert*-butyl-4-methyl-pyridine (DTBMP) 328

Diels-Alder reactions
– cyclopentadienes 70
– dienes 68, 69, *68*
– dienes attached to auxiliary 83–93
– dienophiles attached to auxiliary 65–86
– – acrylic esters as dienophile 65–73
– – α-chloronitroso compounds 75–80, **76**, **77**
– – enol and allenyl ethers as dienophile 80
– – imines as dienophiles 74–80
– heterodienes 90–93
– heterodienophiles 74–80, **76**
dienes
– Diels-Alder reactions 68, 69, *68*
– epoxidation *324*, **343**
– hydrocyanation 206
diethylzinc 254, 255, 294–302, **298**
4,7-dihydro-1,3-dioxepine **196**, *246*
2,3-dihydrofuran 190, **193**, **194**, 246, *246*, *247*, 249
2,5-dihydrofuran 190, **191**, **192**
3,4-dihydroisoquinoline 17
2,3-dihydropyrrole 246, *246*
dimethyl itaconate 170, **171**
dimethyl malonate 237
1,3-diphenylprop-2-enyl acetate 238, 239
discodermolide 61, *61*
Dufthaler-Hafner reaction 144
dynamic kinetic resolution (DKR) of alpha-chlorocarboxylic esters 22–24

e
enones 244, 255, 266
(+)-enshuol 327
enynes 323, *324*, 342, **343**
(–)-epinegamycin 97
epoxidation 113–116
– allylic glycosides 114, *115*
– benzylidenecyclobutanes 341
– γ-butyrolactones 341, *343*
– chromenes 340, 341, *340*
– dienes *324*
– diorinanes 325, 326
– enynes *324*
– ketones
– olefins 323–325, *324*, *329*, **333**, **334**
– 1-phenylcyclohexone 335, 336, *336*
– styrene derivatives 337–341, **339**
– tetrasubstituted cyclobutylidenes *342*
– vinyl silanes *324*

f
fructose derivatives 48
– in 1,2- cycloadditions 302

– cyclopropanation 109, 110
– in epoxidations 322
– as lingands in hydrocyanation reactions 207, *208*, **209**, **210**
fucose derivatives 20
– in allylations 14

g
N-(β-D-galactopyranosyl)-2-pyridone 43, *44*
N-(β-D-galactopyranosyl)-4-pyridone 43, *43*
galactose derivatives 5, 17, *41*, *42*, 43
– in 1,4- nucleophilic additions 41, 43
– in allylations 13, *14*
– derivatives, radical chain reactions 365
– as ligands in hydrogenation reactions 179, 180, *180*
– in phosphite addition reactions 21
Gilbertson's ligand 248
glabrescol 328, *328*
glucose and glucosamine derivatives 24
– in cyclopropanation reactions 108
– in hydrogenation reactions 166
– in 1,4- nucleophilic addition reactions 258–261, *259*, *262*, 286, 287
– in aldol condensations 56–59
– as Diels-Alder dienophiles 74
– in Diels-Alder reactions 66
– in hydride reductions **134**
– in hydroformylations 187, *188*, 190, 195
– Mannich reaction 356
– in Mannich reactions 356, 357
– in nucleophilic substitutions 23, 24
– 1,2-oxazolidin-2-ones 29, 30
– in radical chain reactions 365
– in Tsuji-Trost reactions 220, 221, *220*
gulonic acid 58

h
Heck reaction 245
– ligands 247–250
– substrates 245, 246
hex-1-en-2-yl benzoate **170**, 171
hydride reduction 127
– aluminohydrides 128, 129
– borohydrides 129–134
hydrocyanation 206
– diphosphinite ligands 206–211
– imines 352
hydroformylation 183
– dihydrofurans 190
– diphosphine ligands 200, 201
– diphosphinite ligands 198–200
– disphosphite ligands 194–198
– phosphorus donor ligands 201

hydrogenation   157
– C=C bonds   158, 159
– C=N bonds   158, 159
– P-N donors   177, *178*
– P-S donors   178–180
– phosphines with mixed donor centers 173–177
– – phosphine-phosphite   175, 176
– – phosphine-phosphoamidate   176, 177
– phosphinite ligands   160–166
– phosphite ligands   166–173
hydrosilylation   211
– ketones   213, *213*, *214*
– P-S donor ligands   212–214
– phosphite ligands   211, 212
hydrovinylation   202, 203
– phosphinite ligands   203, 204

i
ibuprofen   204
imines
– nucleophilic 1,2-addition   306, 307, **308**
– amino acid   6
– as Diels-Alder dienophiles   75–81
– as organo catalysts   352–355, *353*, *354*, **354**
– Tsuji-Trost reactions   238–240, *241*
(–)-(*S*)-ipsenol   144

j
Jacobsen-Katsuki epoxidation   113

k
ketones
– 1,2- nucleophilic addition   308–310
– as epoxidation organocatalysts   326–330, *330–332*, **339**
– – spiro ring substitutions   336, 337
– hydrosilylation   213, *214*

l
β-lactams   101–103
(–)-lasiol   35, 36, 59, 60
levoglucoseone   73, 74
lithium enolates   50–52, 57, 58

m
Mannich-type reactions   15–19, 355, 356
– nitro-   356, 357
mannose derivatives   *100*, 365
– in Diels-Alder reactions   75, **76**, **77**
– in Tsuji-Trost reactions   221
– in radical chain reactions   365

methyl 6-deoxy-2,3, di-*O*-(*t*-butyldimethylsilyl)-α-D-glucopyranoside 32–35, *33*, *59*, *60*
trans-5-methyl-3-hexen-2-one   282
Michael additions   358–364
– acetylacetone to olefins   **354**
– bifunctional primary amino-thiourea catalysts   358–361
– bifunctional secondary amino-thiourea catalysts   362, 363
– bifunctional tertiary amino-thiourea catalysts   363, 364
Mizoroki-Heck reaction, *see* Heck reaction
Mukaiyama aldol reactions   59, *60*

n
ent-nakorone   *327*
(+)negamycin   97
nikkomycin Bz   98, 99
nitrodienes   360
nitrones   94, 96–98
– *N*-glycosyl   93–101, *100*
nojirimycin derivatives   78–80, *78*
trans-non-3-en-2-one   281, 282, **283–285**
nucleophilic addition
– 1,2- to carbonyl compounds   135–140, 266–268, 293, 294
– – bidentate phosphorus donors 257–260
– – bidentate phosphorus-nitrogen donors 260
– – bidentate phosphorus-sulfur donors 260–263
– – copper catalysed   253–257
– – cyclopentenone and cycloheptenone substrates   281
– – diethylzinc to aldehydes   294–302
– – ketones   308–310
– – ligands with peripheral donor centers 265
– – organometallic ligands   266–280
– – rhodium-catalysed   286–288
– – stoichiometric reagents   149–153
– – trialkylaluminiums to aldehydes 302–304
– 1,4- to acrylic esters and amides, organometallics   32–34, *34*
– 1,4- to carbonyl compounds
– – monodentate P- and S- ligands   280
– – rhodium-catalysed   286, 287
– 1,4- to pyridones   40–45
– acrylic amides and esters   27–40
– phosphites   19–22
– pyridones   40–45

o

olefins
- in 1,4-nucleophilic additions   289–290
- epoxidation   323–325, *324*, *329*, **333**, **334**, **338**
- hydroformylation   *200*
oxazolidinones   61
- aldol alkylations involving   56–58
- aldol reactions involving   51, 52, *52*, 61
- *N*-acylated   28, 29, *28*
oxazolines   236, 237
- in cyclopropanations
- in 1,2- nucleophilic additions   303, 304, *303*
- in 1,4- nucleophilic additions   263, 274, 288
- in Heck reactions   248, *248*
- in hydrogenations ca.   177–179
- in Tsuji-Trost reactions ca.   238–240
oxidation
- *vic*-diols   345, 346
- *see also* epoxidation
dioxiranes   325, 326

p

Paterno-Büchi reactions   103, 104
phenyl triflate   245, 246, *247*
phenylacetylene   306, 307, **306**
1-phenylcyclohexone   335–337, *336*
phenylglyoxilate esters   59, *60*
phosphinates   22, 121
phosphines
- construction of chiral phosphorus centers ca.   120
- in Heck reactions   247
- in hydroformylations   200, 201
- in hydrogenation reactions   174, 175
- phosphine-oxazoline   236
- phosphine-phosphite   175, 176
- in Tsuji-Trost reactions ca.   222–223, 240
phosphinites
- in 1,4- nucleophilic additions   259–263, 265, 269–278, 284, 286–287, 289–290
- in Heck reactions   248
- in hydrocyanations   206–211
- in hydroformylations   199, 200
- in hydrogenation reactions   162–168, 179, 180
- in hydrovinylations   203, 204
- in Tsuji-Trost reactions   223–225
phosphites
- in 1,2- nucleophilic additions   302–304, *303*, 304
- in 1,4- nucleophilic additions ca.   259–263, ca.   269–278
- in Heck reactions   249, *249*
- in hydroformylations   185–198
- in hydrogenation reactions ca.   168–175
- in hydrosilylations   211, 212
- in hydrovinylations   203, 204
- in Tsuji-Trost reactions   228, *230*, *231*, 236–238, *238*
phosphoroamidites   176, 177
- in 1,2- nucleophilic additions   302–304, *303*
- in Tsuji-Trost reactions   227–230
phosphorus compounds
- *see also* phosphines; phosphites; phosphoroamidites
pumilotoxin   *41*, *42*
pyridones
- 1,4 nucleophilic addition   40–45
- – for prochiral desymmetrization   42, 43
- 1,4- nucleophilic addition   43

r

radical-chain reactions   365, 366
Reformatsky reactions   136, 137, *137*
rhodium diphosphinates   162, *163*, 169, 179, 190
rhodium-catalysed reactions   160, 164, 173, 174, 286–288
ribose derivatives   94, 95
- in 1,4- nucleophilic additions   257, 258, 259, 261, *262*, 265

s

Sharpless-Katsuki epoxidation   113
silicon compounds, allyl   151–153
Simmons-Smith cyclopropanation   107–110
Strecker reactions   3–8, 352–354, *353*, *354*, **354**
styrenes
- cyclopropanation   313–315, *314*
- epoxidation   337–341
- hydroformylation   *185*, *186*, 190, *200*
- hydrovinylation   204, *205*, *206*
sulfoxides   346
chiral sulfur compounds
- via the DAG-method
- via oxidation using chiral dioxiranes
sulfur donor ligands   231–233

t

talose derivatives
- in hydroformylations   187
2,3,4,6-tetra-*O*-pivaloyl-β-D-galactopyranosylamine   40–42

thioethers
- 1,4- nucleophilic addition   263, 265, 274–276, 278
- in hydrogenations ca.   179–180
- in hydrosilylations   212, *213, 214*
- in Tsuji-Trost reactions   231–233, *232*
thioglucosides   233, 234, 240, 241
thiols   **366**
- radical-chain reactions   365
thiourea catalysts   *361*
titanium compounds
- allyl   149–152
- enolates   *146*
- Lewis acids   143–149
trialkylaluminum reagents   253, 254
- 1,2- addition to aldehydes   302–304
triethylaluminium   279, 304
Tsuji-Trost reaction
- heterodonor ligands   233–235
- - N-S   240, *241*
- P-donor ligands   220, 221
- P-N donor ligands   236–240
- P-O ligands   2425
- phosphinite ligands   221–223
- phosphite ligands   223–226
- phosphoroamidite ligands   227
- sulfur donor ligands   231–233

u
Ugi reaction   8–12

v
vinyl arenes
- hydrocyanation   204, 205, *208, 209*
- hydroformylation   *199*
vinyl cuprates   38, 39, *39*
vinyl silanes, epoxidation   *324*

x
xylose derivatives   30, *31*
- in 1,2- nucleophilic additions   295
- in 1,4- nucleophilic additions   27, 257, *258*, 261, *262, 263*
- in aldol reactions   51, 52
- as borohydride reagents   **134**
- in cyclopropanations   110
- in Diels-Alder reactions   67, 80
- in hydrogenation reactions   166–169
- phosphine-phosphite   175, 176
- in Tsuji-Trost reactions   234, 235
- in Ugi reactions   12

z
zinc triflate   305–307